HANDBOOK ON THE PHYSICS AND CHEMISTRY OF THE ACTINIDES

VOLUME 5

HANDBOOK ON THE PHYSICS AND CHEMISTRY OF

THE ACTINIDES

VOLUME 5

EDITORS:

A.J. FREEMAN

Department of Physics and Astronomy
Northwestern University
Evanston, Illinois 60201, USA

G.H. LANDER

European Institute for Transuranium Elements
Postfach 2340, D-7500 Karlsruhe, Fed. Rep. Germany

N·H
P·C

NORTH-HOLLAND

1987 AMSTERDAM · OXFORD · NEW YORK · TOKYO

ISBN: 0 444 87056 3

Published by:

North-Holland Physics Publishing

a division of

Elsevier Science Publishers B.V.
P.O. Box 103
1000 AC Amsterdam
The Netherlands

Sole distributors for the USA and Canada:

Elsevier Science Publishing Company, Inc.

52 Vanderbilt Avenue
New York, NY 10017
USA

Library of Congress Cataloging-in-Publication Data
(Revised for Vol. 5)

Handbook on the physics and chemistry of the actinides.

Vol. 3– : Editors, A.J. Freeman, C. Keller.
Includes bibliographies and indexes.
1. Actinide elements. I. Freeman, Arthur J.
II. Lander, G. H. III. Keller, Cornelius.
QD172.A3H36 1984 546'.4 84-20698
ISBN 0-444-86903-4 (V. 1)

Printed in The Netherlands

PREFACE

A.J. FREEMAN

It is now generally accepted that the actinides are a unique set of the elements. This uniqueness stems not only from their nuclear properties, which have led from the 1940's on to their important role in nuclear energy and nuclear technology, but also from their unusual chemical and physical properties which have added new excitement and discoveries to both these disciplines. To many scientists, the actinide elements—along with their sisters, the rare earths—have appeared as out of place add-ons to the Periodic Table. That a large number of synethetic transuranium elements were literally added to the Periodic Table is well-known and has enhanced the sense that the actinides represented the "last frontier" for research. Indeed, actinide research has presented unique challenges both for experimentalists and theorists: As is made clear in these volumes, these elements have "unfortunate properties"—scarcity, radioactivity, toxicity and reactivity—which cause severe complications for their preparation and handling. The unusual behavior of their 5f electrons—which causes questions such as localization versus delocalization to come to the forefront—has added complexity not found in the other elements. The fruits of these extensive and greatly accelerated experimental and theoretical investigations include not only a deeper understanding of these unique elements but of the rest of the elements in the Periodic Table as well.

It is the purpose of these volumes to describe in detail the present understanding of the actinides by means of comprehensive, critical, broad and up to date reviews covering both the physics and chemistry of these exotic elements. They are intended to serve as introduction to the subject for the non-specialist, as a convenient reference work for the specialist, and as guides for charting pathways for future research. The rapid accelerated pace of research in the last decade continues and carries with it new vigor, ferment and excitement to a field in a state of transition.

Volumes 1 and 2 (co-edited by G.H. Lander) consist of chapters dealing with aspects of the physics of the actinides. Volumes 3 and 4 (co-edited by C. Keller) cover topics in chemistry. The present Volume 5 (also co-edited by G.H. Lander) continues with topics in physics. In particular, it includes a review of the so-called "heavy-electron" actinide systems. These were first discovered in 1983 in the actinides and demonstrate, once again, how full of surprises are the actinide elements and their compounds.

CONTENTS

CONTENTS OF VOLUMES 1–4

VOLUME 1

VOLUME 2

VOLUME 3

VOLUME 4

Handbook on the Physics and Chemistry of the Actinides, edited by A.J. Freeman and G.H. Lander

Chapter 1

RELATIVISTIC STUDIES IN ACTINIDES

P. WEINBERGER
Institut für Technische Elektrochemie, Technical University of Vienna, Vienna, Austria
and
Center for Materials Science, Los Alamos National Laboratory, Los Alamos, NM 87545, USA

A. GONIS
Department of Physics, Northwestern University, Evanston, IL 60201, USA
and
Center for Materials Science, Los Alamos National Laboratory, Los Alamos, NM 87545, USA

Contents

Notation

$\boldsymbol{r}$	vector
$\mathbf{A}$	matrix: $\mathbf{A}=\{A_{ij}\}$. Occasionally the dimension is indicated by an index such as in $\mathbf{I}_4$ (four-dimensional unit matrix)
$\boldsymbol{A}$	supermatrix: $\boldsymbol{A}=\{A_{ij\|kl}\}$
$\hat{\Psi}(\boldsymbol{r})$; $\hat{n}$	field operator; number operator
$\boldsymbol{\alpha}; \beta, \gamma$	Dirac matrices: $\boldsymbol{\alpha}=(\alpha_1, \alpha_2, \alpha_3)$; β or $\gamma=\gamma_\mu$, $\mu=0,1,2,3$
$\boldsymbol{\sigma}$	Pauli spin vector, Pauli spin matrices: $\boldsymbol{\sigma}=\{\sigma_1, \sigma_2, \sigma_3\}$

The Dirac notation of scalar products and projection operators will be used whenever possible and useful.

1. Introduction

The 'relativistic studies of actinides' involve a complexity of problems related not only to demands of material sciences but also to a variety of theoretical approaches dealing with '5f'-electrons. Quite clearly by 'relativistic' a proper use of the correct energy–momentum equation is meant, i.e., the application of the usual postulates of quantum mechanics under the premises of the special theory of relativity (e.g. Rose 1961). This in particular implies the formulation of a fully relativistic single-particle theory, based on the well-accepted density functional theory such that (a) different aspects of material sciences, and (b) present and near-future theoretical aspects are contained in a straightforward manner. The diversity of problems in material sciences requires a description of the electronic structure of ordered, semi-ordered and statistically disordered systems in such a way that the final quantity is representative for a variety of experiments including, e.g., NMR or photoemission measurements.

The desired theoretical description for actinide systems has to include ordered phases like UPt_3, substoichiometric systems or mixed crystals like UN_x or USb_xAs_{1-x} and alloys like Pu_xAl_{1-x}. Local ordering (short range order) has to be part of this theoretical description just as well as dislocations (as for example in cases of interstitial hydrogen in actinide metals). In addition to all these different problems of actinide metallurgy, attempts have to be made to tackle the problem of magnetism in actinide systems.

This is of course a very pretentious program. The main emphasis of this review chapter will be to outline the theoretical background regardless of whether the discussed methods are already well-used in actinide theory, or whether it is only to be expected that they will be applied within the next few years. Many theoretical approaches becoming now standard for d-electron systems will indeed be the state-of-the-art for actinide systems in the near future.

For matters of completeness this chapter takes off with a short introduction to the density functional theory. A section on the Dirac Hamiltonian is followed by a very brief summary on group theoretical concepts. Single site scattering is presented such that formal extensions to the case of the presence of an internal (external) magnetic field and/or anisotropic scattering are evident. Multiple scattering is discussed such that it can readily be applied also to the problem of dislocations. In connection with the problem of selfconsistency particular attention is drawn to the use of complex energies. Finally, in a section dealing with applications the various theoretical aspects discussed are illustrated through the results of numerical calculations.

2. Local density functional theories

Although density functional theory seems to be well-discussed in the literature, it is useful to repeat some of its concepts before entering a discussion of its various practical manifestations. The aim of all density functional theories is to describe – at least at the beginning – the ground state properties of an N-particle system accurately in a realistic and practical manner.

2.1. *The ground state density functional*

Suppose the Hamiltonian for an N-particle system is defined in terms of a system of interacting fields (Dirac fields in the case of a relativistic description) in the presence of a scalar non-quantised external field $\Phi(\boldsymbol{r})$ (Kohn and Vashishta 1983, MacDonald and Vosko 1979):

$$\hat{\mathscr{H}} = \hat{\mathscr{H}}_{\text{sys}} + \int \mathrm{d}\boldsymbol{r}\, \hat{\rho}(\boldsymbol{r})\, \Phi(\boldsymbol{r}), \tag{2.1}$$

where $\hat{\rho}(\boldsymbol{r})$ is the charge density operator,

$$\hat{\rho}(\boldsymbol{r}) = -e\hat{\Psi}^{+}(\boldsymbol{r})\, \hat{\Psi}(\boldsymbol{r}), \tag{2.2}$$

and $\hat{\Psi}^{+}(\boldsymbol{r})$ and $\hat{\Psi}(\boldsymbol{r})$ are field operators. $\hat{\mathscr{H}}_{\text{sys}}$ is the Hamiltonian in the absence of external fields. In terms of the number operator, $\hat{n}(\boldsymbol{r}) = \hat{\Psi}^{+}(\boldsymbol{r})\hat{\Psi}(\boldsymbol{r})$, and a scalar potential, $V(\boldsymbol{r}) = -e\Phi(\boldsymbol{r})$, eq. (2.1) can be rewritten as

$$\hat{\mathscr{H}} = \hat{\mathscr{H}}_{\text{sys}} + \int \mathrm{d}\boldsymbol{r}\, \hat{n}(\boldsymbol{r})\, V(\boldsymbol{r}) = \hat{\mathscr{H}}_{\text{sys}} + \hat{V}. \tag{2.3}$$

The Hohenberg–Kohn theorem (Hohenberg and Kohn 1964) states that (1) for nondegenerate ground states the ground state $|\Psi\rangle$ of the system, and therefore in turn all related ground state properties, are unique functionals of the density $n(\boldsymbol{r})$,

$$n(\boldsymbol{r}) = \langle \Psi | \hat{n}(\boldsymbol{r}) | \Psi \rangle; \tag{2.4}$$

(2) the ground state energy functional

$$E[n] = \langle \Psi[n] | \hat{\mathscr{H}}_{\text{sys}} | \Psi[n] \rangle + \int \mathrm{d}\boldsymbol{r}\, n(\boldsymbol{r})\, V(\boldsymbol{r}) \tag{2.5}$$

is a minimum over densities corresponding to a given total charge at the correct density in the presence of $V(\boldsymbol{r})$.

First the internal energy functional corresponding to $\hat{\mathscr{H}}_{\text{sys}}$,

$$F[n] = \langle \Psi[n] | \hat{\mathscr{H}}_{\text{sys}} | \Psi[n] \rangle, \tag{2.6}$$

shall be considered in terms of its various contributions:

$$F[n] = E_{\mathrm{H}}[n] + T_{\mathrm{s}}[n] + E_{\mathrm{xc}}[n], \tag{2.7}$$

where $E_{\mathrm{H}}[n]$ is the classical Coulomb contribution:

$$E_{\mathrm{H}}[n] = \frac{e^2}{2} \iint \mathrm{d}\boldsymbol{r}\, \mathrm{d}\boldsymbol{r}'\, \frac{n(\boldsymbol{r})\, n(\boldsymbol{r}')}{|\boldsymbol{r} - \boldsymbol{r}'|}, \tag{2.8}$$

and T_{s} is the internal energy functional for an independent single-particle system, denoted by the suffix s,

$$T_{\mathrm{s}}[n] = \langle \Psi_{\mathrm{s}}[n] | \hat{\mathscr{H}}_0 | \Psi_{\mathrm{s}}[n] \rangle. \tag{2.9}$$

In eq. (2.9), is $\hat{\mathscr{H}}_0$ the internal part of the Hamiltonian for the single-particle system,

$$\hat{\mathscr{H}}_0 = \int \mathrm{d}\boldsymbol{r}\, \hat{\Psi}^{+}(\boldsymbol{r})\, t_{\mathrm{s}} \hat{\Psi}(\boldsymbol{r}), \tag{2.10}$$

and t_s the single-particle operator for the kinetic energy of independent single-particles,

$$t_s = \begin{cases} -i\hbar e\gamma\cdot\nabla + mc^2, & \text{relativistic,} \\ -\frac{1}{2}\nabla^2, & \text{non-relativistic.} \end{cases} \tag{2.11}$$

For a discussion of t_s in the relativistic case see section 3. $E_{xc}[n]$, finally, is determined by eq. (2.7).

The functional to be minimized is therefore given by

$$E[n] = T_s[n] + E_H[n] + E_{xc}[n] + \int \mathrm{d}\boldsymbol{r}\, n(\boldsymbol{r})\, V(\boldsymbol{r}) \tag{2.12}$$

and is subject to the condition that the number of particles remains constant (see also corresponding footnote in the article by Kohn and Vashishta 1983):

$$\int n(\boldsymbol{r})\, \mathrm{d}\boldsymbol{r} = N, \tag{2.13}$$

$$\frac{\delta T_s[n]}{\delta n(\boldsymbol{r})} + \int \frac{n(\boldsymbol{r}')}{|\boldsymbol{r}-\boldsymbol{r}'|}\,\mathrm{d}\boldsymbol{r}' + \frac{\delta E_{xc}[n]}{\delta n(\boldsymbol{r})} + V(\boldsymbol{r}) - \mu = 0. \tag{2.14}$$

The Lagrange parameter μ in eq. (2.14) is associated with condition (2.13). Denoting by $V_{xc}([n], \boldsymbol{r})$ the functional derivative for $E_{xc}[n]$,

$$V_{xc}([n], \boldsymbol{r}) = \frac{\delta E_{xc}[n]}{\delta n(\boldsymbol{r})}, \tag{2.15}$$

abbreviating by $V_{eff}([n], \boldsymbol{r})$ the terms

$$V_{eff}([n], \boldsymbol{r}) = V(\boldsymbol{r}) + \int \frac{n(\boldsymbol{r}')}{|\boldsymbol{r}-\boldsymbol{r}'|}\,\mathrm{d}\boldsymbol{r}' + V_{xc}([n], \boldsymbol{r}) \tag{2.16}$$

in eq. (2.14), and applying a variational procedure to eq. (2.10), a set of selfconsistent single-particle equations is derived, which can be used to determine the density $n(\boldsymbol{r})$. In the case of a relativistic Hamiltonian these equations are given by (MacDonald and Vosko 1979)

$$\{-i\hbar c\boldsymbol{\alpha}\cdot\nabla + \beta mc^2 + V_{eff}([n], \boldsymbol{r})\}\, \Psi_i(\boldsymbol{r}) = \varepsilon_i\, \Psi_i(\boldsymbol{r}), \tag{2.17}$$

$$n(\boldsymbol{r}) = \sum_i \Psi_i^*(\boldsymbol{r})\, \Psi_i(\boldsymbol{r})\, \theta(\mu - \varepsilon_i), \tag{2.18}$$

$$N = \sum_i \theta(\mu - \varepsilon_i), \qquad Q = -eN, \tag{2.19}$$

where Q is the total charge. Any set of equations such as (2.17)–(2.19) is traditionally called 'Kohn–Sham equations'.

The energy itself is given by

$$E[n] = \sum_i \varepsilon_i \theta(\mu - \varepsilon_i) + E_{xc}[n] - \tfrac{1}{2} E_H[n] - \int \mathrm{d}\boldsymbol{r}\, n(\boldsymbol{r}) \frac{\partial E_{xc}[n]}{\partial n(\boldsymbol{r})}. \tag{2.20}$$

The sums over i in eqs. (2.18) and (2.19) are over a complete set of single-particle states $\langle \boldsymbol{r}|\Psi_i\rangle = \Psi_i(\boldsymbol{r})$. In the following discussion these sums will be replaced by the more conventional sums over occupied states Σ_i^{occ}. If the Hamiltonian is relativistic as in eq. (2.17) the sum is restricted to positive energy states only (MacDonald and Vosko 1979).

For a more rigorous discussion of density functionals the reader is refered to work by Lieb (1982) and Englisch and Englisch (1983).

In principle therefore the ground state energy and the density $n(\boldsymbol{r})$ can be calculated exactly. In practice, however, an approximation must be made for the functional $E_{\text{xc}}[n]$. The most prominent approximation being the local density approximation (LDA).

2.2. *The local density functional approximation*

The operator $V_{\text{xc}}([n],\boldsymbol{r})$ in eq. (2.15) is usually termed the exchange-correlation operator. Within a local approximation to the density functional method the functional $E_{\text{xc}}[n]$ (eqs. 2.7 and 2.12) is given by

$$E_{\text{xc}}[n] \sim \int \mathrm{d}\boldsymbol{r}\, n(\boldsymbol{r})\, \varepsilon_{\text{xc}}([n]), \tag{2.21}$$

where $\varepsilon_{\text{xc}}([n])$ is the exchange-correlation energy per particle of a uniform electron gas of density n. In the LDA the exchange-correlation operator is given by

$$V_{\text{xc}}([n],\boldsymbol{r}) = \frac{\mathrm{d}}{\mathrm{d}n(\boldsymbol{r})}\{\varepsilon_{\text{xc}}[n]\, n(\boldsymbol{r})\} = \mu_{\text{xc}}([n],\boldsymbol{r}), \tag{2.22}$$

where $\mu_{xc}[n]$ is the exchange and correlation contribution to the chemical potential of a uniform system.

2.2.1. *Non-relativistic local density approximations*

Within the non-relativistic (NR) LDA, which is also used in many cases for actual relativistic band structure calculations, fits to $\varepsilon_{\text{xc}}([n])$ have been devised, most of which are of comparable accuracy (1 − 2%). Most frequently the fit described by Gunnarsson and Lundqvist (1976) to $\varepsilon_{\text{xc}}[n]$ is applied:

$$\varepsilon_{\text{xc}}([n]) = -\frac{3}{2\pi\alpha r_{\text{s}}} - 0.0666\, G\!\left(\frac{r_{\text{s}}}{11.4}\right); \tag{2.23}$$

$$\alpha = \left(\frac{4}{9\pi}\right)^{1/3}, \qquad \frac{4\pi}{3} r_{\text{s}}^3 = n^{-1}; \tag{2.24}$$

$$G(x) = [(1+x^3)\ln(1+x^{-1}) + \tfrac{1}{2}x - x^2 - \tfrac{1}{3}]. \tag{2.25}$$

In the Gunnarsson–Lundqvist approximation, $\mu_{\text{xc}}[n]$ is given by

$$\mu_{\text{xc}}([n],r) = \mu_{\text{xc}}(r_{\text{s}}) = \beta(r_{\text{s}})\, \mu_{\text{x}}(r_{\text{s}}), \tag{2.26}$$

$$\mu_{\text{x}}(r_{\text{s}}) = -\frac{2}{\pi\alpha r_{\text{s}}}, \tag{2.27}$$

$$\beta(r_s)=1+0.0545\, r_s \ln\left(1+\frac{11.4}{r_s}\right), \tag{2.28}$$

and the total energy (2.20) becomes

$$E_{\mathrm{tot}}^{\mathrm{LDA}}=\sum_i^{\mathrm{occ}} \varepsilon_i-\frac{1}{2}\int \frac{n(\boldsymbol{r})\, n(\boldsymbol{r}')}{|\boldsymbol{r}-\boldsymbol{r}'|}\,\mathrm{d}\boldsymbol{r}\,\mathrm{d}\boldsymbol{r}'$$

$$+\int n(\boldsymbol{r})\{\varepsilon_{\mathrm{xc}}([n])-\mu_{\mathrm{xc}}([n],\boldsymbol{r})\}\,\mathrm{d}\boldsymbol{r}. \tag{2.29}$$

2.2.2. *Relativistic local density approximations*

Relativistic corrections to the exchange part of $V_{\mathrm{xc}}[n]$ have been suggested by several authors (Ramana and Rajagopal 1979, MacDonald and Vosko 1979, Rajagopal 1978), and were first applied in an actual calculation by MacDonald et al. (1981). For the exchange-correlation potential $V_{\mathrm{xc}}[n]$ the following form was suggested by MacDonald and Vosko (1979):

$$V_{\mathrm{xc}}([n],\boldsymbol{r})=\left(\frac{81 n(\boldsymbol{r})}{4\pi}\right)^{1/3} \alpha(n(\boldsymbol{r})), \tag{2.30}$$

where $\alpha[n]$ is given by

$$\alpha(n(\boldsymbol{r}))=\tfrac{2}{3}\{1-\tfrac{3}{2}[(\beta\eta-\ln\delta)/\beta^2]^2+0.0545\, r_s \ln\left(1+\frac{11.4}{r_s}\right), \tag{2.31}$$

where

$$\beta=\left(\frac{9\pi}{4}\right)^{1/3}\Big/ 137\, r_s,\quad \eta=(1+\beta^2)^{1/2},\quad \delta=\beta+\eta,$$

the factor 1/137 being the familiar fine-structure constant.

2.3. *Non-relativistic density functional theory for magnetic systems*

If diamagnetic effects are neglected the one-particle operator $\hat{V}$ (eq. 2.3) also includes the interaction of a magnetic field $\boldsymbol{B}(\boldsymbol{r})$ with the electron spin. With α and β denoting 'spin indices' we have

$$\hat{V}=\int [V(\boldsymbol{r})\,\hat{n}(\boldsymbol{r})-\boldsymbol{B}(\boldsymbol{r})\cdot\hat{\boldsymbol{m}}(\boldsymbol{r})]\,\mathrm{d}\boldsymbol{r}=\int W(\boldsymbol{r})\,\mathrm{d}\boldsymbol{r}, \tag{2.32}$$

$$\hat{n}(\boldsymbol{r})=\hat{\Psi}_\alpha^+(\boldsymbol{r})\,\hat{\Psi}_\alpha(\boldsymbol{r}), \tag{2.33}$$

$$\hat{\boldsymbol{m}}(\boldsymbol{r})=-\mu_0\sum_{\alpha,\beta}\boldsymbol{\sigma}_{\alpha\beta}\,\hat{\Psi}_\alpha^+(\boldsymbol{r})\,\hat{\Psi}_\beta(\boldsymbol{r}). \tag{2.34}$$

In eq. (2.34) $\hat{\boldsymbol{m}}(\boldsymbol{r})$ is the magnetisation density operator, μ_0 the Bohr magneton ($e\hbar/2mc$) and $\boldsymbol{\sigma}_{\alpha\beta}$ the Pauli spin matrix operator (see also section 3). The energy functional is now formally equivalent to the functional discussed in section (2.1) and depends on the magnetisation density

$$\boldsymbol{m}(\boldsymbol{r}) = \langle \Psi | \hat{\boldsymbol{m}}(\boldsymbol{r}) | \Psi \rangle. \tag{2.35}$$

2.3.1. *The non-relativistic local spin density approximation* (*NR-LSDA*)

Suppose that $\boldsymbol{B}(\boldsymbol{r})$ and $\boldsymbol{m}(\boldsymbol{r})$ have only one non-vanishing component or can be chosen such that $B_z(\boldsymbol{r}) = B(\boldsymbol{r})$ and $m_z(\boldsymbol{r}) = m(\boldsymbol{r})$. The functional derivatives with respect to $n(\boldsymbol{r})$ and $m(\boldsymbol{r})$ lead again to a set of selfconsistent equations for what is usually called 'spin-up' and 'spin-down' electrons (for details see Gunnarsson and Lundqvist 1976 and Kohn and Vashishta 1983):

$$\left\{ \left[-\tfrac{1}{2}\nabla^2 + V(\boldsymbol{r}) + \int \frac{n(\boldsymbol{r}')}{|\boldsymbol{r} - \boldsymbol{r}'|} \mathrm{d}\boldsymbol{r}' + V_{\mathrm{xc}}(\boldsymbol{r}) \right] \mathbf{I}_2 + \sigma_z \mu_0 [B(\boldsymbol{r}) - W_{\mathrm{xc}}(\boldsymbol{r})] \right\} \boldsymbol{\Psi}_i(\boldsymbol{r}) = \boldsymbol{\varepsilon}_i \boldsymbol{\Psi}_i(\boldsymbol{r}), \tag{2.36}$$

where

$$\boldsymbol{\Psi}_i(\boldsymbol{r}) = \{ \Psi_{i1}(\boldsymbol{r}), \Psi_{i2}(\boldsymbol{r}) \}, \qquad \boldsymbol{\varepsilon}_i = \begin{pmatrix} \varepsilon_{i1} & 0 \\ 0 & \varepsilon_{i2} \end{pmatrix}, \tag{2.37}$$

$$V_{\mathrm{xc}}(\boldsymbol{r}) = \frac{\delta E_{\mathrm{xc}}([n, m])}{\delta n(\boldsymbol{r})}, \qquad W_{\mathrm{xc}}(\boldsymbol{r}) = \frac{\delta E_{\mathrm{xc}}([n, m])}{\delta m(\boldsymbol{r})}, \tag{2.38}$$

with

$$n(\boldsymbol{r}) = \sum_i^{\mathrm{occ}} \operatorname{tr} \boldsymbol{\Psi}_i^*(r) \mathbf{I}_2 \boldsymbol{\Psi}_i(\boldsymbol{r}) \tag{2.39}$$

and

$$m(\boldsymbol{r}) = -\mu_0 \sum_i^{\mathrm{occ}} \operatorname{tr} \boldsymbol{\Psi}_i^*(\boldsymbol{r}) \sigma_z \boldsymbol{\Psi}_i(\boldsymbol{r}). \tag{2.40}$$

Within the NR- local spin density functional approximation, where the plus and minus suffix refer to the upper and the lower part of eq. (2.36), $\mu_{\mathrm{xc}}^{\pm}(\boldsymbol{r})$ is given by Gunnarsson and Lundqvist (1979) as

$$\mu_{\mathrm{xc}}^{\pm}(r) = V_{\mathrm{xc}}(r) \pm \mu_0 W_{\mathrm{xc}}(r), \tag{2.41}$$

$$\mu_{\mathrm{xc}}^{\pm}(r) = \mu_p^x(r_s) \left[\beta(r_s) \pm \tfrac{1}{3}\delta(r_s) \frac{\xi}{1 \pm \gamma\xi} \right], \tag{2.42}$$

$$\mu_p^x(r_s) = -\frac{2}{\pi\alpha r_s}, \qquad \alpha = \left(\frac{4}{9\pi}\right)^{1/3}, \tag{2.43}$$

$$\beta(r_s) = 1 + 0.0545\, r_s \ln\left[1 + \frac{11.4}{r_s}\right], \tag{2.44}$$

$$\delta(r_s) = 1 - 0.036 r_s - 1.36 r_s/(1 + 10 r_s), \tag{2.45}$$

$$\gamma = 0.297, \tag{2.46}$$

$$\xi = \sum_i^{\text{occ}} \frac{\operatorname{tr} \boldsymbol{\Psi}_i^*(\boldsymbol{r})\, \sigma_z \boldsymbol{\Psi}_i(\boldsymbol{r})}{n(\boldsymbol{r})}. \tag{2.47}$$

Before commenting shortly on the implications of eq. (2.36), the back-bone of all NR 'spin-polarized' calculations, it is worthwhile to consider first the relativistic form of the density functional theory for magnetic systems.

2.4. *Relativistic density functional theory for magnetic systems*

The interaction of a system of Dirac electrons with a magnetic field $\boldsymbol{B}(\boldsymbol{r})$ is contained in the coupling of the Dirac current to the vector potential $\boldsymbol{A}^{\text{ext}}(\boldsymbol{r})$ ($\boldsymbol{\nabla} \times \boldsymbol{A}(\boldsymbol{r}) = \boldsymbol{B}(\boldsymbol{r})$). In a similar way as in eq. (2.3) the Hamiltonian $\hat{\mathscr{H}}$ is given by MacDonald and Vosko (1979) as

$$\hat{\mathscr{H}} = \hat{\mathscr{H}}_{\text{sys}} + \int \hat{n}(\boldsymbol{r})\, V(\boldsymbol{r})\, \mathrm{d}\boldsymbol{r} - \frac{1}{c} \int \hat{\boldsymbol{J}}(\boldsymbol{r}) \cdot \boldsymbol{A}^{\text{ext}}(\boldsymbol{r})\, \mathrm{d}\boldsymbol{r}, \tag{2.48}$$

where $\hat{\boldsymbol{J}}(\boldsymbol{r})$ is the spatial part of the Dirac four current density operator

$$\hat{\boldsymbol{J}}(\boldsymbol{r}) = e \hat{\Psi}^+(\boldsymbol{r})\, \gamma\, \hat{\Psi}(\boldsymbol{r}), \tag{2.49}$$

γ being a four vector of Dirac matrices (Messiah 1969). The energy functional corresponds to the following single particle equations:

$$\{-i\hbar c \boldsymbol{\alpha} \cdot \boldsymbol{\nabla} + \beta m c^2 + V_{\text{eff}}([J_\mu], \boldsymbol{r}) + e \boldsymbol{\alpha} \cdot A_{\text{eff}}([J_\mu], \boldsymbol{r})\}\, \Psi_i(\boldsymbol{r}) = \varepsilon_i \Psi_i(\boldsymbol{r}), \tag{2.50}$$

$$\boldsymbol{J}^u(\boldsymbol{r}) = -e \sum_i^{\text{occ}} \Psi_i^*(\boldsymbol{r})\, \gamma^\mu \Psi_i(\boldsymbol{r}), \tag{2.51}$$

$$\boldsymbol{J}^u(\boldsymbol{r}) = (J_0(\boldsymbol{r}),\ J^k(\boldsymbol{r})), \quad k = 1, 2, 3. \tag{2.52}$$

$$V_{\text{eff}}([J_\mu], \boldsymbol{r}) = -e\left[\Phi(\boldsymbol{r}) + \frac{\delta E_{\text{H}}[J_0]}{\delta J_0(\boldsymbol{r})} + \frac{\delta E_{\text{xc}}[J_\mu]}{\delta J_0(\boldsymbol{r})}\right], \tag{2.53}$$

$$A^\mu = (c\Phi, \boldsymbol{A}^{\text{ext}}). \tag{2.54}$$

In eqs. (2.50)–(2.54), A^μ is the four potential and $\boldsymbol{J}^\mu = (\rho, \boldsymbol{J})$ is the four current. Neglecting diamagnetic effects just as in the case of the non-relativistic density functional for magnetic systems (section 2.3) and employing the Gordon decomposition to the current, a Hamiltonian can be defined such that

$$\hat{\mathscr{H}} = \hat{\mathscr{H}}_{\text{sys}} + \int \mathrm{d}\boldsymbol{r}\ [\hat{n}(\boldsymbol{r})\, V(\boldsymbol{r}) - \hat{\boldsymbol{m}}(\boldsymbol{r}) \cdot \boldsymbol{B}(\boldsymbol{r})], \tag{2.55}$$

$$V(\boldsymbol{r}) = -e A_0(\boldsymbol{r})/c, \tag{2.56}$$

$$\hat{\boldsymbol{m}}(\boldsymbol{r}) = -\mu_0 \hat{\Psi}^+(\boldsymbol{r})\, \boldsymbol{\sigma} \hat{\Psi}(\boldsymbol{r}), \tag{2.57}$$

$$\boldsymbol{B}(\boldsymbol{r}) = \boldsymbol{\nabla} \times \boldsymbol{A}(\boldsymbol{r}). \tag{2.58}$$

The corresponding energy functional leads to one-particle equations, which are formally very similar to their non-relativistic counterparts

$$\{-i\hbar c\boldsymbol{\alpha}\cdot\boldsymbol{\nabla} + \beta mc^2 + V_{\text{eff}}([n, m], \boldsymbol{r}) + \beta\boldsymbol{\sigma}\cdot\boldsymbol{B}_{\text{eff}}([n, m], \boldsymbol{r})\}\,\Psi_i(\boldsymbol{r}) = \varepsilon_i\Psi_i(\boldsymbol{r}). \tag{2.59}$$

The magnetisation density $\boldsymbol{m}(\boldsymbol{r})$ and the effective potential $V_{\text{eff}}([n, m], \boldsymbol{r})$ are given by

$$\boldsymbol{m}(\boldsymbol{r}) = \sum_i^{\text{occ}} \operatorname{tr} \Psi_i^*(\boldsymbol{r})\,\beta\boldsymbol{\sigma}\Psi_i(\boldsymbol{r}), \tag{2.60}$$

$$V_{\text{eff}}([n, m], \boldsymbol{r}) = -\left\{V(\boldsymbol{r}) + e^2\int\frac{n(\boldsymbol{r}')}{|\boldsymbol{r}-\boldsymbol{r}'|}\,\mathrm{d}\boldsymbol{r}' + \frac{\delta E_{\text{xc}}([n, m])}{\delta n(\boldsymbol{r})}\right\}, \tag{2.61}$$

and $\boldsymbol{B}_{\text{eff}}([n, m], \boldsymbol{r})$ is a fictitious external magnetic field coupling to the 'spin-only' current (see also Strange et al. 1984)

$$\boldsymbol{B}_{\text{eff}}([n, m], \boldsymbol{r}) = -\mu_0\left(\boldsymbol{B}(\boldsymbol{r}) + \frac{\delta E_{\text{xc}}([n,m])}{\delta\boldsymbol{m}(\boldsymbol{r})}\right). \tag{2.62}$$

In particular choosing the z-axis in spin space such that only the z-component of the coupling between $\boldsymbol{\sigma}$ and $\boldsymbol{B}_{\text{eff}}(\boldsymbol{r})$ remains in eq. (2.59), the one particle equations are reduced to

$$\{-i\hbar c\boldsymbol{\alpha}\cdot\boldsymbol{\nabla} + \beta mc^2 + V_{\text{eff}}([n, m], \boldsymbol{r}) + \beta\sigma_z B_{\text{eff}}(\boldsymbol{r})\}\,\Psi_i(\boldsymbol{r}) = \varepsilon_i\Psi_i(\boldsymbol{r}). \tag{2.63}$$

Equation (2.63) is now completely analogous to eq. (2.36). As compared to eq. (2.36), eq. (2.63) presents a much more subtle problem, since in eq. (2.36) the two equations – one for 'spin-up', and one for 'spin-down' electrons – can be solved separately and are only coupled by the conditions (2.39) and (2.40). It is worthwhile to note that in both cases σ_z is involved as a 'key-quantity', but in the relativistic case it is further complicated by a Dirac matrix β in front (see also sections 3.4 and 5.3).

2.5. *Physical interpretation of the Kohn–Sham eigenvalues*

The energy levels ε_i appearing in the one-particle equations correspond to occupied or unoccupied levels, depending whether they contribute to $n(\boldsymbol{r})$ (e.g., eq. 2.13) or not. These energies are real and correspond in general not to the real part of excitation energies (see e.g., Kohn and Vashishta 1983). For an infinite system, however, the highest occupied eigenvalue does have a physical meaning, since this eigenvalue equals the physical chemical potential μ. In terms of Fermi surfaces, it can be expected that the physical Fermi surface as defined by the Dyson equation differs only very little from the Kohn–Sham Fermi surface.

In the following chapters the suffix 'effective' for $V_{\text{eff}}(\boldsymbol{r})$ will be dropped. It will be assumed that the effective one-particle potential $V(\boldsymbol{r})$ is constructed in the framework of the local density functional approximation or a related approximation such as, e.g., the Slater exchange approximation (including schemes like the so-called $X-\alpha$ method).

3. The Dirac Hamiltonian

The one-particle Dirac equation for $c=\hbar=m=1$ and $\mathbf{I}_4$ the 4×4 unit matrix is given by

$$H|\Psi\rangle = E|\Psi\rangle, \tag{3.1}$$

where H is the Dirac Hamiltonian

$$H = \boldsymbol{\alpha}\cdot\boldsymbol{p} + \beta + V\mathbf{I}_4 \tag{3.2}$$

and $E=p_0$ is the total energy of the electron satisfying the energy–momentum equation,

$$E^2 = p^2 + 1 \tag{3.3}$$

with $\boldsymbol{p}$ being the momentum and V being a scalar field (potential).

The wavefunction $|\Psi\rangle$ is a four component column vector and $\boldsymbol{\alpha}$ and β are 4×4 matrices:

$$\boldsymbol{\alpha} = \begin{pmatrix} 0 & \boldsymbol{\sigma} \\ \boldsymbol{\sigma} & 0 \end{pmatrix}; \quad \beta = \begin{pmatrix} I_2 & 0 \\ 0 & -I_2 \end{pmatrix}, \tag{3.4}$$

where $\boldsymbol{\sigma}$ are the Pauli spin matrices:

$$\sigma_1 = \begin{pmatrix} 0 & 1 \\ 1 & 0 \end{pmatrix}, \quad \sigma_2 = \begin{pmatrix} 0 & -i \\ i & 0 \end{pmatrix}, \quad \sigma_3 = \begin{pmatrix} 1 & 0 \\ 0 & -1 \end{pmatrix}, \tag{3.5}$$

and $\mathbf{I}_2$ is the 2×2 unit matrix.

3.1. *Intrinsic symmetries of the Dirac Hamiltonian*

Any transformation of H in eq. (3.2) has to be of the form $(\alpha|v)\mathbf{S}(\alpha)$ (Jansen and Boon 1967), where α is a pure space point operation, v a pure space translation and $\mathbf{S}(\alpha)$ is a 4×4 matrix performing the transformation of the Dirac matrices. From the fundamental theorem for the Dirac matrices follows that $\mathbf{S}(\alpha)$ is a non singular matrix such that

$$\begin{aligned} &\mathbf{S}(\alpha)^{-1}\gamma_i\mathbf{S}(\alpha) = \gamma_i', \quad \gamma_i = \alpha_i, \quad i=1,2,3, \\ &\mathbf{S}(\alpha)^{-1}\gamma_i\mathbf{S}(\alpha) = \gamma_i', \quad \gamma_i = \beta, \quad i=4. \end{aligned} \tag{3.6}$$

Since $\gamma_4 = \beta$ is block-diagonal, the transformation $\mathbf{S}(\alpha)$ has to be of the following type:

$$\mathbf{S}(\alpha) = \begin{pmatrix} \mathbf{P}(\alpha) & 0 \\ 0 & \mathbf{R}(\alpha) \end{pmatrix}, \tag{3.7}$$

where $\mathbf{P}(\alpha)$ and $\mathbf{R}(\alpha)$ are 2×2 matrices. A transformation of the wavefunction $|\Psi\rangle$ in eq. (3.1) is therefore of the type

$$(\alpha|v)\,\mathbf{S}(\alpha)|\Psi\rangle=(\alpha|v)\,\mathbf{S}(\alpha)|\Psi_1,\Psi_2,\Psi_3,\Psi_4\rangle$$
$$=(\alpha|v)\begin{pmatrix}\mathbf{P}(\alpha) & 0\\ 0 & \mathbf{R}(\alpha)\end{pmatrix}\begin{pmatrix}|\Phi\rangle\\ |\chi\rangle\end{pmatrix}=(\alpha|v)\begin{pmatrix}\mathbf{P}(\alpha)|\Phi\rangle\\ \mathbf{R}(\alpha)|\chi\rangle\end{pmatrix}, \tag{3.8}$$

i.e., the wavefunction $|\Psi\rangle$ transforms like a 'bi-spinor'. The functions $|\Phi\rangle$ and $|\chi\rangle$ are usually called upper and lower component or large and small component, respectively.

3.2. *The muffin-tin approximation*

Suppose the operator V in eq. (3.1) is given by a superposition of spin-independent, 'effective' one-electron potentials, centred at positions $\boldsymbol{R}_i$,

$$\langle\boldsymbol{r}|V=\sum_i\langle\boldsymbol{r}-\boldsymbol{R}_i|V_i^{\text{eff}}=\sum_i\langle\boldsymbol{r}_i|V_i, \tag{3.9}$$

such that the domains D_i of an arbitrary pair of such individual operators V_i are disjoint in the tensorial space of spin and configuration. In terms of scattering theory, discussed in section 6, the domains D_i are called 'scattering regions'. If $\langle\boldsymbol{r}|V$ is given by a system of central fields,

$$V_i^{\text{eff}}(\boldsymbol{r}_i)=V_i(\boldsymbol{r}_i)=\begin{cases}V_i(r_i), & r_i=|\boldsymbol{r}_i|\leqslant R_s^i,\\ 0, & \text{otherwise,}\end{cases} \tag{3.10}$$

then the Hamilton operator in each domain D_i, H_i, is the well-known Dirac Hamiltonian for a central field, which in terms of polar coordinates (Rose 1961) is given by

$$H(\boldsymbol{r})=\mathrm{i}\gamma_5\sigma_r\left(\frac{\partial}{\partial r}+\frac{1}{r}-\frac{\beta}{r}K\right)+V(r)+\beta, \tag{3.11}$$

where

$$\gamma_5=\begin{pmatrix}0 & -\mathbf{I}_2\\ -\mathbf{I}_2 & 0\end{pmatrix},\qquad \sigma_r\boldsymbol{\sigma}\cdot\boldsymbol{L}=\mathrm{i}\boldsymbol{\sigma}\cdot\hat{\boldsymbol{r}}\times\boldsymbol{L}. \tag{3.12}$$

In each domain D_i the Dirac Hamiltonian H_i has the following constants of the motion:

$$[H_i,J^2]_-=[H_i,J_z]_-=[H_i,S^2]_-=[H_i,K]_-=0, \tag{3.13}$$

where J^2 is the total angular momentum operator, $\boldsymbol{J}=\boldsymbol{L}+\boldsymbol{S}$, J_z its z-component, $\boldsymbol{S}$ the spin momentum operator, $\boldsymbol{S}=\frac{1}{2}\boldsymbol{\sigma}$, $\boldsymbol{L}$ the angular momentum operator and in eq. (3.11) as well as in eq. (3.13) $K=\beta(\boldsymbol{\sigma}\cdot\boldsymbol{L}+1)$.

Denoting the bi-spinor $|\Psi\rangle$ in eq. (3.11) by $|\Phi,\chi\rangle$, the constants of motion imply that

$$J^2\begin{pmatrix}|\Phi\rangle\\ |\chi\rangle\end{pmatrix}=j(j+1)\begin{pmatrix}|\Phi\rangle\\ |\chi\rangle\end{pmatrix}, \tag{3.14}$$

$$J^2|\Phi\rangle = j(j+1)|\Phi\rangle, \quad J^2|\chi\rangle = j(j+1)|\chi\rangle, \quad j = \tfrac{1}{2}, \tfrac{3}{2}, \tfrac{5}{2}, \ldots;$$

$$J_z \begin{pmatrix} |\Phi\rangle \\ |\chi\rangle \end{pmatrix} = \mu \begin{pmatrix} |\Phi\rangle \\ |\chi\rangle \end{pmatrix}, \tag{3.15}$$

$$J_z|\Phi\rangle = \mu|\Phi\rangle, \quad J_z|\chi\rangle = \mu|\chi\rangle, \quad -j \leqslant \mu \leqslant j;$$

$$K \begin{pmatrix} |\Phi\rangle \\ |\chi\rangle \end{pmatrix} = -\kappa \begin{pmatrix} |\Phi\rangle \\ |\chi\rangle \end{pmatrix}, \tag{3.16}$$

$$(\boldsymbol{\sigma}\cdot \boldsymbol{L}+1)|\Phi\rangle = -\kappa|\Phi\rangle, \quad (\boldsymbol{\sigma}\cdot \boldsymbol{L}+1)|\chi\rangle = \kappa|\chi\rangle, \quad \kappa = \pm 1, \ \pm 2, \ \pm 3, \ \ldots.$$

This means that the spinors $|\Phi\rangle$ and $|\chi\rangle$ are proportional to spin-spherical harmonics:

$$\langle \boldsymbol{r}|\Psi\rangle = \begin{pmatrix} \langle \boldsymbol{r}|\Phi\rangle \\ \langle \boldsymbol{r}|\chi\rangle \end{pmatrix} = \begin{pmatrix} g_\kappa(r)\,\langle \hat{r}|\kappa\mu\rangle \\ \mathrm{i} f_\kappa(r)\,\langle \hat{r}|-\kappa\mu\rangle \end{pmatrix} \tag{3.17}$$

$$|\kappa\mu\rangle = |Q\rangle = \sum_{s=\pm 1/2} c(l\,j\,\tfrac{1}{2};\, \mu - s,\, s)|l,\, \mu - s\rangle\, \Phi(s) \tag{3.18}$$

$$\langle \hat{r}|lm\rangle = Y_{lm}(\hat{r})$$

where the spinors $\Phi(s)$ are given by

$$\Phi(\tfrac{1}{2}) = \tbinom{1}{0}; \quad \Phi(-\tfrac{1}{2}) = \tbinom{0}{1} \tag{3.19}$$

and the expansion coefficients $c(l\,j\,\frac{1}{2};\, \mu - s,\, s)$ are the well-known Clebsch–Gordan coefficients (table 1), l and m being the eigenvalues of $\boldsymbol{L}$ and L_z, respectively.

The radial amplitudes $g_\kappa(r)$ and $f_\kappa(r)$ in eq. (3.17) are the solutions of the 'radial Dirac equation' and are given by the following system of coupled differential equations:

$$\frac{\mathrm{d} f_\kappa(r)}{\mathrm{d}r} = \frac{(\kappa - 1)}{r} f_\kappa(r) - [E - 1 - V(r)]\, g_\kappa(r), \tag{3.20a}$$

$$\frac{\mathrm{d} g_\kappa(r)}{\mathrm{d}r} = [E - V(r) + 1]\, f_\kappa(r) - \frac{\kappa + 1}{r}\, g_\kappa(r). \tag{3.20b}$$

TABLE 1

Expansion coefficients $c(l\,j\,\frac{1}{2};\, m-s, s)$.

	$s=\frac{1}{2}$	$s=-\frac{1}{2}$
$j=l+\frac{1}{2}$	$\left[\dfrac{l+m+\frac{1}{2}}{2l+1}\right]^{1/2}$	$\left[\dfrac{l-m+\frac{1}{2}}{2l+1}\right]^{1/2}$
$j=l-\frac{1}{2}$	$-\left[\dfrac{l-m+\frac{1}{2}}{2l+1}\right]^{1/2}$	$\left[\dfrac{l+m+\frac{1}{2}}{2l+1}\right]^{1/2}$

TABLE 2
Quantum numbers.

	j	κ	l	$\bar{l}$	2μ
$s^{1/2}$	1/2	−1	0	1	−1, 1
$p^{3/2}$	3/2	−2	1	2	−3, −1, 1, 3
$d^{5/2}$	5/2	−3	2	3	−5, −3, −1, 1, 3, 5
$f^{7/2}$	7/2	−4	3	4	−7, −5, −3, −1, 1, 3, 5, 7
$p^{1/2}$	1/2	1	1	0	−1, 1
$d^{3/2}$	3/2	2	2	1	−3, −1, 1, 3
$f^{5/2}$	5/2	3	3	2	−5, −3, −1, 1, 3, 5

For matters of convenience a quantum number $\bar{l}$ shall be defined:

$$\bar{l}=\begin{cases} l+1, & \kappa<0, \\ l-1, & \kappa>0. \end{cases}$$

The quantum numbers involved are summarized in table 2.

3.3. *The elimination method*

Writing the Dirac Hamiltonian explicitely with respect to the velocity of light, c,

$$H|\Psi\rangle=E|\Psi\rangle$$

$$H=c\boldsymbol{\alpha}\cdot\boldsymbol{p}+(\beta-\mathbf{I}_4)c^2+V\mathbf{I}_4, \tag{3.21}$$

and the wavefunction $|\Psi\rangle$ as a bi-spinor $|\Psi\rangle=|\Phi,\chi\rangle$, eq. (3.21) yields the following equations:

$$\begin{aligned} &c\boldsymbol{\sigma}\cdot\boldsymbol{p}|\chi\rangle+V|\Phi\rangle=E|\Phi\rangle \\ &c\boldsymbol{\sigma}\cdot\boldsymbol{p}|\Phi\rangle+[V-2c^2]|\chi\rangle=E|\chi\rangle. \end{aligned} \tag{3.22}$$

The spinor $|\chi\rangle$ can then be expressed in terms of the spinor $|\Phi\rangle$:

$$|\chi\rangle=\frac{1}{2c}B^{-1}\boldsymbol{\sigma}\cdot\boldsymbol{p}|\Phi\rangle, \tag{3.23}$$

$$B=1+\frac{1}{2c^2}[E-V], \tag{3.24}$$

such that instead of eq. (3.2) only one equation, namely for $|\Phi\rangle$, has to be solved:

$$D|\Phi\rangle=E|\Phi\rangle, \tag{3.25}$$

$$D=[\tfrac{1}{2}\boldsymbol{\sigma}\cdot\boldsymbol{p}\,B^{-1}\boldsymbol{\sigma}\cdot\boldsymbol{p}+V]. \tag{3.26}$$

It should be noted that in eq. (3.23) the spinor $|\chi\rangle$ is only formally eliminated and will appear again in the norm of the wavefunction

$$\langle\Psi|\Psi\rangle=\langle\Phi|\Phi\rangle+\langle\chi|\chi\rangle$$

$$=\langle\Phi|1+\frac{1}{4c^2}\boldsymbol{\sigma}\cdot\boldsymbol{p}\,B^{-2}\,\boldsymbol{\sigma}\cdot\boldsymbol{p}|\Phi\rangle. \tag{3.27}$$

For a central field $\langle \boldsymbol{r}|V=V(r)$, eq. (3.25) has the same constants of motion as eq. (3.11) and can therefore be separated with respect to the radial and angular variables. The differential equation for the radial amplitude of $|\Phi\rangle$, $P_\kappa(r)$, however, is now quite complicated (Rosicky et al, 1976):

$$\left[\frac{1}{2}\left(-\frac{\mathrm{d}^2}{\mathrm{d}r^2}+\frac{l(l+1)}{r^2}\right)+V(r)-E\right]P_\kappa(r)$$

$$=\left\{\frac{1}{2c^2}B^{-2}\frac{\mathrm{d}V(r)}{\mathrm{d}r}\frac{\kappa}{r}P_\kappa(r)+\frac{1}{2c^2}\left[(E-V(r))B^{-1}\left(-\frac{\mathrm{d}^2}{\mathrm{d}r^2}+\frac{l(l+1)}{r^2}\right)\right]P_\kappa(r)\right.$$

$$\left.+\frac{1}{2c^2}B^{-2}\frac{\mathrm{d}V(r)}{\mathrm{d}r}\frac{\mathrm{d}}{\mathrm{d}r}P_\kappa(r)\right\}. \tag{3.28}$$

Although eq. (3.28) is of no practical use, this equation is very suitable for purposes of discussing the various levels of 'relativistic approximations' used in the literature. For $c=\infty$, eq. (3.28) reduces to the radial Schrödinger equation. By approximating $\boldsymbol{B}$ in eq. (3.24) by unity, eq. (3.28) reduces to the Pauli–Schrödinger equation with the terms on the right-hand side being in turn the 'spin–orbit interaction', the 'mass–velocity term' and the 'Darwin shift'. From eq. (3.28) it can easily be seen that even within this approximation for $\boldsymbol{B}$ it is not sufficient just to consider only the 'mass–velocity' term and the 'Darwin shift', however attractive this might be in practical terms.

It should be noted that in eq. (3.28) higher terms than c^{-4} enter via the potential, i.e., via the normalization for the wavefunction $|\Psi\rangle$.

3.4. *Central field solutions in the presence of an external (internal) field*

In order to discuss central field solutions in the presence of an external (internal) field, one has to go back to the Kohn–Sham equations (2.62) in section 2. In analogy to eq. (3.10) both the effective potential $V_{\mathrm{eff}}(\boldsymbol{r})=V(\boldsymbol{r})$ and the effective field $\boldsymbol{B}_{\mathrm{eff}}(\boldsymbol{r})=\boldsymbol{B}(\boldsymbol{r})$ shall be considered to be spherically symmetrical. The z-axis in spin space is taken to be along B. In terms of polar coordinates the ('effective') Hamiltonian is given now by (Rose 1961, Strange et al. 1984)

$$\left[\mathrm{i}\gamma_5\sigma_r\left(\frac{\partial}{\partial r}+\frac{1}{r}-\frac{\beta}{r}K\right)+V(r)+\beta+\tfrac{1}{2}e\beta\sigma_z B(r)\right]=\boldsymbol{H}(\boldsymbol{r}) \tag{3.29}$$

$$V(\boldsymbol{r})=\begin{cases}V(r), & r\leqslant R_s,\\ 0, & \text{otherwise},\end{cases} \tag{3.30a}$$

$$B(\boldsymbol{r})=\begin{cases}B(r), & r\leqslant R_s,\\ 0, & \text{otherwise}.\end{cases} \tag{3.30b}$$

The Hamiltonian in eq. (3.29) no longer commutes with K or J^2. The only remaining constant of motion is J_z, leaving quite the formidable task of formulating the boundary conditions at the atomic sphere radius $r=R_s$ (see section 5.3).

4. Double group symmetry

4.1. *Centred O_h^*-states*

The spin spherical harmonics (3.18) form a basis in the tensorial space of spin and configuration, which is reducible under point group operations (Messiah 1961, Onodera and Okazaki 1966a, Pyykko and Toivonen 1983, Altmann and Dirl 1984). In many cases it is therefore desirable not to use the spin spherical harmonics as such, but symmetry adapted spin spherical harmonics, just as in non-relativistic calculations the use of, e.g., cubic harmonics is frequently of decisive importance. In particular the calculation of energy bands along principal directions in *k*-space are a well-known tool for 'characterizing' the energy band structure of ordered solid systems.

In table 3 the decomposition of the reducible spin spherical harmonics basis set, centred at a particular site such as the origin of the lattice, into basis sets, irreducible under transformations of some 'prominent' point groups (O_h and subgroups of O_h) are listed in terms of the irreducible representations involved. For a comparison, the irreducible 'vector' representations, corresponding to the decomposition of the reducible basis of the spherical harmonics are given. In table 4 the compatibility relations can be found for a cubic system such as for example an fcc metal. From this table, one can immediately see most of the 'hybridisation effects' characteristic for the 'sd'-valence band of an fcc metal such as Pd or Au.

If in the following chapters 'double group cubic harmonics' are used, nothing but a transformation of the spin spherical harmonic basis is implied. In order to distinguish the double group from the single group frequently an asterisk is used as a superscript (e.g. O_h^*). Quite clearly the present chapter cannot provide a proper group theoretical overview, but rather a short summary of some of the notations used.

4.2. *Shifted O_h^* states*

If symmetry adapted sets of spin spherical harmonics are needed centred on sites other than the origin of the lattice (centre of symmetry), the problem of symmetrization is more subtle. This is in particular the case when a cluster approach to the solid state is applied, where because of the size of the resulting matrices the use of symmetry is of crucial importance. Any symmetry operation of a given point group will cause a permutation of sites belonging to a particular shell of neighbours. One is therefore confronted with the problem of dealing with spin, configuration and permutations. For cubic systems such 'generalized cubic harmoncis' were discussed by Dirl et al. (1984) to which a group theoretically minded reader is referred.

TABLE 3
Splitting of $d^l \otimes d^{1/2}$: two-dimensional: $\Gamma_6^\pm$, $\Gamma_7^\pm$, Δ_6, Δ_7, Σ_5, $\Lambda_{4,5}$[a], $L_{4,5}^\pm$[a], $X_6^\pm$, $X_7^\pm$; four-dimensional: $\Gamma_8^\pm$.

$P(\boldsymbol{k})$	$l=0$	$l=1$	$l=2$	$l=3$	$\boldsymbol{k}$
O_h	Γ_6^+	$\Gamma_6^- + \Gamma_8^-$	$\Gamma_7^+ + 2\Gamma_8^+$	$\Gamma_6^- + 2\Gamma_7^- + 2\Gamma_8^-$	Γ
C_{4v}	Δ_6	$2\Delta_6 + \Delta_7$	$2\Delta_6 + 3\Delta_7$	$3\Delta_6 + 4\Delta_7$	Δ
C_{2v}	Σ_5	$3\Sigma_5$	$5\Sigma_5$	$7\Sigma_5$	Σ
C_{3v}	Λ_6	$\Lambda_{4,5} + 2\Lambda_6$	$2\Lambda_{4,5} + 3\Lambda_6$	$2\Lambda_{4,5} + 5\Lambda_6$	Λ
D_{3d}	L_6^+	$L_{4,5}^- + 2L_6^-$	$2L_{4,5}^+ + 3L_6^+$	$2L_{4,5}^- + 5L_6^-$	L
D_{4h}	X_6^+	$X_7^- + 2X_6^-$	$2X_6^+ + 3X_7^+$	$3X_6^- + 4X_7^-$	X

[a]Time degenerate: $\Lambda_4 = \Lambda_5^*$; $L_4^\pm = (L_5^\pm)^*$.

5. Single site scattering

Single site scattering can be viewed either as a matching problem or more generally in terms of a Lippmann–Schwinger type equation. For a spin-independent ('effective') potential of finite range the matching procedure seems to provide a more physical insight to the problem. One has to realize, however, that by solving the Lippmann–Schwinger equation properly, one is not confined to spin-independent potentials or muffin-tin type potentials.

5.1. *Resolvents and Green's functions*

For matters of convenience a few relations connecting resolvents, *t*-matrices, etc., are repeated at the beginning of this section. For a detailed discussion the reader is referred to standard textbooks on scattering theory (Newton 1966, Rodberg and Thaler 1967, Taylor 1972, Calogero 1967).

Suppose the Hamiltonian H is given by

$$H = H_0 + V, \tag{5.1}$$

TABLE 4
Compatibility relations.

	$L \longleftarrow [111] \text{——} \Gamma \text{——} [100] \longrightarrow X$				
	D_{3d}	C_{3v}	O_h	C_{4v}	D_{4d}
s,d	L_6^+	Λ_6	Γ_6^+	Δ_6	X_6^+
p,f	L_6^-	Λ_6	Γ_6^-	Δ_6	X_6^-
s,d	L_6^+	Λ_6	Γ_7^+	Δ_7	X_7^+
s,d	$L_{4,5}^+, L_6^+$	$\Lambda_{4,5}, \Lambda_6$	Γ_8^+	Δ_6, Δ_7	X_6^+, X_7^+
p,f	L_6^-	Λ_6	Γ_7^-	Δ_7	X_7^-
p,f	$L_{4,5}^-, L_6^-$	$\Lambda_{4,5}, \Lambda_6$	Γ_8^-	Δ_6, Δ_7	X_6^-, X_7^-

H_0 being the operator of the kinetic energy. The resolvent of H and H_0, $G(z)$ and $G_0(z)$, respectively, are defined as

$$\left.\begin{aligned} G(z)&=(z-H)^{-1}, \\ G_0(z)&=(z-H_0)^{-1}, \end{aligned}\right\} \quad z=\varepsilon+\mathrm{i}\delta, \quad \delta>0 \tag{5.2}$$

The Hamiltonian operator and its resolvent have the same eigenfunctions.

$$H|\Psi_i\rangle=E_i|\Psi_i\rangle, \tag{5.3}$$

$$G(z)|\Psi_i\rangle=(z-E_i)^{-1}|\Psi_i\rangle. \tag{5.4}$$

If P_i is a projection operator associated with the ith eigenvalue, then

$$G(z)=\sum_i \frac{P_i}{(z-E_i)}. \tag{5.5}$$

For matters of convenience the argument z will be dropped in the following. The solutions of H and H_0, $|\Psi\rangle$ and $|\Phi\rangle$, respectively, are related to each other in the following way:

$$H|\Psi\rangle=E|\Psi\rangle, \tag{5.6}$$

$$H_0|\Phi\rangle=E_0|\Phi\rangle, \tag{5.7}$$

$$\left.\begin{aligned} |\Psi\rangle&=\Omega|\Phi\rangle, \\ \Omega&=(1+GV)=(1+VG), \end{aligned}\right\} \tag{5.8}$$

$$|\Psi\rangle=(1+G_0T)|\Phi\rangle \tag{5.9}$$

where Ω ('Møller operator'), G and T are defined by the recursion relations

$$\Omega=1+G_0V\Omega, \tag{5.10}$$

$$T=V(1+GV)=V\Omega=\Omega V, \tag{5.11}$$

$$G=G_0(1+GV)=G_0\Omega=\Omega G_0. \tag{5.12}$$

Relations of the type (5.10)–(5.12) are called a Lippmann–Schwinger equation. Any representation of a resolvent G or G_0 is called a Green's function.

For the non-relativistic Hamiltonian H_0 the configuration space representative of G_0,

$$\langle \boldsymbol{r}|G_0|\boldsymbol{r}'\rangle=-\frac{1}{4\pi}\frac{\mathrm{e}^{\mathrm{i}p|\boldsymbol{r}-\boldsymbol{r}'|}}{|\boldsymbol{r}-\boldsymbol{r}'|}, \quad p=\sqrt{E}, \tag{5.13}$$

is a rather well-known quantity, which occasionally is termed (NR) 'free space Green's function' or 'free space propagator'. In the relativistic case the eigenfunctions of G_0, $|\boldsymbol{p}\rangle$ ($|\boldsymbol{p}|>0$), and the angular momentum representation of G_0 are given by

$$\langle \boldsymbol{r}|\boldsymbol{p}\rangle=\sum_Q \frac{4\pi}{p(p_0+1)r}\begin{pmatrix} j_l(pr)\,\langle \hat{r}|\kappa\mu\rangle\,\langle\kappa\mu|\hat{p}\rangle \mathrm{i}^l \\ \dfrac{pS_\kappa}{p_0+1}\, j_{\bar{l}}(pr)\,\langle \hat{r}|-\kappa\mu\rangle\,\langle -\kappa\mu|\hat{p}\rangle \mathrm{i}^{\bar{l}} \end{pmatrix}, \tag{5.14}$$

$$\langle \boldsymbol{r}|G_0|\boldsymbol{r}'\rangle = \mathrm{i}p(p_0+1)\sum_Q \{\langle \boldsymbol{r}|\Phi_Q\rangle\langle\Phi_Q^{\mathrm{out}}|\boldsymbol{r}'\rangle\,\theta(\boldsymbol{r}'-\boldsymbol{r}) + \langle \boldsymbol{r}|\Phi_Q^{\mathrm{out}}\rangle\langle\Phi_Q|\boldsymbol{r}'\rangle\,\theta(\boldsymbol{r}-\boldsymbol{r}')\}, \tag{5.15}$$

$$\theta(\boldsymbol{r}-\boldsymbol{r}') = \begin{cases}1, & r\leqslant r',\\ 0, & \text{otherwise},\end{cases} \qquad \theta(\boldsymbol{r}'-\boldsymbol{r}) = 1-\theta(\boldsymbol{r}-\boldsymbol{r}'), \tag{5.16}$$

$$\langle \boldsymbol{r}|\Phi_Q\rangle = \begin{pmatrix} j_l(pr)\langle\hat{r}|\kappa\mu\rangle \\ \dfrac{\mathrm{i}pS_\kappa}{p_0+1}\, j_{\bar{l}}(pr)\langle\hat{r}|-\kappa\mu\rangle\end{pmatrix}, \tag{5.17}$$

$$\langle \boldsymbol{r}|\Phi_Q^{\mathrm{out}}\rangle = \begin{pmatrix} h_l^+(pr)\langle\hat{r}|\kappa\mu\rangle \\ \dfrac{\mathrm{i}pS_\kappa}{p_0+1}\, h_{\bar{l}}^+(pr)\langle\hat{r}|-\kappa\mu\rangle\end{pmatrix}, \tag{5.18}$$

where $j_l(pr)$, $h_l^+(pr)=j_l(x)+\mathrm{i}n_l(x)$ and $n_l(x)$ are spherical Bessel, Hankel and Neumann functions, respectively, and $S_\kappa=\kappa/|\kappa|$ (Shen 1974, Onodera and Okazaki 1966b, Takada 1966).

The Green's function $\langle \boldsymbol{r}|G_0|\boldsymbol{r}'\rangle$ is a 2×2 matrix $\mathbf{G}^0$ which for $r>r'$, for example, has the following elements (Rose 1961):

$$G_{11}^0 = \mathrm{i}p(p_0+1)\sum_{\kappa\mu} h_l^+(pr)\,j_l(pr')\,\langle\hat{r}|\kappa\mu\rangle\,\langle\kappa\mu|\hat{r}'\rangle, \tag{5.19}$$

$$G_{12}^0 = -p^2\sum_{\kappa\mu} S_\kappa h_l^+(pr) j_l(pr)\langle\hat{r}|\kappa\mu\rangle\langle-\kappa\mu|\hat{r}'\rangle, \tag{5.20}$$

$$G_{21}^0 = -G_{12}^0, \tag{5.21}$$

$$G_{22}^0 = \frac{p_0-1}{p_0+1}\,G_{11}^0. \tag{5.22}$$

5.2. *Matching of central field spinors – the reactance*

Suppose V in eq. (5.1) is a finite range spherical well (single muffin-tin potential, see eqs. 3.9 and 3.10). The componentwise matching of wave functions at the sphere boundary (R_s) is given by $[p(p_0+1)\cong p]$

$$\begin{aligned} g_\kappa(R_s,E) &= \cos\delta_\kappa(E)\,j_l(pR_s)-\sin\delta_\kappa(E)\,n_l(pR_s),\\ f_\kappa(R_s,E) &= pS_\kappa\{\cos\delta_\kappa(E)\,j_{\bar{l}}(pR_s)-\sin\delta_\kappa(E)\,n_{\bar{l}}(pR_s)\},\end{aligned} \tag{5.23}$$

and defines the κ-like phaseshift $\delta_\kappa(E)$ (c speed of light, Rydberg units)

$$\tan\delta_\kappa(E) = \frac{cf_\kappa(R_s,E)\,j_l(pR_s)-pS_\kappa j_{\bar{l}}(pR_s)\,g_\kappa(R_s,E)}{cf_\kappa(R_s,E)\,n_l(pR_s)-pS_\kappa n_{\bar{l}}(pR_s)\,g_\kappa(R_s,E)}. \tag{5.24}$$

In eq. (5.24), $g_\kappa(r,E)$ and $f_\kappa(r,E)$ are the solutions of the radial Dirac equation

(3.20), regular at the origin. In terms of the κ-like reactance $K_\kappa(E)$,

$$K_\kappa(E) = -\frac{1}{p(p_0+1)} \tan\delta_\kappa(E) \cong -\frac{1}{p}\tan\delta_\kappa(E), \tag{5.25}$$

eq. (5.23) can be written as

$$\begin{aligned} G_\kappa(R_S, E) &= [pn_l(pR_s) + K_\kappa^{-1}(E) j_l(pR_s)], \\ F_\kappa(R_s, E) &= pS_\kappa[pn_{\bar{l}}(pR_s) + K_\kappa^{-1}(E) j_{\bar{l}}(pR_s)], \end{aligned} \tag{5.26}$$

whereby eq. (5.26) is nothing but the normalization condition for 'scattering solutions' (Faulkner 1982)

$$Z_\kappa(\boldsymbol{r}, E) = \begin{pmatrix} G_\kappa(r, E)\langle \hat{r}|\kappa\mu\rangle \\ \mathrm{i}F_\kappa(r, E)\langle \hat{r}|-\kappa\mu\rangle \end{pmatrix}. \tag{5.27}$$

Equivalent forms of matching can be defined using on the rhs of eq. (5.23) Bessel and Hankel functions. The resulting quantity $t_\kappa(E)$ is usually termed the single site t-matrix and can be calculated directly from the reactance

$$t_\kappa(E) = K_\kappa(E)[1 + \mathrm{i}p(p_0+1)\, K_\kappa(E)]^{-1}, \tag{5.28}$$

$$t_\kappa(E) = -\frac{1}{p(p_0+1)} \sin\delta_\kappa(E)\, \mathrm{e}^{\mathrm{i}\delta_\kappa(E)} = -\frac{1}{p(p_0+1)} f_\kappa(E). \tag{5.29}$$

5.3. *Lippmann–Schwinger equation and the single site t-matrix for a spherical symmetrical finite range potential*

From eq. (5.9) it is evident that in general the Lippmann–Schwinger equation has to be solved, whereby G_0 is defined by eqs. (5.15) and (5.16) and $\langle \boldsymbol{r}|\boldsymbol{p}\rangle$ is given in terms of eq. (5.14):

$$|\Psi_p\rangle = |\boldsymbol{p}\rangle + G_0 V|\Psi_p\rangle, \tag{5.30}$$

$$\langle \boldsymbol{r}|\Psi_p\rangle = \langle \boldsymbol{r}|\boldsymbol{p}\rangle + \langle \boldsymbol{r}|G_0 V|\Psi_p\rangle = \langle \boldsymbol{r}|\boldsymbol{p}\rangle + \langle \boldsymbol{r}|G_0 t|\boldsymbol{p}\rangle. \tag{5.31}$$

In analogy to $\langle \boldsymbol{r}|\boldsymbol{p}\rangle$ we make the following Ansatz for $\langle \boldsymbol{r}|\Psi_p\rangle$:

$$\begin{aligned} \langle \boldsymbol{r}|\Psi_p\rangle &= \sum_Q \frac{A_Q}{p(p_0+1)r} \langle \boldsymbol{r}|\Psi_Q\rangle \begin{pmatrix} \mathrm{i}^l\langle\kappa\mu|\hat{p}\rangle & 0 \\ 0 & \mathrm{i}^{\bar{l}}\langle -\kappa\mu|\hat{p}\rangle \end{pmatrix} \\ &= \sum_Q \frac{A_Q}{p(p_0+1)r} \begin{pmatrix} g_\kappa(r, E)\langle \hat{r}|\kappa\mu\rangle\langle\kappa\mu|\hat{p}\rangle \mathrm{i}^l \\ \mathrm{i}f_\kappa(r, E)\langle \hat{r}|-\kappa\mu\rangle\langle -\kappa\mu|\hat{p}\rangle \mathrm{i}^{\bar{l}} \end{pmatrix}, \end{aligned} \tag{5.32}$$

from which the asymptotic form of the Lippmann–Schwinger equation can be guessed using eq. (5.13):

$$|\boldsymbol{r}-\boldsymbol{r}'| \xrightarrow[r \gg r']{} r - \frac{(\boldsymbol{r}\cdot\boldsymbol{r}')}{r} + O(r^{-2}), \tag{5.33}$$

$$\frac{\mathrm{e}^{\mathrm{i}p|\boldsymbol{r}-\boldsymbol{r}'|}}{|\boldsymbol{r}-\boldsymbol{r}'|} \xrightarrow[r \gg r']{} \frac{\mathrm{e}^{\mathrm{i}pr}}{r} \mathrm{e}^{\mathrm{i}(\boldsymbol{r}\cdot\boldsymbol{r}')/r}(1 + O(r^{-1})), \tag{5.34}$$

$$\langle r|\Psi_Q\rangle \xrightarrow[r\to\infty]{} \langle r|\Phi_Q\rangle + p\,e^{ipr} f_Q(p). \tag{5.35}$$

In eq. (5.35), $f_Q(p)$ is the Q-(κ-)like scattering amplitude. Using the asymptotic form of the Bessel and Neumann functions,

$$\begin{aligned} j_l(x) &\xrightarrow[|x|\to\infty]{} \sin(x - l\pi/2), \\ n_l(x) &\xrightarrow[|x|\to\infty]{} -\cos(x - l\pi/2), \end{aligned} \tag{5.36}$$

it can be shown from eq. (5.35) that $\langle \boldsymbol{r}|\Psi_Q\rangle$ is of the form

$$\langle r|\Psi_Q\rangle \xrightarrow[r\to\infty]{} e^{i\delta_\kappa(E)} \begin{pmatrix} \sin(pr - l\pi/2 + \delta_\kappa(E)) \\ \dfrac{ip}{p_0+1} S_\kappa \sin(pr - \bar{l}\pi/2 + \delta_\kappa(E)) \end{pmatrix}, \tag{5.37}$$

where $\delta_\kappa(E)$ is the κ-like phaseshift (see also the discussion by Rose 1961). Equation (5.37) shows that at long distances the radial amplitudes $g_\kappa(r, E)$ and $f_\kappa(r, E)$ in eq. (5.32) behave like the free radial functions except that the phase of oscillations is shifted by δ_κ. At the origin, i.e., for $r\to 0$, the radial amplitudes $g_\kappa(r, E)$ and $f_\kappa(r, E)$ are regular.

It is important to note that multiplication of eq. (5.30) with $G_0^{-1} = (z - H_0)$ from the left, yields the differential equation(s) corresponding to the integral equation(s) (5.31). The differential equations are nothing but the Kohn–Sham equations for a spin-independent potential as discussed in eqs. (2.27) and (3.11).

The amplitudes A_Q in eq. (5.32) can be determined from the fact that the radial amplitudes in eq. (5.32)

$$\langle r|\Psi_Q\rangle = \Psi_Q(r) = \begin{pmatrix} g_\kappa(r, E) \\ if_\kappa(V, E) \end{pmatrix} \tag{5.38}$$

have to satisfy the Volterra equations (Calogero 1963, 1967), since the integral kernel in eq. (5.31) can be split up in terms of $\theta(\boldsymbol{r} - \boldsymbol{r}')$ (eq. 5.16) into a Volterra type kernel and a separable kernel

$$\Psi_Q(r) = \boldsymbol{j}_l(pr) - \int_0^r dr'\, V(r')[\boldsymbol{N}_\kappa(r, r') - \boldsymbol{N}_\kappa(r', r)]\, \boldsymbol{P}\Psi_Q(r'). \tag{5.39}$$

In eq. (5.39), $\boldsymbol{N}_\kappa(r, r')$ and $\boldsymbol{P}$ are 2×2 matrices:

$$\boldsymbol{N}_\kappa(r, r') = (j_l(pr), S_\kappa j_{\bar{l}}(pr)) \otimes (n_l(pr'), S_\kappa n_{\bar{l}}(r')) = \boldsymbol{j}_l(pr) \otimes \boldsymbol{n}_l(pr'), \tag{5.40}$$

$$\boldsymbol{N}_\kappa(r, r') = \begin{pmatrix} j_l(pr)\, n_l(pr') & S_\kappa j_l(pr)\, n_l(pr') \\ S_\kappa j_{\bar{l}}(pr)\, n_l(pr') & j_l(pr)\, n_l(pr') \end{pmatrix}, \tag{5.41}$$

$$\boldsymbol{j}(x) \otimes \boldsymbol{n}(x) = \begin{pmatrix} 0 & 1 \\ -1 & 0 \end{pmatrix}, \tag{5.42}$$

$$\boldsymbol{P}=\begin{pmatrix} \dfrac{p_0+1}{p} & 0 \\ 0 & \dfrac{p}{p_0+1} \end{pmatrix}, \tag{5.43}$$

where ⊗ denotes a direct (Kronecker) product.

Equation (5.39) can be solved numerically (Calogero 1963) by writing $\Psi_Q(r)$ as linear combination of $\boldsymbol{j}_\kappa(pr)$ and $\boldsymbol{n}_\kappa(pr)$ ('variable phase approach'):

$$\Psi_Q(r)=C_\kappa(p,r)\boldsymbol{j}_\kappa(pr)-S_\kappa(p,r)\boldsymbol{n}_\kappa(pr). \tag{5.44}$$

The functions $S_\kappa(p,r)$ and $C_\kappa(p,r)$ not only define the phaseshift, but also the amplitudes A_Q in eq. (5.32):

$$\tan\delta_\kappa(p,r)=\frac{S_\kappa(p,r)}{C_\kappa(p,r)}, \tag{5.45}$$

$$\tan\delta_\kappa(p,r)\xrightarrow[r\to\infty]{}\frac{S_\kappa(p,\infty)}{C_\kappa(p,\infty)}=\tan\delta_\kappa(E)=-p(p_0+1)\,K_\kappa(E), \tag{5.46}$$

$$A_Q=A_\kappa=[C_\kappa(p,\infty)-\mathrm{i}S_\kappa(p,\infty)]^{-1}. \tag{5.47}$$

It should be noted that both the matching procedure (5.23) as well as the 'variable phase approach' (5.46) are standard numerical procedures to calculate the single site t-matrix $t_\kappa(E)$ (eq. 5.28) for a spin-independent spherical symmetrical potential of finite range.

5.4. *Single site t-matrix for 'spin-dependent' spherical symmetrical potentials*

In the previous sections 5.2 and 5.3 we dealt with the scattering problem for a single spin-independent, spherical symmetrical potential of finite range, in particular as given by eq. (2.26) or (2.30). In this section we will concentrate on the Lippmann–Schwinger equation for a spin-dependent potential as corresponding to the Kohn–Sham eqs. (2.63), where both the effective potential $V_{\mathrm{eff}}(r)$ and the effective magnetic field $B_{\mathrm{eff}}(r)$ are spherically symmetrical and of finite range. The Lippmann–Schwinger equation is given by (Newton 1966, Strange et al. 1984)

$$\langle\boldsymbol{r}|\Psi_{p,m_s}\rangle=\langle\boldsymbol{r}|\Psi^0_{m_s}\rangle+\int \mathrm{d}\boldsymbol{r}'\,\langle\boldsymbol{r}|G_0|\boldsymbol{r}'\rangle[V(r')+B(r')\beta\sigma_z]\langle\boldsymbol{r}'|\Psi_{p,m_s}\rangle. \tag{5.48}$$

In eq. (5.48), m_s is the eigenvalue of σ_z with respect to the spin eigenfunctions $\Phi(s)$ (eq. 3.19),

$$\sigma_z\Phi(s)=m_s\Phi(s)=m_s\Phi_{m_s}(s). \tag{5.49}$$

The unperturbed wave function $|\Psi^0_{m_s}\rangle$ is given similarly to eq. (5.14) as

$$\langle\boldsymbol{r}|\Psi^0_{m_s}\rangle=\left[\frac{p(p_0+1)}{\pi}\right]^{1/2}$$

$$\times \sum_{Q} \begin{pmatrix} j_l(pr)\langle \hat{r}|\kappa\mu\rangle\langle\kappa\mu|\hat{p}\rangle \mathrm{i}^l \Phi(s) \\ \dfrac{\mathrm{i}p}{p_0+1} j_{\bar{l}}(pr)\langle \hat{r}|-\kappa\mu\rangle\langle -\kappa\mu|\hat{p}\rangle \mathrm{i}^{\bar{l}}\Phi(s) \end{pmatrix}, \tag{5.50}$$

whereas $|\Psi_{p,m_s}\rangle$ is expanded similarly to eq. (5.32):

$$\langle \boldsymbol{r}|\Psi_{p,m_s}\rangle = \sum_{Q,Q'} A_{QQ'} \begin{pmatrix} g_{QQ'}(r,E)\langle \hat{r}|\kappa\mu\rangle\langle\kappa'\mu'|\hat{p}\rangle \mathrm{i}^l\Phi(s) \\ \mathrm{i} f_{QQ'}(r,E)\langle \hat{r}|-\kappa\mu\rangle\langle -\kappa'\mu'|\hat{p}\rangle \mathrm{i}^l\Phi(s) \end{pmatrix}. \tag{5.51}$$

Using the Ansatz (5.51), it can be shown (Feder et al. 1983, Strange et al. 1984) that the radial amplitudes $g_{QQ'}(r, E)$ and $f_{QQ'}(r, E)$ satisfy a set of coupled integral equations, which in turn corresponds to the following coupled set of differential equations:

$$\left[-\frac{\partial}{\partial r}+\frac{(\kappa'-1)}{r}\right] f_{QQ'}(r,E)+[V(r)-E+1]g_{QQ'}(r,E)$$

$$+B(r)\sum_{\kappa''} G(\kappa'',\kappa',\mu)\, g_{Q''Q'}(r,E)=0, \tag{5.52a}$$

$$\left[\frac{\partial}{\partial r}+\frac{(\kappa'+1)}{r}\right] g_{QQ'}(r,E)+[V(r)-E-1]f_{QQ'}(r,E)$$

$$-B(r)\sum_{\kappa''} G(-\kappa'',-\kappa',\mu)\, f_{Q''Q'}(r,E)=0, \tag{5.52b}$$

$$G(\kappa,\kappa',\mu)=\mathrm{tr}\,\{\langle \hat{r}|\kappa\mu\rangle\sigma_z\langle\kappa'\mu'|\hat{r}\rangle\}. \tag{5.53}$$

The single site t-matrix, formally defined by the asymptotic Lippmann–Schwinger equation can be calculated solving either a coupled set of Volterra-like equations or by the matching method, discussed below.

The coupling in eq. (5.53) occurs only for non-vanishing coefficients $G(\kappa, \kappa', \mu)$,

$$G(\kappa,\kappa',\mu)\neq 0 \begin{cases} \Delta l=0, & \Delta j=\pm 1, \quad (5.54\text{a}) \\ \Delta l=\pm 2, & \Delta j=\pm 1. \quad (5.54\text{b}) \end{cases}$$

Case (a) implies that $\kappa=-l-1$ and $\kappa'=l$ (and vice versa) and μ runs over the common eigenvalues of J_z. For case (b) it was shown by Feder et al. (1983) that this kind of coupling is by the order of c smaller than the first type of coupling.

Neglecting in the following the second type of coupling, the coefficients $G(\kappa, \kappa', \mu)$ in eq. (5.53) are given by the following simple product of Clebsch–Gordan coefficients (eq. 3.18, table 1):

$$G(\kappa,\kappa',\mu)=c(l\,j\tfrac{1}{2};\mu-s,s)\,c(l\,j'\tfrac{1}{2};\mu-s,s)-c(l\,j\tfrac{1}{2};\mu+s,s)\,c(l\,j'\tfrac{1}{2};\mu+s,s). \tag{5.55}$$

For $r>R_s$, i.e., in the region of zero scalar and magnetic field, solutions of the differential equations (5.53) are of the following form:

$$F^0_\nu(r,E)=\begin{bmatrix} j_{l_1}(pr)\,\delta_{\kappa_1\kappa}-\mathrm{i}ph^+_{l_1}(pr)\,t_{\kappa_1\mu,\,\kappa\mu}(E) \\ \dfrac{\mathrm{i}pS_{\kappa_1}}{p_0+1}\,[j_{\bar{l}_1}(pr)\,\delta_{\kappa_1\kappa}-\mathrm{i}ph^+_{\bar{l}_1}(pr)\,t_{\kappa_1\mu,\,\kappa\mu}(E)] \\ j_{l_2}(pr)\,\delta_{\kappa_2\kappa}-\mathrm{i}ph^+_{l_2}(pr)\,t_{\kappa_2\mu,\,\kappa\mu}(E) \\ \dfrac{\mathrm{i}pS_{\kappa_2}}{p_0+1}\,[j_{\bar{l}_2}(pr)\,\delta_{\kappa_2\kappa}-\mathrm{i}ph^+_{\bar{l}_2}(pr)\,t_{\kappa_2\mu,\,\kappa\mu}(E)] \end{bmatrix}, \tag{5.56}$$

where κ describes the quantum number of the component of the incoming beam and $\nu=1,2$ covers the cases $\kappa=\kappa_1$ and $\kappa=\kappa_2$, respectively. At the sphere boundary, $r=R_s$, the 'outside' solution is matched to the 'interior' solution $\Psi_{\kappa\mu}(r,E)$,

$$\Psi_{\kappa\mu}(r,E)=\sum_{\nu=1,2} A_{\kappa\mu,\nu}\begin{pmatrix} \bar{g}_{\kappa_1\mu,\nu}(r,E) \\ \mathrm{i}\bar{f}_{\kappa_1\mu,\nu}(r,E) \\ \bar{g}_{\kappa_2\mu,\nu}(r,E) \\ \mathrm{i}\bar{f}_{\kappa_2\mu,\nu}(r,E) \end{pmatrix}, \tag{5.57}$$

whereby at the origin the radial amplitudes in eq. (5.57) are determined by the ratios of the corresponding functions g and f (see also Loucks 1967, Feder et al 1983):

$$(\bar{g}_{\kappa_1\mu,\nu},\ \mathrm{i}\bar{f}_{\kappa_1\mu,\nu},\ \bar{g}_{\kappa_2\mu,\nu},\ \mathrm{i}\bar{f}_{\kappa_2\mu,\nu})=\begin{cases}(1,p^{(\kappa_1)},0,0)\,r^{S(\kappa_1)}, & \nu=1,\\ (0,0,1,p^{(\kappa_2)})\,r^{S(\kappa_2)}, & \nu=2,\end{cases} \tag{5.58}$$

$$S(\kappa_i)=[\kappa_i^2-(2Z/c)^2]^{1/2}, \tag{5.59}$$

$$p(\kappa_i)=\frac{\kappa_i+S(\kappa_i)}{2Z/c}\,. \tag{5.60}$$

As in the case of a spin-independent potential (Loucks 1967) the actual values of the radial amplitudes at $r=0$ can be found by normalization. For the calculation of the single site t-matrix, however, only the ratios $f_{\kappa\mu,\nu}/g_{\kappa\mu,\nu}$ are necessary (Strange et al. 1984):

$$t_{\kappa_1\mu,\kappa_1\mu}(E)=\left[\Delta^{(1)}j_{l_1}(pR_s)-\frac{\mathrm{i}p}{p_0+1}S_{\kappa_1}j_{\bar{l}_1}(pR_s)\right]$$
$$\times\left[\mathrm{i}p\Delta^{(1)}h^+_{l_1}(pR_s)-\frac{\mathrm{i}p}{p_0+1}S_{\kappa_1}\mathrm{i}p\,h^+_{\bar{l}_1}(pR_s)\right]^{-1}, \tag{5.61}$$

$$t_{\kappa_1\mu,\kappa_2\mu}(E)=[j_{l_1}(pRs)-\mathrm{i}pt_{\kappa_1\mu,\kappa_1\mu}(E)\,h^+_{l_1}(pR_s)]$$
$$\times[F^{(1)}_1\bar{g}_{\kappa_1\mu,1}(E,R_s)+F^{(1)}_2\bar{g}_{\kappa_1\mu,2}(E,R_s)]^{-1}, \tag{5.62}$$

$$\Delta^{(1)}=[F^{(1)}_1\mathrm{i}\bar{f}_{\kappa_1\mu,1}(E,R_s)+F^{(1)}_2\mathrm{i}\bar{f}_{\kappa_1\mu,2}(E,R_s)]$$
$$\times[F^{(1)}_1\bar{g}_{\kappa_1\mu,1}(E,R_s)+F^{(1)}_2\bar{g}_{\kappa_1\mu,2}(E,R_s)]^{-1}, \tag{5.63}$$

$$F_1^{(1)} = -\mathrm{i}p\left[\frac{\mathrm{i}p}{p_0+1} S_{\kappa_2} h_{\bar{l}_2}^{+}(pR_s) - a_{\kappa_2\mu,2} h_{\bar{l}_2}^{+}(pR_s)\right]$$

$$\times [\mathrm{i}\bar{f}_{\kappa_2\mu,1}(E, R_s) - \bar{g}_{\kappa_2\mu,1}(E, R_s)\, a_{\kappa_2\mu,2}]^{-1}, \tag{5.64}$$

$$F_2^{(2)} = -\mathrm{i}p\left[\frac{\mathrm{i}p}{p_0+1} S_{\kappa_2} h_{\bar{l}_2}^{+}(pR_s) - a_{\kappa_2\mu,1} h_{l_2}^{+}(pR_s)\right]$$

$$\times [\mathrm{i}\bar{f}_{\kappa_2\mu,2}(E, R_s) - \bar{g}_{\kappa_2\mu,2}(E, R_s)\, a_{\kappa_2\mu,1}]^{-1}, \tag{5.65}$$

$$a_{\kappa_2\mu,\nu} = \frac{\mathrm{i}\bar{f}_{\kappa_2\mu,\nu}(E, R_s)}{g_{\kappa_2\mu,\nu}(E, R_s)}. \tag{5.66}$$

For the calculation of $t_{\kappa_2\mu,\kappa_2\mu}(E)$ and $t_{\kappa_2\mu,\kappa_1\mu}(E)$ the suffixes (1) in eqs. (5.61–5.65) are replaced by (2), i.e., the case $\nu=2$ in eq. (5.58) applies.

It should be noted that the single site t-matrix $t_{QQ'}(E)$ defines a scattering amplitude $f_{QQ'}(E)$ just as in the case of a spin-independent potential (5.29),

$$t_{QQ'}(E) = -\frac{1}{p(p_0+1)} f_{QQ'}(E). \tag{5.67}$$

6. Multiple scattering: the RKKR, RKKR–GF and RKKR–CPA methods

6.1. *Fundamental equations of multiple scattering*

We now return to the Dirac equation (2.63) and the muffin-tin approximation (3.9) and (3.10):

$$[\boldsymbol{\alpha}\cdot\boldsymbol{p} + \beta + V\mathbf{I}_4 + \beta\sigma_z B]\Psi = E\Psi, \tag{6.1}$$

$$\langle \boldsymbol{r}|V = \sum_i \langle \boldsymbol{r} - \boldsymbol{R}_i|V_i^{\text{eff}}, \tag{6.2}$$

$$\langle \boldsymbol{r}|B = \sum_i \langle \boldsymbol{r} - \boldsymbol{R}_i|B_i^{\text{eff}}, \tag{6.3}$$

$$\langle \boldsymbol{r}_i|V_i^{\text{eff}} = \begin{cases} V_i(r_i), & r_i \leqslant R_s^i, \\ 0, & \text{otherwise}, \end{cases} \tag{6.4}$$

$$\langle \boldsymbol{r}_i|B_i^{\text{eff}} = \begin{cases} B_i(r_i), & r_i \leqslant R_s^i, \\ 0, & \text{otherwise}. \end{cases} \tag{6.5}$$

The representative of the total T-matrix in the tensorial space of spin and configuration for an arbitrary array of scatterers can be written as a sum of scattering path operators (Gyorffy and Stott 1973, Staunton et al. 1980, Feder et al.

1983, Strange et al. 1984)

$$\langle \boldsymbol{r}|T|\boldsymbol{r}'\rangle = \sum_{ij} \tau^{ij}, \tag{6.6}$$

where i and j refer to the scatterers centred at $\boldsymbol{R}_i$ and $\boldsymbol{R}_j$. The angular momentum representative of the 'on the energy shell' components of T satisfy the following equation of the motion:

$$\tau^{ij}_{QQ'}(E) = t^{i}_{QQ'}(E)\,\delta_{ij} + \sum_{k\neq i}\sum_{Q_1,Q_2}\{t^{i}_{QQ_1}(E)\,G^{ik}_{Q_1Q_2}(E)\,\tau^{kj}_{Q_2Q'}(E)\}, \tag{6.7}$$

where $G^{ik}_{QQ'}(E)$ is the QQ'th element of the angular momentum representation of $\langle \boldsymbol{r}|G_0|\boldsymbol{r}'\rangle$ with r and r' centred at the sites $\boldsymbol{R}_i$ and $\boldsymbol{R}_j$, and $t^{i}_{QQ'}(E)$ refers to the single site t-matrix of the scatterer centred at $\boldsymbol{R}_i$. The matrices $G^{ij}_{QQ'}(E)$ are frequently termed structure constants. In matrix notation, where $\mathbf{A}$ denotes as angular momentum representation, and by dropping the obvious parameter E, eq. (6.7) can be written as

$$\boldsymbol{\tau}^{ij} = \mathbf{t}^{i}\delta_{ij} + \sum_{k\neq i}\mathbf{t}^{i}\mathbf{G}^{ik}\boldsymbol{\tau}^{kj}. \tag{6.8}$$

Equations (6.6–6.8) are derived from a regrouping of terms in the 'traditional' expansion for the T-matrix,

$$\mathbf{T} = \sum_{i}\mathbf{t}^{i} + \sum_{i\neq j}\mathbf{t}^{i}\mathbf{G}^{ij}\mathbf{t}^{j} + \sum_{i\neq j}\sum_{j\neq k}\mathbf{t}^{i}\mathbf{G}^{ij}\mathbf{t}^{j}\mathbf{G}^{jk}\mathbf{t}^{k} + \cdots. \tag{6.9}$$

If the sites i and j are translationally invariant, i.e., all sites have the same single site t-matrix $\mathbf{t}$, then eq. (6.7) can be solved by means of a lattice Fourier transformation

$$\begin{aligned}\boldsymbol{\tau}^{ij} &= \frac{1}{\Omega_{\mathrm{BZ}}}\int [\mathbf{t}^{-1} - \mathbf{G}(\boldsymbol{k})]^{-1}\exp[\boldsymbol{k}\cdot(\boldsymbol{R}_i - \boldsymbol{R}_j)]\,\mathrm{d}\boldsymbol{k} \\ &= \frac{1}{\Omega_{\mathrm{BZ}}}\int \boldsymbol{\tau}(\boldsymbol{k})\exp[\mathrm{i}\boldsymbol{k}\cdot(\boldsymbol{R}_i - \boldsymbol{R}_j)]\,\mathrm{d}\boldsymbol{k},\end{aligned} \tag{6.10}$$

where the elements of $\mathbf{G}(\boldsymbol{k})$ are the lattice structure constants (see the following section).

In many cases of interest only the site diagonal scattering path operator $(i=j)$ is needed:

$$\boldsymbol{\tau}^{nn} = \mathbf{t}^{n}\left[1 + \sum_{k\neq n}\mathbf{G}^{nk}\boldsymbol{\tau}^{kn}\right], \tag{6.11}$$

which can also be written as (Gonis et al. 1985a,b)

$$\boldsymbol{\tau}^{nn} = [\mathbf{m}^{n} - \boldsymbol{\Delta}^{n}]^{-1}, \tag{6.12}$$

$$\mathbf{m}^{n} = (\mathbf{t}^{n})^{-1}, \tag{6.13}$$

$$\begin{aligned}\boldsymbol{\Delta}^{n} &= \sum_{k\neq n}\mathbf{G}^{nk}\mathbf{m}^{k}\left[\delta_{km} + \sum_{i\neq n}\mathbf{G}^{ki}\mathbf{m}^{i}(\delta_{im} + \cdots)\right]\mathbf{G}^{mn} \\ &= \sum_{k\neq n}\sum_{m\neq n}\mathbf{G}^{nk}\boldsymbol{\Gamma}^{(n)}_{km}\mathbf{G}^{mn}.\end{aligned} \tag{6.14}$$

In eq. (6.14) $\boldsymbol{\Gamma}_{km}^{(n)}$ is the scattering path operator corresponding to a material from which site n has been removed. For translationally invariant materials $\boldsymbol{\Delta}^n$ can be explicitly obtained from eq. (6.12):

$$\boldsymbol{\Delta}^n = \mathbf{m}^n - [\boldsymbol{\tau}^{nn}]^{-1}. \tag{6.15}$$

The quantity $\boldsymbol{\Delta}^n$, which has been termed 'renormalized interactor' will prove to be very useful in the discussion of Bloch spectral functions (section 6.6).

Finally, it should be noted that if the underlying lattice is a complex lattice, i.e., consists of n primitive lattices, the scattering path operator in eq. (6.10) is an $n \times n$ supermatrix with i and j refering to the ith and jth unit cell, respectively,

$$\boldsymbol{\tau}^{ij} = \begin{pmatrix} \boldsymbol{\tau}_{11}^{ij} & \boldsymbol{\tau}_{12}^{ij} & \cdots & \boldsymbol{\tau}_{1n}^{ij} \\ \boldsymbol{\tau}_{21}^{ij} & \cdots & & \\ \vdots & & & \\ \boldsymbol{\tau}_{n1}^{ij} & \cdots & & \boldsymbol{\tau}_{nn}^{ij} \end{pmatrix}, \tag{6.16}$$

$$\boldsymbol{t} = \begin{pmatrix} \mathbf{t}_{11} & 0 & \cdots \\ 0 & \mathbf{t}_{22} & \vdots \\ \vdots & & \mathbf{t}_{nn} \end{pmatrix}, \tag{6.17}$$

$$\boldsymbol{G}(k) = \begin{pmatrix} \mathbf{G}_{11}(k) & \mathbf{G}_{12}(k) & \cdots & \mathbf{G}_{1n}(k) \\ \vdots & & & \\ \mathbf{G}_{n1}(k) & \cdots & \cdots & \mathbf{G}_{nn}(k) \end{pmatrix}. \tag{6.18}$$

A 'cell-diagonal' scattering path operator is then defined by the following lattice Fourier transformation:

$$\boldsymbol{\tau}^{ii} = \frac{1}{\Omega_{\mathrm{BZ}}} \int [\boldsymbol{t}^{-1} - \boldsymbol{G}(k)]^{-1} \, \mathrm{d}k. \tag{6.19}$$

(see also the discussion by Klima et al. 1985 and Schadler et al. 1985).

The Green's function can be obtained directly from the scattering path operator (see eq. 5.11)

$$\langle \boldsymbol{r}_i | G | \boldsymbol{r}'_j \rangle = \sum_{QQ'} \{ Z_Q^i(\boldsymbol{r}_i)\, \tau_{QQ'}^{ij}\, Z_{Q'}^j(\boldsymbol{r}'_j) - \delta_{ij} Z_Q^i(\boldsymbol{r}_i)\, J_{Q'}^j(\boldsymbol{r}'_j)\, \delta_{QQ'} \}, \tag{6.20}$$

where in the case of $B^{\mathrm{eff}}(r) = 0$ the $Z_Q^i(\boldsymbol{r}_i)$ are the 'scattering solutions' defined in eq. (5.27) and $J_Q^i(\boldsymbol{r}_i)$ is a solution of the radial Dirac equation (3.20), which in general is not regular at the origin and joins smoothly $\langle r_i | \Phi_Q \rangle$ (eq. 5.17) for $r > R_s^i$. In eq. (6.20), $Z_Q^i(\boldsymbol{r}_i)$ and $J_Q^i(\boldsymbol{r}_i)$ are real functions.

6.2. *Structure constants*

From eqs. (6.7) and (6.8) it is evident that the only quantities remaining to be discussed are the structure constants $\mathbf{G}^{ij}$ and/or $\mathbf{G}(k)$.

6.2.1. *Real space structure constants*

In real space scattering the relativistic structure constants are given by $(p \cong (p_0+1)p)$

$$G^{ij}_{QQ'} = \sum_{s=\pm 1/2} c(l\,j\,\tfrac{1}{2};\mu-s,s)\, g^{ij}_{l\mu-s,l'\mu'-s}\, c(l'\,j'\,\tfrac{1}{2};\mu'-s,s), \tag{6.21}$$

where $g^{ij}_{LL'}$ are the non-relativistic structure constants,

$$g^{ij}_{LL'} = g_{LL'}(\boldsymbol{R}_i - \boldsymbol{R}_j) = -4\pi p \sum_{L''} \mathrm{i}^{l-l'+l''} C^{L''}_{LL'} h^{+}_{l''}\,(p|\boldsymbol{R}_i - \boldsymbol{R}_j|)\, Y_{L''}(\widehat{\boldsymbol{R}_i - \boldsymbol{R}_j}), \tag{6.22}$$

with

$$C^{L''}_{LL'} = \int \mathrm{d}\Omega\, Y_L(\Omega)\, Y_{L'}(\Omega)\, Y^{*}_{L''}(\Omega).$$

The structure constants in eq. (6.21) can be directly used for cluster calculations. More appropriate, however, is the use of symmetry adapted structure constants, i.e., the use of shifted 0^{*}_{h} states in the case of cubic lattices (see section 4). It should be noted that from eq. (6.22) structure constants for displaced atoms (Gonis et al. 1985a) can readily be calculated. If $\boldsymbol{\rho}$ is a vector shifting $\boldsymbol{R}_i - \boldsymbol{R}_j$ by $-\boldsymbol{\rho}$ from the origin, then by choosing $\boldsymbol{R}_i = 0$ (reference site), the displaced atom structure constants are given by

$$g_{LL'}(\boldsymbol{R}_j - \boldsymbol{\rho}) = \sum_{L_1 L_2} \mathrm{i}^{l-l'+l_1}\, C^{L_1}_{LL'}\, g_{L_1 L_2}(\boldsymbol{R}_j)\; j_{L_2}(-\boldsymbol{\rho}), \tag{6.23}$$

$$j_L(\boldsymbol{r}) = j_l(pr)\langle \hat{r}|lm\rangle. \tag{6.24}$$

6.2.2. *Reciprocal space structure constants*

More frequently the lattice Fourier transform of G_0, $\mathbf{G}(\boldsymbol{k})$ (eq 6.10) or $\boldsymbol{G}(\boldsymbol{k})$ (eq. 6.18) is needed. If i and j denote sublattices, then an element of a submatrix $\mathbf{G}^{ij}(\boldsymbol{k})$ in eq. (6.18) is given by (Onodera and Okazaki, 1966b; Takada, 1966; Shen, 1970)

$$G^{ij}_{QQ'}(\boldsymbol{k}) = [B^{ij}_{QQ'}(\boldsymbol{k}) + \mathrm{i}p\delta_{ij}\,\delta_{QQ'}](p_0+1). \tag{6.25}$$

If we agree to ignore terms of the order c^{-2} for the plane wave expansion ($V=0$ in eq. 3.24), then the matrix $\mathbf{B}^{ij}$ can be calculated from the non-relativistic structure constants $\mathbf{A}^{ij}$ (Kohn and Rostoker 1954, Segall 1957, Ham and Segall 1961):

$$B^{ij}_{QQ'}(\boldsymbol{k}) = \sum_{s=\pm 1/2} c(l\,j\,\tfrac{1}{2};\mu-s,s)\, A^{ij}_{l\mu-s,l'\mu'-s}(\boldsymbol{k})\, c(l'\,j'\,\tfrac{1}{2};\mu'-s,s), \tag{6.26}$$

$$A^{ij}_{LL'}(\boldsymbol{k}) = 4\pi \sum_{L''} C^{L''}_{LL'} D^{ij}_{L''}(\boldsymbol{k}), \tag{6.27}$$

$$D^{ij}_{lm}(\boldsymbol{k}) = D^{ij(1)}_{lm}(\boldsymbol{k}) + D^{ij(2)}_{lm}(\boldsymbol{k}) + \delta_{ij}\delta_{l0}\,\delta_{m0}\, D^{(3)}_{00}, \tag{6.28}$$

$$D_{lm}^{ij(1)}(\boldsymbol{k}) = -\frac{4\pi}{\Omega} \mathrm{i}^l \exp[p^2/\eta]\, p^{-l}$$

$$\times \sum_n \frac{|\boldsymbol{K}_n + \boldsymbol{k}|^l \exp[-(\boldsymbol{K}_n + \boldsymbol{k})^2/\eta]}{(\boldsymbol{K}_n + \boldsymbol{k})^2 - p^2}$$

$$\times \exp[\mathrm{i}(\boldsymbol{K}_n + \boldsymbol{k})\cdot(\boldsymbol{a}_i - \boldsymbol{a}_j)]\, Y_{lm}^*(\widehat{\boldsymbol{K}_n + \boldsymbol{k}}) \tag{6.29}$$

$$D_{lm}^{ij(2)}(\boldsymbol{k}) = -\frac{2^{l+1}}{\sqrt{\pi}\, p^l} \sum_{\substack{n\neq 0\\ \text{for } i=j}} |\boldsymbol{R}_n - \boldsymbol{a}_i + \boldsymbol{a}_j|^l$$

$$\times \exp[\mathrm{i}(\boldsymbol{k}\cdot \boldsymbol{R}_n)]\, Y_{lm}^*(\widehat{\boldsymbol{R}_n - \boldsymbol{a}_i + \boldsymbol{a}_j})$$

$$\times \int_{1/2\sqrt{\eta}}^{\infty} \xi^{2l} \exp\left[-\xi^2(\boldsymbol{R}_n - \boldsymbol{a}_i + \boldsymbol{a}_j)^2 + \frac{p^2}{4\xi^2}\right] \mathrm{d}\xi \tag{6.30}$$

$$D_{00}^{(3)} = -\frac{\sqrt{\eta}}{2\pi} \sum_{n=0}^{\infty} \frac{(p^2/\eta)^n}{(2n-1)n!}. \tag{6.31}$$

In eqs. (6.29) and (6.30), $\boldsymbol{R}_n$ and $\boldsymbol{K}_n$ are the real and reciprocal lattice vectors, respectively, and η ('Ewald parameter') is an arbitrary positive constant which determines the convergence of the series. The vectors $\boldsymbol{a}_i$ and $\boldsymbol{a}_j$ are the non-primitive translations, spanning the supermatrix in eq. (6.18) and Ω is the volume of the unit cell. A calculation of structure constants by summation over real lattice vectors is discussed by Korringa (1947). For a discussion of practical aspects of numerical calculations of the structure constants (eqs. 6.29 and 6.30), see in particular the work of Stocks et al. (1979). Particular attention is also drawn to the structure constants for two-dimensional periodicity (Kambe 1967a–c), which have to be used in the case of crystals with surfaces (semi-infinite crystals).

As in the case of the real space structure constants, reciprocal structure constants for displaced atoms can be defined (Gonis et al. 1985a). In the limit of infinitesimal small displacements for example, the reciprocal space displaced atom structure constants $\hat{g}_{LL'}(\boldsymbol{k})$ are given by

$$\hat{g}_{LL'}(\boldsymbol{k}) = g_{LL'}(\boldsymbol{k}) + \mathrm{i}\frac{4\pi}{3} p\rho \sum_{L''\mu} \mathrm{i}^{-l+l''} C_{LL'}^{L''} g_{L'',1\mu}(\boldsymbol{k}) Y_{1\mu}(-\hat{\rho}), \tag{6.32}$$

yielding the relativistic displaced atom structure constants in terms of a transformation with Clebsch–Gordan coefficients (eq. 6.26).

6.3. *The RKKR equations*

From eq. (6.19) it is evident that for a particular k-vector the following matrix equation has to be solved:

$$\boldsymbol{\tau}(\boldsymbol{k})[\boldsymbol{t}^{-1} - \boldsymbol{G}(\boldsymbol{k})] = \boldsymbol{1}, \tag{6.33}$$

where the matrix multiplying $\boldsymbol{\tau}(\boldsymbol{k})$ in eq. (6.33) usually is called the KKR-matrix (Kohn and Rostoker 1954, Ham and Segall 1961, Onodera and Okazaki 1966b). $\boldsymbol{\tau}(\boldsymbol{k})$ has poles of the type $(x-x_0)^{-1}$, whenever the determinant of the KKR-matrix vanishes:

$$\det[\boldsymbol{t}^{-1}-\boldsymbol{G}(\boldsymbol{k})]=0. \tag{6.34}$$

A particular energy for which eq. (6.34) applies is an eigenvalue ε_k of eq. (6.1), whereby the subscript $\boldsymbol{k}$ refers to the kth irreducible representation of the translational group. It should be noted in this context, that the BZ integral in eq. (6.19) is nothing but the resolution of the identity in terms of projections operators of the translational group (Weinberger 1983). For a given direction (k varying along a ray in k-space) an energy band is determined by compatible eigenvalues ε_k along this direction (see also table 4). Compatibility relations are subduction or induction relations for irreducible representations of the point group of the wave vector $\boldsymbol{k}$ with $\boldsymbol{k}$ varying along a ray. Equation (6.34) can be used to calculate the bandstructure of ordered phases fully relativistically.

6.4. *Single impurities, the RKKR–GF method*

For a simple lattice the site-diagonal scattering path operator $\boldsymbol{\tau}^{A,00}$ for an unrelaxed, isolated impurity of type A at the origin of the lattice ($i=j=0$) of an otherwise perfect host can be calculated from the site-diagonal scattering path operator of the host metal (eq. 6.10), $\boldsymbol{\tau}^{00}$, in the following way (Holzwarth 1975, Faulkner 1982, Weinberger 1982a; for a lengthy derivation see also Ehrenreich and Schwartz 1976)

$$\boldsymbol{\tau}^{A,00}=\mathbf{D}^{A,00}\,\boldsymbol{\tau}^{00}. \tag{6.35}$$

In eq. (6.35), $\mathbf{D}^{A,00}$ ('impurity matrix') is given by

$$\mathbf{D}^{A,00}=[\mathbf{1}+\mathbf{m}^{A}\boldsymbol{\tau}^{00}]^{-1}, \tag{6.36}$$

$$\mathbf{m}^{A}=\mathbf{t}_A^{-1}-\mathbf{t}^{-1}, \tag{6.37}$$

where $\mathbf{t}_A$ and $\mathbf{t}$ are the single site t-matrices of the impurity atom and the host, respectively. In the case of complex lattices the cell-diagonal scattering path operator for a single impurity on the sublattice s is given by

$$\boldsymbol{\tau}^{A,00}=\boldsymbol{D}_s^{A,00}\boldsymbol{\tau}^{00}, \tag{6.38}$$

$$\boldsymbol{D}_s^{A,00}=[\boldsymbol{1}+\boldsymbol{M}_s^{A}\boldsymbol{\tau}^{00}]^{-1}, \tag{6.39}$$

$$\boldsymbol{M}_s^{A}=\begin{pmatrix} \mathbf{m}^A\delta_{1s} & 0 & 0 & 0 \\ 0 & \mathbf{m}^A\delta_{2s} & & \\ \vdots & & & \mathbf{m}^A\delta_{ns} \end{pmatrix}, \tag{6.40}$$

where $\boldsymbol{\tau}^{00}$ is given by eq. (6.19). The supermatrix $\boldsymbol{M}_s^{A}$ has only one non-vanishing block, namely in the sth row and sth column.

Equations (6.35)–(6.37) can be used to describe impurities in metals such as

δ-PuAl, whereas the electronic structure of a single vacancy in pnictides and chalcogenides such as CeN or US can be calculated from eqs. (6.38)–(6.40).

In particular it should be noted that in eqs. (6.35) and (6.38) the site- or cell-diagonal scattering path operator for the host is needed, implying that also the host can be described properly (RKKR–Green's function method).

6.5. *The RKKR–CPA method*

The aim of the coherent potential approximation (CPA) is to find an ordered lattice of effective scatterers such that the motion of the electron through the effective lattice resembles the averaged motion of an electron in the random lattice. For a simple lattice of a binary substitutional alloy $A_x B_{1-x}$ the effective scatterers are described by the effective single site matrix $\mathbf{t}_C$

$$\boldsymbol{\tau}^{C,00} = \frac{1}{\Omega_{BZ}} \int [\mathbf{t}_C^{-1} - \mathbf{G}(\boldsymbol{k})]^{-1} \, d\boldsymbol{k}. \tag{6.41}$$

Within the 'single site approximation' for the CPA medium the unknown effective single site t-matrix is determined by the condition that the average scattering of impurities of the type A and B vanishes (Gyorffy and Stocks 1979, Faulkner 1982, Staunton et al. 1980). This condition usually is called the CPA condition:

$$x\boldsymbol{\tau}^{A,00} + (1-x)\boldsymbol{\tau}^{B,00} = \boldsymbol{\tau}^{C,00}, \tag{6.42}$$

or equivalently,

$$x\mathbf{m}^A\mathbf{D}^{A,00} + (1-x)\mathbf{m}^B\mathbf{D}^{B,00} = 0, \tag{6.43}$$

$$\mathbf{m}^\alpha = \mathbf{t}_\alpha^{-1} - \mathbf{t}_C^{-1}. \tag{6.44}$$

It implies that the average site-diagonal scattering path operator $\boldsymbol{\tau}^{C,00}$ equals the sum of restricted averages of the site-diagonal scattering path operator, whereby 'restricted' means (Faulkner 1982) that in the unit cell at the origin the single site t-matrix is fixed to $\mathbf{t}_A(\mathbf{t}_B)$ and the averaging is performed over $N-1$ unit cells (see also fig. 1). Restricting the averaging to one unit cell or subcell of a sublattice of a complex lattice (Klima et al. 1985), however, implies also that short range order is explicitely excluded (see also section 8).

For simple lattices eq. (6.42) can be rewritten as (Staunton et al. 1980)

$$\mathbf{t}_C^{-1} = xt_A^{-1} + (1-x)t_B^{-1} + \mathbf{m}^A\boldsymbol{\tau}^{C,00}\mathbf{m}^B. \tag{6.45}$$

Since $\mathbf{t}_C$ appears on both sides of this equation the effective single site t-matrix has to be determined self-consistently, whereby in each iteration the BZ integral (6.41) has to be performed. Since τ^{ii} is invariant under point group transformations, in the case of cubic systems double group cubic harmonics can be used for the angular momentum representation of τ^{ii}. A sublattice diagonal block has the matrix structure shown in fig. 2. From this figure one can see that for d-electron systems eq. (6.45) consists of 7 equations, whereas for f-electron systems already 17 equations have to be solved simultaneously. It is worthwhile to note that off-diagonal terms occur in fig. 2. In particular the off-diagonal terms due to $d_{3/2}/d_{5/2}$ and $f_{5/2}/f_{7/2}$

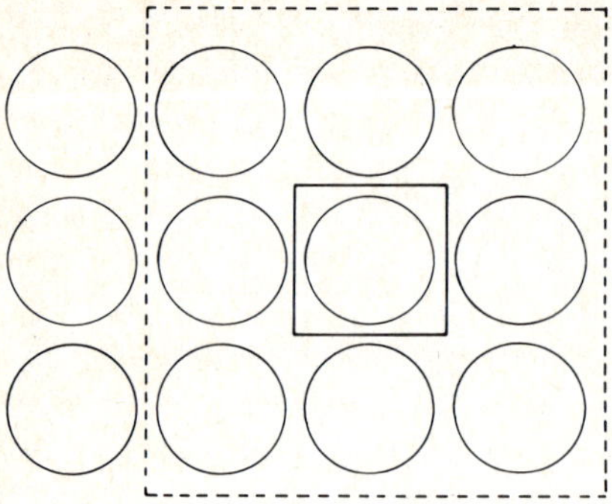

Fig. 1. Arrangement of scatterers and averaging regions.

interactions can be responsible for additional disorder, not present in a non-relativistic approach (Staunton et al. 1980). For complex lattices with one sublattice alloyed randomly, the CPA condition is similar to eq. (6.42), namely

$$x\boldsymbol{\tau}^{\mathrm{A},00}+(1-x)\boldsymbol{\tau}^{\mathrm{B},00}=\boldsymbol{\tau}^{\mathrm{C},00}, \tag{6.46}$$

where the $\boldsymbol{\tau}^{\alpha,00}$ are cell-diagonal scattering path operators and x is the concentration of components on the alloyed sublattice (Klima et al. 1985).

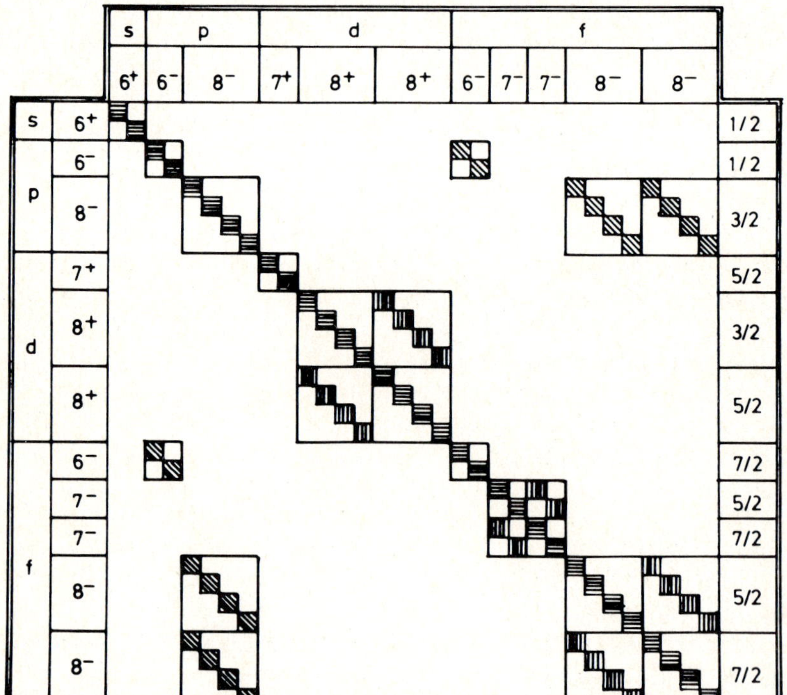

Fig. 2. Matrix structure of τ^{ii}_{ss} for cubic systems.

Equation (6.42) allows one to calculate the electronic structure of substitutional binary alloys such as δ-Pu_xAl_{1-x}, whereas by means of eq. (6.46) the electronic structure of mixed crystals such as UAs_xSb_{1-x} or $U_xTh_{1-x}Sb$ or of substoichiometric compounds such as UN_x can be obtained.

6.6. *Spectral quantities: densities of states and charge densities*

Once the site- or cell-diagonal scattering path operator is determined, spectral quantities such as densities of states (DOS), charge densities and Bloch spectral functions can be calculated. From eq. (5.5) it follows immediately that the first-order density matrix ('Dirac density matrix') $\gamma(\boldsymbol{r},\boldsymbol{r}',E)$ is given by

$$\gamma(\boldsymbol{r},\boldsymbol{r}',E) = -\pi^{-1}\,\mathrm{Im}\langle \boldsymbol{r}|G(E)|\boldsymbol{r}'\rangle, \tag{6.47}$$

since

$$\lim_{\delta\to 0}\frac{1}{x-x_0\pm \mathrm{i}\delta} = \mathrm{PP}\frac{1}{x-x_0}\mp\pi\delta(x-x_0), \tag{6.48}$$

where PP stands for principal part ('Hauptwert') and $\delta(x-x_0)$ is the Dirac δ-function (see in particular the very useful appendix A in the work by Messiah 1969). Quite clearly the diagonal first-order density matrix gives directly the density of states $n(E)$ and the charge density $\rho(r)$:

$$n(E) = \int_{\Omega}\gamma(\boldsymbol{r},\boldsymbol{r},E)\,\mathrm{d}\boldsymbol{r}, \tag{6.49}$$

$$\rho(\boldsymbol{r}) = \int_{-\infty}^{E_\mathrm{F}}\gamma(\boldsymbol{r},\boldsymbol{r},E)\,\mathrm{d}E, \tag{6.50}$$

whereby Ω is the volume of the unit cell.

6.6.1. *Densities of states*

Using eqs. (6.47), (6.49) and (6.20), the density of states (DOS) is in general given by the cell-diagonal scattering path operator

$$n(E) = -\pi^{-1}\int \mathrm{Im}\,G(\boldsymbol{r}_i,\boldsymbol{r}_i)\,\mathrm{d}\Omega_i = -\pi^{-1}\,\mathrm{Im}\,\mathrm{tr}\,\boldsymbol{F}^{ii}\boldsymbol{\tau}^{ii}, \tag{6.51}$$

where in the case of complex lattices as used in eq. (6.51) the unit cell Ω_i is a sum of subcells Ω_{is} corresponding to the sublattices s:

$$\Omega_i = \sum_s \Omega_{is}. \tag{6.52}$$

The matrix $\boldsymbol{F}^{ii}$ is block-diagonal:

$$\boldsymbol{F}^{ii} = \begin{pmatrix} \mathbf{F}^{ii}_{11} & 0 & 0 & 0 & \dots \\ 0 & \mathbf{F}^{ii}_{22} & 0 & 0 & \\ \vdots & & & & \\ 0 & & \dots & & \mathbf{F}^{ii}_{nn} \end{pmatrix}, \tag{6.53}$$

each block consisting of a matrix of integrals $F^{ii}_{ss,QQ'}$,

$$F^{ii}_{ss,QQ'} = \int Z^{is}_{Q}(\boldsymbol{r})\, Z^{is}_{Q'}(\boldsymbol{r})\, \mathrm{d}\Omega_{is}, \tag{6.54}$$

where the functions $Z^{is}_{Q}(\boldsymbol{r})$ are the scattering solutions (5.27) centred on the sth sublattice in the ith unit cell.

The integral (6.54) can be split up into a contribution coming from the integration over the volume of the corresponding muffin-tin $\Omega^{\mathrm{mt}}_{is}$ and the remainder $\bar{\Omega}_{is} = \Omega_{is} - \Omega^{\mathrm{mt}}_{is}$,

$$\begin{aligned} F^{ii}_{ss,QQ'} &= \int_{\Omega^{\mathrm{mt}}_{is}} Z^{is}_{Q'}(\boldsymbol{r})\, Z^{is}_{Q'}(\boldsymbol{r})\, \mathrm{d}\boldsymbol{r} + \int_{\bar{\Omega}_{is}} Z^{is}_{Q'}(\boldsymbol{r})\, Z^{is}_{Q'}(\boldsymbol{r})\, \mathrm{d}\boldsymbol{r} \\ &= S^{ii}_{ss,QQ'}\,\delta_{QQ'} + T^{ii}_{ss,QQ'}, \end{aligned} \tag{6.55}$$

whereby because of the spherical symmetry within the muffin-tin sphere the first term is diagonal in $Q = (\kappa\mu)$. Restricting the integration to the muffin-tins yields the so-called muffin-tin density of states, $n^{\mathrm{mt}}(E)$.

For simple lattices, eq. (6.51) reduces to

$$n(E) = -\pi^{-1}\,\mathrm{Im\,tr}\,\mathbf{F}^{ii}\boldsymbol{\tau}^{ii} \tag{6.56}$$

and the muffin-tin density of states to

$$n^{\mathrm{mt}}(E) = -\pi^{-1}\,\mathrm{Im\,tr}\,\mathbf{S}^{ii}\boldsymbol{\tau}^{ii}. \tag{6.57}$$

Using now the formulations of sections 6.4 and 6.5 the host, single impurity DOS and the DOS for concentrated alloys for simple lattices is given by (Staunton et al. 1980, Weinberger 1982a,b)

$$n(E) = \begin{cases} -\pi^{-1}\,\mathrm{Im\,tr}\,\mathbf{F}^{00}\boldsymbol{\tau}^{00}, & \text{host} \\ -\pi^{-1}\,\mathrm{Im\,tr}\,\mathbf{F}^{\mathrm{AA}}\mathbf{D}^{\mathrm{A},00}\boldsymbol{\tau}^{00}, & \text{single impurity of type A,} \\ -\pi^{-1}\{x\,\mathrm{Im\,tr}\,\mathbf{F}^{\mathrm{AA}}\mathbf{D}^{\mathrm{A},00}\boldsymbol{\tau}^{\mathrm{C},00} & \\ \quad +(1-x)\,\mathrm{Im\,tr}\,\mathbf{F}^{\mathrm{BB}}\mathbf{D}^{\mathrm{B},00}\boldsymbol{\tau}^{\mathrm{C},00}\} & \\ \quad = x n^{\mathrm{A}}(E) + (1-x) n^{\mathrm{B}}(E), & \text{concentrated alloy } \mathrm{A}_x\mathrm{B}_{1-x}. \end{cases} \tag{6.58}$$

In eq. (6.58), $n^{\mathrm{A}}(E)$ and $n^{\mathrm{B}}(E)$ are the so-called component densities of states, associated with atoms of type A and B, respectively. For pseudobinary systems $(\mathrm{A}_x\mathrm{B}_{1-x})\mathrm{X}$ with X being the component on the pure sublattice ($s=2$) and A and B the components on the alloyed sublattice ($s=1$), the DOS is given by (Klima et al. 1985)

$$\begin{aligned} n(E) = -\pi^{-1}\{& x\,\mathrm{Im\,tr}\,\mathbf{F}^{\mathrm{AA}}_{11}\mathbf{D}^{\mathrm{A},00}_{1,11}\boldsymbol{\tau}^{\mathrm{C},00}_{11} + (1-x)\,\mathrm{Im\,tr}\,\mathbf{F}^{\mathrm{BB}}_{11}\mathbf{D}^{\mathrm{B},00}_{1,11}\boldsymbol{\tau}^{\mathrm{C},00}_{11} \\ & + \mathrm{Im\,tr}\,\mathbf{F}^{\mathrm{XX}}_{22}\boldsymbol{\tau}^{\mathrm{C},00}_{22}\}. \end{aligned} \tag{6.59}$$

It should be noted that since for the muffin-tin DOS (6.57) the integrals $S^{ii}_{QQ'}$ are diagonal in Q, $n^{\mathrm{mt}}(E)$ can be written as a sum of κ-like densities of states $n_{\kappa}(E)$. In the case of a simple lattice, $n^{\mathrm{mt}}(E)$ is then simply given by

$$n^{\mathrm{mt}}(E) = \sum_{\kappa} n_{\kappa}(E), \tag{6.60}$$

$$n_\kappa(E) = -\pi^{-1} S_\kappa^{00} \sum_\mu \operatorname{Im} \tau^{00}_{\kappa\mu,\kappa\mu}, \tag{6.61}$$

where S_κ^{00} is a radial integral of the type

$$S_\kappa^{00} = \int_{\Omega_0^{\mathrm{mt}}} [g_\kappa^2(r) + f_\kappa^2(r)] r^2 \, \mathrm{d}r, \tag{6.62}$$

and $g_\kappa(r)$ and $f_\kappa(r)$ are the solutions of the radial Dirac equation (3.20), normalized to single site scattering (eq. 5.26).

6.6.2. *Charge densities*

Considering for the sake of simplicity a simple lattice, eq. (6.50) can be used to calculate the charge density from the site-diagonal scattering path operator:

$$\begin{aligned}\rho(\mathbf{r}) &= -\pi^{-1} \int_{-\infty}^{E_\mathrm{F}} \operatorname{Im}\left[\sum_{QQ'} Z_Q^i(\mathbf{r})\, \tau^{ii}_{QQ'}\, Z_{Q'}^i(\mathbf{r}) \right] \mathrm{d}E \\ &= -\pi^{-1} \int_{-\infty}^{E_\mathrm{F}} \operatorname{Im}\left[\sum_{QQ'} \tau^{ii}_{QQ'} \{ g_\kappa^i(r)\, g_{\kappa'}^i(r) \langle \hat{r} | \kappa\mu \rangle \langle \kappa'\mu' | \hat{r} \rangle \} \right. \\ &\qquad \left. + f_\kappa^i(r)\, f_{\kappa'}^i(r) \langle \hat{r} | -\kappa\mu \rangle \langle -\kappa'\mu' | \hat{r} \rangle \} \right] \mathrm{d}E,\end{aligned} \tag{6.63}$$

where $g_\kappa^i(r)$ and $f_\kappa^i(r)$ are the radial amplitudes of the scattering solutions (5.26). The product of the spin spherical harmonics is given in terms of the addition theorem of spherical harmonics as

$$\langle \hat{r} | \kappa\mu \rangle \langle \kappa'\mu' | \hat{r} \rangle = \sum_{s=\pm 1/2} c(l\, j\, \tfrac{1}{2};\ \mu - s, s)\, c(l'\, j'\, \tfrac{1}{2};\ \mu' - s, s) \sum_{LM} C^{LM}_{l\mu - s, l'\mu' - s} \langle \hat{r} | LM \rangle. \tag{6.64}$$

In the case of cubic systems the use of double group cubic harmonics results in an enormous reduction of terms (see also fig. 2). The so-called muffin-tin charge density $\rho(r)$ (spherical average of $\rho(\mathbf{r})$) is defined by taking only the $L = M = 0$ term in eq. (6.64). However, since

$$\sum_{s=\pm 1/2} c(l\, j\, \tfrac{1}{2};\ \mu - s, s)\, c(l'\, j'\, \tfrac{1}{2};\ \mu' - s, s)\, C^{00}_{l\mu - s, l'\mu' - s} \langle \hat{r} | 00 \rangle = \frac{1}{\sqrt{4\pi}}\, \delta_{ll'} \delta_{\mu\mu'} \delta_{jj'} \tag{6.65}$$

the muffin-tin charge density is simply given by

$$\rho^i(r) = -\frac{1}{4\pi^2} \sum_{\kappa\mu} \int_{-\infty}^{E_\mathrm{F}} \operatorname{Im} \tau^{ii}_{\kappa\mu,\kappa\mu} \{ |g_\kappa^i(r)|^2 + |f_\kappa^i(r)|^2 \} \, \mathrm{d}E. \tag{6.66}$$

Using the corresponding expressions for the site-diagonal scattering path operator for the host, for a single impurity or for concentrated alloys, charge densities or muffin-tin charge densities are found in a similar way as discussed for the density of states. In particular for a binary alloy the muffin-tin charge density is given by

$$\rho(r) = x \rho^{\mathrm{A}}(r) + (1 - x) \rho^{\mathrm{B}}(r), \tag{6.67}$$

where $\rho^{\alpha}(r)$ is calculated by replacing in eq. (6.66) $\boldsymbol{\tau}^{ii}$ by $\boldsymbol{\tau}^{\alpha,ii}$. In eq. (6.67), $\rho^{A}(r)$ and $\rho^{B}(r)$ are the component muffin-tin charge densities.

6.7. *Bloch spectral functions*

Bloch spectral functions are defined by the imaginary part of a representation of the Green's function diagonal in momentum space (Faulkner and Stocks 1980, Faulkner 1982, Weinberger et al. 1982b):

$$\begin{aligned} A_{B}(\boldsymbol{k},E) &= -\pi^{-1}\,\mathrm{Im}\langle \boldsymbol{k}|G|\boldsymbol{k}\rangle \\ &= -\pi^{-1}\,\mathrm{Im}\left\{\sum_{n} e^{i\boldsymbol{k}\cdot\boldsymbol{R}_n}\int_{\Omega_0} G(\boldsymbol{r}_0,\boldsymbol{r}_0+\boldsymbol{R}_n)\,d\boldsymbol{r}_0\right\} \\ &= -\pi^{-1}\left\{\mathrm{Im}\int_{\Omega_0} G(\boldsymbol{r}_0,\boldsymbol{r}_0)\,d\boldsymbol{r}_0+\sum_{n\neq 0} e^{ik\cdot \boldsymbol{R}_n}\int_{\Omega_0} G(\boldsymbol{r}_0,\boldsymbol{r}_0+\boldsymbol{R}_n)\,d\boldsymbol{r}_0\right\}. \end{aligned} \tag{6.68}$$

For a pure (ordered) system the Bloch spectral functions are nothing but a sum of Dirac δ-functions:

$$A_{B}(\boldsymbol{k},E)=\sum_{n}\delta(E-\varepsilon_{\boldsymbol{k}}^{n}), \tag{6.69}$$

where the label n ('band index') is a numbering index for the eigenvalues at a given wavevector $\boldsymbol{k}$. From eq. (6.68) one realizes that (a) the periodicity of unit cells is implied, and (b) non-site (non-cell) diagonal elements of the Green's function are needed. From eq. (6.20) and in analogy to eq. (6.51) follows that in general for a complex lattice we have (Schadler et al. 1985)

$$\mathrm{Im}\int G(\boldsymbol{r}_i,\boldsymbol{r}_j)\,d\Omega_i=\mathrm{Im}\,\mathrm{tr}\,\boldsymbol{F}^{ij}\boldsymbol{\tau}^{ij}. \tag{6.70}$$

The supermatrix $\boldsymbol{F}^{ij}$ is block-diagonal (see eqs. 6.53 and 6.54), each sublattice diagonal block $\mathbf{F}^{ij}_{ss}$ consists of integrals of the type

$$F^{ij}_{ss,QQ'}=\int Z^{is}_{Q}(\boldsymbol{r})\,Z^{js}_{Q'}(\boldsymbol{r})\,d\Omega_{is}. \tag{6.71}$$

For a simple lattice there is of course only one such block. The averaging of the Green's function is now slightly more tedious, since restricted averages keeping the single site t-matrix at is and js fixed to $\mathbf{t}_A$ or $\mathbf{t}_B$ are needed (Faulkner, 1982, Schadler et al. 1985). Summing over all restricted averages implies therefore a quadratic concentration dependence for the Bloch spectral functions.

For a simple lattice of a binary alloy A_xB_{1-x} the Bloch spectral functions are given by (Faulkner and Stocks 1980, Weinberger et al. 1982b)

$$\begin{aligned} A_{B}(\boldsymbol{k},E) = -\pi^{-1}\Bigg\{ &\sum_{\alpha,\beta=A,B} c_\alpha c_\beta\,\mathrm{Im}\,\mathrm{tr}\,\mathbf{F}^{\alpha\beta}\mathbf{D}^{\alpha,00}\,\hat{\boldsymbol{\tau}}(k)\mathbf{D}^{\beta,00} \\ &+\sum_{\alpha=A,B} c_\alpha\,\mathrm{Im}\,\mathrm{tr}\,\mathbf{F}^{\alpha\alpha}\mathbf{D}^{\alpha,00}\boldsymbol{\tau}^{C,00}\Bigg\}, \end{aligned} \tag{6.72}$$

whereby

$$\hat{\tau}(\boldsymbol{k}) = \boldsymbol{\tau}^{\mathrm{C}}(\boldsymbol{k}) - \boldsymbol{\tau}^{\mathrm{C},00}, \tag{6.73}$$

$$\boldsymbol{\tau}^{\mathrm{C}}(\boldsymbol{k}) = [\mathbf{t}_{\mathrm{C}}^{-1} - \mathbf{G}(\boldsymbol{k})]^{-1}. \tag{6.74}$$

For pseudobinary alloys of the type $(A_x B_{1-x})X$ with X being the atomic species on the ordered sublattice ($s=2$) the Bloch spectral functions are given by (Schadler et al. 1985)

$$A_{\mathrm{B}}(\boldsymbol{k}, E) = -\pi^{-1}\left\{\sum_{\alpha=\mathrm{A,B}} \mathrm{Im\,tr}\, \boldsymbol{W}_1^{\alpha\alpha} + \sum_{\alpha,\beta=\mathrm{A,B}} \mathrm{Im\,tr}\, \boldsymbol{U}_1^{\alpha\beta}\right\}, \tag{6.75}$$

$$W_{1,ss}^{\alpha\alpha} = \begin{cases} c_\alpha \mathbf{F}_{11}^{\alpha} \mathbf{D}_{1,11}^{\alpha,00} \boldsymbol{\tau}_{11}^{\mathrm{C},00}, & s=1, \\ \mathbf{F}_{22}^{\mathrm{XX}} \boldsymbol{\tau}_{22}^{\mathrm{C},00}, & s=2, \end{cases} \tag{6.76}$$

$$U_{1,ss}^{\alpha\beta} = \begin{cases} c_\alpha c_\beta \mathbf{F}_{11}^{\alpha\beta} \mathbf{D}_{1,11}^{\alpha,00} \hat{\boldsymbol{\tau}}_{11}^{\mathrm{C}}(\boldsymbol{k}) \mathbf{D}_{1,11}^{\beta,00}, & s=1, \\ \mathbf{F}_{22}^{\mathrm{X}} \hat{\boldsymbol{\tau}}_{22}^{\mathrm{C}}(\boldsymbol{k}), & s=2, \end{cases} \tag{6.77}$$

$$\hat{\boldsymbol{\tau}}^{\mathrm{C}}(\boldsymbol{k}) = \boldsymbol{\tau}^{\mathrm{C}}(\boldsymbol{k}) - \boldsymbol{\tau}^{\mathrm{C},00}, \tag{6.78}$$

$$\boldsymbol{\tau}^{\mathrm{C}}(\boldsymbol{k}) = [\boldsymbol{t}_{\mathrm{C}}^{-1} - \boldsymbol{G}(\boldsymbol{k})]^{-1}. \tag{6.79}$$

In eqs. (6.76) the $\mathbf{D}_{1,ss}^{\alpha,00}$ are the blocks of the supermatrix defined in eq. (6.39).

6.7.1. *Projected Bloch spectral functions*

Rewriting eq. (6.41) by adding and subtracting $\langle \tau_{00} \rangle^{-1}$ to the integrand (Gonis et al. 1985a)

$$\begin{aligned} \langle \boldsymbol{\tau}_{00} \rangle = \boldsymbol{\tau}^{\mathrm{C},00} &= \frac{1}{\Omega_{\mathrm{BZ}}} \int [\langle \boldsymbol{\tau}_{00} \rangle^{-1} - \langle \boldsymbol{\tau}_{00} \rangle^{-1} + \mathbf{t}_{\mathrm{C}}^{-1} - \mathbf{G}(\boldsymbol{k})]^{-1}\, \mathrm{d}\boldsymbol{k} \\ &= \frac{1}{\Omega_{\mathrm{BZ}}} \int [\langle \boldsymbol{\tau}_{00} \rangle^{-1} + \boldsymbol{\Delta} - G(\mathbf{k})]^{-1}\, \mathrm{d}\boldsymbol{k} \end{aligned} \tag{6.80}$$

one gains the advantage that the 'renormalized interactor' $\boldsymbol{\Delta}$ (eq. 6.15) and $\mathbf{G}(\boldsymbol{k})$ are 'structural quantities' only, which are by definition not affected by disorder. The tensorial product between projections $P_{i\alpha}|i\rangle$ on two different sites, $|i\rangle$ representing a site (cell, cluster) dependent state

$$P_{i\alpha}|i\rangle = \delta_{i\alpha}|i\rangle \tag{6.81}$$

$$\delta_{i\alpha} = \begin{cases} 1, & \text{for } |i\rangle \text{ occupied by } \alpha \\ 0, & \text{otherwise} \end{cases}$$

is represented by the following 'projection' matrix (Gonis et al. 1985a, Schadler et al. 1985)

$$\begin{aligned} \mathscr{P}_{\mathrm{A,B}}^{ij} &= \begin{pmatrix} \delta_{i\mathrm{A}}\delta_{j\mathrm{A}} & \delta_{i\mathrm{A}}\delta_{j\mathrm{B}} \\ \delta_{i\mathrm{B}}\delta_{j\mathrm{A}} & \delta_{i\mathrm{B}}\delta_{j\mathrm{B}} \end{pmatrix} \\ &= \begin{pmatrix} \delta_{i\mathrm{A}}\delta_{j\mathrm{A}} & 0 \\ 0 & 0 \end{pmatrix} + \begin{pmatrix} 0 & \delta_{i\mathrm{A}}\delta_{j\mathrm{B}} \\ 0 & 0 \end{pmatrix} + \begin{pmatrix} 0 & 0 \\ \delta_{i\mathrm{B}}\delta_{j\mathrm{A}} & 0 \end{pmatrix} + \begin{pmatrix} 0 & 0 \\ 0 & \delta_{i\mathrm{B}}\delta_{j\mathrm{B}} \end{pmatrix} \\ &= \mathbf{P}_{\mathrm{AA}}^{ij} + \mathbf{P}_{\mathrm{AB}}^{ij} + \mathbf{P}_{\mathrm{BA}}^{ij} + \mathbf{P}_{\mathrm{BB}}^{ij}, \end{aligned} \tag{6.82}$$

such that

$$\mathscr{P}^{ii}_{\mathrm{A,B}} \otimes \tau^{ii} = \begin{pmatrix} \tau^{ii}_{\mathrm{AA}} & 0 \\ 0 & \tau^{ii}_{\mathrm{BB}} \end{pmatrix} \equiv \begin{pmatrix} \tau^{ii}_{\mathrm{A}} & 0 \\ 0 & \tau^{ii}_{\mathrm{B}} \end{pmatrix},$$
$$\mathscr{P}^{ij}_{\mathrm{A,B}} \otimes \tau^{ij} = \begin{pmatrix} \tau^{ij}_{\mathrm{AA}} & \tau^{ij}_{\mathrm{AB}} \\ \tau^{ij}_{\mathrm{BA}} & \tau^{ij}_{\mathrm{BB}} \end{pmatrix}, \tag{6.83}$$

where $\otimes$ denotes the Kronecker product. The Kronecker product of $\mathscr{P}^{00}_{\mathrm{A,B}}$ with $\mathbf{\Delta} - \mathbf{G}(\boldsymbol{k})$ is always a supermatrix of the form

$$\mathscr{P}^{00}_{\mathrm{A,B}} \otimes [\mathbf{\Delta} - \mathbf{G}(\boldsymbol{k})] = \Delta - \boldsymbol{G}(\boldsymbol{k}) = \begin{pmatrix} \mathbf{\Delta} - \mathbf{G}(\boldsymbol{k}) & \mathbf{\Delta} - \mathbf{G}(\boldsymbol{k}) \\ \mathbf{\Delta} - \mathbf{G}(\boldsymbol{k}) & \mathbf{\Delta} - \mathbf{G}(\boldsymbol{k}) \end{pmatrix}. \tag{6.84}$$

In particular the Kronecker product of $\mathscr{P}^{00}_{\mathrm{A,B}}$ with $\boldsymbol{\tau}(\boldsymbol{k})$ gives

$$\mathscr{P}^{00}_{\mathrm{A,B}} \otimes \boldsymbol{\tau}(\boldsymbol{k}) = \boldsymbol{\tau}(\boldsymbol{k}) = \begin{pmatrix} \tau_{\mathrm{AA}}(\boldsymbol{k}) & \tau_{\mathrm{AB}}(\boldsymbol{k}) \\ \tau_{\mathrm{BA}}(\boldsymbol{k}) & \tau_{\mathrm{BB}}(\boldsymbol{k}) \end{pmatrix}$$
$$= \begin{pmatrix} \dfrac{1}{c_{\mathrm{A}}}\boldsymbol{\tau}_{\mathrm{A}}^{-1} + \mathbf{\Delta} - \mathbf{G}(\boldsymbol{k}) & \mathbf{\Delta} - \mathbf{G}(\boldsymbol{k}) \\ \mathbf{\Delta} - \mathbf{G}(\boldsymbol{k}) & \dfrac{1}{c_{\mathrm{B}}}\boldsymbol{\tau}_{\mathrm{B}}^{-1} + \mathbf{\Delta} - \mathbf{G}(\boldsymbol{k}) \end{pmatrix}. \tag{6.85}$$

It is worthwhile to note that the Kronecker product in eq. (6.83) is well-defined, regardless whether τ^{ij} is itself a supermatrix or not. Furthermore, a ternary alloy simply implies the Kronecker product with a 3×3 'projection matrix'.

From eqs. (6.80), (6.82) and (6.85) follows that the Bloch spectral functions for a simple lattice of a binary alloy $\mathrm{A}_x\mathrm{B}_{1-x}$ can be written as a sum over projected, integrated traces of $\boldsymbol{\tau}(k)$:

$$A_{\mathrm{B}}(\boldsymbol{k},E) = -\pi^{-1} \sum_{\alpha,\beta=\mathrm{A,B}} \mathrm{Im\,tr}\,\mathbf{F}^{\alpha\beta}\,\boldsymbol{\tau}_{\alpha\beta}(\boldsymbol{k}) = \sum_{\alpha,\beta=\mathrm{A,B}} A^{\alpha\beta}_{\mathrm{B}}(\boldsymbol{k},E), \tag{6.86}$$

where $\mathbf{F}^{\alpha\beta}$ is given by eq. (6.71). For pseudobinary alloys, i.e., for complex lattices, the cell-diagonal scattering path operator and $\boldsymbol{\tau}(k)$ are already supermatrices (eqs. 6.16 and 6.19) and $\mathscr{P}^{00}_{\mathrm{A,B}} \otimes \boldsymbol{\tau}(\boldsymbol{k}) = \boldsymbol{\tau}(\boldsymbol{k})$ is therefore a supermatrix where the elements are themselves supermatrices:

$$\boldsymbol{\tau}(\boldsymbol{k}) = \begin{pmatrix} \boldsymbol{\tau}_{\mathrm{AA}}(\boldsymbol{k}) & \boldsymbol{\tau}_{\mathrm{AB}}(\boldsymbol{k}) \\ \boldsymbol{\tau}_{\mathrm{BA}}(\boldsymbol{k}) & \boldsymbol{\tau}_{\mathrm{BB}}(\boldsymbol{k}) \end{pmatrix}. \tag{6.87}$$

For pseudobinary alloys the Bloch spectral functions are therefore given by (Schadler et al. 1985)

$$A_{\mathrm{B}}(\boldsymbol{k},E) = -\pi^{-1} \sum_{\alpha,\beta=\mathrm{A,B}} \mathrm{Im\,tr}\,\boldsymbol{F}^{\alpha\beta}\,\boldsymbol{\tau}_{\alpha\beta}(\boldsymbol{k}) = \sum_{\alpha,\beta=\mathrm{A,B}} A^{\alpha\beta}_{\mathrm{B}}(\boldsymbol{k}) \tag{6.88}$$

whereby the supermatrix $\boldsymbol{F}^{\alpha\beta}$ is defined in eq. (6.71). The quantities $A_{\mathrm{B}}^{\alpha\beta}(k,E)$ in eqs. (6.86) and (6.88) are 'component projected' Bloch spectral functions.

6.7.2. *Conceptual significance of Bloch spectral functions*

The Bloch spectral functions constitute the bandstructure along a given direction in $\boldsymbol{k}$-space. For an ordered system one traditionally plots an ε_k^n versus $\boldsymbol{k}$ diagram, which of course is nothing but plotting the positions of the Dirac δ-functions in eq. (6.69) along this direction as a function of the energy parameter E. For disordered systems states correspond to peaks in the Bloch spectral functions, which, represented at a given value of $\boldsymbol{k}$ and varying the energy E, can show considerable broadening in the parameter space of E. Keeping the energy fixed and varying $\boldsymbol{k}$ along a set of directions generates sheets of peaks ('bands') broadened in momentum space. In particular at the Fermi energy these sheets are nothing but cuts through the Fermi surface. A disordered system is said to have a well-defined Fermi surface if the halfwidth of the peaks in the Bloch spectral functions with constant E is small as compared to the scale of the Brillouin zone (for an interpretation of the broadening in terms of life times, see, for example, Ehrenreich and Schwartz 1976).

Bloch spectral functions have to be considered as a general concept for the term 'bandstructure', since they define the evolution of states in the parameter space of the energy and in momentum space for both ordered and disordered systems. The component projected Bloch spectral functions (6.86) and (6.88) are in this context nothing but a more general concept for what is usually called for ordered systems the 'character of bands'.

7. Self-consistent RKKR methods

For a given set of scattering potentials $V_i^{\mathrm{eff}}(r_i)$ (and fields $B_i^{\mathrm{eff}}(r_i)$) sections 5 and 6 provide schemes for calculating fully relativistically the electronic structure for ordered systems – with or without single impurities or vacancies – and for substitutionally disordered materials. Since one of the results is the charge density, the scattering potentials can be determined self-consistently, providing thus an ab initio type description for the electronic structure of ordered and disordered matter. The step of self-consistency for the scattering potentials, however, seems to be more than a technicality and requires a somewhat detailed discussion.

The RKKR-matrix or its inverse $\boldsymbol{\tau}(\boldsymbol{k})$ (eq. 6.33) depends on the wave vector $\boldsymbol{k}$ and the energy parameter E. In order to determine the site- (cell-)diagonal scattering path operator $\boldsymbol{\tau}^{00}$, in principle, two approaches are possible: (a) the energy parameter E is sampled discretely and $\boldsymbol{k}$ is varied continuously, and (b) E varies continuously and $\boldsymbol{k}$ is sampled discretely. The first approach is sometimes called 'constant energy mode', whereas the second usually is referred to as 'constant $\boldsymbol{k}$-mode'. The 'constant $\boldsymbol{k}$ mode' implies that the translational group is mapped homomorphically onto a 'substitutional translation group' of some finite order ${}^0T$. This is important to note, since sum rules such as the number of electrons per unit cell are nothing but sum rules for the irreducible representations of the 'substitutional translation group'. In particular, this implies that the BZ integral in eq. (6.19), for example, is replaced by a sum, since the resolution of the identity of the projection operators P_k of the 'substitutional translation group' is given by

$$\sum_{k=1}^{0_T} P_k = 1. \tag{7.1}$$

Within the 'constant $\boldsymbol{k}$ mode', however, the energy varies continuously, implying e.g., that for a given order of the 'substitutional translational group' the eigenvalues ε_k^n of an ordered system can be determined in principle exactly from eq. (6.34).

In the 'constant E mode' the BZ integral (6.19) can in principle be calculated exactly; however, E has to be sampled on a discrete mesh. Choosing even a very fine energy grid, it can in principle happen that in the density of states van Hove singularities do not show up, since E is only very close to the exact eigenvalue for a topological point.

According to these two modes of the RKKR method, two practical methods for calculating self-consistently charge densities, and, consequently, the scattering potentials have to be discussed. Since the 'constant-k mode' has been used only for ordered systems and also, as will be shown in section 7.2, makes no use of the analytical properties of the Green's function, this mode will be discussed for ordered systems only. Both schemes will be discussed for the case of $B_i^{\text{eff}}(r_i)=0$.

7.1. '*Constant-$\boldsymbol{k}$*' *self-consistency*

Having found an energy ε ('eigenvalue'), where the RKKR determinant (6.34) vanishes, one has to determine the wavefunction or directly its contribution to the charge density. Expanding the wavefunction within the ith muffin-tin sphere $\Psi^i(\boldsymbol{r}_i,\varepsilon)$ as

$$\Psi^i(\boldsymbol{r}_i,\varepsilon)=\sum_Q C^i_Q(\varepsilon)\begin{pmatrix} g^i_\kappa(r_i,\varepsilon) & \langle\hat{r}_i|\kappa\mu\rangle \\ \mathrm{i}f^i_\kappa(r_i,\varepsilon) & \langle\hat{r}_i|-\kappa\mu\rangle \end{pmatrix}, \tag{7.2}$$

where $g^i_\kappa(r_i,\varepsilon)$ and $f^i_\kappa(r_i,\varepsilon)$ are the solutions of the radial Dirac equation at the energy of the found eigenvalue ε_k^μ. The expansion coefficients $C^i_Q(\varepsilon)$ are related to the RKKR-matrix as follows (Onodera and Okazaki 1966b):

$$\sum_j\sum_Q \Lambda^{ij}_{QQ'}(\varepsilon)\, C^j_{Q'}(\varepsilon)=0, \tag{7.3}$$

$$\Lambda^{ij}_{QQ'}(\varepsilon)=W^i_\kappa(\varepsilon)\,[B^{ij}_{QQ'}(\boldsymbol{k},\varepsilon)-\delta_{ij}\delta_{QQ'}K^i_\kappa(\varepsilon)]\;W^j_{\kappa'}(\varepsilon). \tag{7.4}$$

In eq. (7.4) the $B^{ij}_{QQ'}(\boldsymbol{k},\varepsilon)$ are the structure constants defined in eqs. (6.26)–(6.31), $K^i_\kappa(\varepsilon)$ is the reactance for the ith scatterer (5.25) and $W^i_\kappa(\varepsilon)$ is given by

$$W^i_\kappa(\varepsilon)=\det\begin{pmatrix} g^i_\kappa(R^i_s,\varepsilon) & cf^i_\kappa(R^i_s,\varepsilon) \\ j_l(\sqrt{\varepsilon}\,R^i_s) & \sqrt{\varepsilon}\,S_\kappa j_{\bar{l}}(\sqrt{\varepsilon}\,R^i_s) \end{pmatrix}. \tag{7.5}$$

It should be noted that eqs. (7.3)–(7.5) are nothing but the Korringa picture of scattering, namely in terms of incident and scattered waves (see also Rosicky et al.

1976). The expansion coefficients $C^i_Q(\varepsilon)$ in eq. (7.3) have to be normalized (for a discussion see e.g., Ham and Segall 1961). Once the expansion coefficients are normalized the muffin-tin charge density within the ith muffin-tin sphere is given by (Weinberger et al. 1979)

$$\rho^i(r_i) = \sum_{\varepsilon \leqslant E_F} |C^i_Q(\varepsilon)|^2 [g^i_\kappa(r,\varepsilon)^2 + f^i_\kappa(r,\varepsilon)^2]. \tag{7.6}$$

It should be noted that in eq. (7.6) a sum over discrete energies is involved. The sum over the degeneracies of these eigenvalues determines the Fermi energy. If N is the number of electrons per unit cell and $n(\varepsilon^n_k)$ is the degeneracy of the nkth eigenvalue, then the Fermi energy is defined by

$$N = \sum_{\varepsilon^n_k \leqslant E_F} n(\varepsilon^n_k). \tag{7.7}$$

The Fermi energy is therefore by definition the highest occupied discrete one-electron energy level in a one-electron energy spectrum where the eigenvalues are classified according to the irreducible representations of the 'substitutional translation group'. Quite clearly, the accuracy of eqs. (7.6) and (7.7) is determined by the order of the 'substitutional translation group', i.e., by the 'quality' of the homomorphic mapping.

7.2. *'Constant-E' self-consistency*

7.2.1. *Complex energies*

From eq. (6.66) one can see that in a 'constant E mode' for the charge density an energy integral is involved, which is not easy to calculate since τ^{ii} has $(x-x_0)^{-1}$ type poles. However, as suggested by Zeller and Dederichs (1982) and discussed in quite some detail by Johnson et al. (1984) all 'intrinsic KKR quantities' such as the single site t-matrix and the structure constants can be calculated at complex energies such that instead of an integration along the real axis, a contour integration in the complex plane can be performed. If $z = \varepsilon + \mathrm{i}\delta$ then the reactance $K_\kappa(z)$ (see eq. 5.25) is given by

$$\begin{aligned} K_\kappa(z) = &-z^{1/2}[cf_\kappa(R_s,z)\, j_l(\sqrt{z},R_s) - \sqrt{z}\, S_\kappa j_{\bar{l}}(\sqrt{z},R_s)\, g_\kappa(R_s,z)] \\ &\times [cf_\kappa(R_s,z)\, n_l(\sqrt{z},R_s) - \sqrt{z}\, S_\kappa n_{\bar{l}}(\sqrt{z},R_s)\, g_\kappa(R_s,z)]^{-1}, \end{aligned} \tag{7.8}$$

whereby $f_\kappa(r,z)$ and $g_\kappa(r,z)$ are now the radial amplitudes for a complex energy parameter in eq. (3.20), and the single site t-matrix is given by

$$t_\kappa(z) = K_\kappa(z)[1 + \mathrm{i}\sqrt{z}\, K_\kappa(z)]^{-1}. \tag{7.9}$$

The Bessel and Neumann functions for complex arguments in eq. (7.8) are of course well-defined (see e.g. Abramowitz and Stegun 1970). For the structure constants in eqs. (6.26)–(6.31) only powers of z are involved in order to be applicable also for complex energies. Since the scattering solutions are now complex functions the inhomogenity term in eq. (6.20) has to be taken into account. In terms of a contour

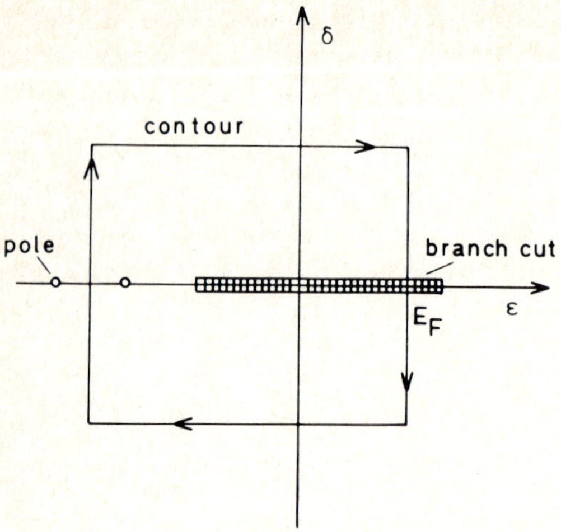

Fig. 3. Schematic paths for contour integrations.

integral $[G(z^*)=G^*(z)$; see fig. 3] the charge density is given as (eq. 6.63)

$$\rho^i(\boldsymbol{r}_i)=-\pi^{-1}\,\mathrm{Im}\int_c \sum_{QQ'}[Z^i_Q(\boldsymbol{r}_i,z)\,\tau^{ii}_{QQ'}(z)\,Z^i_{Q'}(\boldsymbol{r}_i,z)-Z^i_Q(\boldsymbol{r}_i,z)\,J^i_{Q'}(\boldsymbol{r}_i,z)\,\delta_{QQ'}]\,\mathrm{d}z. \quad (7.10)$$

Using complex energies not only broadens the pole structure in the parameter space of energies, but also in momentum space, implying thus that (a) less $\boldsymbol{k}$-values are needed to perform the BZ integral

$$\boldsymbol{\tau}^{ii}(z)=\frac{1}{\Omega_{\mathrm{BZ}}}\int[\mathbf{t}^{-1}(z)-\mathbf{G}(\boldsymbol{k},z)]^{-1}\,\mathrm{d}\boldsymbol{k} \quad (7.11)$$

satisfactorily, and (b) in eq. (7.10) a 'smooth integrand' is obtained (see e.g. Zeller and Dederichs 1982). By means of complex energies, eqs. (7.10) and (7.11) can be calculated numerically for ordered and disordered systems ($\mathbf{t}(z)=\mathbf{t}_{\mathrm{C}}(z)$).

7.2.2. *Riemann–Cauchy continuation*

Exploiting the analyticity of the Green's function $G(z)$ and the scattering path operators $\boldsymbol{\tau}^{ii}(z)$ analytical continuations to the real axis can be obtained numerically from the Riemann–Cauchy conditions. Let $f(z)$ be a complex function that has a derivative at $z_0=x_0+\mathrm{i}y_0$ (see e.g. Churchill 1960):

$$f(z)=u(x,y)+\mathrm{i}v(x,y) \quad (7.12)$$

then according to the Riemann–Cauchy conditions at the point z_0 we have

$$\frac{\partial u}{\partial x}=\frac{\partial v}{\partial y},\qquad \frac{\partial u}{\partial y}=-\frac{\partial v}{\partial x}, \quad (7.13)$$

$$f'(z)=\frac{\partial u}{\partial x}+\mathrm{i}\frac{\partial v}{\partial x}=\frac{\partial v}{\partial y}-\mathrm{i}\frac{\partial u}{\partial y}. \quad (7.14)$$

Calculating therefore $\tau^{ii}(z)$ parallel to the real axis in the complex plane, eq. (7.13) can be used to find the values of the matrix elements of τ^{ii} on the real axis and thus the corresponding density of states (section 6.6.1) (Klima et al. 1985). In principle therefore at no stage of the calculation real energies are required. However, since the analytical continuation is a numerical procedure, numerical difficulties can arise in the vicinity of very flat bands.

In the 'constant E mode' the Fermi energy is found either by a contour integration of $n(z)$ (compare to eq. 6.56) using a couple of different branches of the path leading back to the real axis (see fig. 3):

$$n(z) = -\pi^{-1} \int_{\Omega_i} \mathrm{tr}[\mathbf{Z}^i(r,z)\,\boldsymbol{\tau}^{ii}(z)\,\mathbf{Z}^i(r,z) - \mathbf{Z}^i(r,z)\,\mathbf{J}^i(\boldsymbol{r},z)]\,\mathrm{d}\boldsymbol{r}, \tag{7.15}$$

$$N = \int_{\mathrm{c}} n(z)\,\mathrm{d}z, \tag{7.16}$$

or by using the so-called Lloyd integral density of states (Lloyd and Smith 1972, Gyorffy and Stocks 1979, Staunton et al. 1980)

$$N(E) = N_0(E) - (\pi\Omega_{\mathrm{BZ}})^{-1} \int \mathrm{Im}\,\ln\det[\mathbf{t}_{\mathrm{C}}^{-1} - \mathbf{G}(\boldsymbol{k})]\,\mathrm{d}\boldsymbol{k}$$
$$- \frac{(1-c)}{\pi}\,\mathrm{Im}\,\ln\det\frac{[\mathbf{t}_{\mathrm{A}}^{-1} - \langle\mathbf{t}^{-1}\rangle]}{[\mathbf{t}_{\mathrm{A}}^{-1} - \mathbf{t}_{\mathrm{C}}^{-1}]} - \frac{c}{\pi}\,\mathrm{Im}\,\ln\det\frac{[\mathbf{t}_{\mathrm{B}}^{-1} - \langle\mathbf{t}^{-1}\rangle]}{[\mathbf{t}_{\mathrm{B}}^{-1} - \mathbf{t}_{\mathrm{C}}^{-1}]}, \tag{7.17}$$

$$\langle\mathbf{t}^{-1}\rangle = c\mathbf{t}_{\mathrm{A}}^{-1} + (1-c)\mathbf{t}_{\mathrm{B}}^{-1}. \tag{7.18}$$

Equation (7.17) gives the number of electrons $N(E)$ at energy E, $N_0(E)$ being the contribution of the free electrons (see Lloyd and Smith 1972).

7.3. *Poisson equation and self-consistency for the scattering potentials*

Once the muffin-tin charge densities in the various atomic spheres of the unit cell are found, standard procedures can be used to solve the Poisson equation (see e.g. Slater and DeCicco 1967). Since these procedures are fairly well-used in ordinary band theory they need not be discussed in detail. It should be noted, that quite frequently the 'frozen core approximation' is applied whereby the muffin-tin charge density in the ith muffin-tin is given as a sum over an atomic-like charge density for the core states $\rho_{\mathrm{c}}^i(r_i)$ and the valence charge density $\rho_{\mathrm{v}}^i(r_i)$ (eqs. 7.6 and 7.10), calculated by means of the 'constant E', or 'constant k' mode:

$$\rho^i(r_i) = \rho_{\mathrm{c}}^i(r_i) + \rho_{\mathrm{v}}^i(r_i). \tag{7.19}$$

In terms of complex energies ($z = \varepsilon + \mathrm{i}\delta$), however, an integration along a contour parallel to the imaginary axis and passing through the Fermi energy on the real axis can be performed by closing the path at $\delta = \pm\infty$, avoiding thus the frozen core approximation (Stocks 1985). This particular approach offers very interesting numerical applications. For simple lattices it can be almost of the speed of the so-called 'linearized bandstructure methods'.

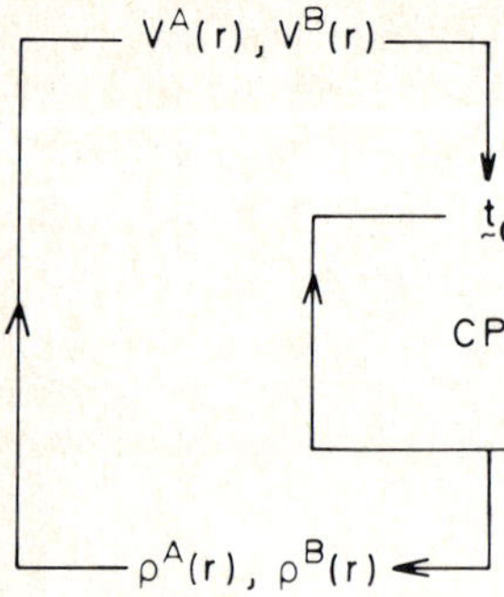

Fig. 4. Self-consistency scheme for disordered materials.

Although the problem of self-consistency sounds fairly technical and doable, for disordered systems containing heavy elements it is still a very major computational task, since, as is shown in fig. 4, two loops of self-consistency have to be performed; one for the single site (cell) effective t-matrix and one for the scattering potentials.

7.4. *Total energies*

Since at present the discussion about the implications of a total energy concept for disordered systems is still going on, this section has to be restricted to ordered systems only.

According to eqs. (2.7) and (2.8) the total energy is given as

$$E[\rho]=T_s[\rho]+U[\rho]+E_{xc}[\rho]. \tag{7.20}$$

For ordered systems (see in particular Weinert et al. 1982) the kinetic energy $T_s[\rho]$ is defined as

$$\begin{aligned} T_s[\rho] &= \sum_i \varepsilon_i - \int_\Omega V_{\text{eff}}(\boldsymbol{r})\,\rho(\boldsymbol{r})\,\mathrm{d}\boldsymbol{r} \\ &= \sum_i \varepsilon_i - \int_\Omega V_C(\boldsymbol{r})\,\rho(\boldsymbol{r})\,\mathrm{d}\boldsymbol{r} - \int_\Omega \mu_{xc}(\boldsymbol{r})\,\rho(\boldsymbol{r})\,\mathrm{d}\boldsymbol{r}, \end{aligned} \tag{7.21}$$

where $\rho(\boldsymbol{r})$ is the charge density and the ε_i correspond to the Kohn–Sham eigenvalues of the occupied one-particle states. In eq. (7.21), $\mu_{xc}(\boldsymbol{r})$ is the exchange-correlation potential (e.g. eq. 2.26) and V_C $(\boldsymbol{r})$ is the Coulomb potential,

$$V_C(\boldsymbol{r})=\int \frac{\rho(\boldsymbol{r}')}{|\boldsymbol{r}-\boldsymbol{r}'|}\,\mathrm{d}r' - \sum_\alpha \frac{Z_\alpha}{|\boldsymbol{r}-\boldsymbol{R}_\alpha|}, \tag{7.22}$$

for a set of scatterers with atomic numbers Z_α and centred at $\boldsymbol{R}_\alpha$.

The total energy per unit cell Ω, E, is then given as

$$E=\sum_i \varepsilon_i - \tfrac{1}{2}\int_\Omega \rho(\boldsymbol{r})[V_C(\boldsymbol{r})+2\mu_{xc}(\boldsymbol{r})]\,\mathrm{d}\boldsymbol{r} - \tfrac{1}{2}\sum_\nu Z_\nu V_M(\boldsymbol{R}_\nu)+E_{xc}[\rho], \tag{7.23}$$

where the sum over ν covers all sites centred at $\boldsymbol{R}_\nu$ in the unit cell Ω and $E_{xc}[\rho]$ is determined by the local density functional used (e.g. eq. 2.23):

$$E_{xc}[\rho] \cong \int_\Omega \rho(\boldsymbol{r})\, E_{xc}(\boldsymbol{r})\, d\boldsymbol{r}. \tag{7.24}$$

The quantities $V_M(\boldsymbol{R}_\nu)$ are generalized Madelung potential terms:

$$V_M(\boldsymbol{R}_\nu) = \int \frac{\rho(\boldsymbol{r})}{|\boldsymbol{r} - \boldsymbol{R}_\nu|}\, d\boldsymbol{r} - \sum_{\alpha \neq \nu} \frac{Z_\alpha}{|\boldsymbol{R}_\alpha - \boldsymbol{R}_\nu|}, \tag{7.25}$$

which can be calculated using the method suggested by Weinert (1981). The charge density, finally, needed to calculate the total energy E is defined in eq. (6.63) for the case of simple lattices. For complex lattices again a summation over subcells can be used in eq. (7.21), implying that only sublattice-diagonal blocks of $\boldsymbol{\tau}^{ii}$ have to be evaluated (see also eqs. 6.51 and 6.59). Quite clearly, in order to calculate the total energy E, the charge density $\rho(r)$ has to be calculated self-consistently.

8. Beyond single site approximations for impurities and for the coherent potential

8.1. *Relaxed impurities and short range order*

Sections 6.4 and 6.5 dealt with single, isolated impurities and the single site approximation to the RKKR–CPA medium. The single site approximation applies always when statistical disorder characterizes the disordered systems. In order to describe also local ordering (short range order) for concentrated alloys and to treat relaxed impurities one has to go beyond the single site approximation. From fig. 1 the situation is quite clear. We are now interested in calculating elements of the Green's function for a cluster of lattice sites (cells), i.e., we want to consider the case when the source and/or field point of such an element can be in the first, second, etc. shell of neighbours.

Keeping the lattice still unrelaxed (frozen) from eq. (6.20) the Green's function $G(\boldsymbol{r}_i, \boldsymbol{r}'_j)$ is immediately calculated once the non-site (non-cell) diagonal scattering path operators are known. In principle, we do know these scattering path operators already from eq. (6.10). The difficulties involved, however, are that in eq. (6.10), because of the phase factor, the $\boldsymbol{k}$-integration can no longer be restricted to the irreducible wedge of the Brillouin zone. In the non-relativistic case this problem was discussed by Gonis and Freeman (1984) and by Gonis et al. (1984). In the relativistic case this problem can be solved by using for cubic systems double group (O_h^*) symmetry or equivalently and more elegantly by means of irreducible projective representations of O_h (Altmann 1986).

Restricting ourselves for a moment to cubic systems with only one atom per unit cell, $\tau(\boldsymbol{R}_n)$ is given by

$$\tau(\boldsymbol{R}_n)=\frac{1}{\Omega_{\mathrm{BZ}}}\int \tau(\boldsymbol{k})\, \mathrm{e}^{\mathrm{i}\boldsymbol{k}\cdot\boldsymbol{R}_n}\, \mathrm{d}\boldsymbol{k}=\frac{1}{\Omega_{\mathrm{BZ}}}\sum_{R\in \mathrm{O}_\mathrm{h}^*}\int_{\mathrm{IBZ}} \tau(R\boldsymbol{k})\, \mathrm{e}^{\mathrm{i}R\boldsymbol{k}\cdot\boldsymbol{R}_n}\, \mathrm{d}\boldsymbol{k}, \tag{8.1}$$

where $\boldsymbol{R}_n=\boldsymbol{R}_i-\boldsymbol{R}_j$ and IBZ is the volume of the irreducible wedge of the Brillouin zone. Let $\mathbf{C}$ be the matrix of O_h^* Clebsch–Gordon coefficients, used to form a representation in cubic harmonics. This representation shall be denoted by a superscript $\wedge$ (see also section 4).

$$\mathbf{C}^+\mathbf{A}\mathbf{C}=\mathbf{A}^\wedge, \qquad \mathbf{C}^+\mathbf{C}=\mathbf{1}. \tag{8.2}$$

In eq. (8.1), $\tau(R\boldsymbol{k})$ is of course given by the following similarity transformation

$$\begin{aligned}&\tau(R\boldsymbol{k})=\mathbf{D}(R)^+\, \tau(\boldsymbol{k})\, \mathbf{D}(R),\\ &\tau^\wedge(R\boldsymbol{k})=\mathbf{D}^\wedge(R)^+\, \tau^\wedge(\boldsymbol{k})\, \mathbf{D}^\wedge(R).\end{aligned} \tag{8.3}$$

Using the decomposition of O_h into left cosets of O,

$$\mathrm{O}_\mathrm{h}=\{\mathrm{O}, J\mathrm{O}\}, \tag{8.4}$$

where J is the inversion, eq. (8.1) can be written as

$$\tau^\wedge(\boldsymbol{R}_n)=\frac{1}{\Omega_{\mathrm{BZ}}}\sum_{R\in \mathrm{O}^*}\int_{\mathrm{IBZ}} \{\mathbf{D}^\wedge(R)^+\, \tau^\wedge(\boldsymbol{k})\, \mathbf{D}^\wedge(R)\, \mathrm{e}^{\mathrm{i}R\boldsymbol{k}\cdot\boldsymbol{R}_n} + \mathbf{D}^\wedge(JR)^+\, \tau^\wedge(\boldsymbol{k})\, \mathbf{D}^\wedge(JR)\, \mathrm{e}^{-R\boldsymbol{k}\cdot\boldsymbol{R}_n}\}\, \mathrm{d}\boldsymbol{k}. \tag{8.5}$$

However, since

$$D(JR)=D(J)\, D(R) \tag{8.6}$$

and

$$\mathbf{D}(J)^+\tau(\boldsymbol{k})\, \mathbf{D}(J)=\tau(-\boldsymbol{k}) \tag{8.7}$$

the integral in eq. (8.5) is nothing but

$$\tau^\wedge(\boldsymbol{R}_n)=\frac{1}{\Omega_{\mathrm{BZ}}}\sum_{R\in \mathrm{O}^*}\int_{\mathrm{IBZ}} \mathbf{D}^\wedge(R)^+\, \tau^\wedge(R,\boldsymbol{k})\, \mathbf{D}^\wedge(R)\, \mathrm{d}\boldsymbol{k}, \tag{8.8}$$

where

$$\tau^\wedge(R,\boldsymbol{k})=\tau^\wedge(\boldsymbol{k})\, \mathrm{e}^{\mathrm{i}\boldsymbol{k}\cdot R^{-1}\boldsymbol{R}_n}+\tau^\wedge(-\boldsymbol{k})\, \mathrm{e}^{-\mathrm{i}\boldsymbol{k}\cdot R^{-1}\boldsymbol{R}_n}. \tag{8.9}$$

The BZ integral in eq. (8.8) need not to be calculated for all relevant distance vectors $\boldsymbol{R}_n=\boldsymbol{R}_i-\boldsymbol{R}_j$, since again group theory can be applied. Let P be the smallest subgroup of O_h, such that

$$P=\{R\,|\,R\boldsymbol{R}_n=\boldsymbol{R}_m;\ |\boldsymbol{R}_m|=|\boldsymbol{R}_n|;\ \forall m,n\} \tag{8.10}$$

then the set of non-site diagonal scattering path operators $\tau^\wedge(\boldsymbol{R}_n)$ corresponding to equivalent distance vectors $\boldsymbol{R}_n$ is given by (for accidental degeneracies, see Dirl et al. 1984)

$$\boldsymbol{\tau}^{\wedge}(\boldsymbol{R}_m) = \mathbf{D}^{\wedge}(R)^{+}\, \boldsymbol{\tau}^{\wedge}(\boldsymbol{R}_n)\, \mathbf{D}^{\wedge}(R), \tag{8.11}$$

where

$$\boldsymbol{R}_m = R\boldsymbol{R}_n, \qquad R \in P.$$

For a first neighbour cluster in an fcc lattice, for example, only 4 such BZ integrals have to be performed, all other non-site-diagonal scattering path operators are defined by eq. (8.11). For a complex lattice, BZ integrals have to be performed for each block in the supermatrix of $\boldsymbol{\tau}^{ij}$ (eq. 6.16),

$$\boldsymbol{\tau}^{ij} = \begin{pmatrix} & \vdots & \vdots & \\ \cdots & \tau(\boldsymbol{R}_{is} - \boldsymbol{R}_{js}) & \tau(\boldsymbol{R}_{is} - \boldsymbol{R}_{js'}) & \cdots \\ \cdots & \tau(\boldsymbol{R}_{is'} - \boldsymbol{R}_{js}) & \tau(\boldsymbol{R}_{is'} - \boldsymbol{R}_{js'}) & \cdots \\ & \vdots & \vdots & \end{pmatrix}, \tag{8.12}$$

where s denotes the sublattice index.

Once the matrices $\boldsymbol{\tau}^{ij}$ are known a cluster-diagonal scattering path operator can be defined (Gonis et al. 1985a,b):

$$\boldsymbol{\tau}^{\mathrm{CC}} = [\Delta \boldsymbol{t}^{-1} + \boldsymbol{\tau}^{-1}]^{-1} \tag{8.13}$$

where for concentrated alloys $\Delta \boldsymbol{t}^{-1}$ is given by

$$\Delta \boldsymbol{t}^{-1} = \begin{pmatrix} \mathbf{t}_{\mathrm{C}}^{-1} - \mathbf{t}_{\alpha 0}^{-1} & 0 & \cdots & 0 \\ 0 & \mathbf{t}_{\mathrm{C}}^{-1} - \mathbf{t}_{\alpha 0}^{-1} & \cdots & \vdots \\ \vdots & & & \\ 0 & \cdots & & \mathbf{t}_{\mathrm{C}}^{-1} - \mathbf{t}_{\alpha n}^{-1} \end{pmatrix} \tag{8.14}$$

and for relaxed single impurities as

$$\Delta \boldsymbol{t}^{-1} = \begin{pmatrix} \mathbf{t}^{-1} - \mathbf{t}_{\alpha 0}^{-1} & 0 & 0 & \cdots & 0 \\ 0 & 0 & 0 & & \\ \vdots & & & & \vdots \\ 0 & \cdots & & & 0 \end{pmatrix}. \tag{8.15}$$

In eq. (8.14) $\mathbf{t}_{\mathrm{C}}$ is the averaged single site t-matrix and $\mathbf{t}_{\alpha i}$ the single site t-matrix of component α at the site $\boldsymbol{R}_i$. In eq. (8.15), $\mathbf{t}$ is the single site t-matrix of the host and $\mathbf{t}_{\alpha 0}$ the single site t-matrix of the impurity atom at the origin of the lattice. The supermatrix $\boldsymbol{\tau}$ is nothing but the matrix formed by the $\boldsymbol{\tau}^{ij}$:

$$\boldsymbol{\tau}=\begin{pmatrix} \overbrace{\quad}^{\text{origin}} & \overbrace{\qquad\qquad}^{1^{\text{st}}\text{ neighbours}} & \overbrace{\qquad\qquad}^{2^{\text{nd}}\text{ neighbours}\ldots} \\ \boldsymbol{\tau}^{00} & \boldsymbol{\tau}^{01}\ldots\boldsymbol{\tau}^{0n} & \boldsymbol{\tau}^{0n+1}\ldots\boldsymbol{\tau}^{0m}\ldots \\ \boldsymbol{\tau}^{10} & & \\ \vdots & & \end{pmatrix}. \tag{8.16}$$

From eq. (8.15) one realizes that not only relaxed single impurities can be calculated, but also double impurities, etc, since $\Delta\boldsymbol{t}^{-1}$ can also be of the form

$$\Delta\boldsymbol{t}^{-1}=\begin{pmatrix} \mathbf{t}^{-1}-\mathbf{t}_{\alpha 0}^{-1} & 0 & 0 & 0 & \cdots & 0 \\ 0 & 0 & 0 & 0 & & \\ 0 & 0 & \mathbf{t}^{-1}-\mathbf{t}_{\alpha 2}^{-1} & 0 & \cdots & \vdots \\ 0 & & \cdots & & & 0 \end{pmatrix}. \tag{8.17}$$

This is in particular important for magnetic interactions of nearby atoms (RKKY–interaction).

For concentrated alloys it is important to note that $\boldsymbol{\tau}^{\mathrm{CC}}$ is defined for a configuration of atoms αn. Let $\{\alpha\}$ denote such a particular configuration, then the average over all configurations determines the new effective site diagonal scattering path operator:

$$\langle\boldsymbol{\tau}^{\mathrm{CC}}\{\alpha\}\rangle_{00}=\boldsymbol{\tau}_{\mathrm{C}}^{00}. \tag{8.18}$$

Equation (8.13) allows therefore not only to calculate site diagonal scattering path operators for a particular configuration of neighbours, which indeed can be very important for physical phenomena of alloys depending strongly on the local environment, but also to include short range order to the CPA medium. For complex lattices each element in $\boldsymbol{\tau}$ and $\Delta\boldsymbol{t}^{-1}$ is itself a supermatrix (8.12), the occupation of sites is then according to the positions $\boldsymbol{R}_{is}$ and $\boldsymbol{R}_{js'}$, with s and s' denoting sublattices.

8.2. *Dislocations and interstitial impurities*

In eq. (6.32), displaced atom reciprocal structure constants for an infinitesimal dislocation ρ_- are discussed. The relativistic counterparts are given by

$$\hat{G}_{QQ'}(\boldsymbol{k})=\sum_{s=\pm 1/2} c(l\,j\,\tfrac{1}{2};\mu-s,s)\,\hat{g}_{l\mu-s,l'\mu'-s}(\boldsymbol{k})\,c(l'\,j'\,\tfrac{1}{2};\mu'-s,s). \tag{8.19}$$

In general, the $\boldsymbol{G}(k)$ are given by (Gonis et al. 1985a)

$$\hat{g}_{LL'}(\boldsymbol{k}) = \sum_{L'', L'''} \mathrm{i}^{l-l'+l''} C_{LL'}^{L''} \cdot g_{L''L'''}(\boldsymbol{k})\, j_{L'''}(-\boldsymbol{\rho}), \tag{8.20}$$

and by (eq. 6.24),

$$j_l(\boldsymbol{r}) = j_l(pr)\langle \hat{r} | lm \rangle$$

and are only restricted by the requirement that the scattering regions do not overlap (see also section 3). Using eqs. (8.18) or (8.19) the site-(cell-)diagonal scattering path operator can be calculated:

$$\hat{\tau}^{00} = \frac{1}{\Omega_{\mathrm{BZ}'}} \int [\mathbf{t}^{-1} - \hat{G}(\boldsymbol{k})]^{-1}\, \mathrm{d}\boldsymbol{k}, \tag{8.21}$$

where $\hat{\tau}^{00}$ is defined as

$$\hat{\tau}^{00} = \langle \boldsymbol{R}_0 - \boldsymbol{\rho} | \tau | \boldsymbol{R}_0 - \boldsymbol{\rho} \rangle = \tau(\boldsymbol{R}_0 - \boldsymbol{\rho}, \boldsymbol{R}_0 - \boldsymbol{\rho}) \tag{8.22}$$

and the vector $\boldsymbol{\rho}$ is the displacement vector. Equation (8.20) has to be regarded as the backbone of investigations on interstitial impurities such as in the case of hydrogen in metals (PuH_x, for example). By means of eq. (8.20), however, also temperature dependent relaxation of the underlying lattice can be described (for a detailed discussion see Gonis et al. 1985a,b).

9. Applications to actinide systems

After having discussed in quite some detail relativistic scattering theory for ordered and disordered systems, we want to illustrate some of the theoretical concepts presented. As in a previous article (Weinberger 1982b) we will use non-relativistic and/or 'non-actinide' examples in all cases, where at present no results for actinide systems are available. In particular, since there are already excellent review articles available in this Handbook dealing with the electronic structure of ordered actinide systems (Arko et al. 1984, Brooks et al. 1985) no attempt is made to give a complete 'listing' of results for such systems.

9.1. *Single site scattering*

In many cases of interest it is necessary to have 'cheap and ad hoc models' for the electronic structure of actinide systems. In particular this is necessary for systems with rather complicated crystal structures. Single site scattering provides such an approach. For resonant angular momentum channels the derivative of the corresponding phaseshift $\delta_\kappa^i(E)$ (eq. 5.24) for a particular scattering region i with respect to the energy parameter E is commonly termed the 'Wigner delay time' $t_\kappa^{\mathrm{D}}(E)$ (Wigner 1955, Lloyd and Smith 1972),

$$t_\kappa^{\mathrm{D}}(E) = \frac{\mathrm{d}\delta_\kappa(E)}{\mathrm{d}E}. \tag{9.1}$$

Because of the Lorentzian shape of $t_\kappa^{\mathrm{D}}(E)$ the 'Wigner delays' determine the shape of the single site density of states $n^i(E)$,

$$n^i(E) = n_0^i(E) + \sum_\kappa (2j+1) \frac{d\delta_\kappa^i(E)}{dE}, \tag{9.2}$$

where $(2j+1)$ is the degeneracy of the κth angular momentum channel. The integrated density of states $N^i(E)$ is nothing but the sum of the phaseshifts,

$$N^i(E) = N_0^i(E) + \sum_\kappa (2j+1)\, \delta_\kappa^i(E). \tag{9.3}$$

In eqs. (9.2) and (9.3), $n_0^i(E)$ and $N_0^i(E)$ are the contributions of the free electrons (Lloyd and Smith 1972).

Considering therefore the density of states simply as a sum of single site contributions per unit cell, a first idea about the electronic structure can be obtained:

$$n(E) \cong \sum_i n^i(E), \tag{9.4}$$

$$N(E) \cong \sum_i N^i(E). \tag{9.5}$$

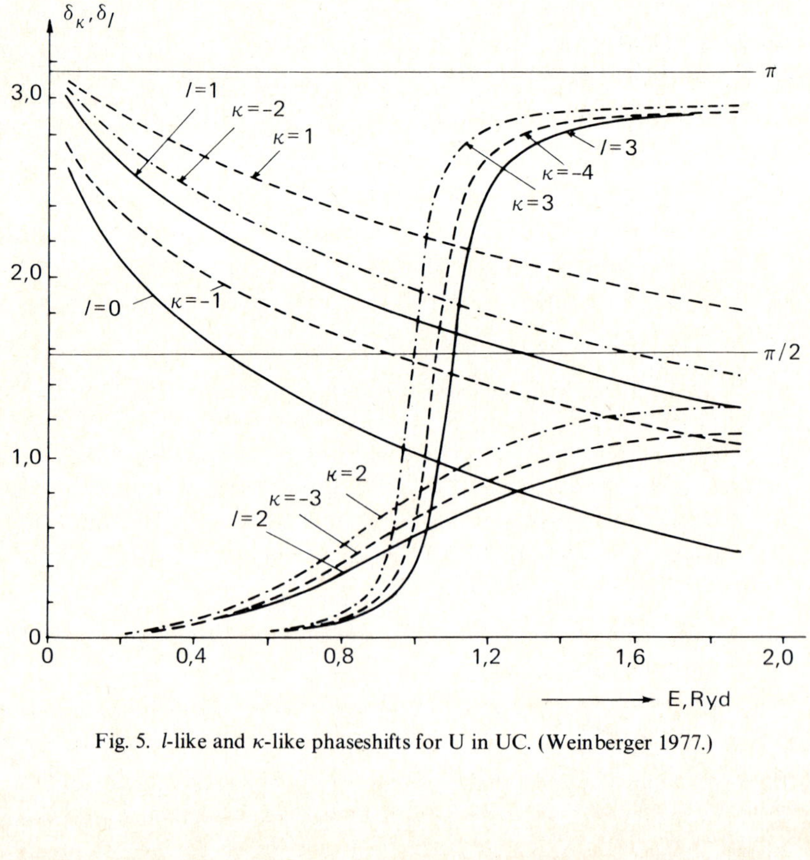

Fig. 5. l-like and κ-like phaseshifts for U in UC. (Weinberger 1977.)

In particular, eq. (9.4) can be used to determine qualitatively the position of the Fermi energy. Since only single site quantities are involved, even an ATA ('averaged t-matrix')-like alloy theory can be applied:

$$\mathbf{t}_C = x\mathbf{t}_A + (1-x)\mathbf{t}_B. \tag{9.6}$$

In eq. (9.6), x is the concentration of species A and $(1-x)$ the concentration of species B on a particular sublattice of a pseudobinary alloy.

It should be noted that for eqs. (9.1)–(9.6) only phaseshifts have to be calculated; a numerical task which can be performed nowadays on the scale of a personal computer.

Figure 5 shows the l-like and κ-like phaseshifts for U in UC (Weinberger 1977).

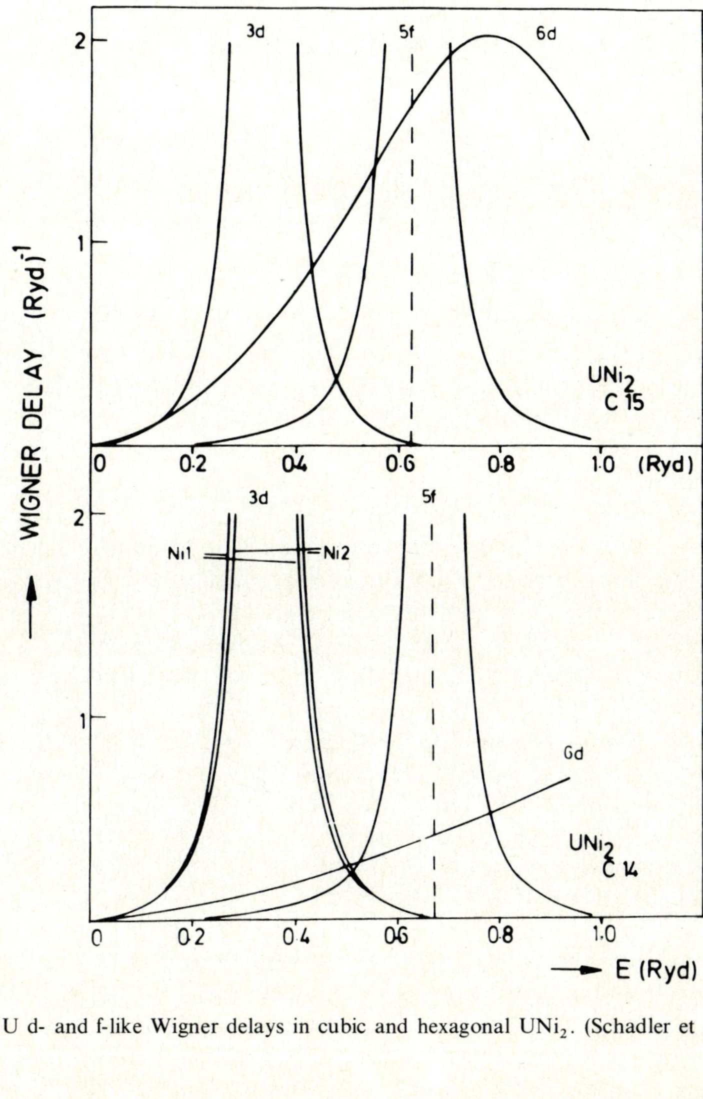

Fig. 6. U d- and f-like Wigner delays in cubic and hexagonal UNi_2. (Schadler et al. 1982.)

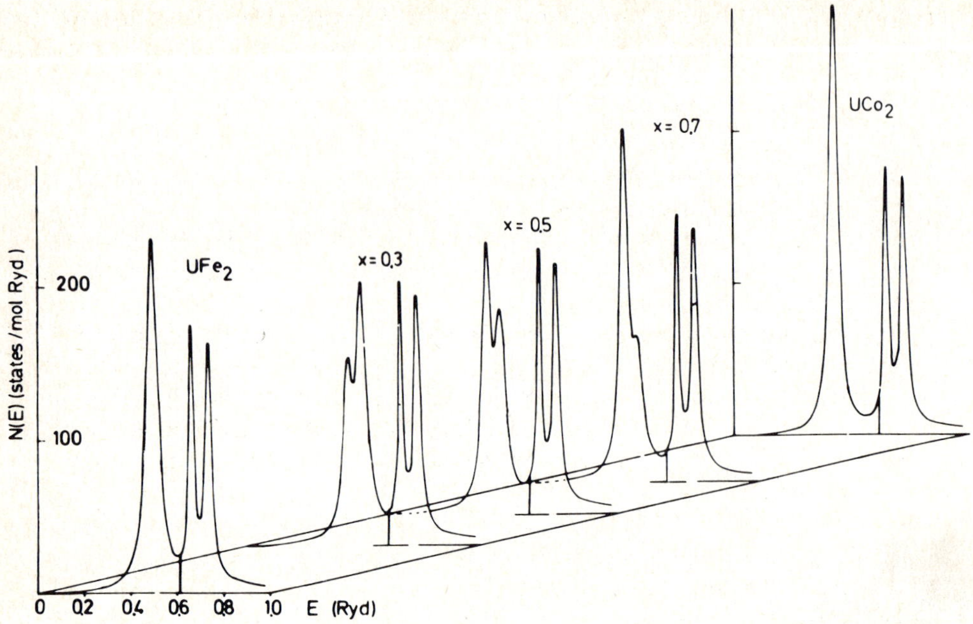

Fig. 7. Density of states for cubic $U(Fe_xCo_{1-x})_2$. (Schadler et al. 1982.)

As one can see from this figure the f-like phaseshifts correspond to sharp resonances, wereas the d-like resonances are rather broad. All other phaseshifts are 'non-resonant'. Sharp resonances correspond to flat bands, whereas weak resonances indicate bands with considerable dispersion.

Figure 6 shows U d- and f-like 'Wigner-delays' for the cubic and hexagonal Laves phases of UNi_2 (Schadler et al. 1982). The cubic Laves phase of UNi_2 is non-magnetic, whereas its hexagonal counterpart is weakly ferromagnetic. From this figure one can see that quite obviously the resonance behaviour of the U 6d-like electrons is quite different in the cubic and the hexagonal phase, indicating that the magnetic properties for UNi_2 are due to a very delicate balance of d- and f-like magnetic moments.

Figure 7 shows the density of states (eqs. 9.5 and 9.6) for $U(Fe_xCo_{1-x})_2$ (pseudobinary cubic Laves phase) for varying concentration x. From this figure one can see the building up of the Co subband and concomitantly the dying out of the Fe subband. The growing and dying out of subbands is at the energetic position of the corresponding impurity peaks. At the Fermi energy, $n(E_F)$ is almost entirely determined by the U subband, the number of electrons moving the Fermi energy up or down. It should be noted that this qualitative alloy description coincides remarkably well with a careful KKR–CPA study for $Ti(Fe_xCo_{1-x})$, which is a pseudobinary alloy with the CsCl structure (Schadler and Weinberger 1986).

Figures 6 and 7 demonstrate that single site scattering not only is a necessary 'ingredient' to multiple scattering, but can also be used for quick, qualitative descriptions of the electronic structure of ordered and disordered materials.

9.2. *Actinide metals and single impurities in actinide metals and related systems*

9.2.1. *Electronic structure of actinide metals*

Figure 8 shows the bandstructure of δ-Pu along [100] for two lattice constants, the smaller of which corresponds to the density of α-Pu, the other to the extended phase (Weinberger et al. 1985a). The results shown in the figure illustrate the use of eq. (6.34). One can see that when the lattice parameter is decreased, the Pu $6s^{1/2}$ band is pushed upwards in energy and penetrates considerably the system of f-like bands. The density of states for δ-Pu at these two lattice constants, shown in fig. 9, was calculated from eq. (6.51), in particular from the 'host entry' of eq. (6.58). From fig. 9 one can see that at the larger lattice constant the width of the f-like bands is smaller then for the smaller lattice constant. Concomitantly the peak structure is considerably 'sharper'.

9.2.2. *Single, isolated impurities in actinide metals*

An important problem in the study of the electronic strucutre of single impurities in Pu is of metallurgical nature: α-Pu is brittle and δ-Pu is ductile. The δ-phase of

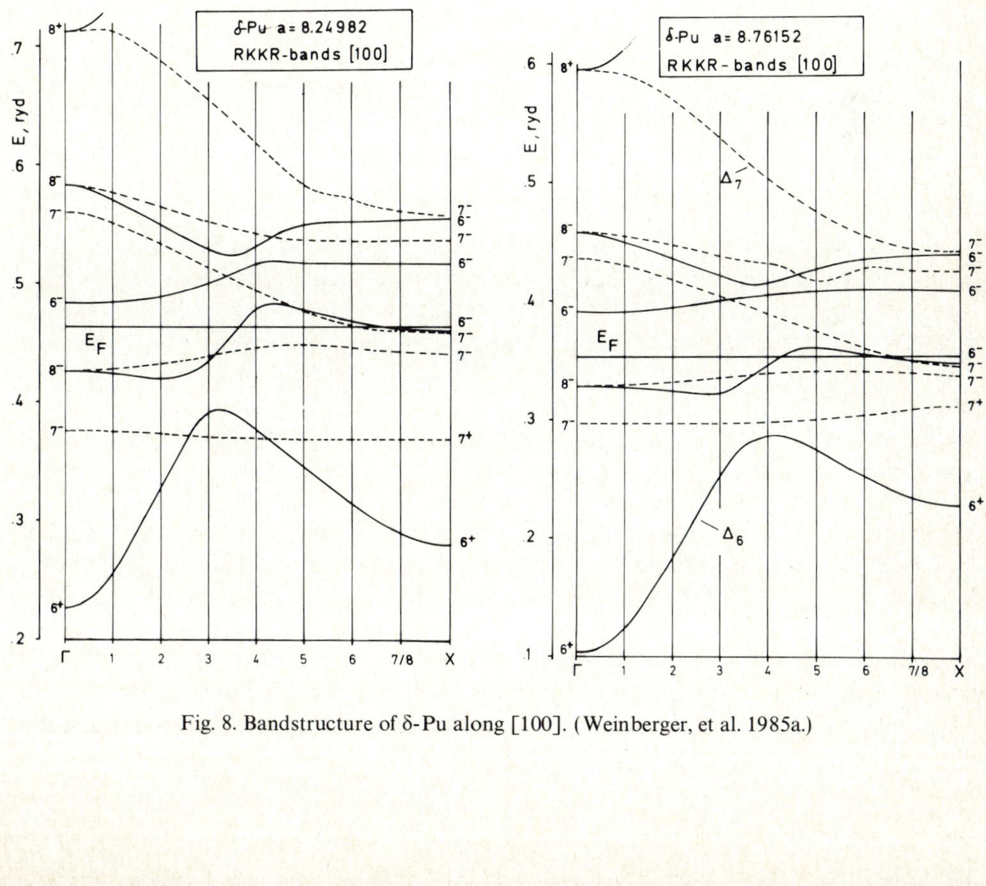

Fig. 8. Bandstructure of δ-Pu along [100]. (Weinberger, et al. 1985a.)

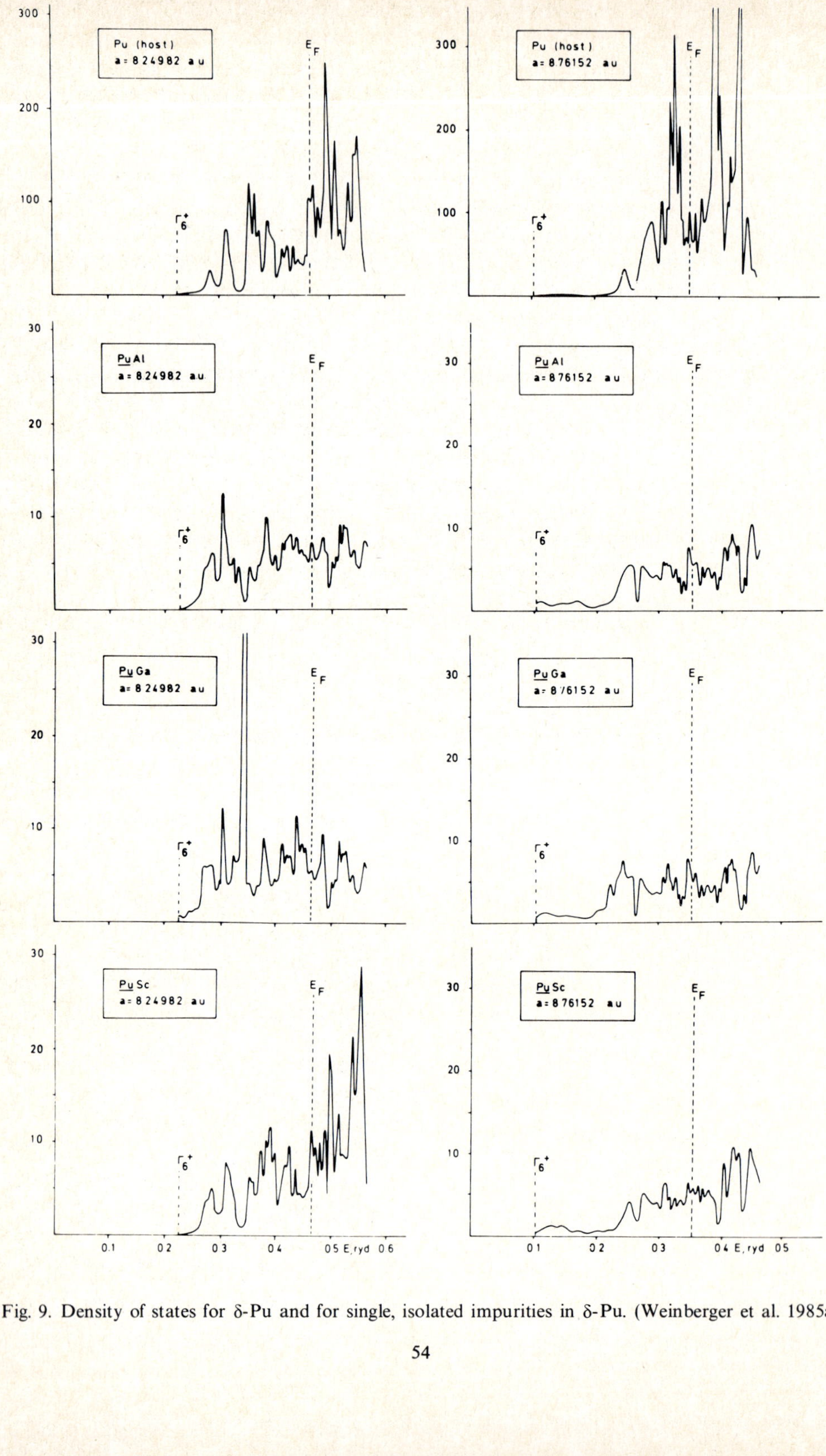

Fig. 9. Density of states for δ-Pu and for single, isolated impurities in δ-Pu. (Weinberger et al. 1985a.)

Pu is stabilized by small concentrations of Al, Ga, Sc, etc. Although a single impurity is only a limiting case of dilute alloy, the calculation of the electronic structure of single impurities is important mainly for two reasons: (a) once the host Green's function is known, it is quite easy (almost free of charge) to calculate the density of states of any impurity of interest, and (b) usually the virtual bound states of the impurity do show up in a similar way in the concentrated alloys for definite, but small, concentrations of the 'impurity component'. In fig. 9 single, isolated impurities of Al, Ga and Sc in δ-Pu (Weinberger et al. 1985a) are shown. Figure 9 illustrates the use of the 'impurity entry' in eq. (6.58). In principle, the charge density of the impurity atom can be calculated self-consistently in the Fermi sea of the host by means of eq. (6.66), whereby $\boldsymbol{\tau}^{ii}$ has to be replaced by $\boldsymbol{\tau}^{\alpha,ii}$ (eq. 6.35). For non-relativistic hosts (3d- and 4d-hosts) already a good deal of self-consistent impurity calculations is available (see for example Braspenning et al. 1982). Figure 9 corresponds to a non-self-consistent calculation.

From fig. 9 one can also see rather broad virtual bound states in the vicinity of the Pu $6s^{1/2}$ band and 'backscattering' to the host at higher energies. In a concentrated alloy the subband of A in A_xB_{1-x} is growing at the energetic position of the virtual bound state of A as impurity in host B. In the figure one can see therefore impurity states which eventually will grow in intensity when the concentration of the 'impurity component' is increased. In the region of 'backscattering', disorder effects have to be expected (for a more detailed discussion of Al in Pu, see also Boring et al. 1985).

Before going ahead it is worthwhile to recall the definition of a virtual bound state. From eq. (6.35) $\boldsymbol{\tau}^{A,00}$ is given as

$$\boldsymbol{\tau}^{A,00} = \frac{\boldsymbol{\tau}^{00}}{[1 + \mathbf{m}^A \boldsymbol{\tau}^{00}]}. \tag{9.7}$$

Suppose that at an energy E_r for a particular irreducible component v of $\boldsymbol{\tau}^{00}$ (see also fig. 2) the denominator in eq. (9.7) tends to zero:

$$[1 + m_v^A(E_r)\, \tau_v^{00}(E_r)] = 0. \tag{9.8}$$

Since

$$m_v^A(E_r) = t_{A,v}^{-1}(E_r) - t_v^{-1}(E_r) = \sqrt{E_r}\,[\cot \delta_v^{\mathrm{host}}(E_r) - \cot \delta_v^A(E_r)]$$

is real, eq. (9.8) corresponds to the following two conditions:

$$\mathrm{Im}\, \tau_v^{00}(E_r) = 0, \tag{9.10}$$

$$\sqrt{E_r}\,[\cot \delta_v^A(E_r) - \cot \delta_v^{\mathrm{host}}(E_r)]\ \mathrm{Re}\, \tau_v^{00}(E_r) = 0. \tag{9.11}$$

If eqs. (9.10) and (9.11) are fulfilled exactly, $\tau_v^{A,00}(E_r)$ grows indefinitely and the density of states at E_r is infinite, i.e., at E_r the density of states has the shape of a Dirac δ-function. In this case eqs. (9.10) and (9.11) are nothing but the occurence of a bound state. If, however, Im τ_v^{00} is only very small, but finite at E_r and eq. (9.11) is fulfilled, a Lorentzian resonance with a finite halfwidth characterizes the impurity density of states at E_r. E_r usually is called resonance energy and is the energy of a virtual bound state. The intensity of this peak is entirely governed by the value of $\mathrm{Im}\,\tau_v^{00}$ at E_r. Of course if only $\mathrm{Im}\,\tau_v^{00}$ is very small, but condition (9.11) is not met, then no virtual bound state exists.

9.3. *Concentrated alloys of actinides*

9.3.1. *Density of states for $Pu_x Al_{l-x}$*

Figure 10 depicts the density of states for $Pu_{0.90}Al_{0.10}$ (Weinberger et al. 1985b) as calculated by the RKKR–CPA method (eqs. 6.41–6.45) and in particular the 'concentrated alloy entry' in eq. (6.58). The black area in this figure corresponds to the contribution of Al to the total density of states. Since this contribution is rather small, it is interesting to investigate the shape of the component density of states of Al in $Pu_{0.90}$ $Al_{0.10}$, in particular the Al $s^{1/2}$, $p^{1/2}$ and $p^{3/2}$ partial local component densities of states (see eq. 6.61). In fig. 11 the Al $p^{1/2}$-like and $p^{3/2}$-like peaks to the left of the dip around 0.3 Ry have to be regarded as the 'virtual bound states' of 10% Al in δ-Pu. Comparing now with the corresponding results for a single, isolated Al impurity in δ-Pu (fig. 9) one can see that the results are indeed very similar. Because of disorder effects in the concentrated alloy, differences occur mainly in the region of 'backscattering'. This implies that, as already mentioned earlier, the Al impurity states grow with increasing Al concentration in the vicinity of the Al p-like virtual bound states. The effect of disorder in the vicinity of the Pu f-bands can be seen by comparing the host density of states in fig. 9 with the alloy density of states in fig. 10. In order to show a very typical growing of impurity-like states, the density of states for the system Au_xNi_{1-x} is shown in fig. 12 (Weinberger 1983b). In this figure one can see the Ni-like component density of states growing with increasing Ni concentration. This behaviour seems to be very typical for substitutionally disordered alloys. The system Pu_x Al_{1-x} and for that matter also other δ-phase Pu alloys behave very similar.

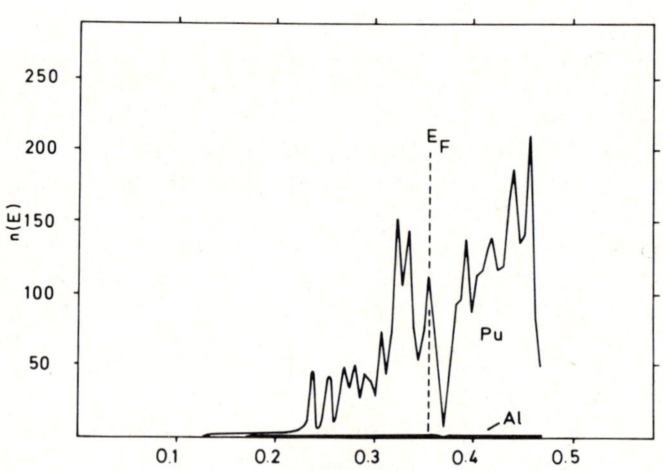

Fig. 10. Density of states for $Pu_{0.90}Al_{0.10}$. (Weinberger et al. 1985b.)

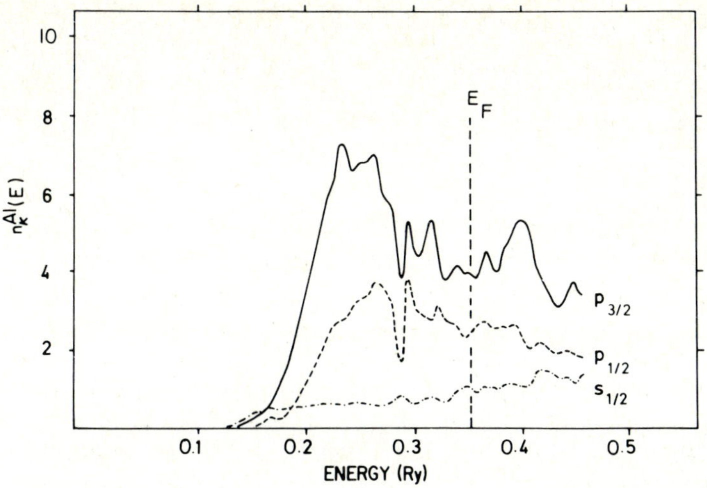

Fig. 11. Partial local densities of states for Al in $Pu_{0.90}Al_{0.10}$. (Weinberger et al. 1985b.)

9.3.2. *Bandstructure for concentrated alloys of actinides*

In fig. 13, constant-$\boldsymbol{k}$ Bloch spectral functions (Weinberger et al. 1985b) are shown for $\boldsymbol{k}=(0,0,0)$ and $\boldsymbol{k}=(0,0,\frac{1}{2})$. For comparison, in the figure a 'traditional' bandstructure along (100) is shown in the energy range of these Bloch spectral functions for the hypothetical case that in $Pu_{0.90}Al_{0.10}$ all sites would be occupied by Pu. From fig. 13 one can see that in the concentrated alloy the peaks do show considerable broadening in the parameter space of energies.

Again it is interesting to see how Bloch spectral functions in a binary substitutionally disordered system change with respect to concentration. In fig. 14 the Bloch spectral functions at Γ are shown for Au_xNi_{1-x} for various values of x (Weinberger 1983b). For pure Au the Bloch spectral functions reduce to a set of Dirac δ-functions, the positions of which are shown as vertical lines. Already at $x=0.9$ an additional peak in the vicinity of the Ni 'virtual bound state' can be seen. This peak grows with increasing Ni concentration and eventually splits up. Figure 14 is a very good example for a typical 'non-rigid-band' behaviour of bands, since within a rigid-band model the number of states at a particular vector $\boldsymbol{k}$ has to remain constant. In this way fig. 14 is also representative for δ-Pu alloys.

Since the dispersion of the bands shown in fig. 13 is rather small, large broadening of peaks in the constant-E Bloch spectral functions has to be expected. In fig. 15 constant-E Bloch spectral functions are shown at the Fermi energy for a few directions in the X–W plane. Because of disorder effects, there is no well-defined Fermi surface in $Pu_{0.90}Al_{0.10}$ (see also section 9.5.1). From fig. 2, showing the matrix structure of the site-diagonal effective scattering path operator (eq. 6.41), one can see that disorder can result either from the diagonal elements of $\boldsymbol{\tau}^{00}$ or the non-diagonal elements, which include the interaction between different values of j belonging to the same value of l. Even more interesting in the case of $Pu_{0.90}Al_{0.10}$

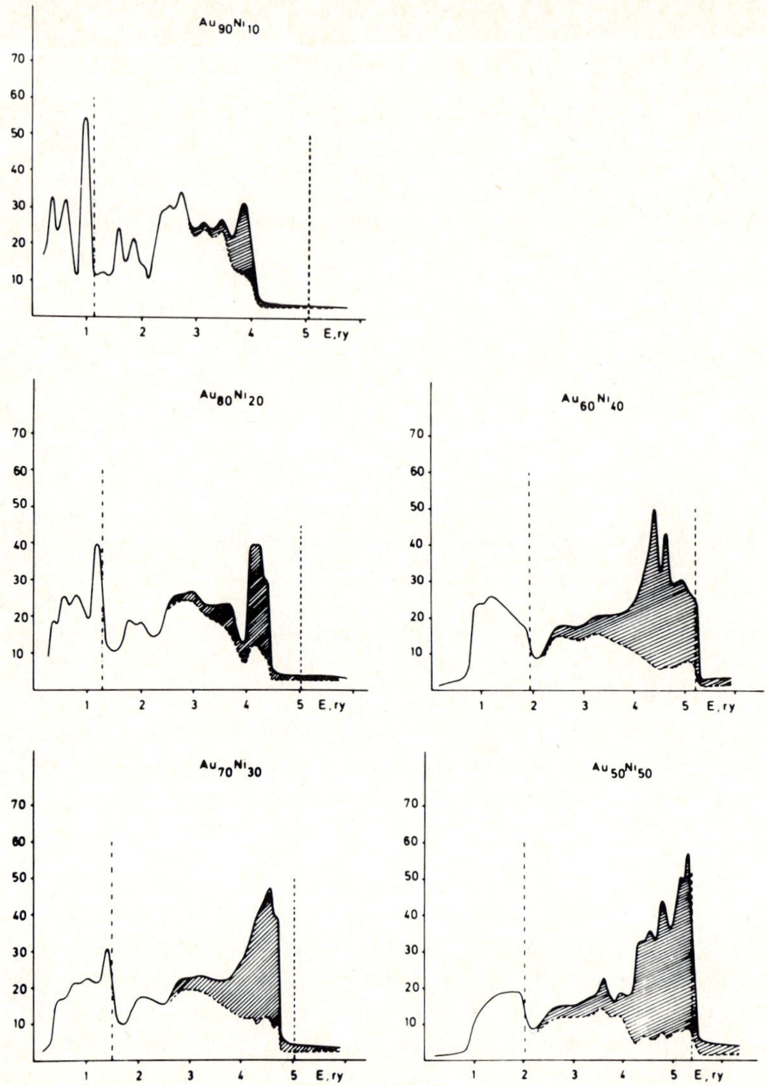

Fig. 12. Density of states for Au_xNi_{1-x}. The shaded area corresponds to the contribution of the Ni component to the total density of states. (Weinberger 1983b.)

are the p–f interactions. For a detailed discussion of figs. 13 and 15, see Weinberger et al. (1985b). Figures 13 and 15 illustrate the use of eqs. (6.72)–(6.74).

Of particular interest are projected Bloch spectral functions. In fig. 16 the total constant-E Bloch spectral functions at E for $Ag_{0.50}Pd_{0.50}$ (Gonis et al. 1985a) are analyzed in terms of the projected Bloch spectral functions (A = Ag, B = Pd). As one can see from this figure the second (diffuse) sheet of the Fermi surface, i.e., the rising

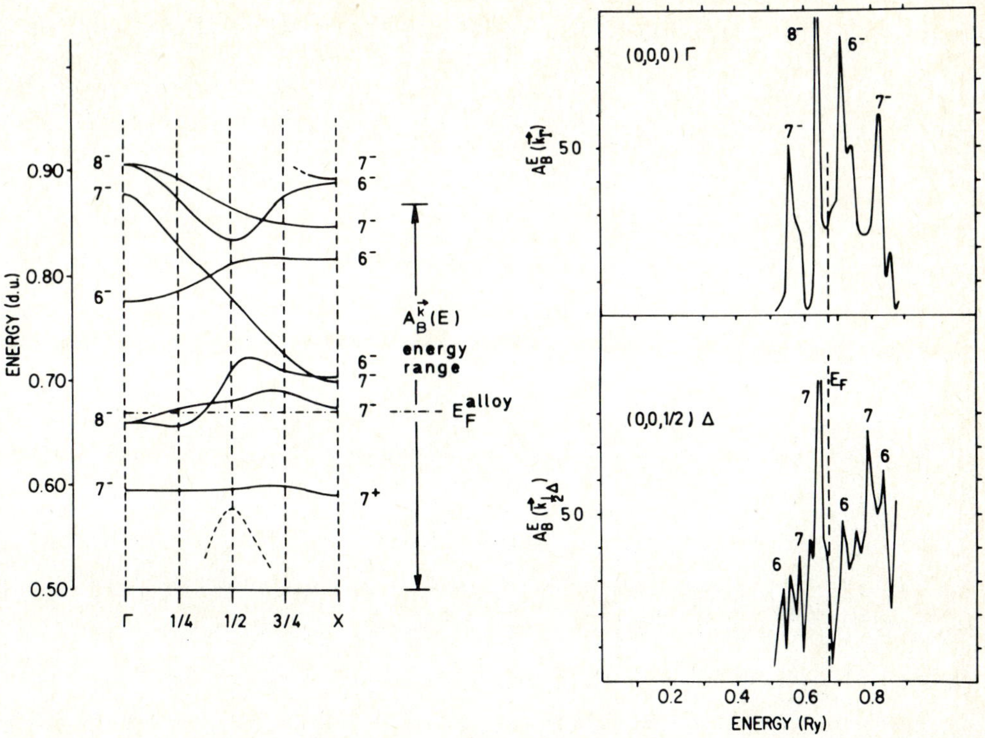

Fig. 13. Constant-$\boldsymbol{k}$ Bloch spectral functions along [100] for $Pu_{0.90}Al_{0.10}$ in the vicinity of the occupied Pu f bands. (Weinberger et al. 1985b.)

intensity near the edge of the BZ, is a remainder of the second, Γ-centred, sheet of the Pd Fermi surface, since the contributions to the total constant-E Bloch spectral functions do come mainly from the BB-like and partially also from the AB-like constant-energy projected Bloch spectral functions. The results shown in fig. 16 are based on the use of eq. (6.86).

9.4. *Ordered and disordered actinide compounds*

9.4.1. *Ordered pnictides and chalcogenides*

The electronic structure of ordered actinide pnictides and chalcogenides has been studied intensively in the past and is reviewed by Brooks et al. (1985). In fig. 17 the bandstructure along (100) is shown for UN (Weinberger et al. 1980). The results displayed in this figure illustrate the use of the 'constant-$\boldsymbol{k}$ self-consistency mode', discussed in section 7.2, and the use of eq. (6.34). In fig. 18 normal photoemission (Baptist 1982) from the UN (001) surface with s and p polarisation is correlated to the bands along Γ to X. The bandstructure shown in fig. 17 seems to be very consistent with the photoemission experiments.

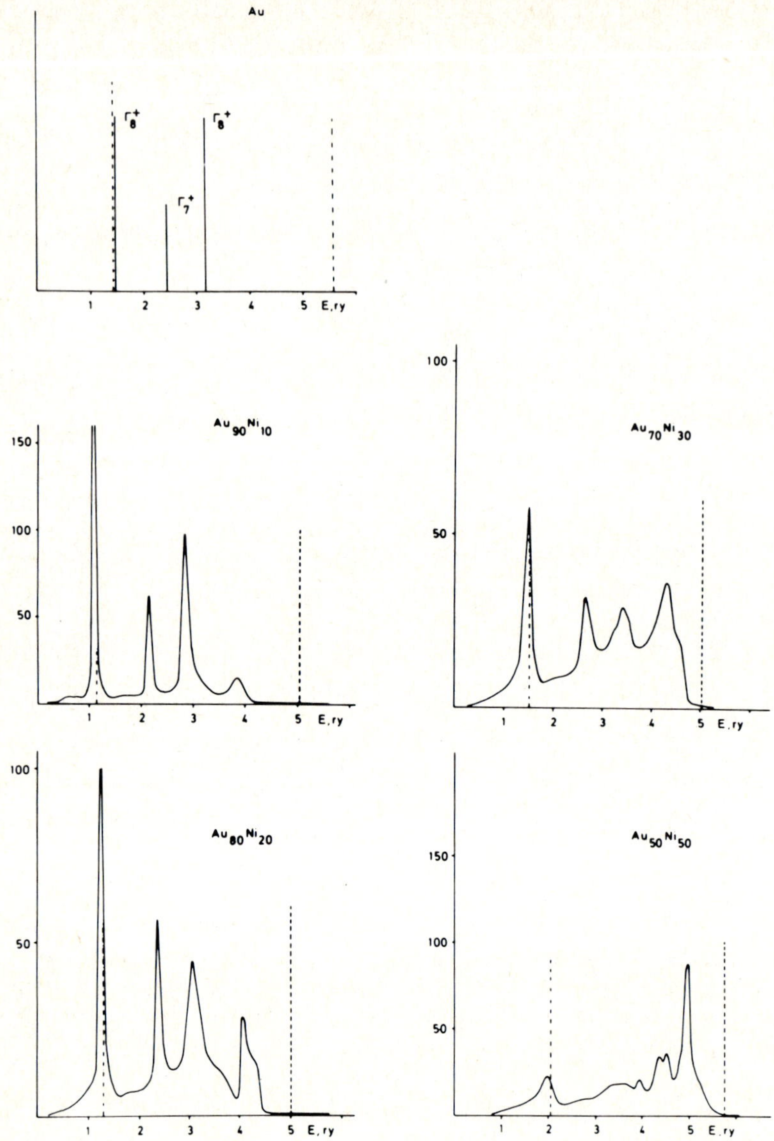

Fig. 14. Bloch spectral functions at $k = (0,0,0)$ for $Au_x Ni_{1-x}$. (Weinberger 1983b.)

Within an isostructural series of compounds it is of considerable interest to see changes in the electronic structure with respect to changes of the metal or the non-metal constituent. In table 5 the top of the non-metal p-like band (see also fig. 17), namely the Γ_6^- and the Γ_8^- state, is analysed for UN and UBi (Weinberger et al. 1982c) in terms of partial charges. As one can see from this table, the Γ_6^- state is mostly a non-metal $p^{1/2}$-like state, whereas the Γ_8^- state is only non-metal $p^{3/2}$-like

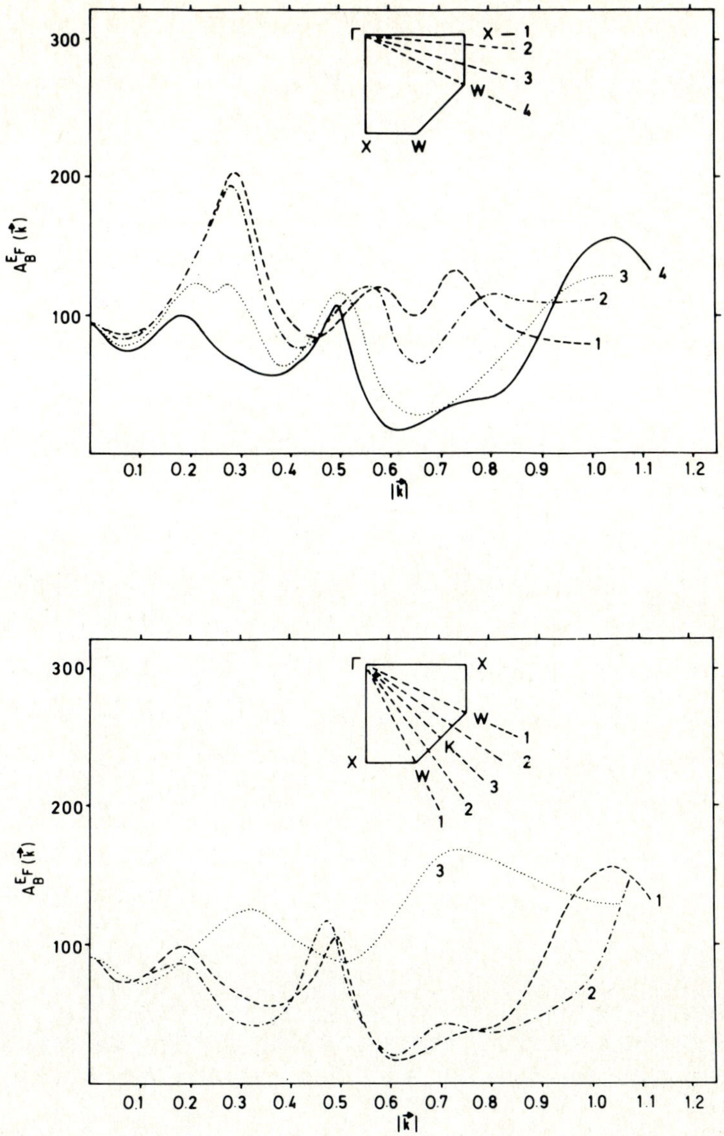

Fig. 15. Constant-E Bloch spectral functions at E_F for $Pu_{0.90}Al_{0.10}$. (Weinberger et al. 1985b.)

in UN. For UBi, Bi being the last element in group 5a, the Γ_8^- state is localized predominantly in the U sphere. Going therefore along the elements of group 5a, the chemical bonding of U pnictides undergoes significant and characteristic changes.

In fig. 19 the bandstructure along (100) is displayed for ThC, PaC, UC and NpC (Mallett 1982). In this figure it is worthwhile to note in particular that the dispersion

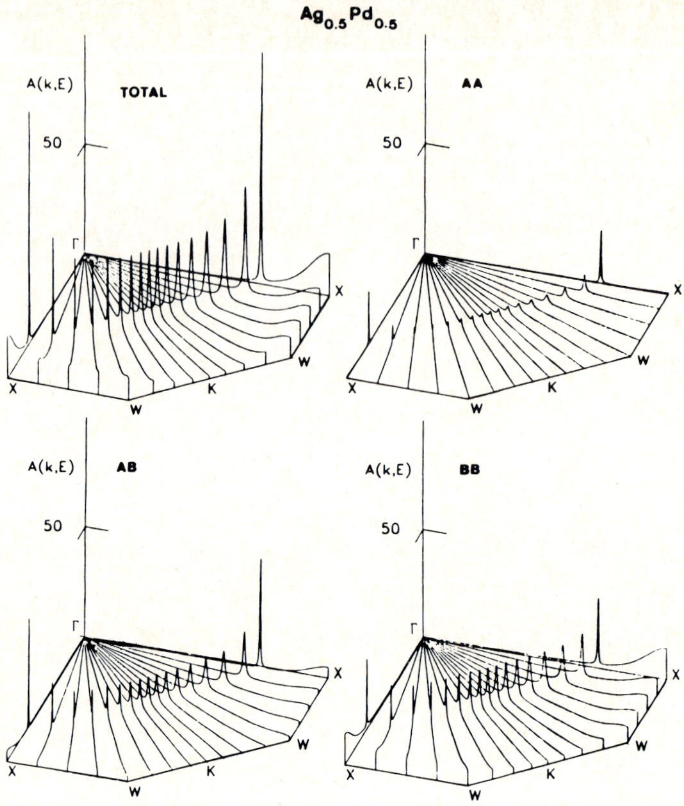

Fig. 16. Projected constant-E Bloch spectral functions at E_F for $Ag_{0.50}Pd_{0.50}$ (A = Ag, B = Pd). (Gonis et al. 1985a.)

of the Δ_7 band coming from above and intersecting the carbon p-like band near X is continously decreased with increasing atomic number of the metal constituent. In ThC the dispersion of this band is typically d-like, whereas in NpC the Δ_7 band is a well-localized f band.

Figure 20 shows contours for the full non-spherical potential for PaC. This figure illustrates beautifully that the muffin-tin approximation (section 3.2) is an excellent approximation even for compounds with considerable 'covalent bonding'.

9.4.2. *Vacancy effects in pnictides and chalcogenides, mixed crystals of actinide compounds*

Most of the NaCl type actinide pnictides and chalcogenides show a non-stoichiometry region, i.e., they are of the form $A(NMe)_x$, A being the actinide component and NMe the non-metal component. For a review on defect structure in actinide compounds see in particular the article by de Novion (1982). Carbides of

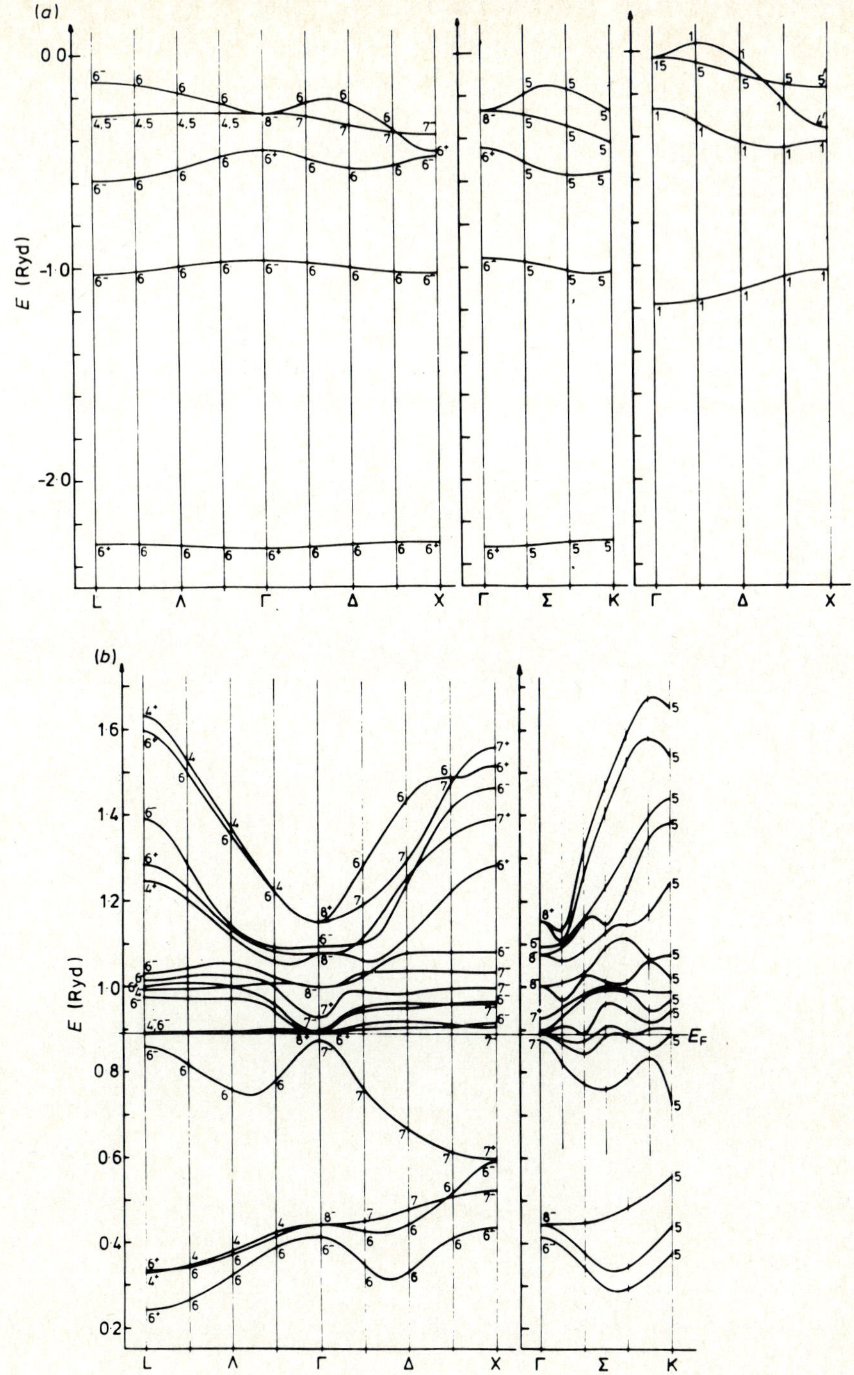

Fig. 17. Energy bandstructure of UN along [100], [111], and [110]. The upper half shows the region of the U $6s^{1/2}$, U $p^{1/2}$, U $p^{3/2}$ and C s bands together with the non-relativistic limit along [100]. (Weinberger et al. 1980.)

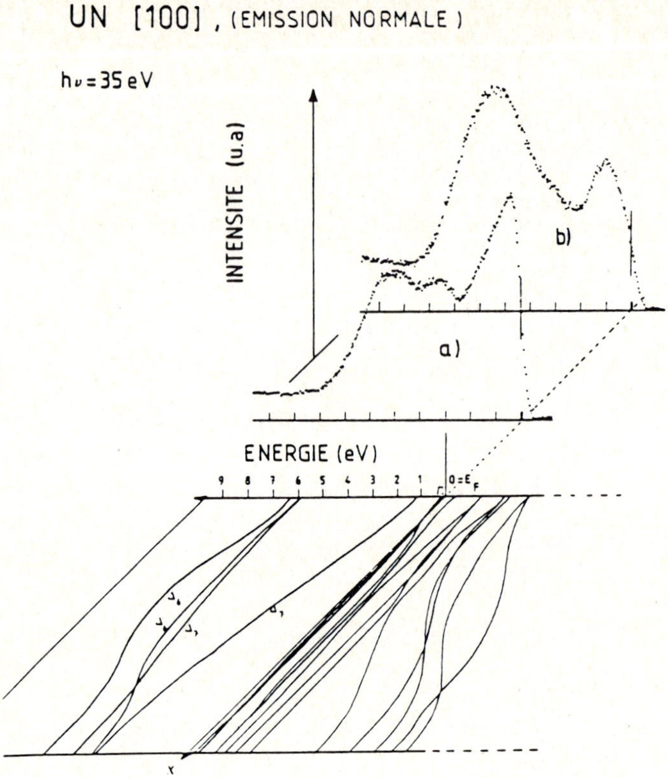

Fig. 18. Normal photoemission from UN (001) for (a) s polarisation, and (b) p polarisation. (Baptist 1982.)

TABLE 5

Partial charges for the $\boldsymbol{k} = (0,0,0)$ states of the 'Non-metal p-band' in UN and UBi.

		E (Ry)	Uranium				Non-metal		out	ΔE (eV)
			$p^{1/2}$	$p^{3/2}$	$f^{5/2}$	$f^{7/2}$	$p^{1/2}$	$p^{3/2}$		
UN[a]	6^-	0.4123	2.53	—	—	19.94	60.61	—	16.92	0.41
	8^-	0.4425	—	8.20	14.89	7.06	—	53.82	16.01	
UBi[b]	6^-	0.2953	1.53	—	—	19.23	56.60	—	22.64	1.46
	8^-	0.4028	—	0.56	50.24	9.93	—	22.91	16.36	

[a] Weinberger et al. (1980).
[b] Weinberger et al. (1982c).

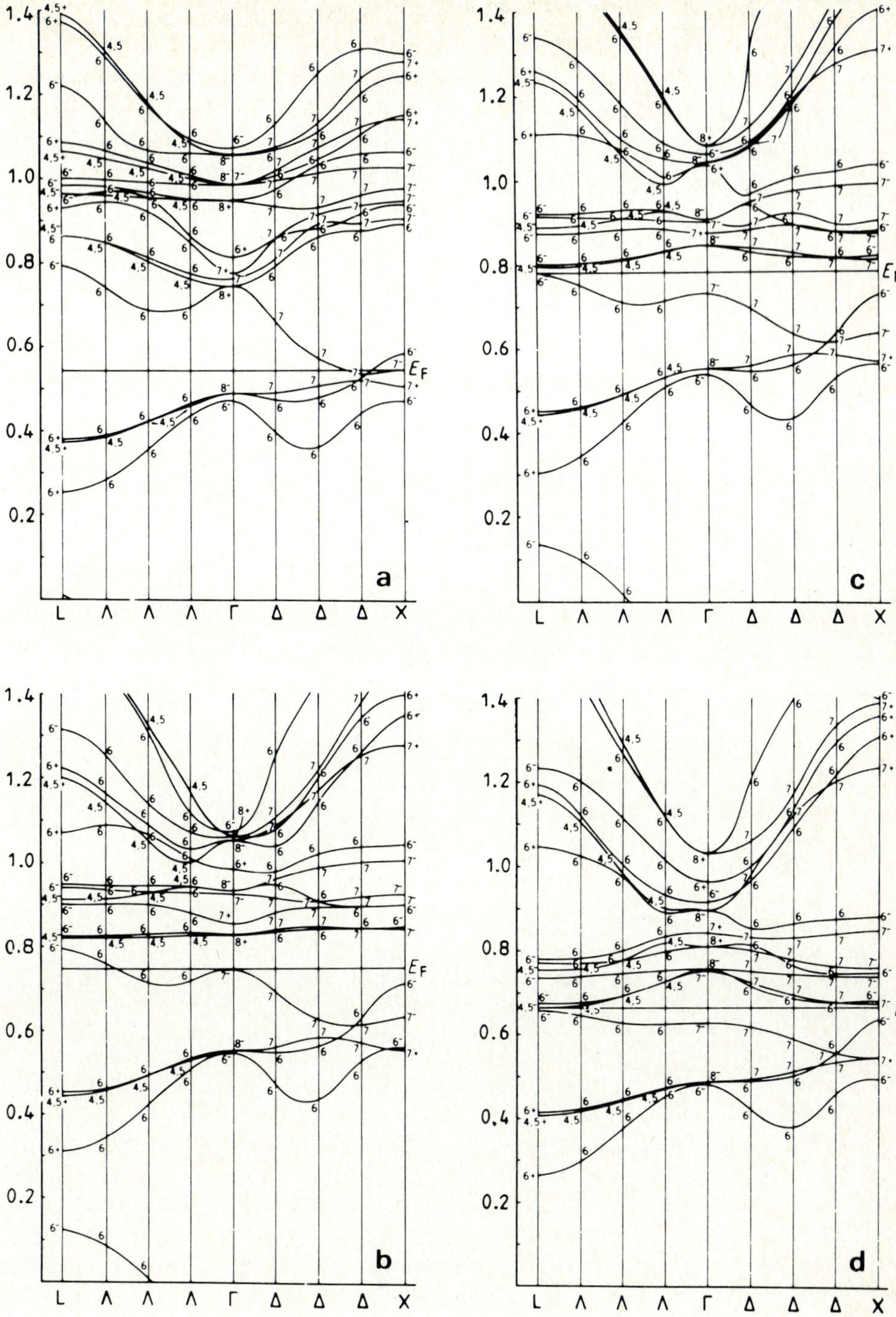

Fig. 19. Energy bandstructure along [100] and [111] for (a) ThC, (b) PaC, (c) UC and (d) NpC. (Mallett 1982.)

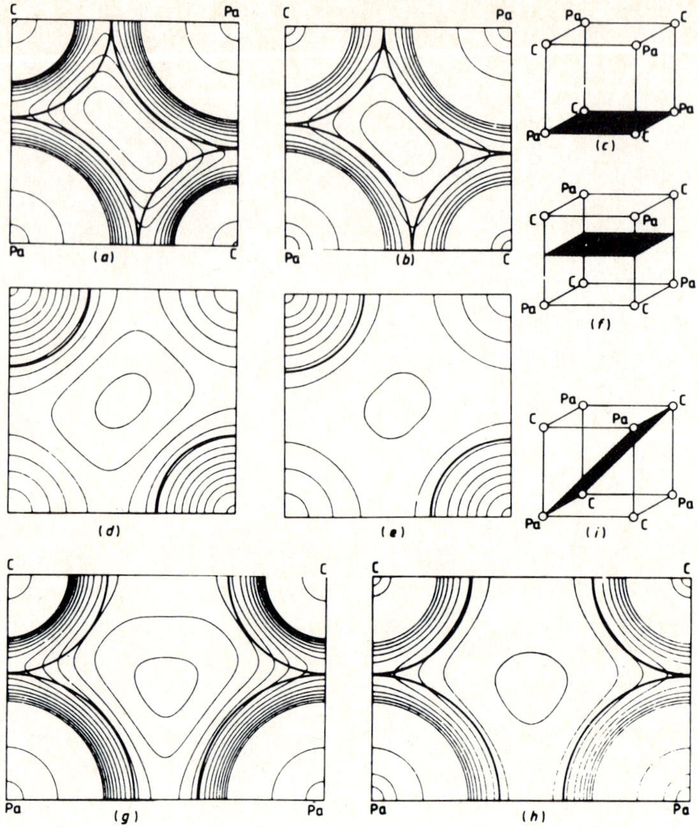

Fig. 20. The full non-spherical potential contours for PaC. (a) and (b) correspond to the Coulomb potential and charge density for the (001) direction (see c). (d)–(f) give the corresponding results for the (001) direction but this time with the plane touching the equipotential intersection for the origin Pa atom. Parts (g)–(h) give the results for the ($\bar{1}$01) plane. Also shown are the muffin-tin sphere radii. (Mallett 1982.)

NaCl structure for example were observed for the following range of concentrations: UC_x $(0.9<x<1.0)$, $NpC_x(0.82<x<0.96)$ and $PuC_x(0.65<x<0.9)$.

It is well-known from the carbides and nitrides of 3d and 4d metals, that the vacancies on the non-metal sublattice introduce new types of states about 1.5–2 eV below the Fermi energy. Since these effects of non-stoichiometry are very similar for Ti, V, Zr, Nb carbides and nitrides (Marksteiner et al. 1986a,b), and since very similar effects have to be expected for actinide (and rare-earth) pnictides and chalcogenides, figs. 21 and 22 illustrate the use of eq. (6.59), applied, however, in its non-relativistic form. In fig. 21 the density of states is shown for TiN_x, $x=1.0$ and $x=0.75$ (Klima et al. 1985). From this figure one can clearly see the effect of disorder on the non-metal sublattice due to vacancies. Figure 22 shows the vacancy component density of states in the non-metal subcell for (a) a single vacancy in an otherwise perfect TiN

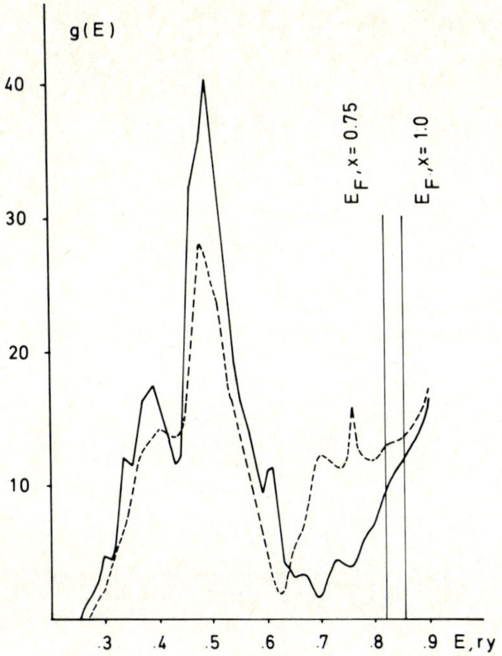

Fig. 21. Density of states for stoichiometric ($x = 1.0$, full line) and non-stoichiometric ($x = 0.75$, dashed line) TiN_x. (Klima et al. 1985.)

host, and (b) for 25% vacancies in TiN_x. In both cases the sharp peak corresponds to an s-like virtual bound state.

In fig. 23 the metal e_g- and t_{2g}-like, the non-metal t_{1u}-like and the vacancy a_{1g}-like component density of states is shown for VC_x for a variety of concentrations x

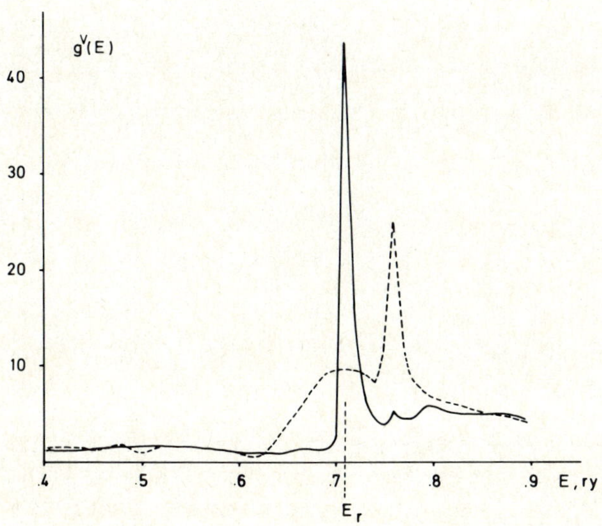

Fig. 22. Component density of states for the vacancies in TiN_x (dashed line) and for a single site vacancy (full line) in a stoichiometric host. (Klima et al. 1985.)

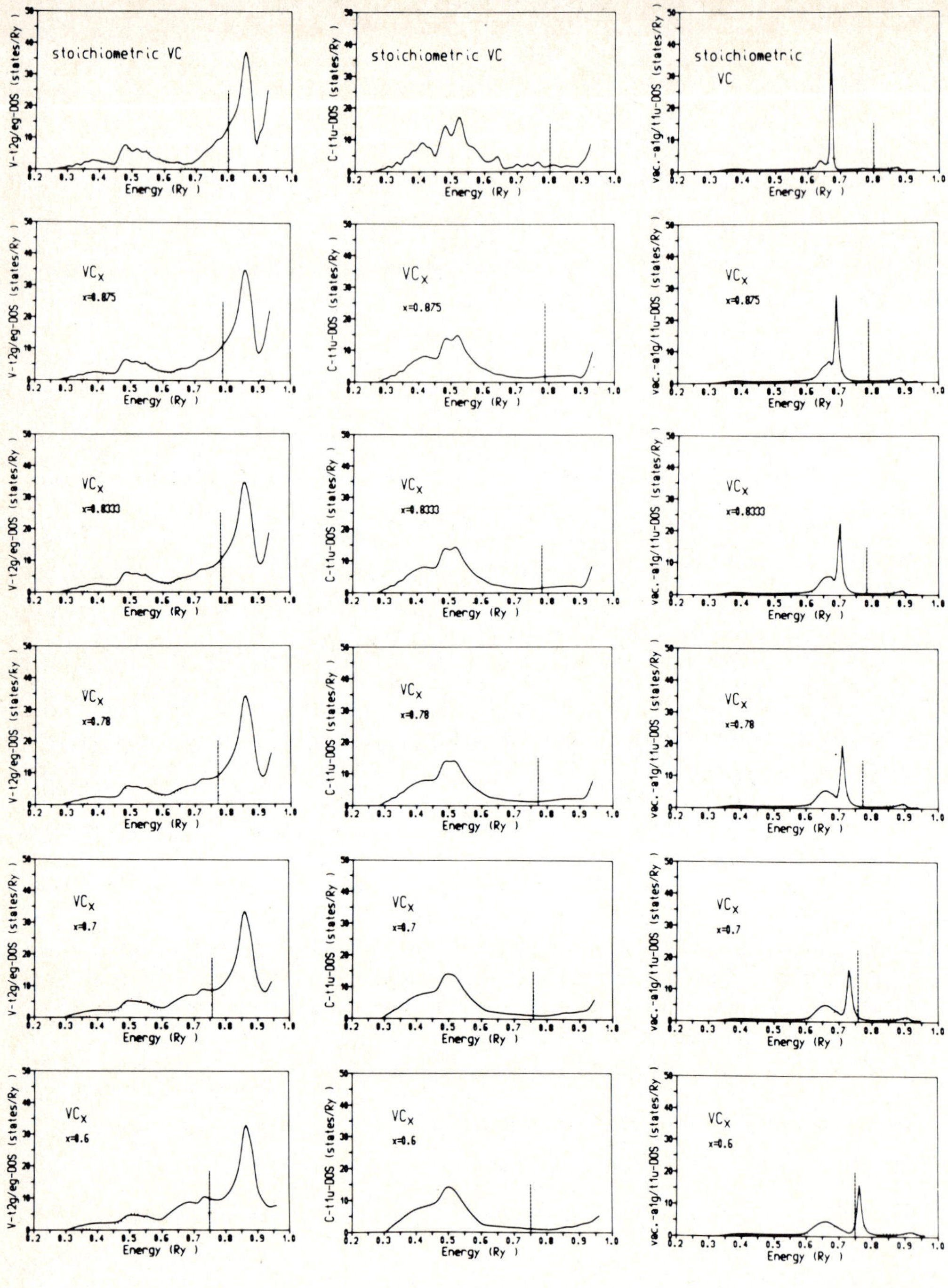

Fig. 23. Partial component densities of states of VC_x. Left column: t_{2g}-like (full line) and e_g-like (dotted line) metal partial density of states; center column: t_{1u}-like non-metal partial density of states; right column: a_{1g}-like (full line) and t_{1u}-like (dotted line) vacancy partial density of states. (Marksteiner et al. 1986a.)

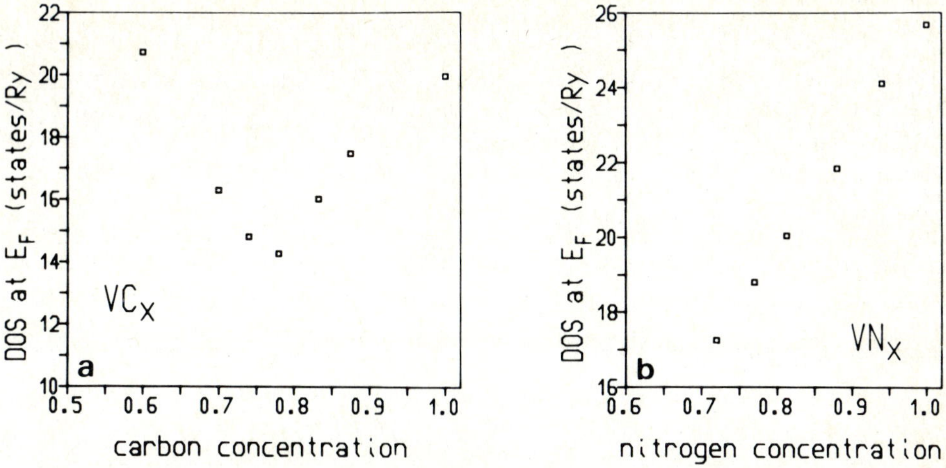

Fig. 24. Density of states at the Fermi energy for (a) VC_x, and (b) VN_x. (Marksteiner et al. 1986a.)

(Marksteiner et al. 1986a). In this figure one can see how profoundly the electronic structure is changed with increasing vacancy concentration. The position of the Fermi energy is in particular interesting for the change of the linear coefficient of the specific heat with respect to the concentration of the vacancies. In fig. 24 the density of states at E_F is displayed for VC_x and VN_x (Marksteiner et al. 1986a). While in VN_x, $n_0(E_F)$ is lowered continuously with increasing vacancy concentration, in VC_x for large vacancy concentrations $n_0(E_F)$ goes up again because the Fermi energy moves into the vacancy induced peak of the density of states. The behaviour shown in figs. 23 and 24 is most likely to be characteristic also for actinide carbides, chalcogenides and pnictides (see also Marksteiner et al. 1986b).

The vacancy induced states change also the topology of the Fermi surface (Schadler et al. 1985) and, consequently, the transport properties. Figures 21 and 24 imply that comparisons to experiments can be quite misleading unless the actual composition of the sample is considered.

Since the same type of KKR–CPA equations (6.46) have to be solved also for mixed crystals such as UAs_xSb_{1-x} or $Th_xU_{1-x}As$, it can be expected that in the near future calculations of the electronic structure for mixed crystals of actinide compounds will be available.

9.5. *Physical properties related to the electronic structure of actinide systems*

In most cases physical properties can be calculated directly from the one-particle Green's functions (6.20). In particular, the unenhanced density of states at the Fermi energy and Fermi surfaces can be obtained rather easily from the equations given in sections 6.6 and 6.7. For other properties reduced matrix elements (Wigner–Eckart theorem) have to be calculated in terms of the one-particle Green's function.

9.5.1. *Fermi surfaces*

Since the Kohn–Sham Fermi surfaces seem to be closely related to the physical Fermi surfaces (see also section 2.5), and since the Fermi surfaces can be measured directly by means of dHvA or positron annihilation experiments, the concept of constant-E Bloch spectral functions (6.68) and (6.72) is of quite some importance. In fig. 25, Fermi surface cuts for δ-Pu at different lattice constants are shown together with a 'contour diagram' for constant Bloch spectral functions at E_F for $Pu_{0.90}Al_{0.10}$ (Weinberger et al. 1985c). From this figure one can see that the third sheet of the Fermi surface in δ-Pu (vertically shaded) overlaps with the second sheet when pressure is applied. For the larger lattice constant the second sheet has a neck around K. $Pu_{0.90}Al_{0.10}$ does not have a well-defined Fermi surface. From the 'contour diagram' of the Bloch spectral functions, however, the features along $\Gamma-X$, for example, show how the building up of pressure is manifested in the alloying process.

For non-stoichiometric compounds drastic changes of the Fermi surface topology have to be expected, as is shown in fig. 26 in the case of TiN_x (Schadler et al. 1985). Similar effects are very likely to occur in substoichiometric actinide compounds such as PuC_x for example. In general, a rather good agreement between measured and calculated Fermi vectors can be expected for ordered and disordered systems. In fig. 27 the (100) Fermi vector is shown for the system Ag_xAu_{1-x} (Ebert et al. 1986). The deviation from experiment is typically about 1–5% of the scale of the BZ in a particular direction.

9.5.2. *Linear coefficient of the specific heat*

For $\mu \gg k_BT$ (μ chemical potential, k_B Boltzmann constant) the linear coefficient of the specific heat is given by

$$\gamma = \frac{\pi^2}{3} k_B n(E_F). \tag{9.12}$$

where $n(E_F)$ is the actual density of states at the Fermi energy, which can be written as

$$n(E_F) = n_0(E_F)[1 + \lambda_{ep} + \lambda_{sf} + \cdots], \tag{9.13}$$

where $n_0(E_F)$ is the value of the single particle density of states at E_F (eqs. 6.56 or 6.59) and λ_{ep} and λ_{sf} are the electron–phonon and the spin fluctuation enhancement factors, respectively.

In fig. 28 the 'unenhanced' density of states $n_0(E_F)$ is used for a comparison (Weinberger et al. 1985b) to experimentally determined linear coefficients of the specific heat in Pu_xAl_{1-x}. Quite clearly, $n_0(E_F)$ can at best only show qualitative trends. However, from fig. 28 it seems quite plausible that with decreasing Al concentration, γ is increased as is predicted from an extrapolation of the experimental data.

In general, it can be expected that the concentration dependence of $n_0(E_F)$ is described quite adequately within the relativistic KKR–CPA method. In fig. 29 for

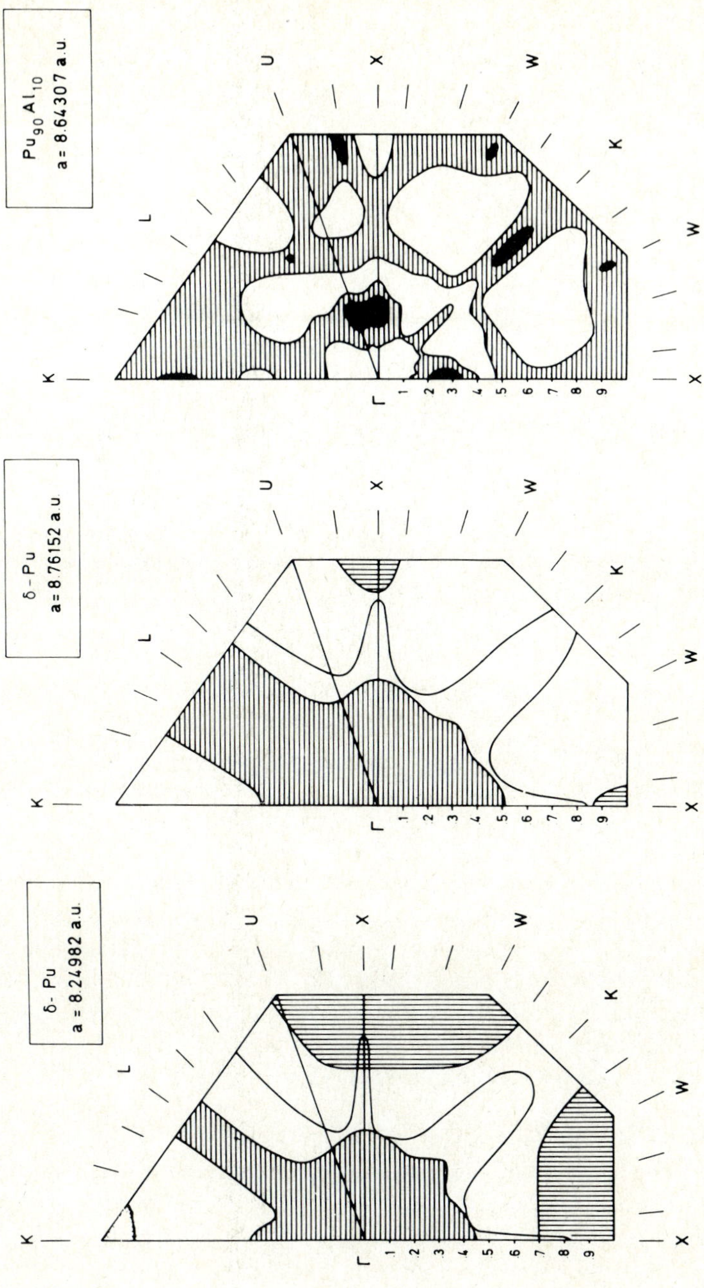

Fig. 25. Fermi surface cuts for δ-Pu and $Pu_{0.90}Al_{0.10}$. (Weinberger et al. 1985c.)

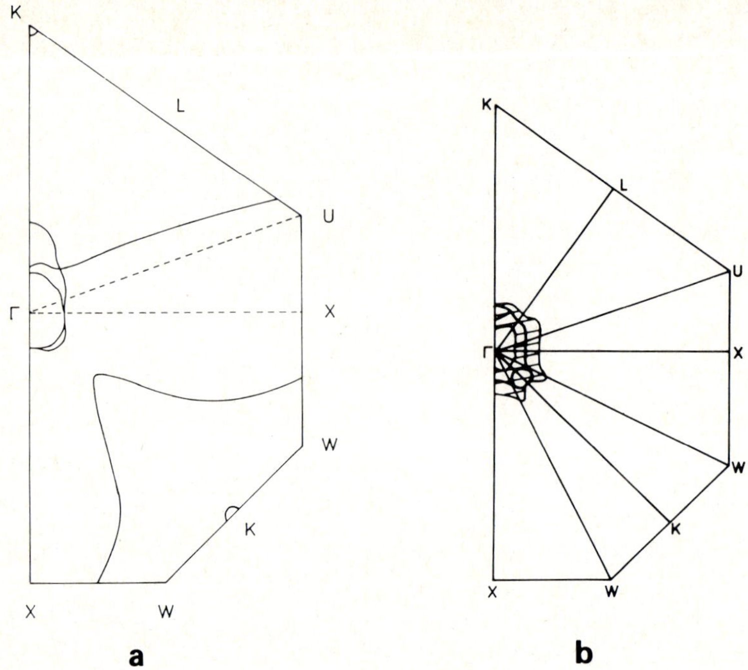

Fig. 26. Fermi surface cuts for (a) stoichiometric ($x=1.0$) and (b) non-stoichiometric ($x=0.75$) TiN_x. (Schadler et al. 1985.)

Au_xNi_{1-x} (Weinberger, 1983b) the calculated value of the density of states at the Fermi energy is compared with experimental values for γ and also with the experimental value of the molar susceptibility. Au_xNi_{1-x} is a rather complicated system. For small concentrations of Ni, Au_x Ni_{1-x} forms a spin glass; with increasing Ni concentration the system changes from 'paramagnetic' to 'superparamagnetic' and finally to ferromagnetic. As one can see from this figure the break in the curve for γ is described correctly in the theoretical calculation up to a few percent. For actinide systems the same kind of numerical accuracy can be expected for $n_0(E_F)$ as a function of concentration either in binary alloys such as Pu_xAm_{1-x} or in pseudobinary alloys such as in PuC_x (see also fig. 24).

9.5.3. *Electron–phonon enhancement*

The calculation of the electron–phonon enhancement factor is rather a tedious and involved problem. In terms of the single site approximation of Gaspari and Gyorffy (1972) λ_{ep} can be estimated as

$$\lambda_{ep}=\frac{\eta^0}{M\langle\omega^2\rangle}, \tag{9.14}$$

where M and $\langle\omega^2\rangle$ are the mass and the averaged square of the phonon frequency, respectively. The parameter η^0, which occasionally is called Hopfield parameter, is

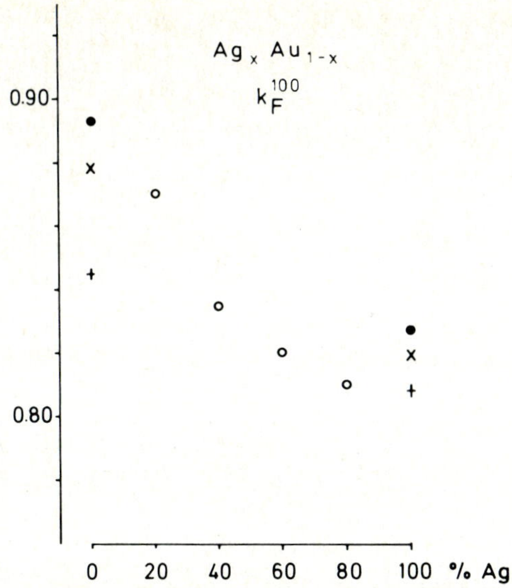

Fig. 27. (100) Fermi vector for $Ag_x Au_{1-x}$; ●: MacDonald et al. (1982); +: Eckardt et al. (1984); ○: Ebert et al. (1986); ×: Experiment, Coleridge and Templeton (1982).

given relativistically within the framework of the Gaspari–Gyorffy Ansatz for metals as (John et al. 1980, Weinberger 1982)

$$\eta^0 = \frac{2E_F^2}{n(E_F)\pi} \sum_{j,j'} (2j+1) \begin{pmatrix} j & 1 & j' \\ \frac{1}{2} & 0 & -\frac{1}{2} \end{pmatrix}^2 \frac{\sin^2[\delta_\kappa(E_F) - \delta_{\kappa'}(E_F)]}{\sin^2 \delta_\kappa(E_F) \sin^2 \delta_{\kappa'}(E_F)}$$

$$\times \operatorname{Im} \tilde{\tau}_{\kappa\kappa}^{00}(E_F) \operatorname{Im} \tilde{\tau}_{\kappa'\kappa'}(E_F), \tag{9.15}$$

$$\operatorname{Im} \tilde{\tau}_{\kappa\kappa}^{00}(E) = \frac{1}{(2j+1)} \sum_\mu \operatorname{Im} \tilde{\tau}_{\kappa\mu,\kappa\mu}^{00}(E). \tag{9.16}$$

In eq. (9.15), $(j\ 1\ j'/\frac{1}{2}\ 0\ -\frac{1}{2})$ are Wigner 3j-coefficients and the $\delta_\kappa(E_F)$ are the κ-like phaseshifts at the Fermi energy.

If the denominator in (9.14) can be averaged as

$$M\langle\omega^2\rangle = xM_A\langle\omega_A^2\rangle + (1-x)M_B\langle\omega_B^2\rangle, \tag{9.17}$$

then eq. (9.4) can be applied also for substitutional alloys such that $\eta^0 = x\eta_A^0 + (1-x)\eta_B^0$. Within the Gaspari–Gyorffy Ansatz the electron–phonon enhancement can be calculated very easily. Equation (9.17), however, has one practical disadvantage. If $j_{max} = l_{max} \pm \frac{1}{2}$ characterizes the resonance with the highest angular momentum quantum number l (e.g. $l=3$ for f electron systems) then usually the $j_{max}/(j_{max}+1)$ interaction is the biggest contribution to η^0. This implies that the matrix elements of $\boldsymbol{\tau}^{00}$ in eq. (9.16) have to be calculated with respect to $j_{max}+1$, which increases drastically the computational effort (size of matrix is more than doubled).

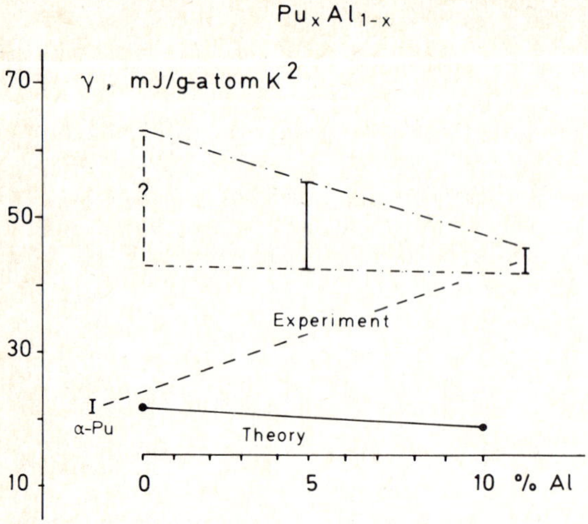

Fig. 28. Linear coefficient of the specific heat in Pu_xAl_{1-x}. (Weinberger et al. 1985b.)

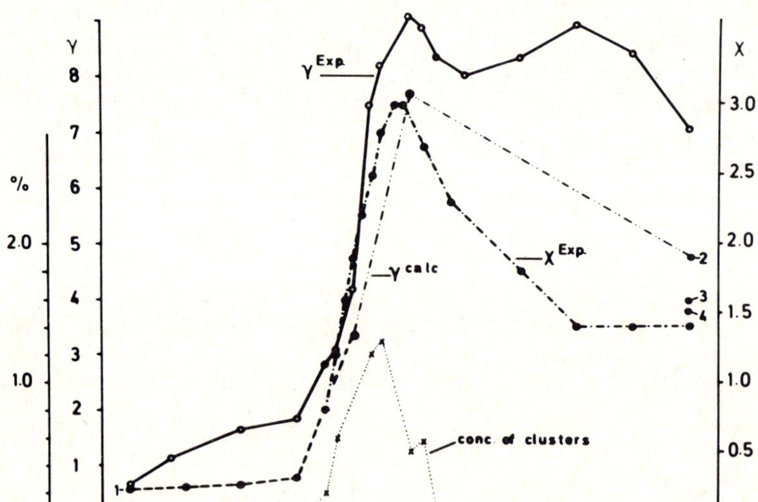

Fig. 29. Experimental linear coefficient γ^{exp} (mJ mol^{-1} K^{-2}) of the electronic specific heat (open circles, full curve); experimental high-field susceptibility χ^{exp} (10^{-6} EMU 9, half full circles, chain curve); cluster concentration (%, crosses, dotted curve) and of calculated value γ (mJ mol^{-1} K^{-2}, full circles, broken curve). The experimental values are taken from Kuentzler and Kappler (1979). For Ni, point 2 refers to the non-spin-polarised calculation of $n(E_F)$ and point 3 (Moruzzi et al. 1978) and 4 (Anderson et al. 1979) refer to the corresponding spin-polarised calculations. For Au, point 1 refers to the value given by Weinberger (1982). (Weinberger 1983b.)

9.5.4. *Soft X-ray emission spectra*

One of the physical observables of considerable interest in the electronic theory of solid matter is the intensity of soft X-ray emission and absorption. Considering only electric dipole transitions in the long wavelength limit, within first-order time-dependent perturbation theory, the intensity for an emission process can be written as (Staunton et al. 1980, Weinberger 1982a, Weinberger et al. 1982b)

$$I_0(\omega) \sim \frac{4\pi}{3}(\Delta\varepsilon)^3 \sum_{j'}(2j+1)\begin{pmatrix} j & 1 & j' \\ \frac{1}{2} & 0 & -\frac{1}{2}\end{pmatrix}^2$$

$$\times R^i_{\kappa\kappa'}(\varepsilon_f, \varepsilon) N^i_{\kappa'}(\varepsilon)^{-1} n^i_{\kappa}(\varepsilon)\delta(\varepsilon_f + \hbar\omega - \varepsilon), \tag{9.18}$$

where $R^i_{\kappa\kappa'}(\varepsilon_f, \varepsilon)$ and $N^i_{\kappa}(\varepsilon)$ are radial integrals,

$$R^i_{\kappa\kappa'}(\varepsilon_f, \varepsilon) = \int_0^{R^i_s} r_i^3 [g^i_{\kappa}(\varepsilon_f, r_i) g^i_{\kappa'}(\varepsilon, r_i) + f^i_{\kappa}(\varepsilon_f, r_i) f^i_{\kappa'}(\varepsilon, r_i)]\, dr, \tag{9.19}$$

$$N^i_{\kappa}(\varepsilon) = \int_0^{R^i_s} r_i^2 [g^i_{\kappa}(\varepsilon, r_i)^2 + f^i_{\kappa}(\varepsilon, r_i)^2]\, dr, \tag{9.20}$$

$(j1j'/\frac{1}{2}0-\frac{1}{2})$ are Wigner $3j$-coefficients, $n^i_x(\varepsilon)$ is the local density of states as given by eq, (6.61) and i denotes the atomic species centred at $\boldsymbol{R}_i$. In the case of a random alloy, for the emission from an α-type constituent, in eq. (9.20) the local component density of states has to be used (see also eq. 6.58). In eq. (9.18), j is the total angular momentum of the corresponding core state (e.g. $j=\frac{1}{2}$ and $\frac{3}{2}$ for L_2 and L_3 emission spectra) with energy ε_f.

Quite recently, eqs. (9.18)–(9.20) have been used to calculate the $N_{6,7}$ emission spectra of Ir, Pt and Au in UIr_3, UPt_3 and UAu_3 (Albers et al. 1985). Since the Pt d-like partial local density of states reflects the very sharp U f-like peak in the density of states for the heavy Fermion system UPt_3, the $N_{6,7}$ spectra can be used for an indirect experimental checking of this peak. Because two emissions are observed in the experiment simultaneously, namely the emission from the 5d metal $4f^{5/2}$ and the $4f^{7/2}$ core hole, this very peak is visible only in the $4f^{5/2}$ emission, reflecting a $\Delta j=0$ transition to the Pt $d^{5/2}$-like and a $\Delta j=-1$ transition to the Pt $d^{3/2}$-like valence states. The comparison of the $N_{6,7}$ spectra for UIr_3, UPt_3 and UAu_3, as shown in fig. 30, makes the recording of these spectra even more interesting, since only for the heavy Fermion system UPt_3 a corresponding shoulder in the vicinity of the Fermi energy occurs. In UIr_3 this shoulder is not visible, whereas in UAu_3 this shoulder is moved out of the vicinity of E_F. For a better understanding of the above arguments and of the spectra in fig. 31 the $d^{3/2}$-like and $d^{5/2}$-like partial local density of states of Ir, Pt and Au in UIr_3, UPt_3 and UAu_3 are displayed.

9.5.5. *Other physical properties*

It is beyond the scope of this review article to discuss in detail all the physical

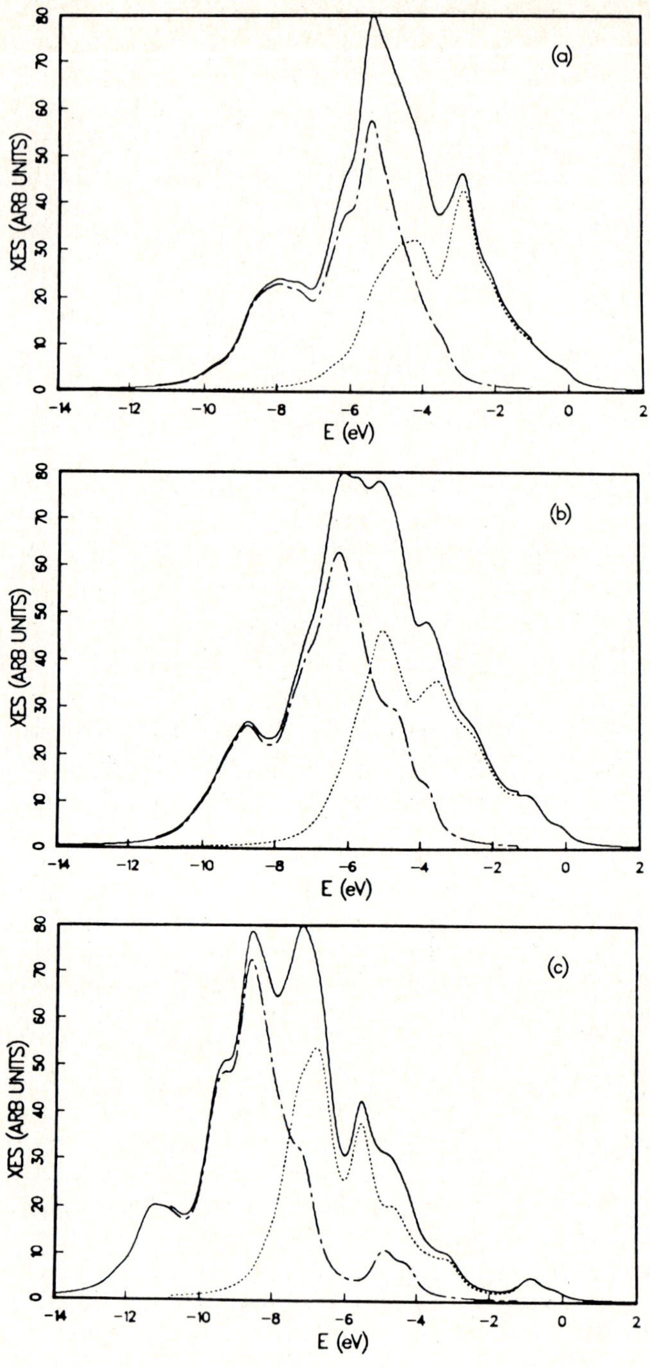

Fig. 30. $N_{6,7}$ soft X-ray emission spectra for (a) Ir in UIr_3, (b) Pt in UPt_3, and (c) Au in UAu_3. (Albers et al. 1985.)

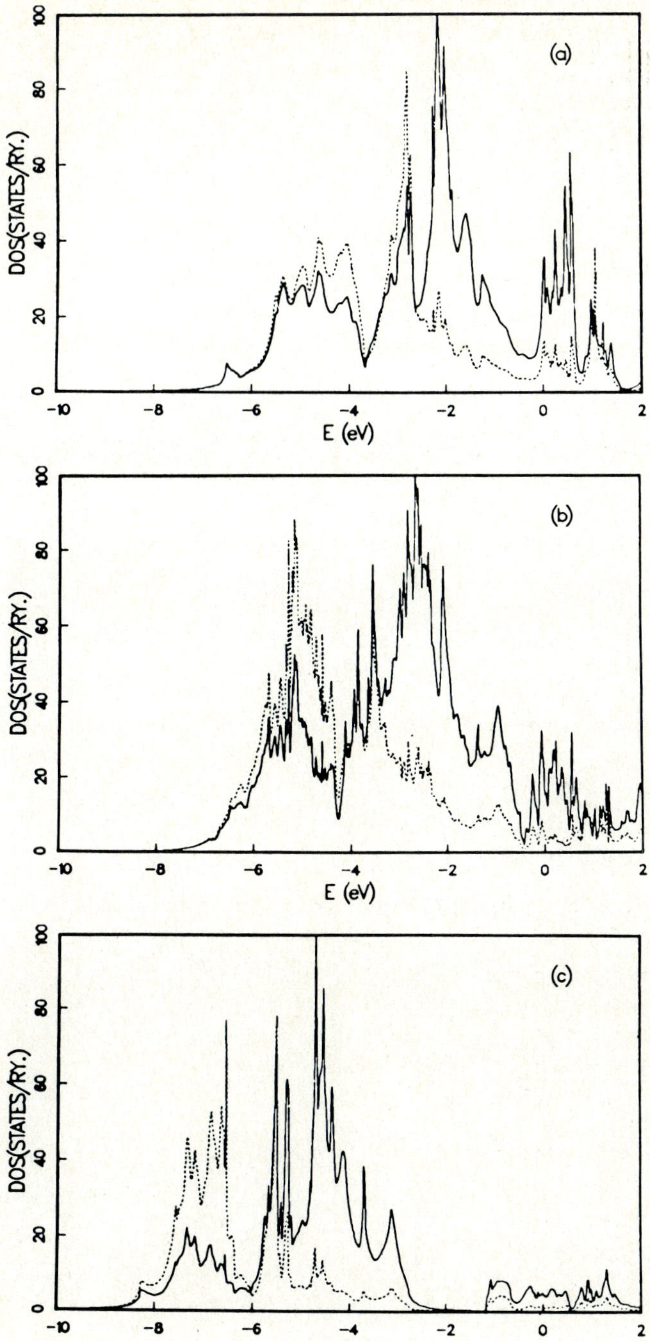

Fig. 31. $d^{3/2}$-like and $d^{5/2}$-like partial local density of states for (a) Ir in UIr_3, (b) Pt in UPt_3 and (c) Au in UAu_3. (Albers et al. 1985.)

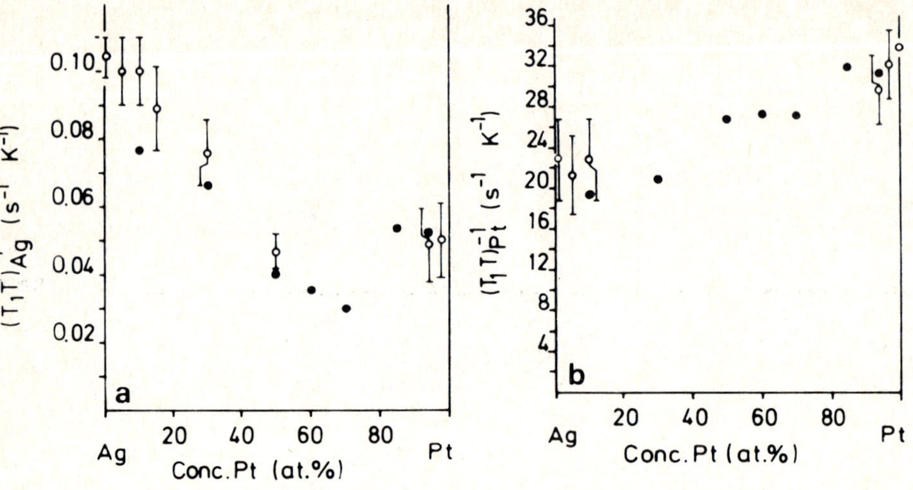

Fig. 32. Nuclear spin lattice relaxation rates for ^{109}Ag and ^{195}Pt in $Ag_x\ Pt_{1-x}$. (Ebert et al. 1985.)

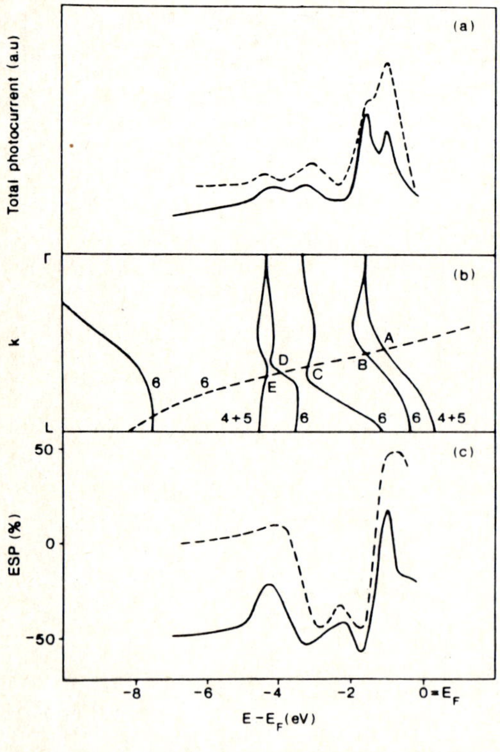

Fig. 33. Normal photoemission from Pt (111). (a) Normal incidence, normal emission spectrum ($h\nu = 14$eV): theory (full line), experiment (dashed line) (Eyers et al. 1984). (b) Energy bands in the (111) direction. The dashed curve is the final state band moved down in energy by 14 eV. Direct transition can occur at the intersection A–E. (c) Electron spin polarisation for right-hand circularly polarised radiation: theory (full line), experiment (dashed line) (Eyers et al. 1984). (Ginatempo et al. 1985.)

properties which can be calculated directly either from the site-(cell-diagonal scattering path operators or in terms of a relativistic multiple scattering approach. As further examples we want to point out that hyperfine field interactions such as the nuclear spin lattice relaxation rate can be calculated fully relativistically in terms of the site-(cell-) diagonal scattering path operators (Ebert et al. 1985), and that the fully relativistic calculation of spin-polarized, angle-resolved photoemission intensities (Ginatempo et al. 1985) is very closely related to the theoretical approach discussed in sections 5 and 6. In fig. 32 the ^{109}Ag and ^{195}Pt nuclear spin lattice relaxation rates are shown for $Ag_x Pt_{1-x}$ (Ebert et al. 1985a). As one can see, the experimental relaxation rates are rather well described by the RKKR-CPA method. For actinide systems, in particular for the case of the heavy Fermion system UPt_3, it would be very useful indeed to have similar measurements, since the nuclear spin lattice relaxation rate directly maps the κ-like decomposition of the density of states at the Fermi energy (Ebert et al. 1985).

In particular, theoretical angle-resolved and angle-integrated photoemission intensities are very likely to be calculated in the very near future also for f-electron systems. As an encouraging example first results for the Pt(111) normal photoemission are shown in fig. 33. In this figure the theoretical calculated photocurrent and the polarisation are shown together with the experimental data. The agreement between experiment and theoretical calculation is very good indeed. It should be noted, that the results shown in the figure correspond to the very first such calculation available.

10. On the problem of magnetism in actinide systems

The problem of magnetism in actinide systems is very tough indeed, not only because the truly interesting features of magnetism, namely the temperature dependence was discussed (non-relativistically) only very recently on the basis of a KKR–CPA approach for the pure metals Fe, Co, Ni and Cr (Gyorffy et al. 1985). Even at $T=0$ K, up to now the description of magnetism in actinide systems is not really satisfactorily (see also the relevant discussion by Brooks et al. 1985).

Assuming that the spin-only current is indeed by far the biggest contribution to the coupling with the fictitious external field $B_i^{\mathrm{eff}}([n,m],\boldsymbol{r}_i)$ (section 2.4), a fully relativistic $T=0$ K theory of magnetism involves the calculation of the single site t-matrix for 'spin-dependent' spherical symmetrical potentials (section 5.4) self-consistently with respect to the single particle density $n(r)$ and the magnetisation density $m(r)$. Computationally this is quite a major task even within the frame-work of a 'complex energy' KKR-scheme and in the age of supercomputers. At present, it seems that only within the next few years first results for the pure metals and perhaps binary alloys of actinide metals will be available. At the moment there is not even such a calculation for a d-only (alloy) system available.

Results from single site scattering (eqs. 5.48–5.67) using the difference between 'quasi-relativistic' self-consistent spin-polarized potentials as internal magnetic field show interesting bits of new actinide physics evolving. In fig. 34 the 'non-magnetic'

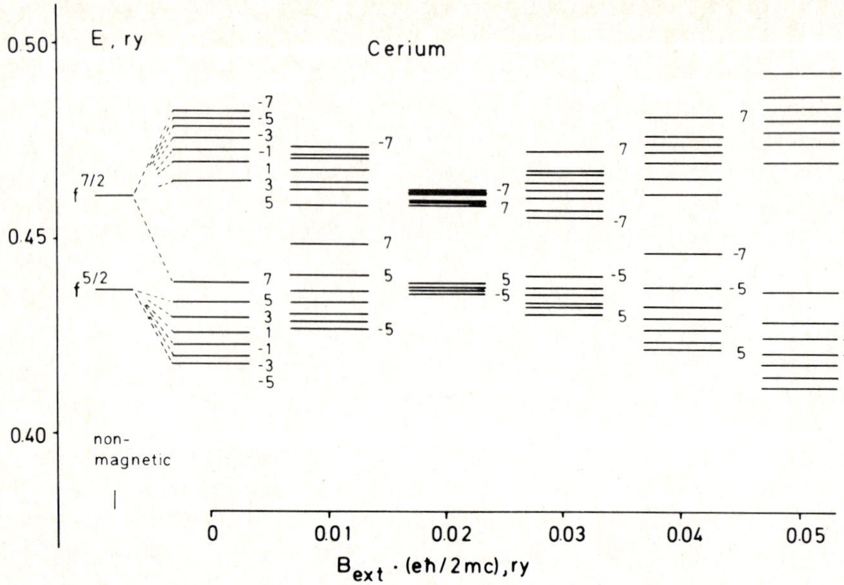

Fig. 34. Resonance energies for f-like single site scattering in Ce. (Schadler et al. 1986.)

$f^{5/2}$- and $f^{7/2}$-like resonance energies are shown for Ce (Schadler et al. 1986, Smith et al. 1986). In the presence of the internal magnetic field these resonances are split up by a kind of internal Zeeman effect. It should be noted that for elastic scattering a resonance energy is defined as the energy at which a particular phaseshift $\delta_\kappa(E)$ (eq. 5.24) passes through $\pi/2$, or equivalently, as the energy at which the real part of the scattering amplitude (5.29) changes sign. For absorptive scattering this is less rigorous, because the scattering amplitude can have loops, i.e., several changes of the sign of its real part can occur. However, by defining for practical purposes the resonance energy in terms of the energy where the diagonal scattering amplitude $f_{QQ}(E)$ (eq. 5.67) passes through the imaginary axis (well) above 0.5, a compact picture of single site scattering can be given. Switching on the external magnetic field, the internal Zeeman splitting is reduced until at a particular external magnetic field the scattering is completely elastic (i.e., non-magnetic). Beyond this external field the ordering of the resonance energies with respect to the eigenvalues of J_z, μ is inverted.

Figure 35 shows the $\mu=\frac{5}{2}$ f-like scattering in the presence of the internal magnetic field. From this figure one can see that the coupling between the $\kappa=-4$ and the $\kappa'=3$ angular momentum channel is quite substantial. In fig. 36 the $\mu=\frac{3}{2}$ f-like off-diagonal scattering amplitude in the presence of the internal magnetic field is shown for different external fields. One can clearly see that at a particular external field this scattering amplitude shrinks to a limiting circle.

In fig. 37 finally the resonance energies for f-like single site scattering in Pu are

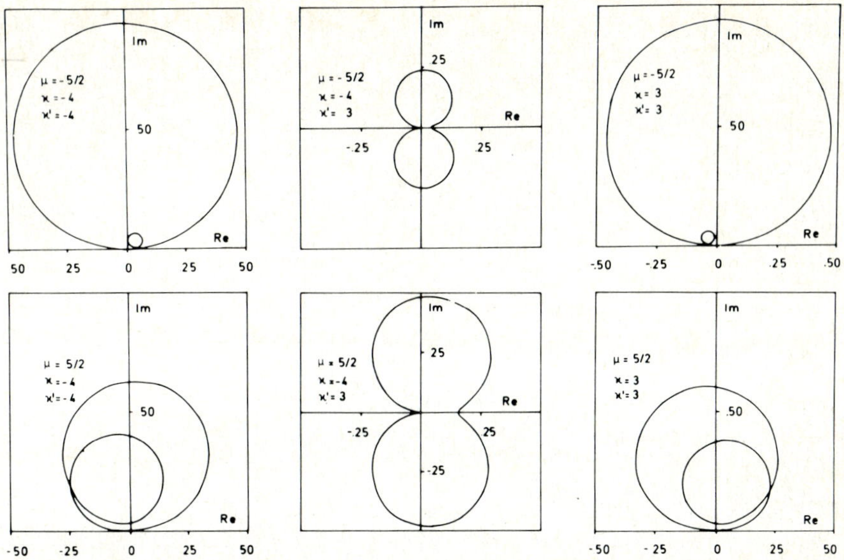

Fig. 35. $\mu = \pm\frac{5}{2}$ f-like scattering amplitudes in Ce. (Schadler et al. 1986.)

displayed. In fig. 34 as well as in fig. 37 the internal magnetic field ($B_{ext}=0$) produces two patterns of 7 resonance energies each. Each of these patterns is to be considered as 'relativistic splitting' of a non-relativistic 'spin-polarized' resonance. In particular, it is interesting to note that the d-like channels in Ce and Pu behave quite differently from the f-like channels (Schadler et al. 1986).

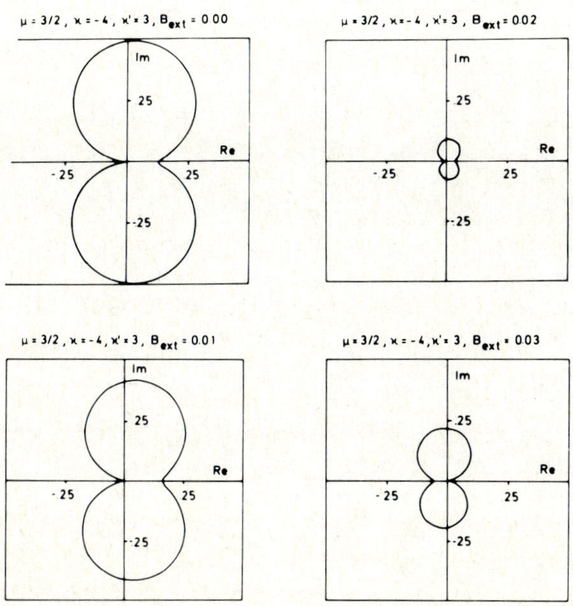

Fig. 36. The $\mu=\frac{3}{2}$ f-like scattering amplitudes in Ce for varying external magnetic field. (Schadler et al. 1986.)

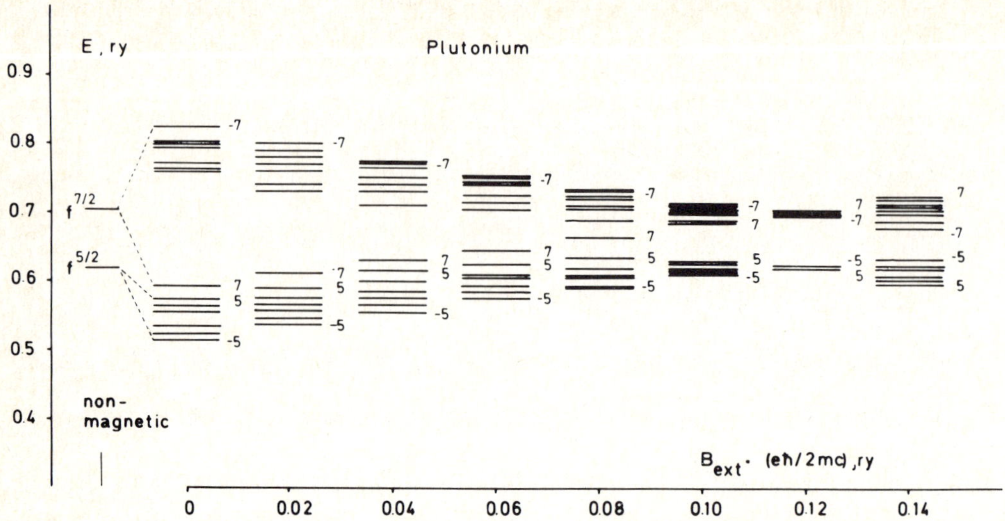

Fig. 37. Resonance energies for f-like single site scattering in Pu. (Schadler et al. 1986.)

Quite clearly figs. 34–37 only summarize the effects to be seen in a 'spin-polarized, relativistic' bandstructure calculation. It is yet to be seen whether the interplay between the crystal field splitting and the splitting due to the internal magnetic field is indeed a driving force for narrowing f-like bands in actinide systems. In particular, however, the $\mu = \pm\frac{7}{2}$ resonance in Ce as well as in Pu seems likely to be in charge of a few future surprises for the electronic structure of actinide systems.

Acknowledgements

This article was partially supported by the Jubiläumsfonds der Österreichischen Nationalbank (Projekt no. 2301) and by the Department of Energy under contract no. DE-FG-84ER45080.

References

Abramowitz, M., and I.A. Stegun, 1970, Handbook of Mathematical Functions (Dover, New York).

Albers, R.C., A.M. Boring, P. Weinberger and N.E. Christensen, 1985, Phys. Rev. B **32**, 7571.

Altmann, S.L., 1977, Induced Representations in Crystals and Molecules (Academic Press, New York).

Altmann, S.L., 1986, Rotations, Quaternions and Double Groups (Clarendon Press, Oxford).

Altmann, S.L., and R. Dirl, 1984, J. Phys. A **17**, 501.

Anderson, J.R., D.A. Papaconstantopoulos and L.L. Boyer, 1979, Phys. Rev. B **20**, 3172.

Arko, A.J., D.D. Koelling and J.E. Schirber, 1984, in: Handbook on the Physics and Chemistry of the Actinides, Vol. 2, eds A.J. Freeman and G.H. Lander (North-Holland, Amsterdam) ch. 3.

Baptist, R., 1982, Thesis (University of Grenoble, France).

Boring, A.M., R.C. Albers and P. Weinberger, 1985, in: High Temperature Alloys: Theory and Design, ed. J.O. Stiegler (The Metallurgical Society of AIME, New York) p. 265.

Braspenning, P.J., R. Zeller, P.H. Dederichs and A. Lodder, 1982, J. Phys. F **12**, 105.

Brooks, M.S.S., B. Johansson and H.L. Skriver, 1985, in: Handbook on the Physics and Chemistry of the Actinides, Vol. 1, eds A.J. Freeman and G.H Lander (North-Holland, Amsterdam) ch. 3.

Calogero, F., 1963, Nuovo Cimento **27**, 1007.

Calogero, F., 1967, The Variable Phase Approach to Potential Scattering (Academic Press, New York).

Churchill, R.V., 1960, Complex Variables and Applications (McGraw-Hill, New York).

Coleridge, P.T., and I.M. Templeton, 1982, Phys. Rev. B **25**, 7818.

Dirl, R., R. Haase, P. Herzig, P. Weinberger and S.L. Altmann, 1984, Phys. Rev. B **32**, 788.

Ebert, H., P. Weinberger and J. Voitlander, 1985, Phys. Rev. B **31**, 7566.

Ebert, H., P. Weinberger and J. Voitlander, 1986, Z. Phys. B **63**, 299.

Eckardt, H., L. Fritsche and J. Noffke, 1984, J. Phys. F **14**, 97.

Ehrenreich, H., and L.M. Schwartz, 1976, Solid State Phys. **31**, 149.

Englisch, H., and R. Englisch, 1983, Physica A **121**, 253.

Eyers, A., F. Schafers, G. Schonhense, U. Heinzmann, H.P. Oepen, K. Hunlich, J. Kirschner and G. Borstel, 1984, Phys. Rev. Lett. **52**, 1559.

Faulkner, J.S., 1982, Prog. Mater. Sci. **27**, 1–187.

Faulkner, J.S., and G.M. Stocks, 1980, Phys. Rev. B **21**, 3222.

Feder, R., F. Rosicky and B. Ackermann, 1983, Z. Phys. B **52**, 31.

Gaspari, G.D., and B.L. Gyorffy, 1972, Phys. Rev. Lett. **28**, 801.

Ginatempo, B., P.J. Durham, B.L. Gyorffy and W.M. Temmermann, 1985, Phys. Rev. Lett. **54**, 1581.

Gonis, A., and A.J. Freeman, 1984, Phys. Rev. B **29**, 4277.

Gonis, A., G.M. Stocks, W.H. Butler and H. Winter, 1984, Phys. Rev. B **29**, 555.

Gonis, A., A.J. Freeman and P. Weinberger, 1985a, Phys. Rev. B **32**, 7713.

Gonis, A., A.J. Freeman and P. Weinberger, 1985b, Phys. Rev. B **32**, 7720.

Gunnarsson, O., and B.I. Lundqvist, 1976, Phys. Rev. B **13**, 4274.

Gyorffy, B.L., and G.M. Stocks, 1979, in: Electrons in Disordered Materials and at Metallic Surfaces, NATO ASI Series B, Vol. 42, eds P. Phariseau, B.L. Gyorffy and L. Scheire (Plenum Press, New York) p. 89.

Gyorffy, B.L., and M.J. Stott, 1973, in: Bandstructure Spectroscopy of Metals and Alloys, eds D.J. Fabian and L.M. Watson (Academic Press, New York) p. 385.

Gyorffy, B.L., A.J. Pindor, J. Staunton, G.M. Stocks and H. Winter, 1985, Phys. Rev. B, submitted for publication.

Ham, F.S., and B. Segall, 1961, Phys. Rev. **124**, 1786.

Hohenberg, P., and W. Kohn, 1964, Phys. Rev. **136**, 864B.

Holzwarth, N.A.W., 1975, Phys. Rev. B **11**, 3718.

Jansen, L., and M. Boon, 1967, Theory of Finite Groups. Applications in Physics (North-Holland, Amsterdam).

John, W., V.V. Nemoshkalenko, V.N. Antonov and Vl.N. Antonov, 1980, Phys. Status Solidi b **99**, 599.

Johnson, D.D., F.J. Pinski and G.M. Stocks, 1984, Phys. Rev. B **30**, 5508.

Kambe, K., 1967a, Z. Naturforsch. **22a**, 322.

Kambe, K., 1967b, Z. Naturforsch. **22a**, 422.

Kambe, K., 1967c, Z. Naturforsch. **22a**, 1280.

Klima, J., G. Schadler, P. Weinberger and A. Neckel, 1985, J. Phys. F **15**, 1307.

Kohn, W., and N. Rostoker, 1954, Phys. Rev. **94**, 1111.

Kohn, W., and P. Vashishta, 1983, General density functional theory, in: Theory of the Electron Gas, eds S. Lundqvist and N.H. March (Plenum Press, New York) p. 79.

Korringa, J., 1947, Physica **13**, 392.

Kuentzler, R., and J.P. Kappler, 1979, J. Phys. F **11**, 457.

Lieb, E.H., 1982, Density functionals for Coulomb systems, in: Physics as Natural Philosophy, eds A. Simon and H. Feshbach (MIT Press, Cambridge).

Lloyd, P., and P.V. Smith, 1972, Adv. Phys. **21**, 69.

Loucks, T.S., 1967, The Augmented Plane Wave Method (Benjamin, New York).

MacDonald, A.H., and S.H. Vosko, 1979, J. Phys. C **12**, 2977.

MacDonald, A.H., J.M. Daams, S.H. Vosko and D.D. Koelling, 1981, Phys. Rev. B **23**, 6377.

MacDonald, A.H., J.M. Daams, S.H. Vosko and D.D. Koelling, 1982, Phys. Rev. B **25**, 713.

Mallett, C.P., 1982, J. Phys. C **15**, 6361.

Marksteiner, P., P. Weinberger, A. Neckel, R. Zeller and P.H. Dederichs, 1986a, Phys. Rev. B **33**, 812.

Marksteiner, P., P. Weinberger, A. Neckel, R. Zeller and P.H. Dederichs, 1986b, Phys. Rev. B **33**, 6709.

Messiah, A., 1969, Quantum Mechanics (North-Holland, Amsterdam).

Moruzzi, V.L., J.F. Janak and A.R. Williams, 1978, Calculated Electronic Properties of Metals (Pergamon, Oxford).

Newton, R.G., 1966, Scattering Theory of Waves and Particles (McGraw-Hill, New York).

Onodera, Y., and M. Okazaki, 1966a, J. Phys. Jpn. **21**, 2400.

Onodera, Y., and M. Okazaki, 1966b, J. Phys. Jpn. **21**, 1273.

Pyykko, P., and H. Toivonen, 1983, Acta Acad. Aboensis B **43**, 1.

Rajagopal, A.K., 1978, J. Phys. C **11**, L943.
Ramana, M.V., and A.K. Rajagopal, 1979, J. Phys. C **12**, L845.
Rodberg, L.S., and R.M. Thaler, 1967, Introduction to the Quantum Theory of Scattering (Academic Press, New York).
Rose, M.E., 1961, Relativistic Electron Theory (Wiley, New York).
Rosicky, F., P. Weinberger and F. Mark, 1976, J. Phys. B **9**, 2971.
Schadler, G., 1985, Thesis (Technical University of Vienna, Austria).
Schadler, G., and P. Weinberger, 1986, J. Phys. F **16**, 27.
Schadler, G., G. Hilscher and P. Weinberger, 1982, J. Magn. & Magn. Mater. **29**, 241.
Schadler, G., P. Weinberger, A. Gonis and J. Klima, 1985, J. Phys. F **15**, 1675.
Schadler, G., P. Weinberger, A.M. Boring and R.C. Albers, 1986, Phys. Rev. B **34**, 713.
Segall, B., 1957, Phys. Rev. **105**, 108.
Shen, A.P., 1974, Phys. Rev. B **9**, 1328.
Slater, J.C., and P.D. DeCicco, 1967, in: Quantum Theory of Solids, Vol. III, ed. J.C. Slater (McGraw-Hill, New York) appendix 4.
Smith, J.L., A.M. Boring and P. Weinberger, 1986, Int. J. Quant. Chem. **XXXIX**, 315.
Staunton, J., B.L. Gyorffy and P. Weinberger, 1980, J. Phys. F **10**, 2665.
Stocks, G.M., 1985, private communication.
Stocks, G.M., W.M. Temmermann and B.L. Gyorffy, 1979, in: Electrons in Disordered Materials and at Metallic Surfaces, NATO ASI Series B, Vol. 42, eds P. Phariseau, B.L. Gyorffy and L. Scheire (Plenum Press, New York).
Strange, P., J. Staunton and B.L. Gyorffy, 1984, J. Phys. C **17**, 3355.
Takada, S., 1966, Progr. Theor. Phys. **36**, 224.
Taylor, J.R., 1972, Scattering Theory (Wiley, New York).
Weinberger, P., 1977, Ber. Bunsenges. Physik. Chemie **81**, 804.
Weinberger, P., 1982a, J. Phys. F **12**, 2171.
Weinberger, P., 1982b, in: Actinides in Perspective, ed. N.M. Edelstein (Pergamon Press, Oxford) p. 145.
Weinberger, P., 1983a, Einführung in die relativistische Quantentheorie der Festkörper, Dresdner Seminar für Theoretische Physik, Sitzungsberichte Nr. 20, ed. P. Ziesche.
Weinberger, P., 1983b, J. Phys. F **13**, 625.
Weinberger, P., R. Podloucky, C.P. Mallett and A. Neckel, 1979, J. Phys. C **12**, 801.
Weinberger, P., C.P. Mallett, R. Podloucky and A. Neckel, 1980, J. Phys. C **13**, 173.
Weinberger, P., J. Staunton and B.L. Gyorffy, 1982a, J. Phys. F **12**, 2229.
Weinberger, P., J. Staunton and B.L. Gyorffy, 1982b, J. Phys. F **12**, L199.
Weinberger, P., R. Podloucky and A. Neckel, 1982c, J. Magn. & Magn. Mater. **29**, 247.
Weinberger, P., M.A. Boring and J.C. Smith, 1985a, Phys. Rev. B **31**, 1964.
Weinberger, P., A. Gonis, A.J. Freeman and A.M. Boring, 1985b, Phys. Rev. B **31**, 1971.
Weinberger, P., A. Gonis, A.J. Freeman and A.M. Boring, 1985c, Physica B **130**, 13.
Weinert, M., 1981, J. Math. Phys. **22**, 2433.
Weinert, M., E. Wimmer and A.J. Freeman, 1982, Phys. Rev. B **26**, 4571.
Wigner, E.P., 1955, Phys. Rev. **98**, 145.
Zeller, R., and P.H. Dederichs, 1982, Solid State Phys. **44**, 993.

Handbook on the Physics and Chemistry of the Actinides, edited by A.J. Freeman and G.H. Lander

Chapter 2

HEAVY-ELECTRON ACTINIDE MATERIALS

H.R. OTT
Laboratorium für Festkörperphysik, ETH-Hönggerberg, 8093 Zürich, Switzerland

Z. FISK
Physics Division, Los Alamos National Laboratory, Los Alamos, NM 87545, USA

Contents

1. Introduction

1.1. *Scope of review*

The defining feature of the so-called heavy-electron or heavy-Fermion materials is a very large and temperature-dependent electronic specific-heat coefficient, γ, near $T=0$ K. It perhaps seems peculiar that the specific-heat behavior at low temperature can provide the basis for meaningfully distinguishing a category of materials. Nevertheless, this appears to be a legitimate procedure, and it is a general goal of this review to show that there appears to be a common physical basis underlying the variety of phenomena observed in the various known heavy-electron materials.

Our attention is focussed on actinide materials particularly those containing uranium, for which the largest body of data exists. The same sort of phenomena are observed, of course, in rare-earth materials and comparisons will be made to these in the course of the review.

A brief outline follows. After introductory remarks, mostly historical, a detailed presentation of the experimental facts concerning the heavy-electron ground state will be given. This is followed by an assessment of the current (1985) theoretical situation. Then, in section 4, the various instabilities of the ground state will be discussed. A section on systematics follows, where we try to assemble some of the wisdom from workers in this field as to where these compounds can be found and what are the pre-conditions for the formation of the heavy-electron ground state. Finally, we give our evaluation of the current situation: where the outstanding problems are and what directions look most promising for new developments. In this respect we consider experimental and theoretical results that were known to us in June 1986.

1.2. *Historical introduction*

The designation 'heavy Fermion' first appeared in the literature in the sense in which we wish to use it in 1979 in the paper by Steglich et al. (1979) on the superconductivity of $CeCu_2Si_2$. In fact, a prime example of this heavy-electron, or heavy-Fermion, behavior had already been reported several years earlier: $CeAl_3$ (Andres et al. 1975). In retrospect, we now realize that there were a number of other earlier examples in the literature, for example UCu_5 (van Daal et al. 1975), $NpSn_3$ (Trainor et al. 1976) and perhaps UAl_2 (Trainor et al. 1975b), where, however, the low-temperature properties were not quite as spectacular as for the two aforementioned Ce compounds and therefore escaped a widespread attention. What was important about $CeCu_2Si_2$ was that the specific-heat anomaly at the superconducting transition temperature, $T_c = 0.6$ K, showed that the superconducting energy gap opens in the heavy-electron band, proving that the low-temperature 'linear' specific-heat γ is what one thinks it is, namely a large electronic density of states at the Fermi energy E_F. This allows one to have some confidence that the same is then true in $CeAl_3$ and others, where we do not find superconductivity. The magnetic anomaly in $NpSn_3$ at $T_N = 9.5$ K had actually been analyzed by Trainor et al. (1976) in these terms: a magnetic instability of itinerant electrons which opens a gap across a large part of the Fermi surface.

The observation of these exceptionally large electronic specific-heat γ's represents one end result of an experimental and theoretical program of long standing. A part of this program can be traced back to the discovery by Lawson and Ting-Yuang Tang (1949) of the isostructural γ to α volume collapse in elemental cerium. Both Zachariasen (1949) and Pauling (1950) immediately interpreted this finding as a change in the f-configuration of the cerium ions from $4f^1$ to $4f^0$. There followed a long period of work on cerium and its intermetallic compounds aided especially by the new availability of high-purity metal from the Ames Laboratory. Some cerium intermetallics were found to be magnetic, some not. Those in the 'not' category were believed to be tetravalent, $4f^0$, cerium compounds, and the magnetic ones, $4f^1$. This attitude was only slightly modified by the parallel development in the field of magnetic impurities in metals and investigations of the Kondo effect. It was clear by 1970 that cerium could behave like a Kondo impurity and often continued to do so in intermetallic compounds of cerium. The atomic size of cerium in these Kondo-type compounds, as deduced from lattice parameter comparisons, led to the conclusion that cerium was intermediate between 3+ and 4+ in these cases, a suggestion that had been made much earlier for α-cerium by Gschneidner and Smoluchowski (1963).

A very different light was shed on the cerium problem when Johansson (1974) interpreted some then recent results as suggesting that in fact cerium was in the $4f^1$ configuration in both the γ and α phase, and that what was involved in the γ–α transition was a Mott localization–delocalization transition of the 4f electron. Further developments in the Kondo problem also led to the conclusion (on different grounds) that the demagnetization of the cerium $4f^1$ state of a cerium impurity in, for example, lanthanum metal involved no change in f-count: the 'valence' of cerium being nearly 3+ in both the high-temperature magnetic and the low-temperature 'compensated' states. This picture was later expanded to describe the γ–α transition as a 'Kondo collapse' (Allen and Martin 1982, Thompson et al. 1983).

The ordered lattice of Kondo impurities presents a much more difficult theoretical problem than the single-impurity one. It seems reasonable to expect that the Kondo energy in some cases will dominate the Ruderman–Kittel–Kasuya–Yoshida (RKKY) magnetic-interaction energy and that a low-temperature ground state could be achieved with a large electronic density of states at the Fermi level resulting from the Kondo screening of the f-moment. This seemed the most likely qualitative explanation of the physics behind the existence of the surprising properties of $CeAl_3$ and $CeCu_2Si_2$.

In parallel with this development of our picture of cerium, considerable effort was being devoted to studying actinide compounds (Brodsky 1978, Fournier and Troć

1985). While a number of similarities presented themselves between uranium and cerium, the former really seemed to occupy a middle ground between the d-transition and the rare-earth elements. The 5f wavefunction is much more spatially extended than the 4f one (Desclaux and Freeman 1984), and direct overlap between actinide 5f wavefunctions seems possible. Hill (1970) introduced a clarifying idea here with his 'Hill plots': he noticed that a uranium–uranium separation of 3.4 Å separated magnetic from non-magnetic uranium compounds. The idea is that 5f atoms spaced closer than 3.4 Å can form bonding and hence non-magnetic f-electron bands. Roughly speaking, at spacings below the Hill limit, uranium in its compounds acts like a transition metal; at spacings above, they behave like the rare earths similar to cerium (Brooks et al. 1984). Still, the parallel with cerium is not complete. For example, in cerium intermetallics one can often observe crystal-field excitations arising from the crystal-field-split $J = \frac{5}{2}$ Hund's rule ground state. Only one observation of definite crystal-field excitations is known for a uranium intermetallic, UPd_3 (Shamir et al. 1978, Buyers and Holden 1985). Further, the high-temperature magnetic susceptibilities of uranium intermetallics above the Hill limit do not clearly correspond to a definite f-configuration: the actinides are expected to fall into a coupling regime intermediate between L–S and jj coupling, making identification even of f-count uncertain. And cases in which uranium impurities act as Kondo impurities have not been clearly documented. U in $LaAl_2$ (Schlabitz et al. 1980), U in Th (Maple et al. 1970) and U in ThSb (Frick et al. 1982) are possible cases and the work on Pu impurities by Elliott and Hill (1970) is also noted. What is happening in the actinide elements is an increasing tendency towards 5f localization going from uranium to neptunium and plutonium (Brooks et al. 1984); in the rare earths, this localization takes place right at the beginning of the series with cerium. It is also interesting that this localization is happening among the actinides only when there is more than one f-electron involved, unlike the case with cerium.

A number of intermetallic compounds with spacings beyond the Hill limit were found not to be magnetic at low temperature. For several Cu_3Au-type compounds, strong 5f-neighboring ligand hybridization was found to account for this (Lander et al. 1980), making the compounds paramagnets with temperature-independent magnetic susceptibilities. In several other compounds, Curie–Weiss-type magnetic susceptibilities were found without low-temperature magnetic ordering: UAl_2 (Buschow and van Daal 1972), UPt_3 (Frings et al. 1983) and UBe_{13} (Troć et al. 1971). All these compounds have large specific heats at low temperature. Of our examples, UBe_{13}, as discussed at length below, is remarkably similar to $CeCu_2Si_2$, and this has raised the hope that the physics underlying the two cases is similar.

Subsequent to the discovery of the superconductivity of UBe_{13} (Ott et al. 1983), one other heavy-electron superconductor of uranium was found, UPt_3 (Stewart et al. 1984c), and several magnetically-ordering uranium compounds having very large low-temperature specific heats. Whereas the high-temperature properties of these uranium compounds would lead one to expect magnetic ordering at sufficiently low-temperature, the development of the large γ value at low temperature can be argued to show that the magnetism which sometimes develops from the heavy-electron state is not of the usual local-moment type, but of an essentially more complicated kind involving high-mass itinerant electrons.

1.3. *Basic facts about heavy-electron systems*

A distinguishing feature of metals is a low-temperature specific heat which varies as $\gamma T + \beta T^3$ at temperatures low compared with their Fermi and Debye temperatures. The linear term is the electronic contribution and the term varying proportional to T^3 is the lattice specific heat; γ is a measure of the electronic density of states at the Fermi level and has a value, for example, of 0.7 mJ/mol K^2 in Cu and Au. In d-transition elements and compounds γ values can be found which are an order of magnitude larger. The explanation for this increased γ is that d-electron states are present at the Fermi level and, being more localized than s- or p-electrons, give rise to narrower electronic bands, hence higher γ's and larger effective masses. Bands deriving from f-orbitals might be expected to be even narrower and, if they intersect the Fermi level, lead to larger γ's. δ-Pu, in fact, has the largest γ reported for any element: 53 mJ/mol K^2 (Stewart and Elliott 1981). 5f electrons are believed to be responsible for this.

It is experimentally observed that a number of interesting physical phenomena occur when γ is high, and the question naturally presents itself as to how large γ might become, and whether some interesting new physics would present itself if substantially larger γ's were found. The first step in this direction occurred in 1975 in the investigation of the properties of $CeAl_3$ at very low temperatures (Andres et al. 1975). What is found in this crystallographically-ordered hexagonal compound is (i) a Curie–Weiss local-moment-type susceptibility above 100 K corresponding to an effective moment of the expected Hund's rule $4f^1$ ground state (Buschow and Fast 1966), but no magnetic order above 15 mK (Andres et al., 1975); (ii) a limiting γ as $T \to 0$ K of 1600 mJ/mol Ce K^2, implying an extremely large electronic effective mass of order 100 m_e, where m_e is the free-electron mass; and (iii) a Kondo-type resistivity with a maximum around 35 K of the order of 150 $\mu\Omega$ cm (Buschow et al., 1971) and which falls to a few $\mu\Omega$ cm between 30 K and 15 mK, ultimately approaching a T^2-dependence at lowest T.

These results pose two related questions: how does this large γ arise, and why is there no cooperative phase transition at low temperature involving these high-effective-mass electrons? A qualitative approach to the first question is offered by a heuristic approach to the Kondo impurity problem. There the idea is that an antiferromagnetic coupling between conduction electrons and a localized d- or f-electron gives rise to a compensation cloud of conduction electrons which screens out the localized-impurity moment, this happening gradually as T decreases below the so-called Kondo temperature, T_K (Varma 1985b). This compensation cloud shows up as a resonance in the electronic energy spectrum at the Fermi level and can dominate γ.

In $CeAl_3$, of course, we do not have an impurity problem, but an ordered lattice of 4f moments, and the appropriate problem is the Kondo-lattice problem, one limit of the more general Anderson-lattice problem. These lattice problems are currently under intense theoretical investigation and only tentative suggestions as to their expected behavior exist. We suspect that the large γ-value results from what is fundamentally a many-body interaction and cannot be treated as a single-electron problem. The interesting conclusion is now that the large γ is not primarily due to hybridized f-derived bands at the Fermi level, but from a many-body resonance at the Fermi level which develops off the f-level which is 1 to 2 eV below E_F. If this conclusion is correct, it means that the large-γ state, which is generally called the heavy-electron state, has not evolved continuously from the γ of transition-metal compounds, some two orders of magnitude smaller, by a simple variation of some kind of hybridization strength.

This Kondo-lattice viewpoint places the heavy-electron materials conceptually in between fully-localized f-systems and the so-called intermediate-valence materials in which the f-configuration is intermediate between two different f-occupation numbers.

Within the Kondo-lattice model we can qualitatively interpret the large drop in resistivity of $CeAl_3$ below 35 K as due to the onset of 'coherence' between lattice sites. The idea is that the f-sites scatter electrons incoherently at higher temperatures where the lattice appears as a set of Kondo impurities. Below the resistivity maximum, phase coherence sets in between the sites, leading at low temperature to a state which can be described as a strongly-interacting Fermi liquid, as can the Kondo-impurity state itself. This view then has two energy scales: T_K, the single-ion Kondo temperature and T_{coh}, the inter-Kondo-ion interaction temperature.

An important point concerning the low-temperature Fermi liquid is that its effective Fermi temperature is small, of order tens to hundreds of degrees Kelvin, for known heavy-electron materials. The large γ, therefore, can only develop at low temperature, and it is to be expected that in some cases a magnetic phase transition might take place due to interactions between the unscreened f-moments at a temperature higher than the effective Fermi temperature, precluding a development of the heavy-electron state.

The heavy-electron state itself might be expected to be unstable in many cases against possible spin-wave, superconducting or charge–density wave transitions, to name the most obvious. Clean examples of the first two possibilities are now known, and it was in fact the surprise of the superconductivity of $CeCu_2Si_2$ reported by Steglich et al. (1979) and the later observation of an analogous behavior in a uranium compound, UBe_{13} (Ott et al. 1983), that launched the research on heavy-electron materials into the forefront of condensed-matter research.

2. Physical properties of heavy-electron actinide materials

As we pointed out in the Introduction, heavy-electron materials still have to be classified as metals. What distinguishes them from ordinary metals is their low-temperature physical properties, some of which may be used to identify a given substance as an example of the class of materials that we are considering here. In this section we first present experimental data of this type for various compounds which should justify why a new classification for these materials is unavoidable. Further data are meant to complete the experimental facts which, taken all together, should finally provide a useful background to compare with results of theoretical models that are supposed to describe the physical background of all these observations.

2.1. *Specific heat*

The low-temperature specific heat of metals is usually analyzed by considering two contributions, namely the specific heat due to thermal excitations of conduction electrons and of lattice vibrations which are supposed to add up to

$$C_{tot}=\gamma T+\beta T^3, \tag{1}$$

where the first and second term on the right describe the electronic and the lattice part, respectively. γ and β are commonly assumed to be temperature-independent parameters and it is clear that this approximation is only valid if $T \ll T_F$, the Fermi degeneracy temperature of the electronic subsystem and also if $T \ll \theta_D$, the Debye temperature associated with the vibrational or phonon spectrum of the lattice. In the simplest applicable models, the free-electron model for the electronic subsystem and the Debye model for the lattice vibrations, the parameters γ and β are given by

$$\gamma=\tfrac{2}{3}k_B^2\pi^2 N(E_F) \tag{2a}$$

and

$$\beta=1944\ n\theta_D^{-3}, \tag{2b}$$

where k_B is the Boltzmann constant, $N(E_F)$ is the density of electronic states per spin direction at the Fermi energy E_F, n is the number of atoms per formula unit and 1944 is the numerical prefactor if the specific heat is given in J/mol K.

What is not considered in eq. (1) is the mutual influence of electrons on phonons or vice versa. Eliashberg (1962) was the first to show that electron–phonon coupling leads to an additional positive term in the low-temperature specific heat of metals of the form

$$C^{el\text{-}ph} \sim T^3 \ln(\theta_D/T). \tag{3}$$

Even for metals with very strong electron–phonon coupling, however, this contribution is less than 1% of the total specific heat at a given temperature and therefore very difficult to observe, at least in the normal state. A more obvious effect of the electron–phonon coupling is the renormalization of the electronic density of states which usually is taken into account by replacing $N(E_F)$, the band-structure density of states, in eq. (2a) by $N(E_F)\ (1+\lambda)$, where λ is the electron–phonon enhancement parameter. In the most general case λ may also be temperature dependent, as mentioned above. From its fundamental definition (Sommerfeld and Bethe 1967) it may be seen that $N(E_F)$ is proportional to the effective mass m^* of the electrons. Therefore the electronic contribution to the specific heat of a metal may also serve as a measure of m^* of its conduction electrons.

One of the most obvious features of heavy-electron materials is the failure of eq. (1), together with the assumption of a temperature-independent value for γ, for describing their specific heat at low temperatures. In contrast to common metallic materials, the temperature dependence of γ is outstanding and the resulting values for γ when T approaches 0 K are very large, two to four orders of magnitude larger than in ordinary metals. It is clear that exactly this latter property led to the nomenclature of the materials considered here. In this context we should like to recall that, as mentioned in the introduction, the first identified examples of this new class of materials were Ce compounds and only about 3 years ago was it realized that some actinide compounds also showed similar features. A compilation of specific-heat data of the presently known heavy-electron actinide materials provides a first look at the anomalous properties of these materials.

2.1.1. UBe_{13}

Extreme heavy-electron behavior in actinide materials was first recognized in UBe_{13}. Figure 1 shows the temperature dependence of the specific heat C_p of UBe_{13} between 0.15 and 13 K (Ott et al. 1984d). The discontinuity of $C_p(T)$ just below 1 K is due to the superconducting transition and we shall consider this part of the curve in section 4. Here we are particularly interested in the normal-state properties, which obviously include a flat maximum of $C_p(T)$ just above 3 K. A C_p/T versus T^2 plot reveals that the data for temperatures above 10 K can fairly well be represented by eq. (1) and the resulting values for the parameters γ and β are 185 mJ/mol K^2 and 0.09 mJ/mol K^4, respectively. This rather small lattice contribution C_p^{latt} to the total specific heat, indicating a Debye temperature of more than 600 K, has recently been confirmed by calculations of C_p^{latt} based on phonon densities of states that were determined experimentally with neutron-scattering experiments (Renker et al. 1985). If we subtract the lattice part, which is almost negligible below 10 K, and plot the remaining specific heat as C_p/T versus T^2 for $T > 1$ K, we obtain what is shown in fig. 2. If interpreted as an *electronic* specific heat C_p^{el}, which is a priori by no means clear, it is obviously very strongly temperature dependent below about 7 K and increases its value by more than a factor of 4 down to 1 K. In this type of plot the maximum of $C_p^{el}(T)$ around 3 K is no longer obvious, but it demonstrates very clearly the difference from ordinary metallic behavior as described by eq. (1). This upturn of C_p^{el}/T with decreasing temperature is a general feature of heavy-electron

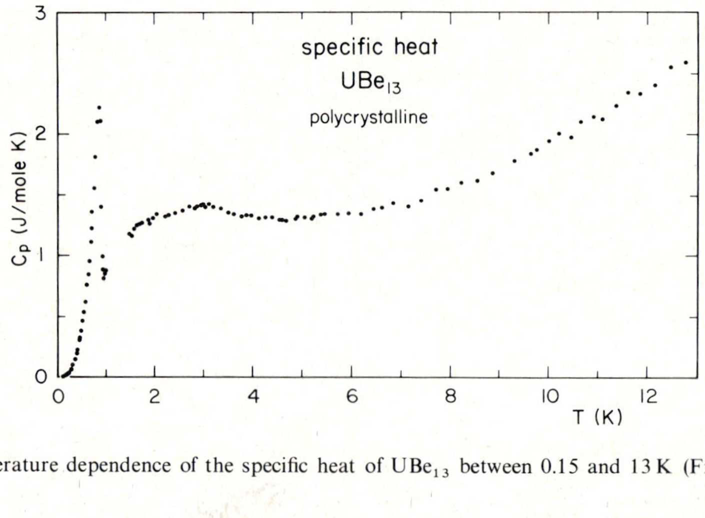

Fig. 1. Temperature dependence of the specific heat of UBe_{13} between 0.15 and 13 K (From Ott et al. 1984d.)

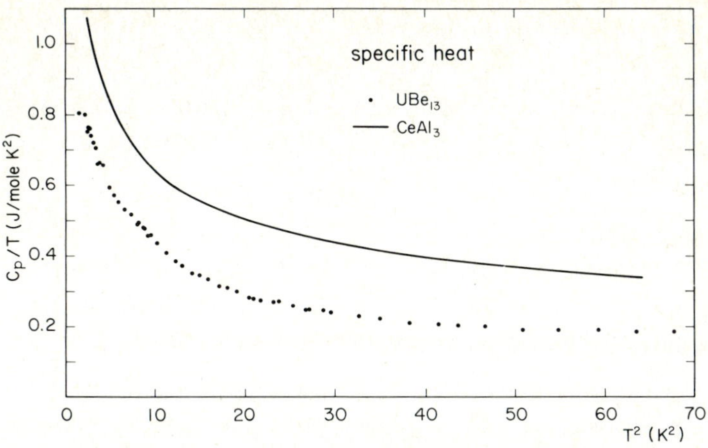

Fig. 2. Electronic part of the low-temperature specific heat of UBe_{13} plotted as C_p^{el}/T versus T^2 between 1.5 and 8 K. For comparison we show the analogous quantity for $CeAl_3$ as a solid line. (Data are taken from Ott et al. 1984d.)

materials and it reflects roughly the temperature dependence of C_p^{el} as shown in fig. 1, if we disregard the anomaly due to the superconducting phase transition. Similar behaviour, without any superconducting phase transition is observed in $CeAl_3$ and $CeCu_6$ (Ott et al. 1984d, Benoit et al. 1981, Stewart et al. 1984e, Ott et al. 1985d). The rapid increase of C_p^{el} with T at very low temperatures is followed by a saturation-type behavior which may even result in a maximum of C_p/T as observed in $CeAl_3$ (Benoit et al. 1981, Bredl et al. 1984). We shall see, however, that this is a general trend only and that different materials show distinctly different details in $C_p^{el}(T)$.

For UBe_{13}, the low-temperature behaviour of C_p^{el} can be changed quite drastically by replacing small amounts of U by non-magnetic impurities like Th and Lu (Ott et al. 1986a). This is shown in fig. 3 where we plotted C_p^{el}/T versus T for Th- and Lu-doped samples of UBe_{13} with approximately the same amount of impurities, in comparison with pure UBe_{13}. Whereas Th impurities shift the upturn of C_p^{el}/T to lower temperatures and also lead to an increase of the saturation value of C_p^{el}/T at very low temperatures, just the opposite is true for Lu impurities. Obvious differences between these two kinds of impurities are that Th has a tetravalent configuration and expands the lattice while Lu is trivalent and leads to a lattice contraction. Further support that the lattice constant has a decisive influence on the specific-heat enhancement is provided by the observation that single crystals of UBe_{13} which systematically have a distinctly larger room-temperature lattice constant than polycrystalline material, indeed also show a larger C_p^{el}/T ratio in the temperature range just above the superconducting transition temperature T_c. In pointing this out we should remember, however, that the respective changes of the lattice parameters are very small indeed ($<0.1\%$) indicating that the observed changes of C_p^{el} result from minute changes of the electronic spectrum. From the

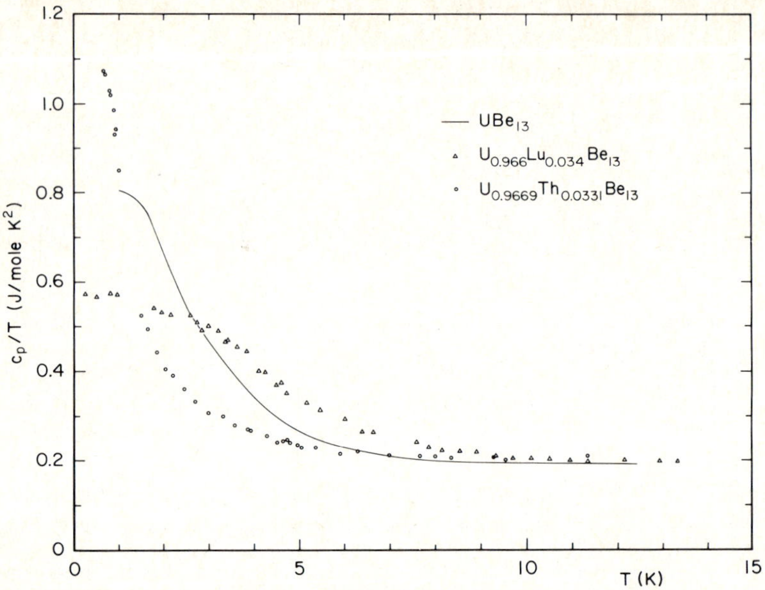

Fig. 3. Comparison of the influence of different impurities on the low-temperature normal-state specific heat of UBe_{13}. (From Ott et al. 1986a.)

experimental data shown in fig. 3, which were obtained from different samples but prepared with always the same procedure (Smith et al. 1985), it seems remarkable that C_p^{el} is almost unaffected above about 7 K, indicating that it is only those interactions which provide the specific-heat enhancement that are strongly affected by disturbing the actinide sublattice. Finally, we want to point out that different measurements of C_p of other workers on different samples of nominally pure UBe_{13} above T_c gave generally similar but not identical results (Bucher et al. 1975, Samorukov et al. 1974, Stewart and Giorgi 1984). In table 1 we list various relevant numerical values in conjunction with these measurements.

It is to be expected that external influences other than impurities may also influence the normal-state specific heat, as for example external magnetic fields or pressure. For UBe_{13} two studies of the influence of magnetic fields on C_p above 1 K have been published. Stewart et al. (1984d) measured C_p of a single crystal of UBe_{13} in a field of 100 kOe. They find an overall increase of C_p of a few percent between 2 and 10 K with a maximum increase of 9% around 3 K. Above 10 K, no magnetic-field-induced shift of C_p is found. Later, Mayer et al. (1986) reported measurements of C_p of polycrystalline UBe_{13} in a field of 130 kOe. This work indicates a negative shift of C_p of a few percent below 3.5 K and an increase of C_p which reaches a maximum of about 10% at 5.5 K, above 3.5 K. The positive shift is very small above 10 K.

2.1.2. *UPt_3, UAl_2*

UPt_3 is also an actinide compound with the characteristic C_p^{el}/T upturn with

TABLE 1
Specific-heat data on UBe_{13}.

Compound	Form	γ(at T_c)	C/T(13 K)	T_c from spec. heat	Remarks	Ref.[a]
UBe_{13}	s.c.	1100		0.85		[1]
	p.c.	850	185	0.9		[2]
	p.c.	805	185	0.92		[3]
	p.c.		150		No spec. heat data below 1.5 K	[4]
	p.c.		123		No spec. heat data below 1.5 K	[5]
	p.c.		128		No spec. heat data below 1.5 K	[6]
	p.c.	760		0.871		[8]
$U_{0.991}Th_{0.009}Be_{13}$	p.c.	1040		0.60		[7]
$U_{0.967}Th_{0.033}Be_{13}$	p.c.	1120	185	0.60		[7]
$U_{0.94}Th_{0.06}Be_{13}$	p.c.	1140	185	0.36		[7]
$U_{0.994}Lu_{0.006}Be_{13}$	p.c.	785		0.625		[7]
$U_{0.984}Lu_{0.016}Be_{13}$	p.c.	750	185	0.25	Broad s.c. transition	[7]
$U_{0.985}Sc_{0.015}Be_{13}$	p.c.		185			[7]
$UBe_{12.94}Cu_{0.06}$	p.c.			<0.05	No C_p data below 2 K	[6]

[a] References:
[1] Ott et al. (1983).
[2] Ott et al. (1984d).
[3] Ott et al. (1986a).
[4] Butcher et al. (1975).
[5] Samorukov et al. (1974).
[6] Stewart and Giorgi (1984).
[7] Ott et al. (1986a).
[8] Mayer et al. (1986).

decreasing temperature in the liquid-helium temperature range and a subsequent transition to a superconducting state below 1 K. Here again we first want to discuss the specific heat of the normal state. In fig. 4 we show a plot of C_p^{el} versus T of UPt_3 between 1 and 17 K. Various workers using single-crystalline and polycrystalline material obtained similar but again not identical results. Here we have chosen to display the data given by Frings (1984) because they allow for the most reliable determination of the lattice contribution which, if subtracted from the total specific heat, leads to the data shown in fig. 4. Again we notice the saturation-type curvature of $C_p^{el}(T)$ with increasing temperature and the smooth transition into a linear-T dependence above 15 K.

The departure from normal metallic behaviour is most obvious in a C_p/T versus T^2 plot as shown in fig. 5. It was Frings et al. (1983) who first pointed to similarities between the low-temperature specific heats of UPt_3 and UAl_2, but it was then Stewart et al. (1984c) who substantiated this conjecture by showing that the total specific heat $C_p(T)$ of UPt_3 could be fitted quite accurately with

$$C_p = \gamma_0 T + \beta^* T^3 + \delta T^3 \ln T \tag{4}$$

between 1 and 17 K. Equation (4) is an approximation to describe the low-

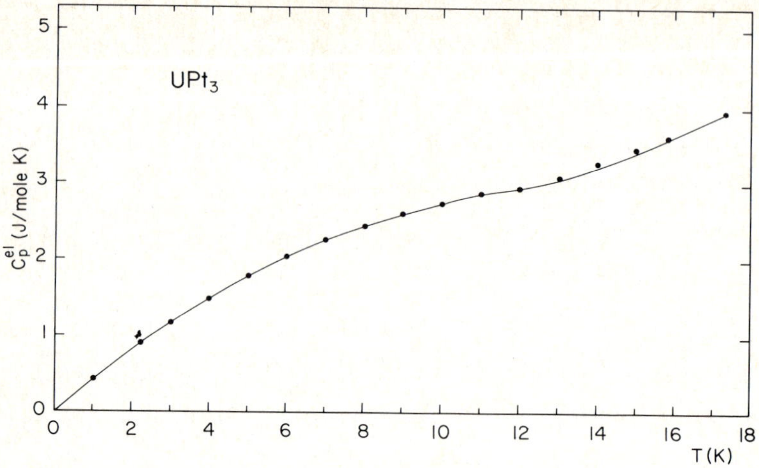

Fig. 4. Temperature dependence of the electronic specific heat of UPt_3 between 1 and 17 K. The data points were obtained as explained in the text, the solid line is to guide the eye. (From Frings 1984.)

temperature specific heat of a metallic system where the energy spectrum of the electrons is noticeably affected by spin fluctuations and originates from calculations by Doniach and Engelsberg (1966), intended to describe some low-temperature properties of a Fermi system with strong spin fluctuations like liquid ^{3}He. γ_0 obviously denotes C_p/T for $T \to 0$ K, $\beta^* = \beta - \delta \ln T^*$ contains the usual lattice term β, δ is the prefactor to the characteristic spin-fluctuation contribution and T^* is a cut-off temperature with regard to the spin fluctuations. At this point we note the important difference in sign between the two seemingly equal terms in eqs. (3) and (4). While the electron–phonon interaction leads to a positive C_p contribution of the

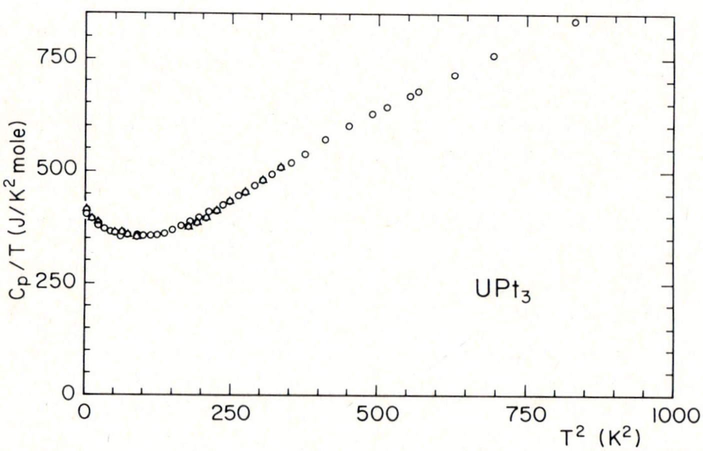

Fig. 5. Low-temperature specific heat of UPt_3 as a C_p/T versus T^2 plot between 1.5 and 28 K. Circles: zero magnetic field; triangles: magnetic field of 50 kOe. (Data are taken from Frings 1984.)

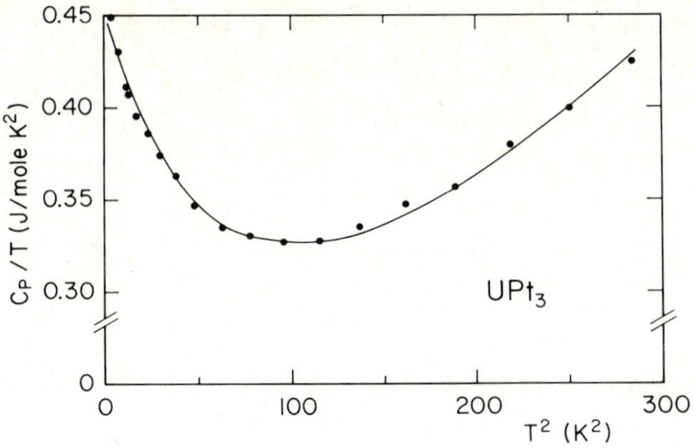

Fig. 6. Comparison of experimental data of C_p/T versus T^2 fo UPt_3 with a fit using eq. (4) by Stewart et al. (1984c).

logarithmic term with increasing temperature below the cut-off temperature, spin fluctuations obviously provoke the opposite effect. For metallic systems, eq. (4) has first been used by Trainor et al. (1975b) to describe the low-temperature specific heat of UAl_2. We show the original data and corresponding fits of $C_p(T)$ for UPt_3 and UAl_2 in figs. 6 and 7. In table 2 we list some of the relevant parameters that result from this fitting procedure as they were obtained by different workers. Note that these parameters may not only change for different samples in different measurements but also, as de Visser et al. (1984a) pointed out, change even for one single set of data, depending on the temperature range chosen for the fitting procedure. Nevertheless, it is clear that experimental data of $C_p(T)$ for both UPt_3 and UAl_2 come much closer to a description by a relatively simple temperature dependence

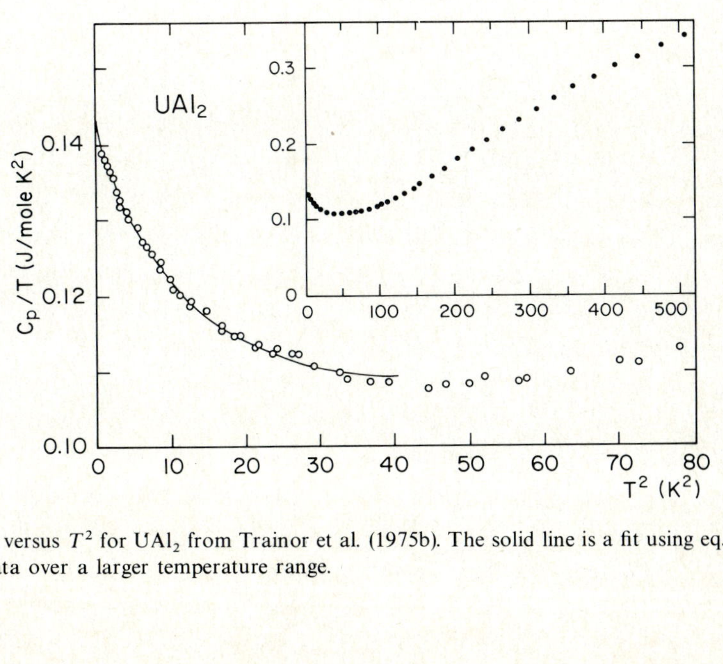

Fig. 7. C_p/T versus T^2 for UAl_2 from Trainor et al. (1975b). The solid line is a fit using eq. (4), the inset shows the data over a larger temperature range.

TABLE 2

Normal-state specific-heat data on UPt_3 and UAl_2. γ and β denote the electronic and lattice contribution, respectively, evaluated at high temperature. β^* and δ are explained in the text.

Compound	Temperature for fitting	γ_0	β^*	δ	T^*	γ	β	Remarks	Ref.[a]
UPt_3									
(s.c.)	3.3–16	452	−4.169	2.247					[1]
(p.c.)	1.3–16.5	422.4(5)	−4.18(5)	1.54(2)	26		0.85		[2]
(p.c.)	7–16					225	0.85		[2]
(s.c.)	1.2–20	422	−3.27	1.38					[3]
(s.c.)	1.2–15	423	−3.80	1.41					[3]
(s.c.)	1.2–10	421	−3.80	1.41					[3]
(s.c.)	6–10	416	−3.13	1.15					[3]
(s.c.)	8–20	413	−3.45	1.30					[3]
UAl_2									
(p.c.)	0.8–6	143	−4.38	1.94	12–16				[4]
(s.c.)	2–23	142.3	−3.644	1.566				Similar results for p.c. material	[5]
(s.c.)	2–7.5	142.1	−3.437	1.413					[5]
(p.c.)	1.3–23	132.7(2)	−3.04(3)	1.35(1)	15		0.65		[2]
	1.3–8	131.3(1)	−2.60(3)	1.11(2)	19		0.65		[2]
	1.3–10	131.1(1)	−2.55(2)	1.08(1)					[2]
	7–16					52	0.65		[2]
	0.3–2	~126			4–7				[6]

[a] References:
[1] Stewart (1984).
[2] Frings (1984).
[3] de Visser et al. (1984a).
[4] Trainor et al. (1975b).
[5] Stewart et al. (1983).
[6] Armbrüster et al. (1979).

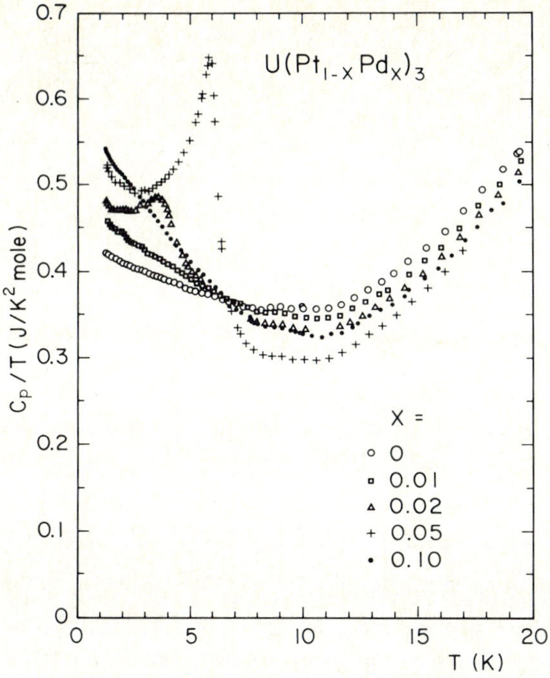

Fig. 8. Influence of Pd substitutions for Pt on the low-temperature specific heat of UPt_3 as shown by de Visser et al. (1986a).

than it seems possible for UBe_{13} and, as a reminder, for other heavy-electron rare-earth compounds such as $CeAl_3$, $CeCu_6$ and $CeCu_2Si_2$. For a further discussion of the validity and the interpretation of eq. (4) for heavy-electron materials we refer to section 3.

As for UBe_{13}, we want also to point out that the low-temperature specific heats of UPt_3 and UAl_2 vary considerably upon changing the chemical constituents on either the cation or the anion site. Data are available for $U(Pt_{1-x}Pd_x)_3$ (de Visser et al. 1986a) and for $U_{1-x}Th_xAl_2$ (Brodsky and Trainor 1977). They are shown in figs. 8 and 9. Up to about 10% Pd for Pt clearly reinforces the enhancement effect on the low-temperature specific heat of UPt_3. Low concentrations of Th on the U sites of UAl_2 systematically suppress the magnitude of the low-temperature upturn.

For UPt_3, application of an external magnetic field of 100 kOe leads to sizeable changes of the low-temperature specific heat. A positive shift of C_p is observed below 7 K which increases with decreasing temperature and reaches about 10% at 2 K. Above 7 K the shift is negative and again of the order of 5 to 10% between 10 and 17 K (Stewart et al. 1984b).

For UAl_2, the magnetic-field dependence of C_p was first investigated by Trainor et al. (1975b) in fields up to 43 kOe. Their measurements between 0.6 and 4 K showed a clear reduction of a few percent of C_p in that temperature range. Later, Stewart et al. (1983) extended the field range to 170 kOe. Their results can be summarized as follows: increasing the magnetic field results in a suppression of the up-

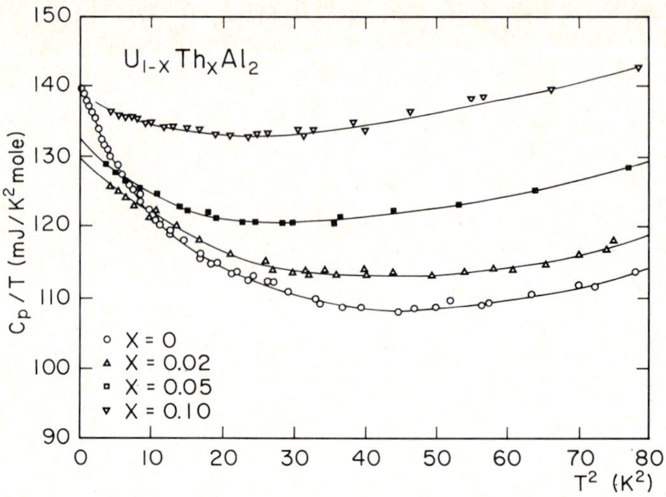

Fig. 9. Influence of Th substitutions for U on the low-temperature specific heat of UAl_2 as reported by Brodsky and Trainor (1977).

turn of C_p/T below 4 K, amounting to about 40% in a field of 125 kOe at 2.1 K. No suppression of C_p is found for temperatures exceeding 4 K, instead the results are compatible with an increase of the specific heat with increasing field. Since an increase of the specific heat with increasing magnetic field is incompatible with conventional views regarding spin-fluctuation contributions to C_p, at least part of the data for UPt_3 and UAl_2 is unexpected. The last mentioned results of UAl_2 are, according to their authors, due to an artifact of the experiment. In this respect one should remember the notorious problems with thermometry in large magnetic fields.

2.1.3. *U_2Zn_{17}*

U_2Zn_{17} orders antiferromagnetically out of a state with enhanced specific heat (Ott et al. 1984b). In distinct contrast to the examples of heavy-electron materials discussed so far, the ratio of C_p^{el}/T is already large and saturated at 10 K, just above the ordering temperature of 9.7 K. Figure 10 shows C_p versus T for both U_2Zn_{17} and Th_2Zn_{17} between 1.5 and 15 K. Ott et al. (1984b) used the latter to calculate the electronic specific heat of U_2Zn_{17} by assuming that the lattice contribution is the same in both compounds. The result for C_p^{el}/T versus T^2 is shown in fig. 11. Although we have not yet mentioned it explicitly, it seems fair, without an elaborate explanation at this point, to presume that the large electronic specific heat is predominately due to the 5f-electrons of the actinide atoms in these compounds. Therefore, the values of C_p^{el} should be compared for the same number of actinide atoms. This implies that the relevant numbers for U_2Zn_{17} have to be divided by 2 in order to be comparable to the same quantities of compounds which contain only one actinide atom per formula unit.

It is, of course, of interest to investigate how this still large C_p^{el}/T-ratio develops at

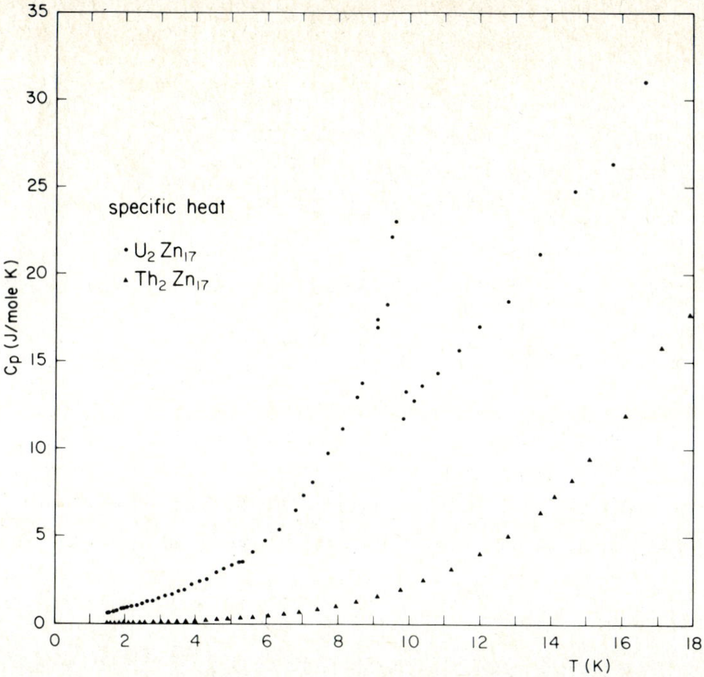

Fig. 10. $C_p(T)$ for U_2Zn_{17} and Th_2Zn_{17} between 1.5 and 17 K. (From Ott et al. 1984b.)

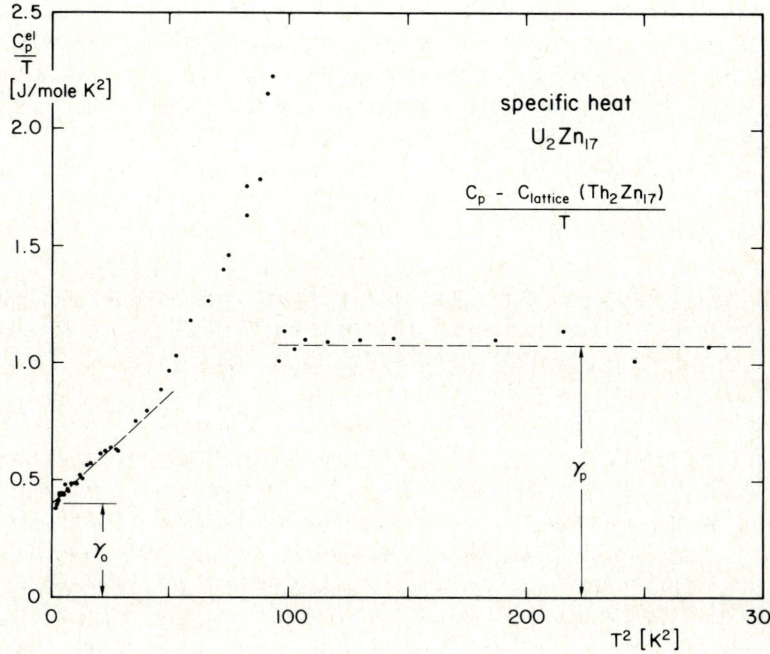

Fig. 11. Specific heat minus the lattice contribution of U_2Zn_{17} plotted as C_p^{el}/T versus T^2. Data are from Ott et al. (1984b). The broken lines indicate the temperature dependences above and below the phase transition.

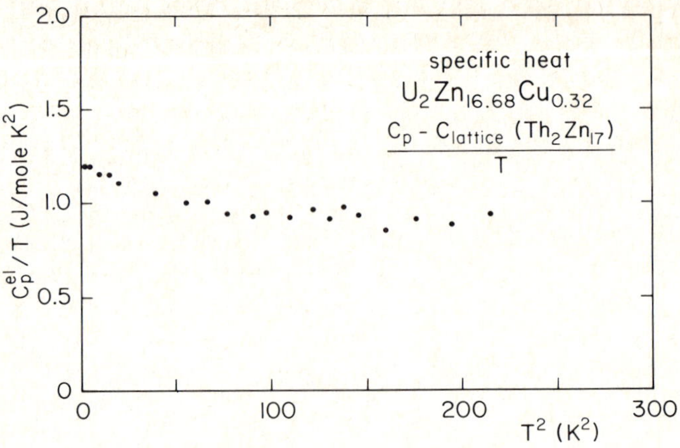

Fig. 12. Influence of Cu substitutions for Zn on the electronic specific heat of U_2Zn_{17} at low temperatures. (Data are from Ott et al. 1985e.)

higher temperatures. Fischer et al. (1986) have measured $C_p(T)$ for U_2Zn_{17} and Th_2Zn_{17} up to 70 K. Again C_p^{el} was obtained by subtracting the lattice contribution of Th_2Zn_{17} from the total specific heat of U_2Zn_{17}. Their results show that C_p^{el}/T rises continuously with decreasing temperature below 70 K by about the same factor of 4 as observed in UBe_{13}, but now in a temperature interval that is 10 times larger. For some unknown reason the temperature dependence of C_p^{el}/T found by Fischer et al. just above T_N is somewhat different from that reported originally by Ott et al. (1984b).

Although we wish to postpone a discussion of all aspects concerning the phase transition to section 4, we show the results of a specific-heat measurement on a sample for which the phase transition is suppressed by a small amount of impurities on the Zn sites in fig. 12 (Ott et al. 1985e). This is to demonstrate that $C_p^{el}(T)$ does not simply follow the characteristic temperature dependence of a Schottky anomaly whose low-temperature side is masked by the anomaly due to the phase transition. In fact, we even observe a slight increase of C_p^{el}/T with decreasing temperature below T_N of U_2Zn_{17}.

2.1.4. *UCd_{11}, $NpSn_3$*

The low-temperature specific heat of UCd_{11} in many ways resembles that of U_2Zn_{17}. We show $C_p(T)$ of UCd_{11} as it was measured by Fisk et al. (1984) between 0.4 and 13 K in fig. 13. Above 8 K, $C_p(T)$ can be well described by eq. (1) and therefore the lattice contribution βT^3 can be determined. The values for γ and β above 8 K are 840 mJ/mol K and 5.6 mJ/mol K^4, respectively. These numbers were obtained from the data plot shown in fig. 14. Unfortunately, no specific-heat data of this compound for temperatures above 13 K have been reported.

A fairly large and constant γ-value above a phase transition was also found for $NpSn_3$, which was earlier claimed to be an itinerant antiferromagnet with

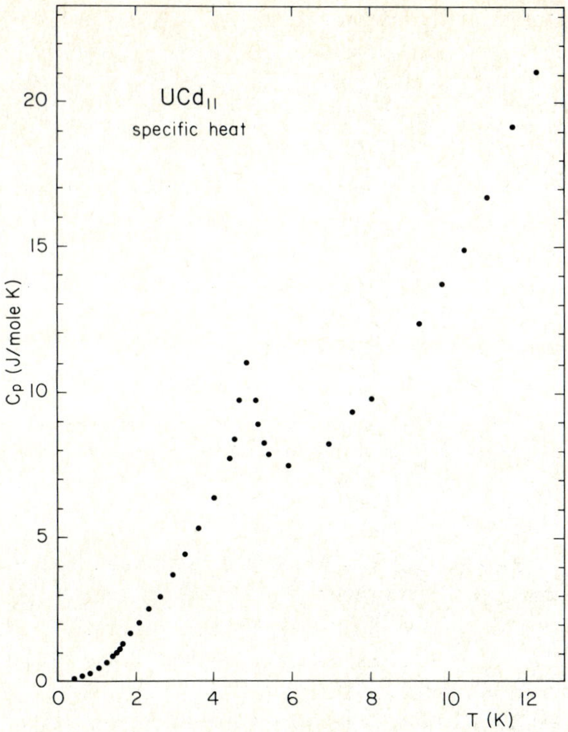

Fig. 13. $C_p(T)$ for UCd_{11} between 0.5 and 12 K as found by Fisk et al. (1984).

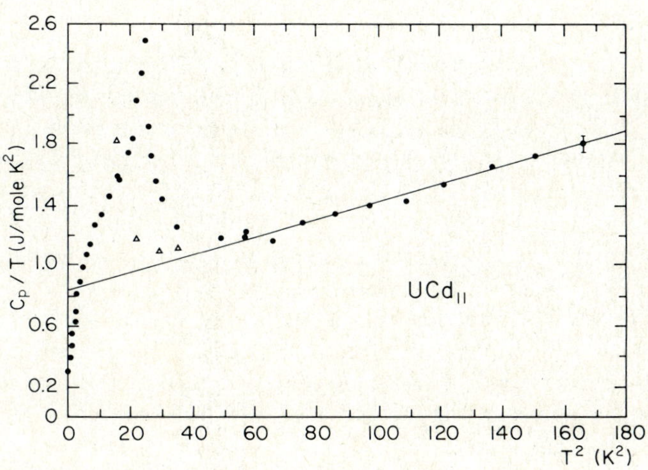

Fig. 14. C_p/T versus T^2 for UCd_{11} as reported by Fisk et al. (1984). The solid line serves to estimate the parameters γ and β of eq. (1). Open triangles are data obtained in an external magnetic field of 110 kOe.

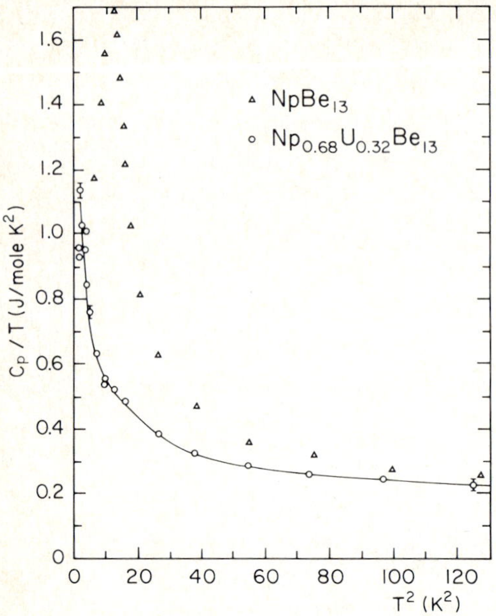

Fig. 15. C_p/T versus T^2 for $NpBe_{13}$ and $Np_{0.68}U_{0.32}Be_{13}$ below 11 K, as reported by Stewart et al. (1984a). The solid line is to guide the eye.

T_N=9.5 K (Trainor et al. 1976). The low-temperature specific heat results in an analogous C_p^{el}/T versus T^2 plot as have been recently observed in U_2Zn_{17} and UCd_{11} (see fig. 114). For $NpSn_3$ the large γ value above T_N of 242 mJ/mol K^2 appears to be independent of temperature up to about 30 K. Again no data exist for $T>30$ K that would show how, and in what temperature range, this still fairly large electronic specific heat is building up.

2.1.5. *NpBe₁₃*

$NpBe_{13}$ is a compound where a phase transition, most probably magnetic, at 3.4 K is preceded by a sizeable increase of C_p/T with decreasing temperature below 10 K (Stewart et al. 1984a). The C_p/T ratio of $NpBe_{13}$ at 10 K is even bigger than that of UBe_{13}. Recalling that the lattice contribution to C_p of UBe_{13} is negligible in this temperature range, it can safely be assumed that the measured specific heat of $NpBe_{13}$ below 10 K is entirely electronic in nature. In fig. 15, C_p/T versus T^2 is shown for $NpBe_{13}$ for $T<11$ K. In this case it is not trivial to separate the increase of specific heat that is due to the phase-transition anomaly from the enhancement of C_p that occurs in heavy-electron materials. As in other cases (see especially section 4) it seems possible to suppress the phase transition by disturbing the actinide sublattice and hence a more distinct observation of the enhancement effect on C_p should be possible. The only available data here are those for $Np_{0.68}U_{0.32}Be_{13}$ (Stewart et al. 1984a) and they are also shown in fig. 15 for $1\ K<T<11$ K. What is not apparent in fig. 15 is a saturation of C_p^{el}/T between 1.5 and 1 K with a subsequent strong increase with decreasing temperature down to the lowest attained temperature T=0.38 K. We therefore also show a plot of data obtained between 0.38 and 1.9 K on a more extended temperature scale in fig. 16. Thus,

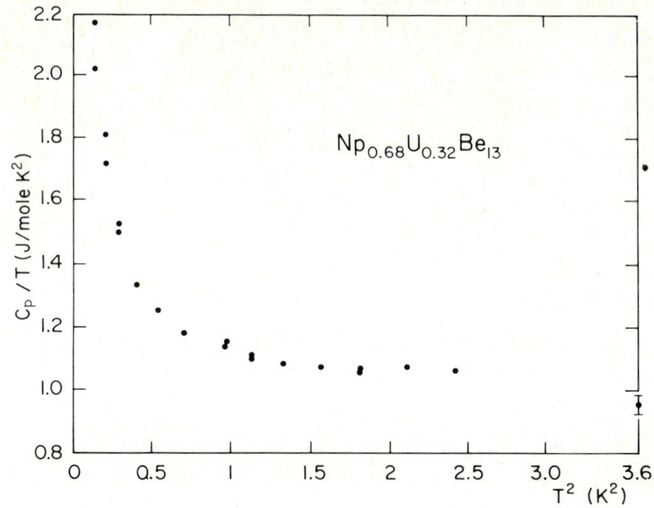

Fig. 16. C_p/T versus T^2 for $Np_{0.68}U_{0.38}Be_{13}$ at temperatures below 1.9 K as found by Stewart et al. (1984a).

$Np_{0.68}U_{0.32}Be_{13}$ shows very similar behaviour in its low-temperature specific heat as pure UBe_{13}. Whether the further anomalous behaviour of C_p below 1 K should be associated with magnetic ordering has not been firmly established yet. By analogy we therefore also classify $NpBe_{13}$ as a typical heavy-electron material, presumably with a magnetically-ordered ground state.

2.1.6. *$PuBe_{13}$*

Another compound with a highly anomalous low-temperature specific heat that may, with some reservations, also be classified as a heavy-electron material is $PuBe_{13}$. We show C_p/T versus T^2 for $PuBe_{13}$ between 2 and 26 K in fig. 17. In their

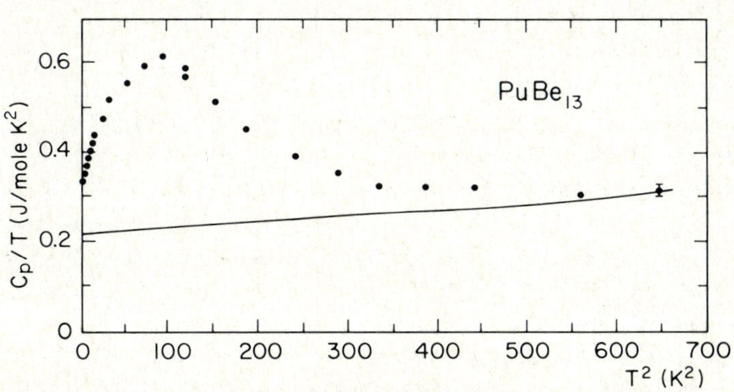

Fig. 17. C_p/T versus T^2 for $PuBe_{13}$ below 25 K as reported by Stewart et al. (1984a). The solid line is an estimate of the background specific heat as defined by eq. (1).

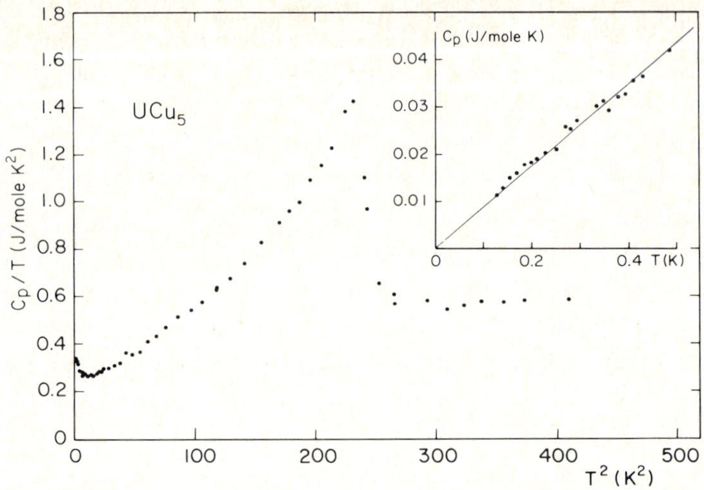

Fig. 18. Low-temperature specific heat of UCu_5 as obtained by Ott et al. (1985c). The inset shows the linear temperature variation of C_p below 0.5 K.

original work, Stewart et al. (1984a) argue that the structure of C_p/T above the extrapolated high-temperature behaviour (solid line in fig. 17) is not due to a phase transition but rather resembles the structure that is obtained for the specific heat of an impurity with a finite moment dissolved in an itinerant-electron system when described with the Coqblin–Schrieffer model (Rajan 1983). Therefore they classified $PuBe_{13}$ as a Kondo system. Even when considering the low concentration of Pu atoms in a compound like $PuBe_{13}$ it is, of course, by no means evident that a theoretical approach which is meant to describe the effect of one single impurity should be relevant for this case. Consequently, we regard the above conclusion as tentative and simply note that the C_p/T ratio for $T \to 0$ K approaches a value of about 330 mJ/mol K^2 which is clearly enhanced above the extrapolated high-temperature γ value of 220 mJ/mol K^2.

2.1.7. *UCu_5*

So far we have shown examples of compounds for which the low-temperature specific heat shows an anomalous enhancement with decreasing temperature. At even lower temperatures they usually undergo phase transitions which, however, can be suppressed by introducing impurities on either the cation or the anion sites of the crystal lattice. Very recently it was found that the specific-heat enhancement effect may also occur in an already magnetically-ordered material, the first such example being UCu_5 (Ott et al. 1985c). That UCu_5 orders antiferromagnetically at $T_N = 15$ K was known for some time. A distinct anomaly in the specific heat (van Daal et al. 1975) confirmed former conclusions from magnetic susceptibility measurements (Misiuk et al. 1973) and neutron-diffraction measurements (Murasik et al. 1974). What was not realised was that an upturn of C_p/T with decreasing temperature below 4 K is associated with the same phenomenon that is characteristic for heavy-electron materials. We show recent specific-heat data of UCu_5 between 0.13 and 21 K in fig. 18 (Ott et al. 1985c). It includes a C_p/T versus T^2 plot of the data that were obtained above 1 K and the inset shows the linear temperature dependence below 0.5 K. Obviously, the resulting $\gamma(0)$ value of 86 mJ/mol K^2 is much smaller than the C_p/T ratio at 1.5 K. This implies that UCu_5 undergoes another phase transition between 0.5 and 1.5 K, as has indeed been observed. Again we postpone the discussion of this phase transition to section 4. In order to preserve the enhancement effect, i.e., to suppress the phase transition, one of the five Cu atoms per formula unit was replaced by Ag. The measured $C_p(T)$ values for $UAgCu_4$ are shown, in the same manner as those of UCu_5, in fig. 19 (Ott et al. 1985c). The resulting $\gamma(0)$ value of 310 mJ/mol K^2 is now almost 4 times larger than that of UCu_5, demonstrating the expected behaviour. For $UAgCu_4$, C_p/T reaches a maximum value of 340 mJ/mol K^2 at 1.4 K. A possible cause for the occurrence of such a maximum is mentioned in the work by Bredl et al. (1984) on $CeAl_3$.

An interesting study by van Daal et al. (1975) revealed a drastic effect on the low-temperature specific heat of UCu_5 if Cu is replaced by Ni. With increasing Ni content $\gamma(0)$, which is obtained by linear extrapolation of a C_p/T versus T^2 plot at temperatures above T_N, decreases rapidly but stays concentration-independent for Ni contents of more than 20%. For UCu_5, van Daal et al. extrapolated a $\gamma(0)$ value of 200 mJ/mol K^2 which reduces to about 50 mJ/mol K^2 for $UNiCu_4$. Clearly, the replacement of Cu by Ni has a quite different effect than when Cu is substituted with Ag.

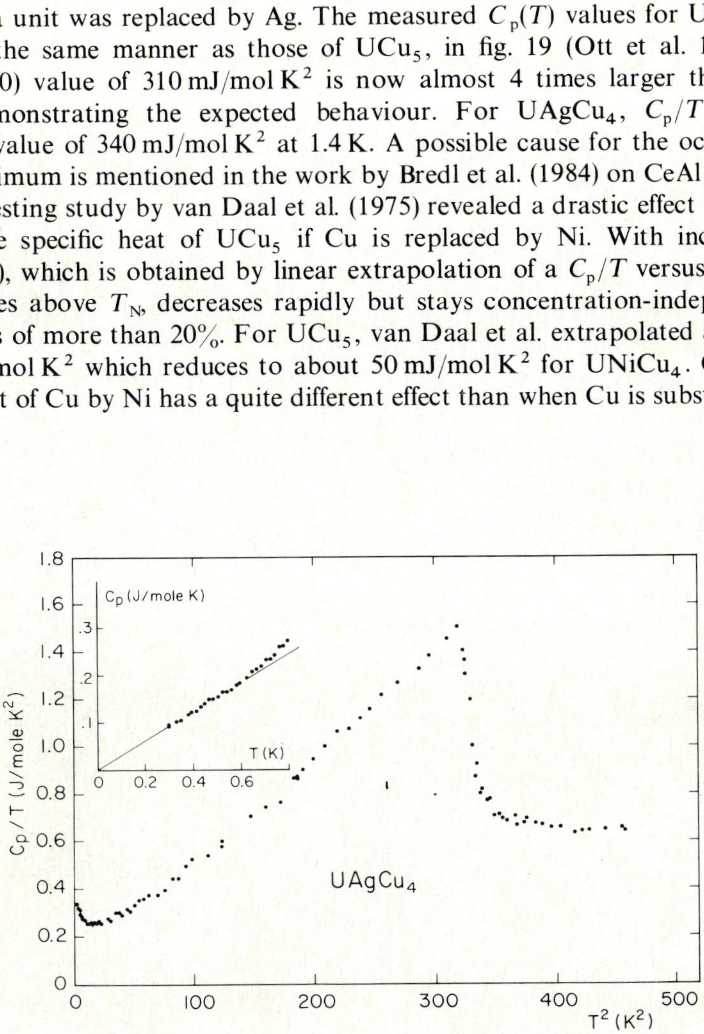

Fig. 19. Low-temperature specific heat of $UAgCu_4$ as obtained by Ott et al. (1985c). The inset demonstrates the linear temperature variation of C_p below 0.6 K. Note the difference in slope with respect to UCu_5.

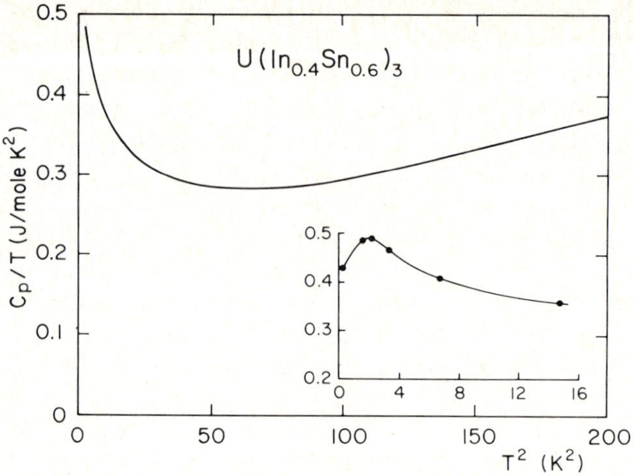

Fig. 20. C_p/T versus T^2 for $U(In_{0.4}Sn_{0.6})_3$ as found by Lin et al. (1986) below 14 K. The inset shows details of the behaviour at temperatures below 4 K.

2.1.8. *$U(Sn_{1-x}In_x)_3$*

A recent study (Lin et al. 1986) of the $U(Sn_{1-x}In_x)_3$ series has shown that the fairly large $\gamma(0)$ value of USn_3 of 170 mJ/mol K^2 which, however, is not preceded by the characteristic C_p/T upturn can be drastically enhanced by replacing Sn with In. The enhancement effect is maximal for $x=0.4$ and in this case the C_p/T versus T^2 curve shows all the features that are ascribed to heavy-electron behaviour. We show the C_p/T versus T^2 plot for $U(Sn_{0.6}In_{0.4})_3$ for the temperature range between 0.3 and 14 K in fig. 20. The inset shows details at very low temperatures, indicating a maximum for the C_p/T ratio of about 500 mJ/mol K^2 before reaching the $\gamma(0)$ value of 430 mJ/mol K^2, appreciably higher than the extrapolated high-temperature $\gamma(0)$ value of 195 mJ/mol K^2. For $x=0.4$, no phase transition is observed down to 0.3 K. This is, however, not surprising if we remember the devastating effect of only tiny amounts of impurities on the anion sites on phase transitions that occur out of a state with enhanced specific heat. For x smaller and larger than 0.4, the enhancement effect disappears fairly rapidly with varying x.

Concluding this subsection we would like to mention the pseudobinary compound $Pu_{0.7}U_{0.3}Al_2$ which seems to have the highest $\gamma(0)$ value of 260 mJ/mol K^2 without the now often mentioned and demonstrated enhancement effect (Trainor et al. 1975a). At temperatures below 15 K, $C_p(T)$ of this compound can be well described with eq. (1) and therefore we do not classify it as a heavy-electron material. Nevertheless, its very high $\gamma(0)$ value makes it a very interesting substance, certainly worthy of future investigations. The same may be said for other actinide compounds like U_2PtC_2, a superconductor (Meisner et al. 1984), or $NpOs_2$, probably an itinerant ferromagnet (Brodsky and Trainor 1977, Fournier and Troć 1985). Especially the Np compounds have γ-values above their respective ordering temperatures that are one order of magnitude larger than those of most metallic materials, not counting, of course, the rare-earth-based heavy-electron substances.

2.2. *Magnetic susceptibility*

The magnetic susceptibility χ of most metals is only weakly temperature dependent and, except for special cases, mostly of paramagnetic nature. Curie–Weiss-type paramagnetic behaviour above the respective transition temperatures is observed for metals which order magnetically. The special feature of the heavy-electron materials is that they show a mixture of these two limiting behaviours which is dictated by the temperature range in which χ is measured. Again, this particular behaviour was first observed in Ce intermetallics and in order to lay out the general features we refer to the temperature dependence of χ of the standard heavy-electron material $CeAl_3$. The original measurements by Buschow and Fast (1966) between 4.2 and 300 K gave a Curie–Weiss-type behaviour

$$\chi \sim P_{\text{eff}}^2/(T+\theta_{\text{p}}) \tag{5}$$

with an effective moment P_{eff} of 2.6 μ_B/Ce ion and some additional temperature dependence below 100 K that could be ascribed to crystal-electric-field effects. The value of the observed effective moment is very close to that expected for a free trivalent Ce ion. Later measurements between 0.05 and 1.5 K by Andres et al. (1975) showed an almost temperature-independent susceptibility in that temperature range with a very flat maximum at about 0.65 K but no magnetic order. The value of this temperature-independent magnetic susceptibility is 3.6×10^{-2} emu/mol, an enormous value for a Pauli-type susceptibility of a metal. However, this value is compatible with the value of the electronic specific heat $\gamma(0)$ of 1.62 J/mol K^2. Using eq. (2a) for γ and the corresponding expression for the Pauli susceptibility of electrons,

$$\chi_{\text{p}} = 2\mu_B^2 N(E_F), \tag{6}$$

the ratio $(\chi_p/\gamma)(\pi^2 k_B^2/3\mu_B^2)$ should be of order unity, if both χ and γ can be ascribed to the same electronic subsystem. For $CeAl_3$, the data given by Andres et al. (1975) provide a value of 1.62 for this ratio. Another fact worth mentioning is that, by varying the temperature, the transition from Curie–Weiss-type to a temperature-independent behaviour of χ is much more abrupt than the development of the enhanced specific heat as seen by the increase of C_p^{el}/T with decreasing temperature. In the following subsections we will show that these general features also appear in the magnitude and the temperature dependence of χ for heavy-electron actinide materials. For the discussion of these experimental data, especially also in connection with other quantities, the reader is again referred to section 3. Phase transition aspects will be dealt with in section 4.

2.2.1. *UBe_{13}*

The earliest investigation of the magnetic susceptibility of UBe_{13} was made by Troć et al. (1971) who measured $\chi(T)$ between 4.2 and 1000 K. Later, Brodsky and Friddle (1975) and Bucher et al. (1975) repeated these measurements, concentrating on $\chi(T)$ below room temperature. In particular, Bucher et al. were searching for evidence of a temperature-independent Van Vleck-type susceptibility that would have provided evidence for a $5f^2$ configuration of the U ions. More recent $\chi(T)$

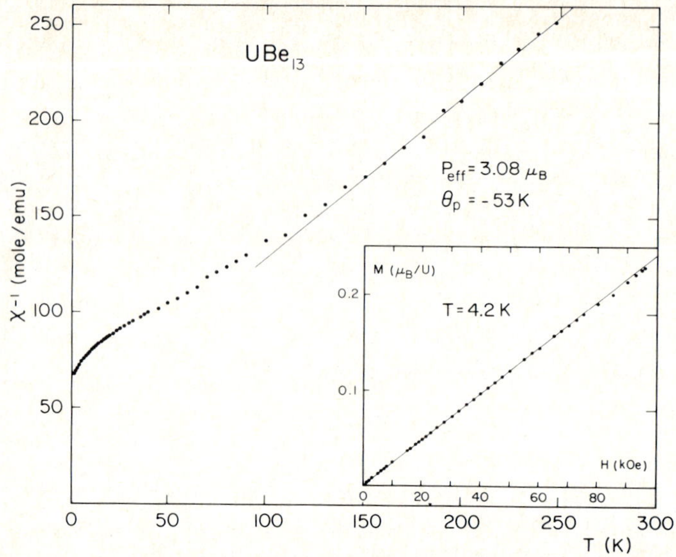

Fig. 21. Inverse magnetic susceptibility of UBe_{13} between 1.5 and 250 K. The inset shows the magnetic-field dependence of the magnetization at 4.2 K up to 100 kOe. (Data are taken from Ott et al. 1984d.)

measurements are due to Ott et al. (1984d) who measured the magnetic susceptibility of single-crystalline UBe_{13} between 1.4 and 300 K. We show their results in fig. 21 as a χ^{-1} versus T plot. At temperatures above 150 K, $\chi^{-1}(T)$ is well described by a Curie–Weiss straight line with an effective moment P_{eff} of 3.08 μ_B/U ion and a negative paramagnetic Curie temperature θ_p of -53 K. At 1.4 K, χ reaches a value of 1.5×10^{-2} emu/mol. These values and those that result from the other mentioned work, including results from recent experiments by Mayer et al. (1986), are listed in table 3.

Although there appear to be considerable differences between measurements, some of them are due to obvious reasons. The values for P_{eff} and θ_p depend very much on the temperature range in which $\chi(T)$ is analyzed, as was first recognized by Brodsky and Friddle (1975). Clearly, all values for P_{eff} and $|\theta_p|$ that are evaluated from $\chi(T)$ below 100 K are too large and misleading when used for conclusions concerning the 5f-electron configuration of the U ions. The temperature dependence of $\chi(T)$ below 100 K may be due to single-ion effects. For example, there is microscopic evidence (e.g. from neutron inelastic scattering) for a crystal-field splitting of 5f-electron energy levels backed by specific-heat results that might explain the deviation of $\chi(T)$ from the Curie–Weiss behavior below 150 K (Felten et al. 1986). Nevertheless, part of the scatter between different data must be traced back to real differences of behaviour in differently prepared samples below 150 K.

In addition, we refer to the inset of fig. 21 where the magnetization $M(H)$ of UBe_{13} at 4.2 K is plotted for magnetic fields between 0 and 100 kOe. Evidently the magnetic susceptibility $\partial M/\partial H$ at these low temperatures is almost field independent. At 100 kOe, χ has decreased only a few percent from its value at $H=0$.

TABLE 3
Experimental data on the magnetic susceptibilities of UBe_{13}, UPt_3 and UAl_2.

	P_{eff} (μm)	θ_p (K)	Temp. range (K)	χ (low temp.) (10^{-3} emu/mol)	Ref.[a]
UBe_{13}					
Powder	3.4	−70	260–1000	14 (at 4.2 K)	[1]
p.c.	2.99	−68	200–300	12 (at 1.5 K)	[2]
p.c.	3.16	−110	1.5–200	12	
p.c.	3.52	−98	1.5–100	16.6 (T=1.5 K, H=1.4 T)	[3]
s.c.	3.08	−53	150–300	15 (T=1.5 K, H=1 T)	[4]
p.c.	3.34	−100	120–300		[5]
UPt_3					
p.c.	2.85		23–110	6.85 (at 1.5 K)	[6]
p.c.	2.61	−95	110–210		[6]
p.c.	2.785	−129	300–1000		[7]
s.c., (*ab*)-plane	2.53	−39	50–300	8.55 (at 1.5 K)	[7]
s.c., *c*-axis	2.37	−151	2–300	4.46 (at 1.5 K)	[8]
UAl_2					
p.c.	3.1	−245	150–300	4.38 (at 1.5 K)	[9]
p.c.	3.38	−335	250–600		[7]

[a] References:
[1] Troć et al. (1971).
[2] Brodsky and Friddle (1975).
[3] Bucher et al. (1975).
[4] Ott et al. (1984d).
[5] Mayer et al. (1986).
[6] Schneider and Laubschat (1981b).
[7] Frings (1984).
[8] Frings et al. (1983).
[9] Brodsky (1979).

Finally, we should like to mention that low-field a.c. susceptibility measurements in the normal state of UBe_{13} reveal an almost temperature-independent χ_{ac} below 1.2 K. Thus again the behaviour of $\chi(T)$ appears to change in a narrow temperature range for UBe_{13} as well.

2.2.2. *UPt_3, UAl_2*

Because UPt_3 has a non-cubic crystal structure (hexagonal Ni_3Sn structure as $CeAl_3$), χ is likely to be anisotropic as was indeed confirmed by measurements using single-crystalline specimens. Measurements of $\chi(T)$ of polycrystalline UPt_3 were reported by Schneider and Laubschat (1981b). As in UBe_{13}, the values of characteristic parameters like P_{eff} and θ_p depend very much on whether they are evaluated at low or at high temperatures. Parameters quoted by the above authors and others mentioned below are listed in table 3. We also note that the observation of a maximum in $\chi(T)$ of UPt_3 at 15 K led Schneider and Laubschat to claim an antiferromagnetic phase transition for this compound. More recent measurements of $\chi(T)$ of poly- and single-crystalline UPt_3 were reported by Frings (1984) and by Franse et al (1984). Figure 22 shows their results of $\chi(T)$ along, and perpendicular

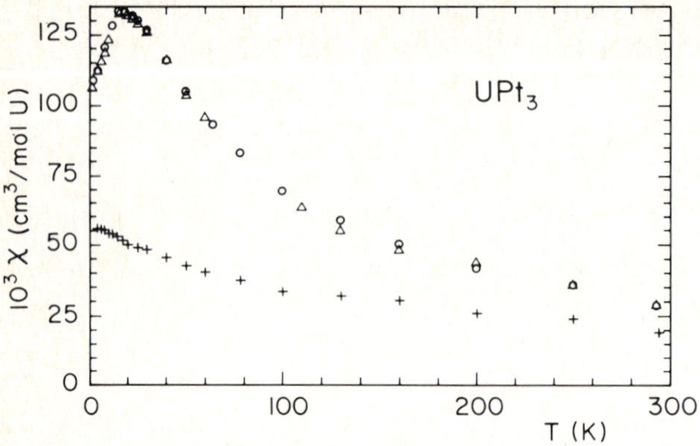

Fig. 22. Temperature dependence of the magnetic susceptibility of UPt_3 between 1.5 and 300 K, measured along different crystallographic directions by Franse et al. (1984). Crosses are data obtained along the hexagonal *c*-axis, circles and triangles are along the *a*- and *b*-axis in the basal plane, respectively.

to, the hexagonal axis of the crystal structure. From fig. 22 it is readily seen that the anisotropy is considerable and that below 100 K, $\chi_\perp$ within the basal plane is at least twice as large as $\chi_{\parallel}$, parallel to the *c*-axis. Moreover, it is evident that the maximum in $\chi(T)$ only occurs within the basal plane. Its occurrence is still a mystery because the phase transition that might readily be inferred from it is not consistent with the specific-heat results that are shown in figs. 4 and 5 in the previous section. So far there is no hint that this particular temperature dependence of $\chi_\perp$ is due to a crystal-field effect. Actually, the sudden slope change around 16 K makes such an interpretation quite unlikely anyway. For polycrystalline UPt_3, Frings (1984) also reported results of measurements between 300 and 1000 K. The important parameters that can be extracted from all these data are also listed in table 3. Contrary to UBe_{13}, the P_{eff} values given by Frings and by Franse et al. are lower when extracted from $\chi(T)$ at lower temperatures. Just the opposite seems to be true for the data by Schneider and Laubschat. Data for the low-field a.c. magnetic susceptibility χ_{ac} are available from work by Smith et al. (1984b) for temperatures between 0.4 and 2.3 K. They indicate that χ_{ac} decreases with decreasing temperature below 2.3 K down to about 1.2 K below which it is temperature independent until the superconducting transition temperature is reached. Obviously the onset of a temperature-independent χ is equally abrupt in UPt_3 as was observed for $CeAl_3$ and UBe_{13}.

Franse et al. (1984) also published data on the field dependence of the a.c. susceptibility in the basal plane up to 320 kOe and of the magnetization $M(H)$ along different crystalline directions and at different temperatures. Strong non-linearities occur in the $M(H)$ curves at helium temperatures when the external field is applied in the basal plane, and for H values that range between 100 and 320 kOe. At temperatures above 20 K these nonlinearities disappear. The same behavior is reflected in the data for the a.c. susceptibility. We also note that $M(H)$ reaches no saturation even in fields of more than 300 kOe.

Since the low-temperature specific heats of UPt_3 and UAl_2 can both be described by eq. (4) it seems reasonable to compare the low-temperature magnetic response of these two compounds as well. Trainor et al. (1975b) pointed out that the magnetic

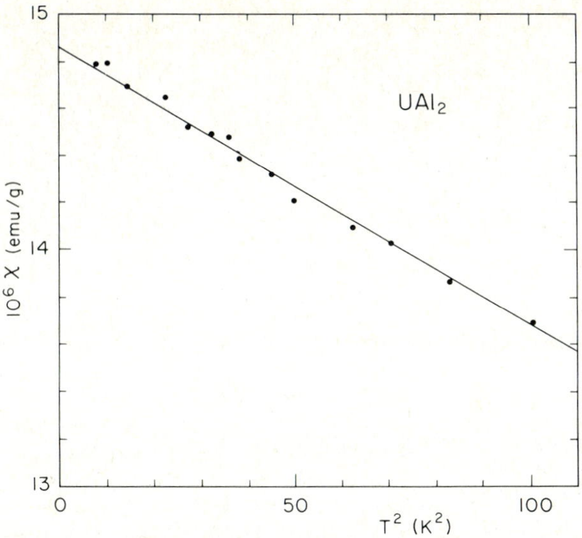

Fig. 23. χ versus T^2 below 10 K for UAl_2. The data are due to Trainor et al. (1975b). The solid line indicates the T^2 dependence of χ in this temperature range.

susceptibility of polycrystalline UAl_2 can be very well fitted by

$$\chi = A(1 - aT^2) \tag{7}$$

in the same temperature range where C_p follows eq. (4). We show their plot in fig. 23. Their numerical values for the parameters in eq. (7) are $A = 4.312 \times 10^{-3}$ and $a = 7.94 \times 10^{-4}$ when χ is given in units of emu/mol. Such a behaviour is clearly not observed in UPt_3 where the anomaly of $\chi_\perp(T)$ dominates $\chi(T)$ of polycrystals. Also $\chi_\parallel$ does not follow a temperature dependence according to eq. (7), as may be seen from fig. 22.

In a later publication, Brodsky and Trainor (1977) showed their results for $\chi(T)$ for UAl_2 between 1.5 and 300 K. We show their results in fig. 24. In this work they claim even two temperature ranges for which eq. (7) is applicable, for different physical reasons, however, and with different values for A and a. For temperatures above 150 K they claim a temperature dependence for χ according to a Curie–Weiss law as in eq. (5). This is, however, in disagreement with more recent results by Frings (1984) to temperatures up to 1000 K which indicate a continuous curvature of χ^{-1} versus T and therefore no real Curie–Weiss fit is satisfactory. The closest reasonable approach to it can be made for the data between 250 and 600 K and the resulting values for P_{eff} and θ_p are given in table 3. The comparison in table 3 underlines the problems connected with the evaluation of Curie–Weiss parameters in most of these compounds. It should always be stated in which temperature range the data analysis was made.

Brodsky (1979) mentioned that small amounts of Th impurities enhance the electronic specific heat of UAl_2 at the lowest temperatures. It is now interesting to note that, on the other hand, with increasing Th content the low-temperature

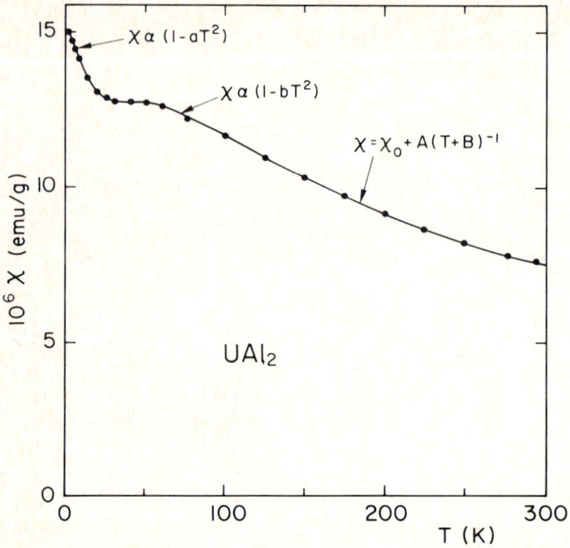

Fig. 24. $\chi(T)$ of UAl_2 between 1.5 and 300 K as reported by Brodsky and Trainor (1977). Emphasized are the different temperature dependences in different temperature ranges.

magnetic susceptibility is being reduced, at least for Th contents of less than 10%. A similar effect was also noted in the case of UBe_{13}, where $\gamma(0)$ clearly increases with small amounts of Th on the U sites (see section 2.1.1.) but χ at low temperature is also reduced (Smith et al. 1984a). Quite different is the effect when diluting U in UAl_2 with Y. Buschow and van Daal (1972) observed a drastic increase of $\chi(T)$ of $U_{1-x}Y_xAl_2$ with increasing x at all temperatures between 1.5 and 300 K.

Measurements of $M(H)$ at different temperatures and in fields up to 350 kOe were claimed to give additional evidence for spin-fluctuation effects in UAl_2 (Franse et al. 1982). At temperatures of 4.2 K and below, $M(H)$ is linear in H up to 150 kOe where the slope changes to a lower value. This slope change was ascribed to the suppression of spin fluctuations in larger magnetic fields. The value for $\partial M/\partial H$ that is observed at low temperatures above 150 kOe is roughly the same as that at temperatures exceeding 20 K in the whole field region covered.

For UPt_3, data for the pressure dependence of the low-temperature magnetization $M(H)$ and of the maximum in $\chi(T)$ are also available. Willis et al. (1985a) measured $M(H)$ of polycrystalline UPt_3 at 4.2 K in fields up to 60 kOe and at 0 and 4.5 kbar pressure. They found an approximately 10% reduction of the linear-in-H slope of M upon application of pressure. The same 4.5 kbar pressure resulted in a 2 K shift of the maximum of $\chi(T)$ to higher temperatures. In this experiment $\chi(T)$ was monitored in an external magnetic field of 53 kOe. These experiments gave a value for the pressure dependence of the low-temperature magnetic susceptibility of $\partial \ln \chi(0)/\partial p = -24\ \text{Mbar}^{-1}$. The influence of pressure on the magnetic susceptibility of UAl_2 was investigated by Fournier and Beille (1979). They measured $\chi(T)$ under hydrostatic pressures up to 7 kbar and at temperatures between 4 and 250 K. They observe that the temperature dependence of χ does not change with increasing pressure but the $\chi(T)$ curves are shifted to lower values as a whole. The overall

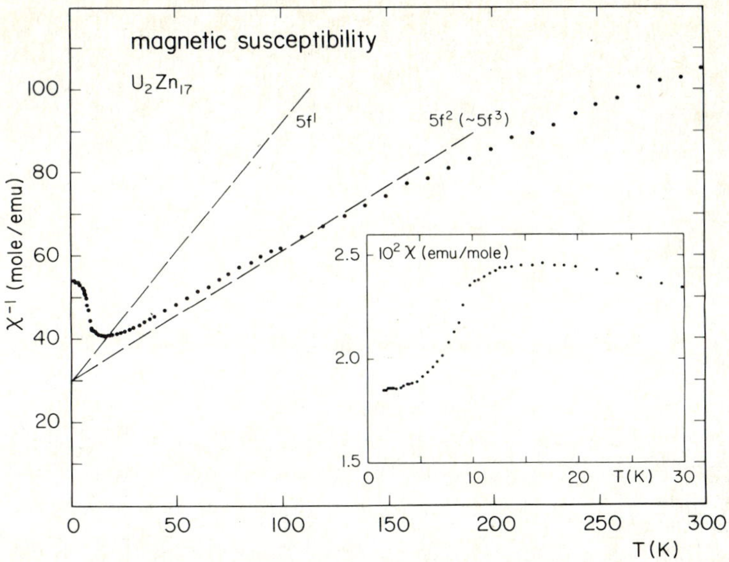

Fig. 25. Temperature dependence of the inverse magnetic susceptibility χ^{-1} of U_2Zn_{17} between 1.5 and 300 K. The inset shows $\chi(T)$ on an extended temperature scale below 30 K. (Data are taken from Ott et al. 1984b.)

pressure-induced change of χ is given as $\partial \ln \chi/\partial p = -25\,\mathrm{Mbar}^{-1}$, obviously very close to the value that was reported for UPt_3.

2.2.3. U_2Zn_{17}

The magnetic susceptibility of U_2Zn_{17} has first been measured by Misiuk et al. (1973) between 4.2 and 900 K on polycrystalline samples with no unambiguously identified crystal structure. They noticed a temperature-independent magnetic susceptibility below 10 K. For temperatures above 400 K they claim an effective moment of 3.6 μ_B/U ion and a θ_p of -270 K although it seems evident from their data, that $\chi^{-1}(T)$ varies with a continuously decreasing slope with increasing temperature. More recent measurements (Ott et al. 1984b) on single-crystalline U_2Zn_{17} with the proper Th_2Zn_{17} structure between 1.5 and 300 K more or less confirmed the magnitude of χ as found in the previous experiment, but they also revealed some distinct differences in the temperature dependence of χ. There newer data are shown in fig. 25. It may be seen that $\chi(T)$ reaches a flat maximum at 17 K and decreases sharply below 9.7 K. The reduction of χ down to the lowest temperatures amounts to about 25%. The χ^{-1} versus T plot shows that there is no temperature region up to 300 K where a reasonable Curie–Weiss fit is appropriate. We therefore refrain from trying to extract any numbers for P_{eff} or θ_p from these data.

Since U_2Zn_{17} has a uniaxal crystal structure it was of interest to study the anisotropy of the magnetic susceptibility. This was most recently done by Willis et al. (1985c) for temperatures between 4.2 and 200 K and their result is shown in fig. 26

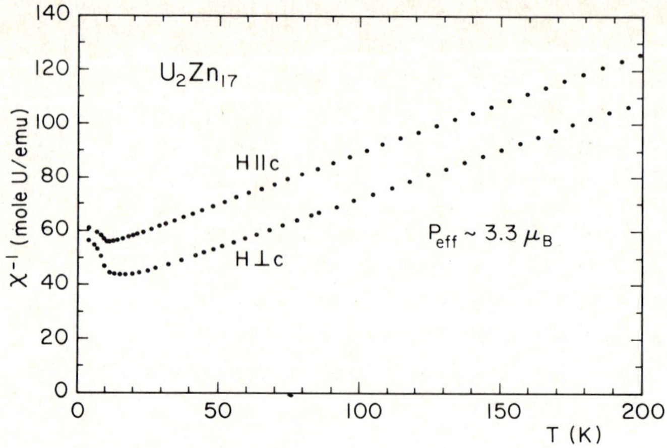

Fig. 26. Anisotropy of the magnetic susceptibility of U_2Zn_{17} as measured by Willis et al. (1985c) along different crystallographic directions.

in the form of χ^{-1} versus T plots. As in UPt_3, $\chi_{\perp}$ within the basal plane is larger than $\chi_{\parallel}$ along the hexagonal axis. The maximum of $\chi(T)$ at 17 K is clearly visible in $\chi_{\perp}$ but $\chi_{\parallel}$ increases until the antiferromagnetic phase transition sets in at 9.7 K. Between 50 and 200 K both $\chi_{\perp}$ and $\chi_{\parallel}$ appear to follow a Curie–Weiss law with the same value for the effective moment of 3.3 μ_B/U ion but with different paramagnetic Curie temperatures of −95 and −130 K, respectively. Again it seems doubtful whether these numbers have any real meaning because they increase with increasing temperature. One possible reason for this kind of behaviour might be single-ion effects.

Impurities on both the cation and the anion sites can drastically alter the low-temperature susceptibility of U_2Zn_{17} (Willis et al. 1986). Some examples are shown in fig. 27. Especially effective are impurities on the Zn sites as may be seen from the

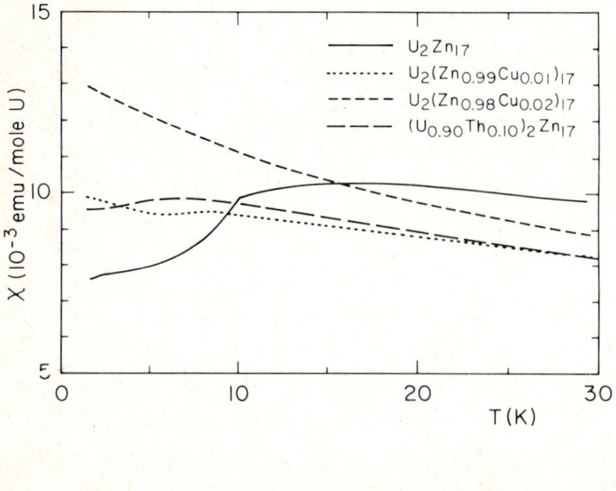

Fig. 27. Influence of various impurities on the low-temperature magnetic susceptibility of U_2Zn_{17} as measured by Willis et al. (1986).

effect of a 2% substitution of Zn by Cu. The suppression of the magnetic transition will be discussed in section 4.

$M(H)$ data were also obtained on single crystals of U_2Zn_{17} with the direction of H either parallel or perpendicular to the hexagonal axis (Willis et al. 1985c). The data were taken at various temperatures between 4.2 and 30 K and in fields up to 50 kOe. For H parallel to the hexagonal axis and temperatures below T_N, M varies linearly with H up to 50 kOe. Distinct upward deviations are observed for $H > 30$ kOe and temperatures exceeding T_N. If the field direction is kept in the basal plane, M is linear in H only up to 20 kOe with clear upturns at higher fields but now for $T < T_N$. For higher temperatures, M is linear in H up to 40 kOe with a tendency to saturation at still higher fields. That the magnetization in the basal plane does not exceed 0.1 μ_B/U ion in 50 kOe seems quite remarkable when considering the value of the ordered moment as obtained by neutron diffraction (see section 4).

2.2.4. *UCd₁₁*, *NpSn₃*

First measurements of the magnetic susceptibility of UCd_{11} were reported by Cafasso et al. (1963) for temperatures between 2 and 300 K. With only 5 data points they established a Curie–Weiss-type behaviour between 5 and 300 K with $P_{eff} = 3.79\ \mu_B$/U ion and $\theta_p = -41.5$ K. The temperature-independent $\chi(T)$ below 5 K was interpreted as a Van Vleck paramagnetic susceptibility hence implying a crystal-field induced singlet ground state of the 5f electrons, compatible only with a $5f^2$ configuration of the U ions. Later measurements by Misiuk et al. (1973) covered the temperature range from 4.2 to 745 K. They confirmed the results of Cafasso et al. but it was again demonstrated that both P_{eff} and $|\theta_p|$ increase considerably when evaluated in a temperature range exceeding 300 K. More recent measurements by Fisk et al. (1984) between 1.5 and 300 K confirmed the low-temperature, (up to 5 K) temperature-independent value of Cafasso et al. The Curie–Weiss-type behaviour between 80 and 300 K, however, resulted in a P_{eff} of 3.45 μ_B/U ion and a θ_p of -20 K. Clear deviations from this behaviour between 5 and 80 K are discernible in fig. 28, where we show the data given by Fisk et al. (1984).

$\chi(T)$ for $NpSn_3$ was measured by Aldred and Lam (1976) and its temperature dependence is described by Trainor et al. (1976). Brodsky and Trainor (1977) show a graph of $\chi(T)$ between 2 and 60 K. $\chi(T)$ rises from its value of the order of 1.5×10^{-2} emu/mol at 2 K, passes over a rounded maximum near 18 K, similar in character as that observed in U_2Zn_{17} (see above) and then steadily drops with increasing temperature to a value of again 1.5×10^{-2} emu/mol at 60 K. No particular feature in $\chi(T)$ indicates the antiferromagnetic transition at $T_N = 9.5$ K. The behaviour of $\chi(T)$ above T_N is interpreted as being due to a narrow-band Pauli-type susceptibility (Trainor et al. 1976).

2.2.5. *NpBe₁₃*

The d.c. magnetic susceptibility of $NpBe_{13}$ was first reported by Brodsky and Friddle (1975) for temperatures between 2 and 300 K. We show their results of $\chi(T)$ in fig. 29. A χ^{-1} versus T plot of these data reveals Curie–Weiss behaviour in two different temperature regions with different parameters P_{eff} and θ_p. The slope change

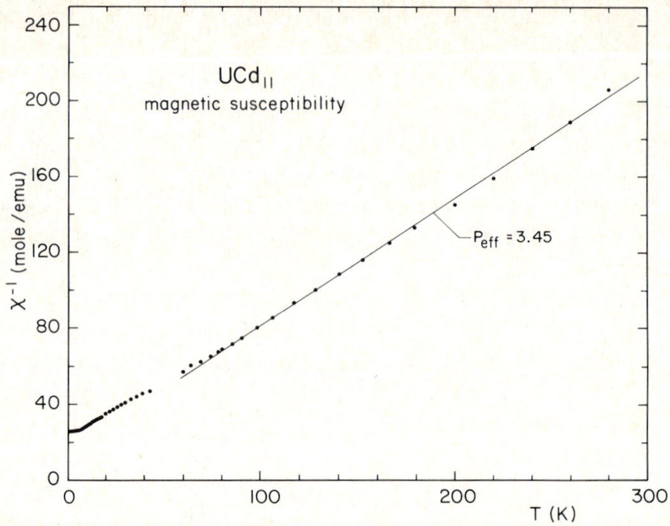

Fig. 28. Temperature dependence of the inverse magnetic susceptibility χ^{-1} of UCd_{11} between 1.5 and 300 K. (Data are from Fisk et al. (1984.)

of $\chi^{-1}(T)$ occurs at about 70 K. At low temperatures $P_{eff}=2.35\,\mu_B$/Np ion and $\theta_p=-12$ K while for $T>70$ K, $P_{eff}=2.76\,\mu_B$/Np ion and $\theta_p=-42$ K. No indication for magnetic order was observed in the whole temperature range.

More recently, Stewart et al. (1984a) re-measured $\chi(T)$ for single-crystalline $NpBe_{13}$ between 1.3 and 15 K. They obtain a 25% higher value for χ at 2 K than that reported by Brodsky and Friddle and they also observe a tiny break in $\chi(T)$ between 3 and 4 K, the same temperature where the specific heat shows its maximum (see above). These latter data are shown in fig. 30. Stewart et al. ascribe the difference between the two $\chi(T)$ measurements to the better quality of their samples.

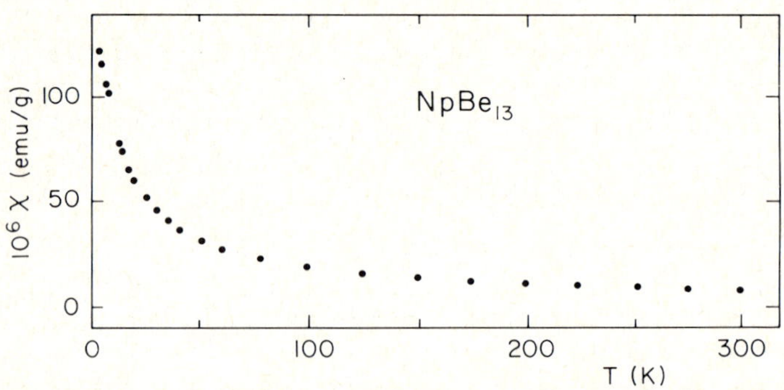

Fig. 29. $\chi(T)$ for $NpBe_{13}$ between 4.2 and 300 K as measured by Brodsky and Friddle (1975).

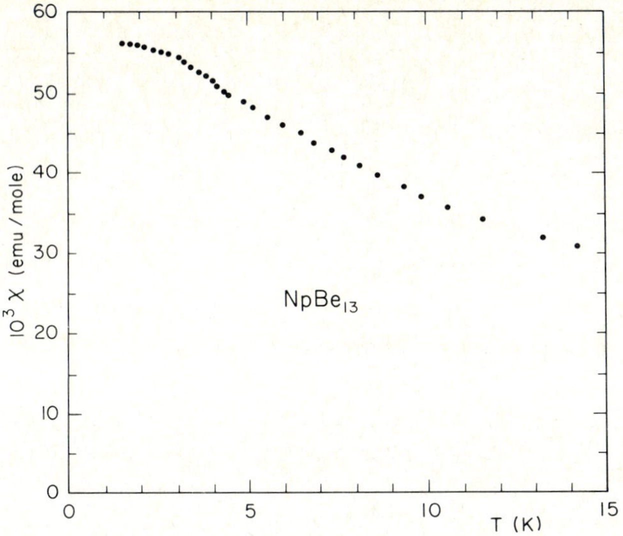

Fig. 30. $\chi(T)$ for $NpBe_{13}$ between 1.5 and 15 K as measured by Stewart et al. (1984a).

2.2.6. *PuBe$_{13}$*

Considerable discrepancy exists between the two available data sets of $\chi(T)$ measurements on $PuBe_{13}$. Again Brodsky and Friddle (1975) reported results obtained using a polycrystalline sample and at temperatures between 2 and 300 K. They observed an undisturbed Curie–Weiss behaviour between 15 and 300 K with $P_{eff}=0.74\ \mu_B$/Pu ion and $\theta_p=+11$ K. A very sharp maximum of $\chi(T)$ at 11.5 K with a subsequent drop of about 65% down to 2 K was interpreted as evidence for antiferromagnetic ordering in this compound.

Stewart et al. (1984a) measured the d.c. susceptibility of single-crystals of $PuBe_{13}$ between 1.3 and 100 K and also report results for a.c. susceptibility of the same sample between 4 and 18 K. We show their d.c. results in fig. 31. They strongly argue against a magnetic transition of $PuBe_{13}$ because of the obvious absence of a peak in $\chi(T)$ and no indication of a phase transition in $\chi_{ac}(T)$. However, their conclusion is also based on combined evidence from measurements of the specific heat (see above) and the electrical resistivity (see below). The different behaviour may originate in a difference of chemical composition between the two sample batches or in radiation-damage effects.

2.2.7. *UCu$_5$*

To our knowledge the first $\chi(T)$ experiments on UCu_5 were made by Misiuk et al. (1973) in the temperature range between 4 and 900 K. Their data are shown in fig. 32 and from the temperature dependence of χ above 400 K it was concluded that $\chi(T)$ obeys a Curie–Weiss law with $P_{eff}=3.6\ \mu_B$/U ion and $\theta_p=-740$ K, a somewhat unlikely value. A maximum in $\chi(T)$ at 15 K was interpreted as the onset of antiferromagnetic ordering.

Brodsky and Bridger (1973) measured $\chi(T)$ of UCu_5 between 2 and 300 K and observed a round maximum around 16 K. They claimed a Curie–Weiss-type

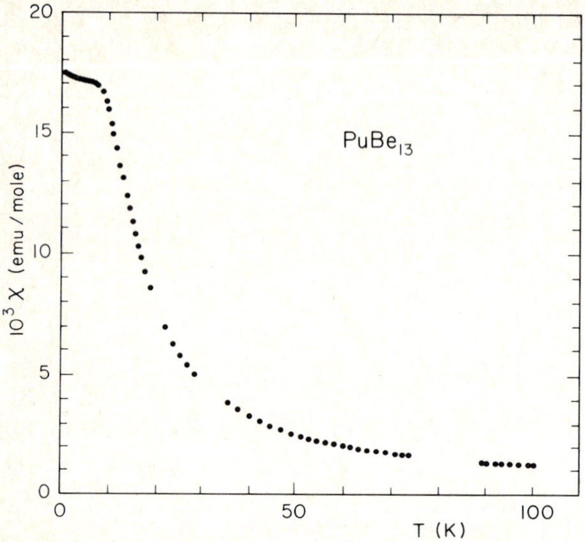

Fig. 31. $\chi(T)$ of $PuBe_{13}$ between 1.5 and 100 K as measured by Stewart et al. (1984a). Note the absence of a peak as was reported previously by Brodsky and Friddle (1975).

behaviour between 23 and 300 K with an effective moment of 2.3 μ_B/U ion and a paramagnetic Curie temperature of −76 K.

Further measurements by van Daal et al. (1975) between 1.5 and 1100 K confirmed the low-temperature values of χ obtained by the previous authors but they observed a rather different temperature dependence of χ. Between 15 and 1100 K they found no temperature range where a linear T dependence of χ^{-1} might be claimed and they doubted the analyses of the data sets put forward by the

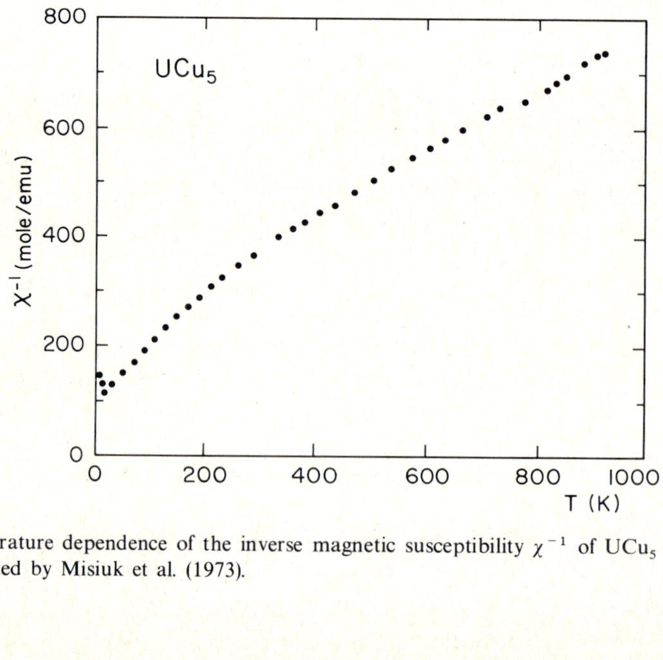

Fig. 32. Temperature dependence of the inverse magnetic susceptibility χ^{-1} of UCu_5 between 4.2 and 900 K as reported by Misiuk et al. (1973).

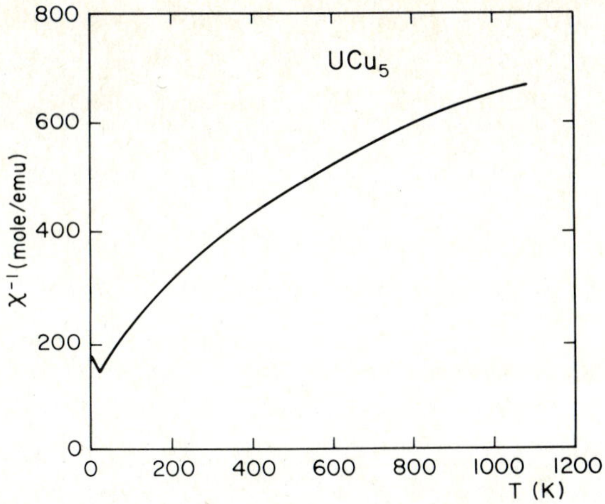

Fig. 33. $\chi^{-1}(T)$ of UCu_5 between 1.5 and 1100 K as measured by van Daal et al. (1975). Note the monotonously varying slope.

previous workers, with good reason as may be seen from fig. 33. For the antiferromagnetic ordering temperature they also obtain 15 K. van Daal et al. also report that replacing Cu by Ni drastically changes the low-temperature behaviour of $\chi(T)$. Small amounts of Ni suppress the magnetic-ordering temperature and an increasing concentration of Ni on the Cu sites finally leads to an overall decrease of χ at all temperatures and a complete loss of the Curie–Weiss-type behaviour. The concomitant decrease of the low-temperature specific heat was mentioned in section 2.1.7.

2.2.8. $U(Sn_{1-x}In_x)_3$

The results for $\chi(T)$ of USn_3 that we show in the usual χ^{-1} versus T plot in fig. 34 were obtained by Buschow and van Daal (1972). Up to about 30 K, χ is almost temperature independent and for temperatures between 50 and 280 K, these authors find a Curie–Weiss behaviour with $P_{eff}=2.56\,\mu_B$/U ion and $\theta_p=-70$ K. The low-temperature value of χ is 10^{-2} emu/mol. It should be mentioned that previous

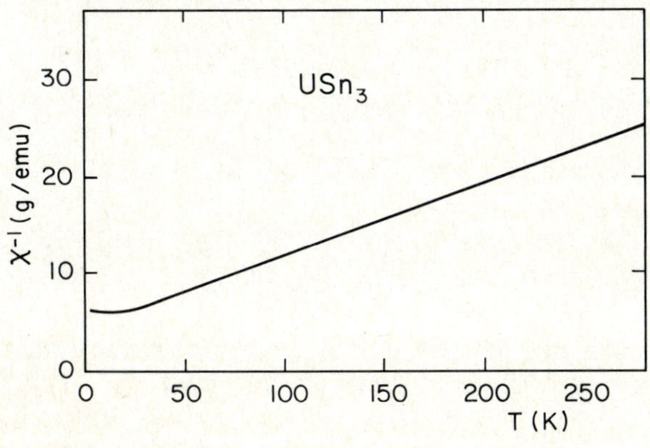

Fig. 34. $\chi^{-1}(T)$ of USn_3 between 4.2 and 300 K from data of Buschow and van Daal (1972).

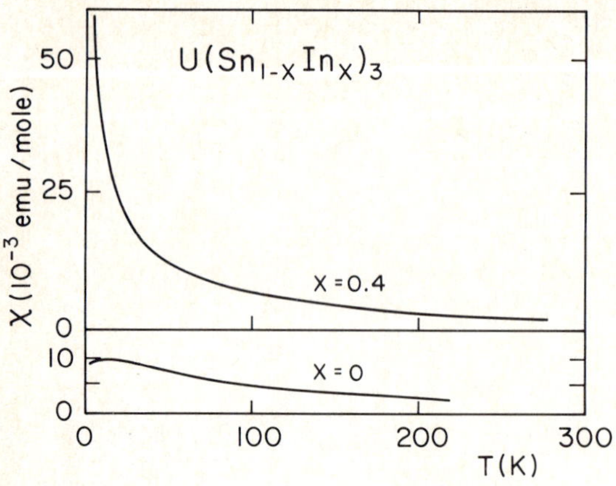

Fig. 35. Comparison of $\chi(T)$ between USn_3 and $U(In_{0.4}Sn_{0.6})_3$ below 300 K. (Data are from Lin et al. 1986.)

authors claimed the absence of any Curie–Weiss-type behaviour for USn_3 in different temperature regions. Rao and Vijayaraghavan (1968) first measured $\chi(T)$ of USn_3 between 100 and 400 K and Mulak and Misiuk (1971) later extended the temperature interval from 4.2 to 900 K. Both experiments showed no linear T-dependence of χ^{-1} at any temperatures.

As was more recently observed (Lin et al. 1986), substitution of In for Sn not only drastically changes the low-temperature specific heat but also the magnetic susceptibility. In fig. 35 we show $\chi(T)$ for two $U(Sn_{1-x}In_x)_3$ compounds with $x=0$ and 0.4. The curve for $x=0$ is essentially the same as that in fig. 34 but here, however, plotted as $\chi(T)$. The resulting Curie–Weiss parameters from this most recent experiment are $P_{\text{eff}}=2.5\ \mu_B$/U ion and $\theta_p=-70$ K. As for the specific heat the largest enhancement effect, now on χ at low temperatures, was observed for $x=0.4$. The curve shown in fig. 35 can be described by a Curie–Weiss-type temperature dependence with $P_{\text{eff}}=2.5\ \mu_B$/U ion and $\theta_p=-9$ K. Thus the enhancement of the low-temperature magnetic susceptibility results mainly in a parallel shift of the $\chi^{-1}(T)$ straight line of USn_3 to lower values, keeping p_{eff} more or less constant and reducing $|\theta_p|$ by almost a factor of 10. As in the case of the specific heat the enhancement effect for χ is suppressed considerably when x is larger or smaller than 0.4.

For the same compound series, $M(H)$ in fields up to 100 kOe was also studied at low temperatures. For $x=0$, M varies linearly with H at all fields, implying a field-independent magnetic susceptibility. For $x=0.4$, however, $M(H)$ increases with H with a continuously decreasing slope, indicating that the large zero-field value of χ is decreasing considerably with increasing field (Lin et al. 1986).

Summarizing all these data for the magnetic susceptibility of actinide heavy-electron or heavy-electron-type materials, the following statements can be made:

(1) Different measurements on the same substance generally give the same salient features for $\chi(T)$, but differences both in the magnitude as well as of the

temperature dependence are encountered. Part of these discrepancies certainly have to be traced back to problems with the calibrations of both the magnetic response signal and that of the thermometers. Another reason for different $\chi(T)$'s is, of course, an intrinsically different behaviour because of different sample quality. Since in almost all cases that we mentioned above, a really elaborate characterization of the specimens with respect to chemical composition or impurity content is lacking, this particular cause for differences in $\chi(T)$ cannot be discussed in detail here.

(2) It seems quite clear that for none of the compounds mentioned above can $\chi(T)$ be described unambiguously with a single Curie–Weiss-type temperature dependence according to eq. (5). For most of them, P_{eff} and the negative paramagnetic Curie temperature increase if the evaluation of these parameters is made at elevated temperatures. One clear exception is UBe_{13} for which P_{eff} and $|\theta_p|$ increase with decreasing temperature, at least above 30 K. These facts are certainly not helpful in determining the f-electron configuration of the actinide ions from $\chi(T)$ measurements.

(3) The change from Curie–Weiss-type behaviour to a temperature-independent χ may occur quite suddenly, as for UBe_{13} and UPt_3 where this change is observed in the vicinity of 1 K, in both cases rather close to the superconducting transition temperature, or also as for $PuBe_{13}$ around 12 K. For UAl_2, U_2Zn_{17} and USn_3 more gradual changes in $\chi(T)$ seem to be observed. Further discussions of $\chi(T)$ in relation with results from other types of measurements will appear in sections 3 and 4.

2.3. *Electrical resistivity, magnetoresistance*

If we take again, as a starting point, the temperature dependence of the electrical resistivity ρ of the standard heavy-electron material, $CeAl_3$, and compare it with $\rho(T)$ that is observed for ordinary metals, we are again confronted with a very anomalous behaviour. At room temperature the electrical resistivity is fairly high, of the order of 10^2 $\mu\Omega$cm. Lowering the temperature results in an only small reduction of ρ which eventually passes through a minimum. With still decreasing temperature, ρ increases substantially, reaching values that exceed the room-temperature value considerably. At about 35 K, $\rho(T)$ passes through a maximum and then decreases with increasing positive slope to lower temperatures (Buschow et al. 1971). At very low temperatures, i.e. below 0.3 K, $\rho(T)$ of $CeAl_3$ can be cast in the form (Andres et al. 1975)

$$\rho(T)=\rho_0+AT^2, \tag{8}$$

where ρ_0 is the residual resistivity determined by impurities and imperfections in the sample and A is a numerical prefactor, whose significance will be discussed in section 3. Here we only mention its remarkably high value of 35 $\mu\Omega\text{cm}/\text{K}^2$. $\rho(T)$ of $CeAl_3$ was studied by various different workers (Buschow et al. 1971, Percheron et al. 1973, Edelstein et al. 1974, Andres et al. 1975, Ott et al. 1984a) and because it shows the characteristic features of $\rho(T)$ of heavy-electron materials particularly

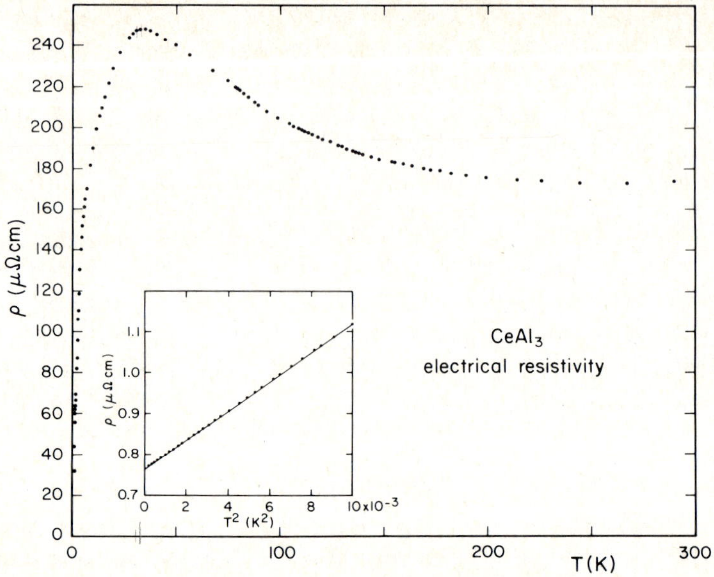

Fig. 36. Temperature dependence of the electrical resistivity of $CeAl_3$ between 1.5 and 300 K as measured by Ott et al. (1984a). The inset shows ρ versus T^2 below 0.3 K as measured by Andres et al. (1975). The solid line indicates the T^2 dependence of ρ in this temperature range.

well, we display it in fig. 36 with an inset that emphasizes its behaviour at very low temperatures.

It was quite early recognized that replacing Ce with non-magnetic substitutions (Buschow et al. 1971) or applying external pressure (Percheron et al. 1973) results in very drastic changes in $\rho(T)$ for $CeAl_3$. Less dramatic are changes induced by external magnetic fields, although a sign change of the magnetoresistance below 1 K was certainly an unexpected feature (Edelstein et al. 1974). In the following subsections we give a compilation of data on the electrical resistivity and its changes upon external influences that were obtained for heavy-electron actinide materials.

2.3.1. UBe_{13}

The temperature dependence of the electrical resistivity of both single-crystalline and polycrystalline UBe_{13} was first studied by Ott et al. (1983) between 0.5 and 300 K. In fig. 37 we show the results for $\rho(T)$ of a polycrystalline sample between 1.2 and 300 K. Although these measurements revealed slight variations in the absolute magnitude of the resistivity between different samples, which probably have to be traced back to differences in the quality and the chemical composition of different specimens, the general features that may be seen in fig. 37 are always the same. The steady increase of ρ with decreasing temperature saturates into a shoulder around 10 K. Below 5 K, $\rho(T)$ rises again rapidly, passes through a distinct maximum at 2.5 K and then decreases with increasing slope down to 1.2 K. This narrow-peak structure at low temperatures is observed in all samples investigated and the peak value at 2.5 K is of the order of 200 μΩcm.

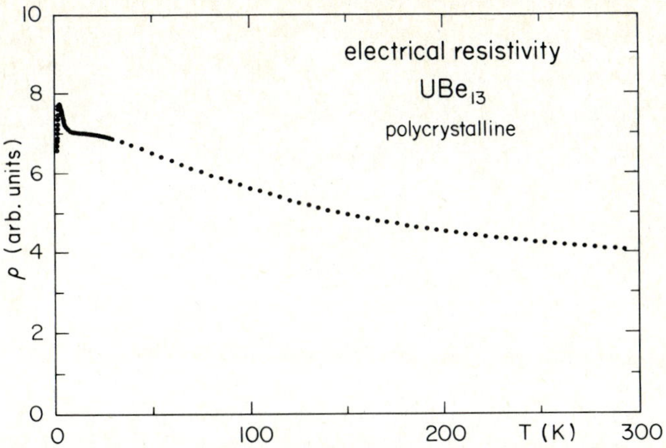

Fig. 37. $\rho(T)$ of UB_{13} between 1.5 and 300 K as measured by Ott et al. (1984d).

Smith et al. (1984a, 1985) investigated the changes in $\rho(T)$ that were caused by substituting U with non-magnetic impurities like Th, Lu and Sc. Particularly drastic is the effect of small amounts of Th atoms on the U sites. In fig. 38 it may be seen that at very low temperatures the electrical resistivity is actually lowered with increasing Th content, a feature that is quite opposite to what usually happens if

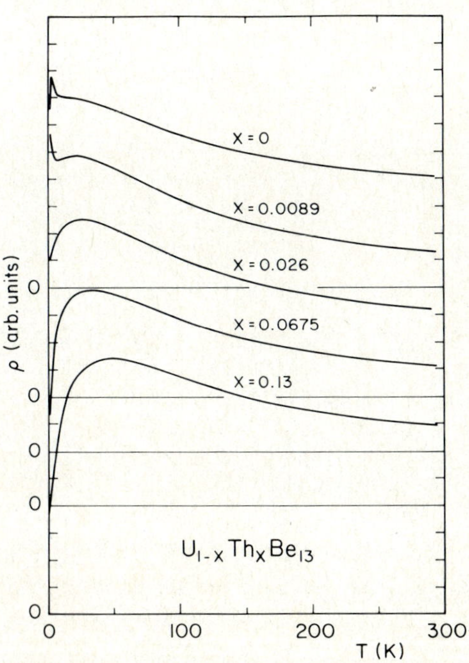

Fig. 38. Influence of Th substitutions for U on the low-temperature electrical resistivity of UBe_{13} as found by Smith et al. (1984a).

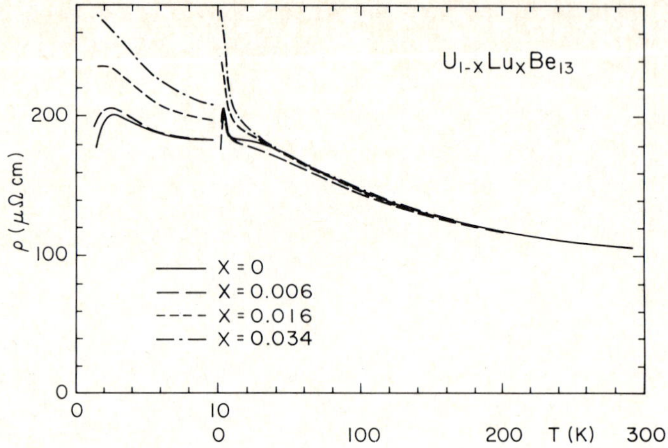

Fig. 39. Influence of Lu substitutions for U on the low-temperature electrical resistivity of UBe_{13} as found by Smith et al. (1985).

impurities are introduced into a metal, and a first indication that UBe_{13} is not an intrinsically dirty system. The shoulder-type structure observed around 10 K in pure UBe_{13} transforms into a maximum and, with increasing x, the maximum value of ρ is reached at higher temperatures. Lu and Sc substitutions, on the other hand, provoke quite a different effect. As an example we show $\rho(T)$ for $U_{1-x}Lu_xBe_{13}$ for various values of x in fig. 39. The increase below 5 K observed in pure UBe_{13} is enforced and for $x=0.034$, $\partial\rho/\partial T$ stays negative down to 20 mK.

Quite surprising was the discovery of a very large negative magneto-resistance of

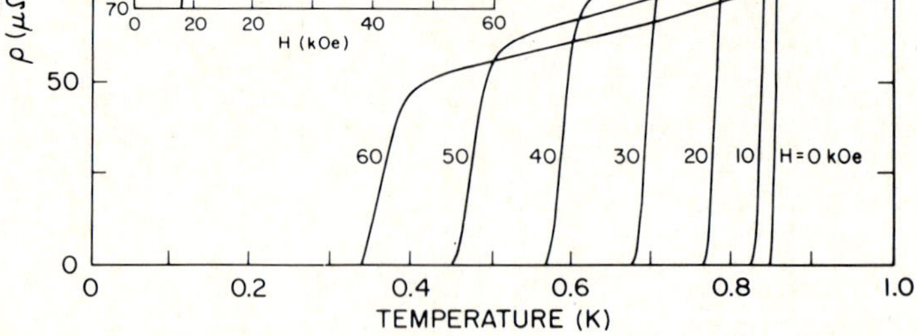

Fig. 40. Magnetic-field dependence of the electrical resistivity of UBe_{13} below 1 K as measured by Maple et al. (1985).

UBe_{13} at low temperatures. In fig. 40 we show data that were published by Maple et al. (1985) and resulted from measurements on a single crystal covering the temperature range from 0.15 to 1.2 K and magnetic fields up to 60 kOe. Although the continuation of $\rho(T)$ towards $T = 0$ K is interrupted by the transition to the superconducting state it seems clear that the residual resisitivity depends on the external magnetic field. This is again an indication that the high resistivity found at T_c in zero field is not primarily due to ordinary impurity scattering. More recently, both the temperature and the magnetic-field range of these measurements were extended. Remenyi et al. (1986), for example, measured $\rho(T)$ of a polycrystalline sample between 0.075 and 4.2 K in fields up to 120 kOe. For selected temperatures, with 0.48 K being the lowest, they measured the variation of ρ even up to 230 kOe. The overall result is a continuous decrease of ρ with decreasing temperature and increasing field. It was also concluded that for fields exceeding 50 kOe and temperatures below about 0.7 K, ρ could be described by eq. (8). Their fitting procedure, however, required both ρ_0 and A to be strongly magnetic-field dependent. Stewart et al. (1984d) studied the magnetoresistance of UBe_{13} between 1.5 and 15 K in fields up to 110 kOe. They found that the negative magnetoresistance persists up to 15 K but decreases in magnitude with increasing temperature. Stewart (1984) later noticed that the temperature dependence of $|\Delta R|/R$, where $\Delta R = R(H) - R(0)$, is very similar to that of C_p^{el}/T which is shown in fig. 2. Fisk (1984) pointed out that all data in this temperature and magnetic-field range fell essentially on one curve if $\Delta R/R$ is plotted versus $H/(T + T^*)$ with $T^* \sim 1$ K. Further data on the temperature and magnetic-field dependence of ρ of UBe_{13} were reported by Batlogg et al. (1985b) for temperatures between 0.4 and 30 K and in fields up to 230 kOe. We show a compilation of their results at low temperatures in fig. 41. Batlogg et al. (1987)

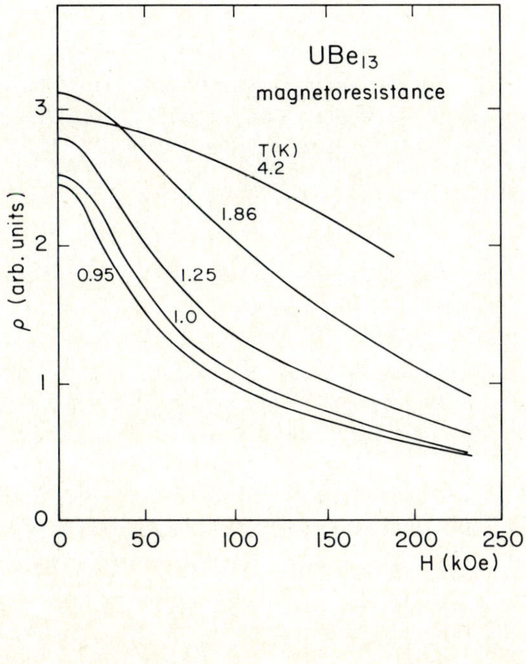

Fig. 41. Magnetic-field dependence of the electrical resistivity of UBe_{13} at various temperatures and in fields up to 230 kOe as measured by Batlogg et al. (1985b).

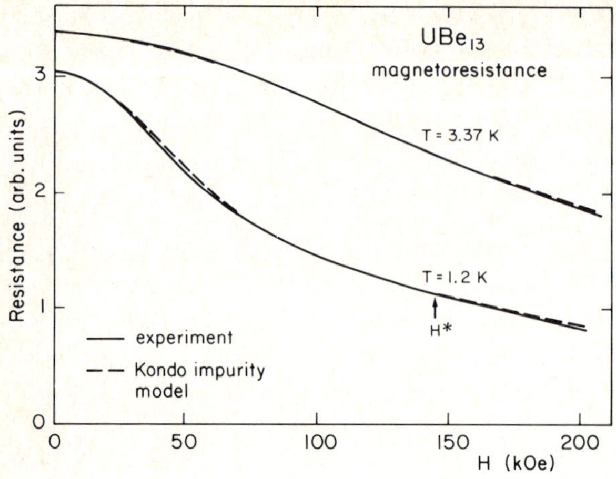

Fig. 42. Comparison of the experimental low-temperature magnetoresistance of UBe_{13} with a Kondo-impurity model calculation by Batlogg et al. (1987). H^* is a model parameter (see Schlottmann 1983).

were able to fit these magnetoresistance data with an $S=\frac{1}{2}$ Bethe-Ansatz solution (Schlottmann 1983) for $\rho(H, T)$ over a wide range of temperatures and magnetic fields. In fig. 42 we show two of these fits at different temperatures to data that are plotted as ρ versus H. These fits imply a temperature-dependent value for T^*, where T^* can also be expressed as an effective magnetic field H^* via the relation $H^*=(k_B/g\mu)T^*$, where μ is the moment calculated from χ. T^* is of the order of 100 K at 25 K and decreases to about 75 K at 10 K. At still lower temperatures, T^* decreases with increasing slope and eventually vanishes linearly with T for temperatures below 1 K. This is shown in the form of $H^*(T)$ in fig. 43. Alekseevskii et al. (1985) also reported magnetoresistance measurements in fields up to 120 kOe at temperatures above 1.5 K. With 120 kOe external field they observed resistivity reductions of 20 and 50% at 4.2 and 1.85 K, respectively.

Magnetoresistance measurements have also been made on a series of $U_{1-x}Th_xBe_{13}$ compounds for $0<x<0.06$ and in small magnetic fields of maximum 4 kOe only. Willis et al. (1985b) found, quite unexpectedly, that the magnetoresistance in these samples was either virtually zero or else positive in this low-field region, clearly different from what had been observed in pure UBe_{13}.

2.3.2. UPt_3, UAl_2

As for $\chi(T)$ discussed above it is to be expected that also the temperature dependence of the electrical resistivity of UPt_3 is anisotropic. This was indeed verified by measurements of $\rho(T)$ on single crystals of UPt_3 by de Visser et al. (1984b) both along and perpendicular to the hexagonal axis of the crystal lattice between 1.5 and 300 K. We show their results in fig. 44. They observed a marked anisotropy which persists over the whole temperature range covered. Within the basal plane the room-temperature value of ρ is 238 $\mu\Omega$ cm while along the hexagonal axis it is only 132 $\mu\Omega$ cm. The main decrease of ρ with decreasing temperature occurs below 50 K both for $\rho_{\parallel}$ and $\rho_{\perp}$. Wire et al. (1984) extended the measurements of $\rho(T)$ of polycrystalline UPt_3 up to 1150 K and the result is shown in fig. 45, together with

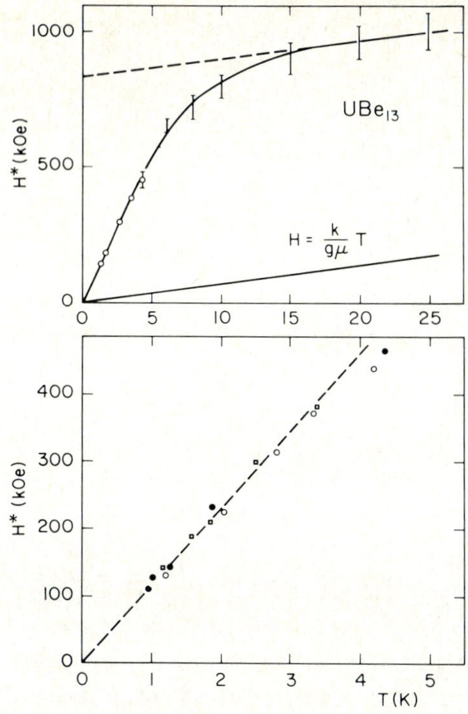

Fig. 43. Temperature dependence of $H^* = k_B/g\mu T^*$ obtained by fitting experimental $\rho(H)$ data to Schlottmann's Bethe-Ansatz calculations for $S=1/2$ by Batlogg et al. (1986b). The lower part of the figure demonstrates how H^* vanishes with decreasing temperature below 5 K.

similar data for UAl_2. It may be seen that $\rho(T)$ of UPt_3 reaches a maximum around 800 K and actually decreases with further increasing temperature. Of interest is also the temperature dependence of ρ at very low temperatures. Different authors claim a T^2-dependence for both $\rho_{\parallel}$ and $\rho_{\perp}$, although there are some discrepancies with regard to the temperature range in which eq. (8) is believed to be valid. de Visser et al. (1984b) showed quite convincingly that eq. (8) is adequate only below about 2 K,

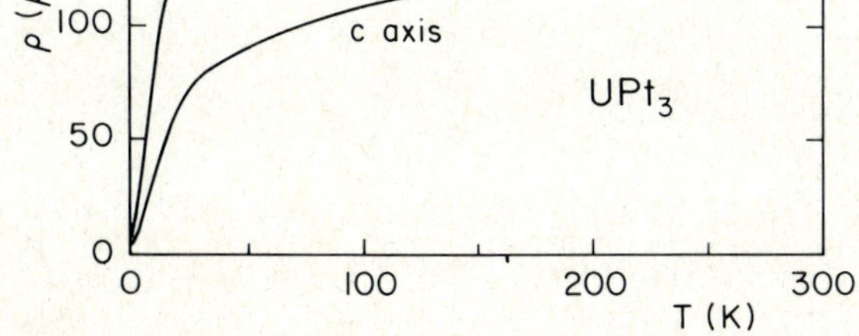

Fig. 44. Temperature dependence and anisotropy of the electrical resistivity of UPt_3 below 300 K as obtained by de Visser et al. (1984b).

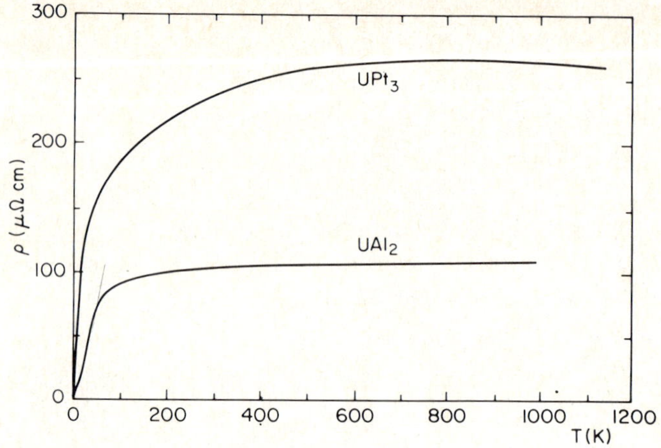

Fig. 45. Comparison of the electrical resistivities of polycrystalline samples of UPt_3 and UAl_2 below 1000 K by Wire et al. (1984).

as demonstrated by their double-logarithmic plot of $\rho_{\parallel}$ and $\rho_{\perp}$ at low temperatures (see fig. 46). The anisotropy of ρ is obviously also reflected in the prefactor A which is $1.6\,\mu\Omega\,cm/K^2$ for $\rho_{\perp}$ and $0.7\,\mu\Omega\,cm/K^2$ for $\rho_{\parallel}$. These values are an order of magnitude smaller than that for $CeAl_3$, mentioned above.

In view of the results of the low-temperature susceptibilities and their magnetic-field dependence, some anomalies in the magnetoresistance had to be expected.

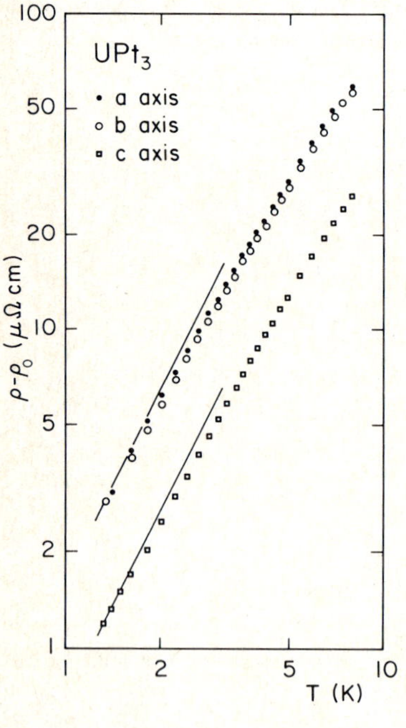

Fig. 46. Logarithmic plot of the temperature-dependent part of the electrical resistivity of UPt_3 along different crystallographic directions between 1.5 and 10 K. The solid lines indicate the expected slope for a T^2 dependence. (Data are from de Visser et al. 1984b.)

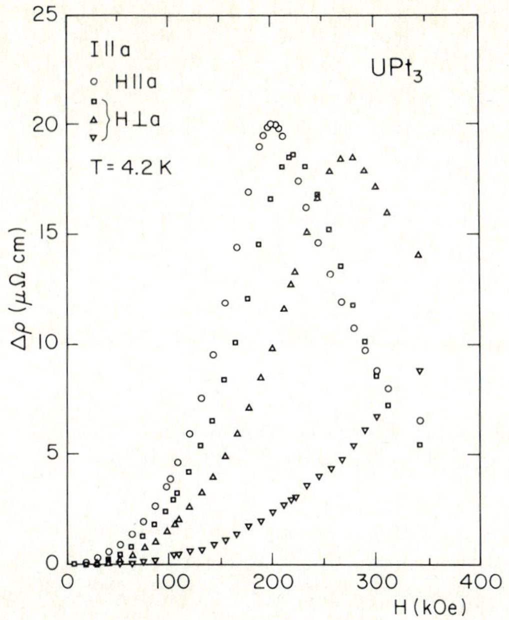

Fig. 47. Magnetoresistance of UPt_3 in fields up to 350 kOe along the *a*-axis of the basal plane with field directions parallel and perpendicular to it at 4.2 K. (Data are from de Visser et al. 1986b.)

What was unexpected, however, was the observation that the low-temperature electrical resistivity increases considerably with increasing magnetic field. de Visser et al. (1986b) and also Franse et al. (1984) demonstrated that this is true for all possible combinations of current- and magnetic-field directions. As an example, we show the low-temperature magnetoresistivity $\Delta\rho=\rho(H)-\rho(0)$ for $\Delta\rho_\perp$ but with H applied either parallel to the current direction or in various directions in a plane perpendicular to it in fig. 47. Similar curves were published for $\Delta\rho_\parallel$. The main result is that $\Delta\rho_\parallel$ and $\Delta\rho_\perp$ show a maximum at 210 kOe whenever the vector of the magnetic-field direction has a non-vanishing component in the basal plane. At higher temperatures, 20 K or above, the magnetoresistivity generally has a negative sign but its magnitude decreases with increasing temperature.

In their investigation of the anisotropy of the electrical resistivity of UPt_3, Wire et al. (1984) also studied the effect of hydrostatic pressure on both $\rho_\parallel$ and $\rho_\perp$ between 1.5 and 300 K and with pressures up to 17 kbar. They noted that, quite generally, ρ decreases with increasing pressure in this whole temperature range. Of particular interest are, of course, the pressure-induced changes of the low-temperature electrical resistivity. Wire et al. note that the decrease of ρ with increasing pressure is accompanied by an extension of the temperature range in which eq. (8) seems to be a good approximation. de Visser et al. (1984b) also mentioned the decrease of A with increasing external pressure. Similar measurements were later extended to temperatures below 1 K. Willis et al. (1985a) report results which cover the temperature range from 0.1 to 2 K and pressures up to 19 kbar. We show these data in fig. 48 which indicate that whereas the prefactor A in eq. (8) changes considerably with pressure, ρ_0 does not.

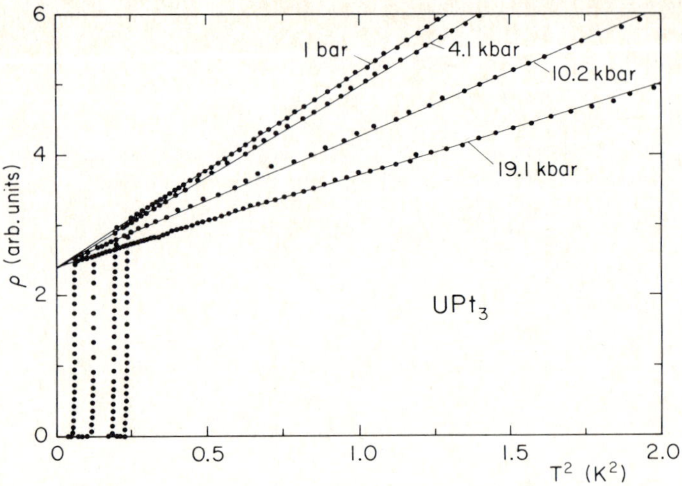

Fig. 48. Pressure dependence of the low-temperature electrical resistivity of UPt_3 as measured by Willis et al. (1985a). Solid lines are to guide the eye.

First measurements of $\rho(T)$ of UAl_2 were reported by Buschow and van Daal (1972) with a room-temperature value of ρ of about 120 $\mu\Omega$ cm and a residual-resistance ratio of 10. Again the main decrease of ρ with decreasing temperature is observed below 50 K. From this work it was concluded that also for UAl_2, eq. (8) was quite suitable to describe $\rho(T)$ at very low temperatures, i.e., below 3.4 K in this case. Arko et al. (1972) published $\rho(T)$ data for UAl_2, obtained on a polycrystalline sample of obviously lesser quality with a room-temperature resistivity of 140 $\mu\Omega$ cm and a resistance ratio of only about 3, and also observed a T^2-dependence of ρ below 2 K.

Wire et al. (1984) have extended these measurements of ρ of UAl_2 between 1.5 and 1000 K. We show the result, together with $\rho(T)$ of UPt_3 in fig. 45. From this graph it may be stated that ρ is virtually constant for UAl_2 at temperatures above 400 K. Again, most attention was paid to the low-temperature region where the enhancement effects in the specific heat and the magnetic susceptibility are observed. Wire et al. (1984) find that the temperature dependence of the as-measured resistivity is never quadratic in T. They argue, however, that in systems with the distinct saturation effect in $\rho(T)$ as observed here, the total resistivity can be described as being due to two resistances that are coupled in parallel and one of them is essentially the saturation resistivity. Within this model they calculate the low-temperature part of the second resistance and find that this varies as T^2 over an appreciable temperature range, from 1.5 up to more than 30 K.

In their early work, Buschow and van Daal (1972) were already interested in changes of $\rho(T)$ upon replacing U with non-magnetic atoms and they showed that substituting Y for U in UAl_2 changes both the magnitude and the temperature dependence of ρ. With increasing Y-concentration ρ rises steadily and looses the low-temperature drop observed in pure UAl_2. When almost two-third of the U

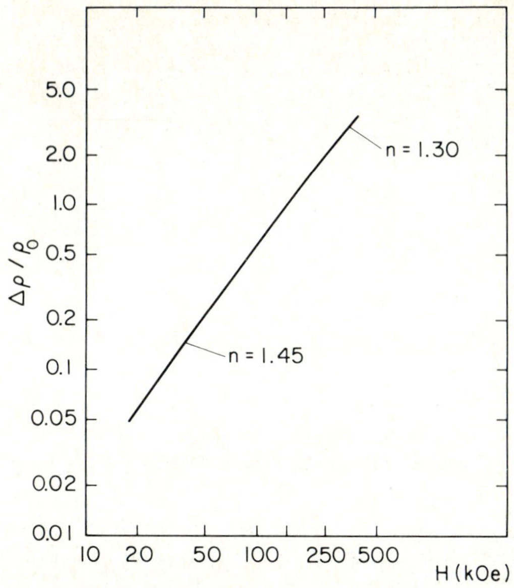

Fig. 49. Low-temperature magnetoresistance of UAl_2 in fields up to 350 kOe. (Data were taken at 4.2 K by Franse et al. 1982.)

atoms are replaced by Y, ρ is almost constant in temperature with a small negative $\partial\rho/\partial T$ and a low-temperature resistivity of about 150 $\mu\Omega$ cm. With even further dilution, $\rho(T)$ changes towards Kondo-type features with a positive $\partial\rho/\partial T$ at high temperatures, a minimum of $\rho(T)$ at an intermediate temperature and a subsequent rise of ρ towards $T=0$ K.

Not much detailed information is available on magnetoresistive effects in UAl_2. Franse et al. (1982) show the relative magnetoresistivity versus field at 1.2 K on a double logarithmic plot as shown in fig. 49. They claim a power-law behaviour $\Delta\rho/\rho \sim H^n$, with $n=1.45$ below 150 kOe and 1.3 in larger fields. That was observed along different crystallographic directions and therefore should not be due to band-structure effects. They argue that from these data substantial suppression (40%) of the spin-fluctuation-induced resistivity in a field of 100 kOe can be derived.

A more detailed study has been made on pressure-induced changes of the resistivity of UAl_2. As for UPt_3, hydrostatic pressure reduces ρ as reported by Wire et al. (1984) at all temperatures between 1.5 and 300 K. Again of special interest was the low-temperature portion of $\rho(T)$ and we show the results in fig. 50. As in UPt_3, increasing pressure sizeably reduces the parameter A of eq. (8) but leaves ρ_0 virtually unchanged.

2.3.3. U_2Zn_{17}

The temperature dependence of the electrical resistivity of U_2Zn_{17} between 1.3 and 300 K is shown in fig. 51 (Ott et al. 1984b). Again, the characteristic increase of ρ with decreasing temperature is observed below room temperature. As well as χ, ρ also reaches a maximum at 17 K and then slowly decreases towards $T_N=9.7$ K. At the antiferromagnetic transition, $\partial\rho/\partial T$ increases almost discontinuously, and be-

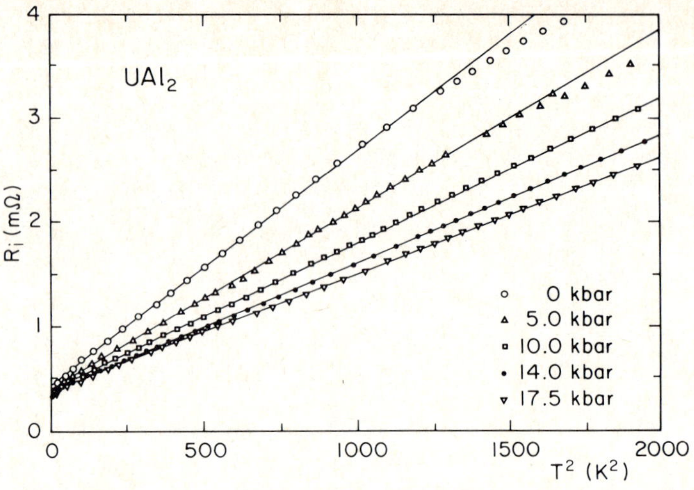

Fig. 50. Pressure dependence of the low-temperature electrical resistivity of UAl_2 as measured by Wire et al. (1984). The solid lines are to guide the eye. The procedure to extract the T^2 part is described in the text.

tween 9.7 and 1.3 K the resistivity decreases by more than an order of magnitude. Below 4 K, the temperature dependence of ρ follows very closely a power law T^n, where $n=2.5$, with some deviations, however, around 1.5 K. The impurity-induced changes of the low-temperature resistivity are just as drastic as those observed for the specific heat and the magnetic susceptibility (Willis et al. 1986). Impurities on both the U and the Zn sites destroy the low-resistance state that is found in pure

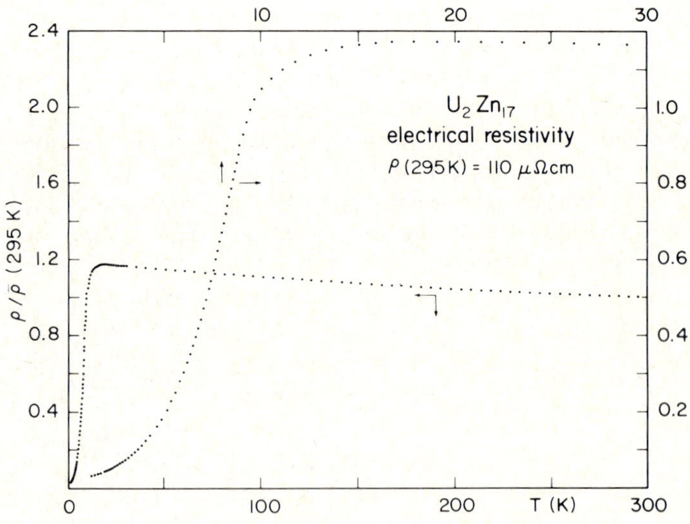

Fig. 51. Temperature dependence of the electrical resistivity of U_2Zn_{17} between 1.2 and 300 K. Note the different scales for different temperature intervals. (Data are from Ott et al. 1984b.)

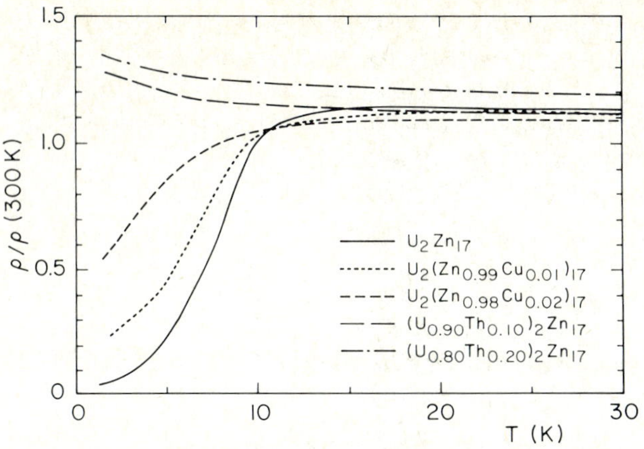

Fig. 52. Influence of different impurities on the low-temperature electrical resistivity of U_2Zn_{17} as found by Willis et al. (1986).

U_2Zn_{17}, as may be seen in fig. 52. So far the most effective scatterer in this respect appears to be Mn (not shown in fig. 52).

An investigation of the influence of magnetic fields on the low-temperature electrical resistivity of U_2Zn_{17} was made by Batlogg et al. (1985b) who measured ρ between 0.4 and 30 K and in fields up to 230 kOe. We show their results of $\rho(T)$ for $H=0$ and 200 kOe in fig. 53. Compared to UBe_{13}, the changes of $\rho(T)$ of U_2Zn_{17} in external magnetic fields are fairly small. We only note that below T_N, ρ increases with increasing H while above T_N, just the opposite is true. The sign change is actually observed at about 2 K above $T_N(0)$. With respect to the phase transition it may be stated that the shift of T_N with increasing H is surprisingly small.

The effect of pressure on the low-temperature resistivity of U_2Zn_{17} is not large in magnitude but $\partial\rho/\partial p$ changes sign around 14 K. Thompson et al. (1986) measured $\rho(T)$ between 1.2 and 300 K at pressures up to 17 kbar. While increasing pressure enhances the resistivity above 14 K, ρ decreases below that temperature. Below 5 K, pressure-induced changes on ρ are very small but a careful analysis reveals a distinct

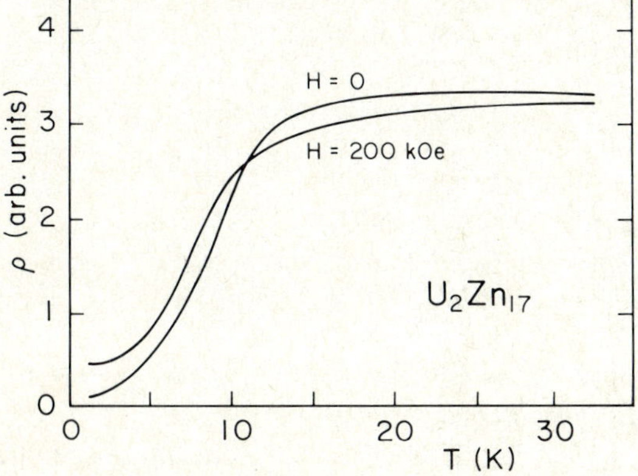

Fig. 53. Low-temperature electrical resistivity of U_2Zn_{17} in zero and 200 kOe magnetic field as measured by Batlogg et al. (1985b).

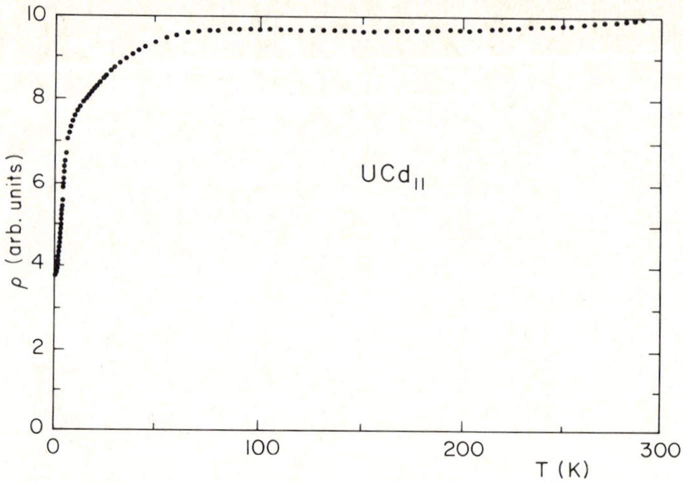

Fig. 54. Temperature dependence of the electrical resistivity of UCd_{11} between 1.5 and 300 K as measured by Fisk et al. (1984).

decrease of the prefactor of the $T^{2.5}$ term established in the zero-pressure data. The deviations from the power-law behaviour below 1.5 K at zero pressure also disappear with increasing pressure.

2.3.4. *UCd_{11}, $NpSn_3$*

Results for the temperature dependence of the electrical resistivity of UCd_{11} were first reported by Fisk et al. (1984) for temperatures between 1.5 and 300 K, as shown in fig. 54. $\rho(T)$ decreases very little with decreasing temperature below 300 K, passes first through a minimum at 170 K and then over a very broad maximum somewhat below 100 K. At still lower temperatures ρ decreases with increasing slope and around 5 K, a distinct break in the $\rho(T)$ curve indicates the occurrence of a phase transition.

While studies of the influence of impurities or external magnetic fields on $\rho(T)$ of UCd_{11} have not yet been reported, Thompson et al. (1986) investigated the changes of $\rho(T)$ that are brought about by application of external pressure up to 17 kbar in the temperature range between 1.2 and 300 K. In contrast to U_2Zn_{17}, $\rho(T)$ of UCd_{11} is modified considerably even with these moderate pressures (see fig. 55). Increasing pressure enhances ρ in the whole temperature range. First, we note that a maximum pressure of 17 kbar enhances ρ at room temperature by about 20% while at helium temperatures this increase amounts to about 70%. At 17 kbar pressure the minimum of $\rho(T)$, which at zero pressure is observed around 170 K, is shifted to above 200 K, and the maximum, which is located at 84 K at ambient pressure, gets more prominent and is observed at 57 K. At this point we should briefly mention that pressure also seems to induce two additional phase transitions at low temperatures but we shall discuss this aspect in more detail in section 4.

The temperature dependence of ρ for $NpSn_3$ was described by Trainor et al.

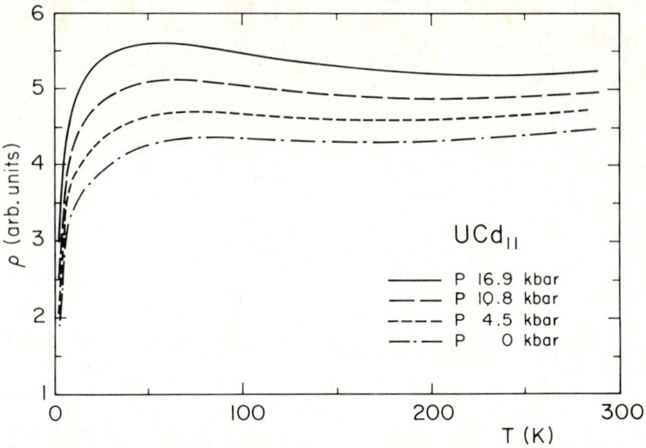

Fig. 55. Influence of external pressure on the electrical resistivity of UCd_{11} between 1.5 and 300 K as measured by Thompson et al. (1986).

(1976). The magnetic ordering is indicated by a maximum in $\partial\rho/\partial T$ rather than in $\rho(T)$ itself, which is a feature of itinerant antiferromagnets like Cr and, as will be shown below, UCu_5. Above 100 K, ρ varies with a strong negative curvature, resulting in a maximum at 175 K and a slow decrease towards room temperature. The residual resistivity at the lowest temperatures is reported to be 3 $\mu\Omega$ cm. The maximum value at 175 K is 125 $\mu\Omega$ cm and the room-temperature value of ρ is 115 $\mu\Omega$ cm.

2.3.5. *$NpBe_{13}$, $PuBe_{13}$*

$\rho(T)$ data for these two compounds between 1.5 and 300 K were published by Stewart et al. (1984a) and are shown in fig. 56. Both data sets reveal the characteristic features that we have discussed in the previous subsections. For $NpBe_{13}$, $\rho(T)$ increases steadily with decreasing temperature below 300 K until a sudden sign change in $\partial\rho/\partial T$ indicates the mentioned phase transition. The slope of $\rho(T)$ close to 300 K is indicative of a minimum in $\rho(T)$ at around that temperature.

For $PuBe_{13}$, $\rho(T)$ is very similar in character to that of $CeAl_3$ without, however, the dramatic drop of more than two orders of magnitude at low temperatures. The minimum of $\rho(T)$ is observed close to 160 K. Below that temperature, ρ increases with progressively increasing negative slope, reaches a rounded maximum that is centered around 13 K and subsequently decreases with still decreasing temperature. The resistivity at the lowest temperatures is, however, still higher than the minimum value at 160 K.

Some data are available on changes of $\rho(T)$ of $NpBe_{13}$ when Np is replaced by U. Stewart et al. (1984a) measured two compositions of $Np_{1-x}U_xBe_{13}$ with $x=0.15$ and $x=0.32$. The conclusion is that $\rho(T)$ changes very little even for these comparatively large values of x. For $x=0.32$, the drop in resistivity that is observed below 3.5 K for $NpBe_{13}$ is no longer observed; $\partial\rho/\partial T$ stays negative down to 1.5 K.

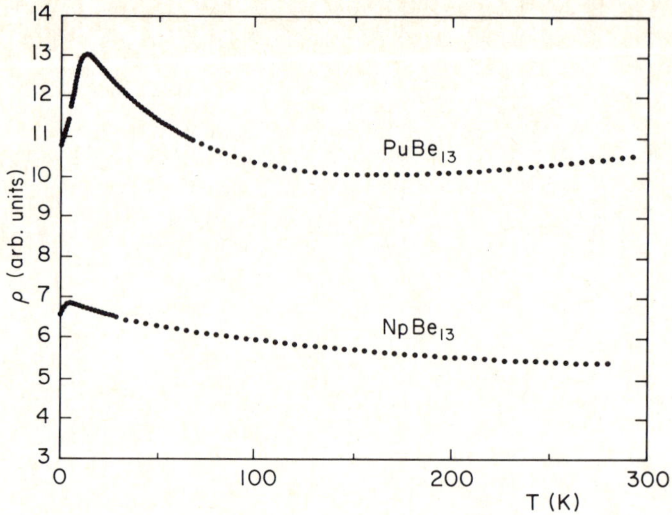

Fig. 56. Temperature dependence of the electrical resistivities of $NpBe_{13}$ and $PuBe_{13}$ between 1.5 and 300 K as reported by Stewart et al. (1984a).

2.3.6. *UCu_5*

$\rho(T)$ measurements on UCu_5 were first published by Brodsky and Bridger (1973) for temperatures between 1.5 and 300 K and they noticed a strong dependence of $\rho(T)$ upon annealing the specimens. They found that annealing lowers the resistivity at all temperatures and also reduces the anomalies that are associated with the phase transition at 15 K. Later measurements by van Daal et al. (1975) confirmed this sensitivity of $\rho(T)$ to annealing procedures but, in fact, they obtained just the opposite effects. Their annealing procedure resulted in an increase of $\rho(T)$ and an enforcement of the structure provoked by the magnetic phase transition and they discussed some of the reasons that might have led to diverging results. In more recent work, Ott et al. (1985c) repeated these measurements and their results for $\rho(T)$ of UCu_5 between 1.0 and 300 K are shown in fig. 57. Before we discuss the data in more detail we should like to mention that preparatory work by Fisk (1985) gave a clue for the reason of the two aforementioned opposite effects with respect to annealing. It is most likely that the annealing temperature of 1210 K chosen by Brodsky and Bridger (1973) was too high, leading to the formation of free Cu in the sample and a Cu deficiency of the compound. This might explain both the reduction of the total resistivity and the diminuation of the phase transition effects.

Regardless of sample preparation it was found that the room-temperature value of ρ is rather high, between 200 and 300 $\mu\Omega$ cm. The data by van Daal et al. (1975) are more or less identical with those shown in fig. 57 except that Ott et al. (1985c) observe a stronger decrease of ρ towards the lowest temperatures. While van Daal et al. (1975) also report the anomaly below 15 K that is, as mentioned above, characteristic for some itinerant antiferromagnets, they missed the observation of

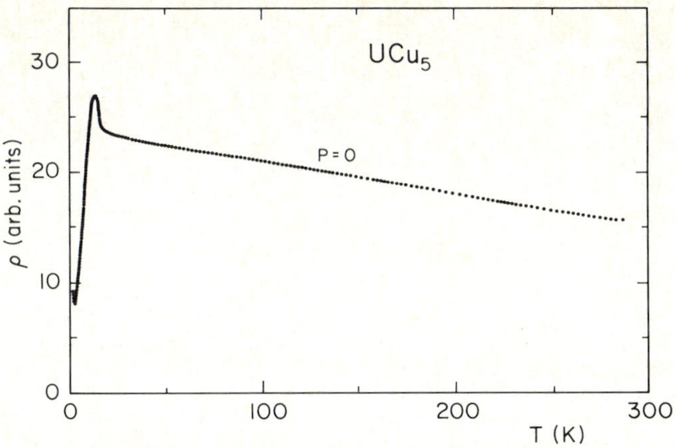

Fig. 57. Temperature dependence of the electrical resistivity of UCu_5 between 1.2 and 300 K. ρ(295 K) = 268 μΩ cm. (Data are due to Ott et al. 1985c.)

the minimum that appears at 1.6 K (see fig. 57). The subsequent increase of ρ below that temperature is associated with another phase transition that will be discussed in detail in section 4.

van Daal et al. (1975) already noted that $\rho(T)$ is not only susceptible to annealing but also to chemical changes. Both Cu-deficiency and replacement of Cu by Ni lead to sizeable changes in $\rho(T)$. Ten percent replacement of Ni for Cu remove the anomaly indicating the magnetic transition at 15 K and also the resistivity decrease below 10 K. ρ at helium temperatures is then as high as 400 $\mu\Omega$ cm. Later measurements by Ott et al. (1985c) showed that replacing Cu by Ag instead of Ni is much less harmful and $\rho(T)$ retains all its features in the temperature range above 2 K for as much as 20% Ag for Cu. For this composition $UAgCu_4$, however, the minimum at 1.6 K is no longer observed, $\partial\rho/\partial T$ is positive down to 1.1 K.

Magnetoresistance measurements on UCu_5 were carried out by Batlogg et al. (1985c). The magnetoresistance of UCu_5 is drastically influenced by the phase transition close to 1 K. Since we will discuss this transition in section 4, we restrict the presentation of $\rho(T, H)$ data to temperatures above 1.5 K. A selection of $\rho(H)$ curves in fields up to 230 kOe is shown in fig. 58. At low fields, $\Delta\rho(H)$ is positive but small. The pronounced maxima that are obviously shifted to higher field values with increasing temperature are still connected with the mentioned phase transition and the negative slope of the $\rho(H)$ curves is characteristic for the low-temperature phase. In the ordinary magnetically-ordered phase below 15 K, $\Delta\rho(H)$ is positive in low fields.

Pressure-induced changes of $\rho(T)$ were studied on both UCu_5 and $UAgCu_4$ by Thompson et al. (1986). Since $\rho(T)$ is not very different for both materials, it is not surprising that also the pressure effects are almost identical and not very large in magnitude. At room-temperature, increasing pressures enhances $\rho(T)$, the effect being reduced by lowering the temperature. Since pressure enhances the magnetic

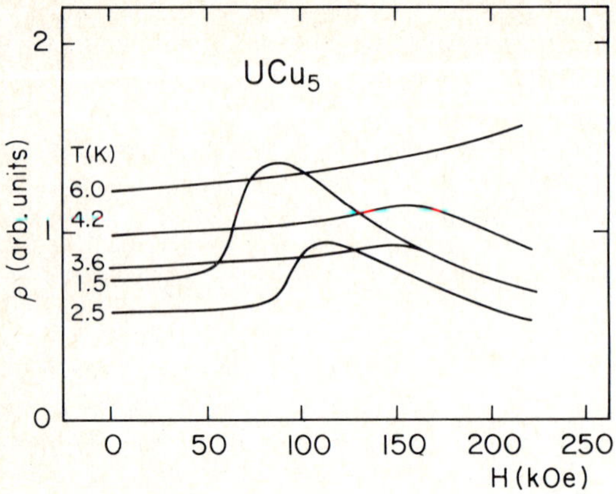

Fig. 58. Influence of magnetic fields on the low-temperature electrical resistivity of UCu_5 in fields up to 230 kOe, as measured by Batlogg et al. (1985c).

ordering temperature, again $\partial\rho/\partial p < 0$ for $T < T_N$. The most distinct effect of external pressure on $\rho(T)$ of UCu_5 is observed below 4 K. We show the $\rho(T)$ curves between 1 and 4 K measured at four different pressures in fig. 59. The shift of the $\rho(T)$-minimum to lower temperatures at these moderate pressures is quite remarkable.

2.3.7. *$U(Sn_{1-x}In_x)_3$*

Results for $\rho(T)$ of USn_3 are available from the work by Buschow and van Daal (1972) and, more recently, from the work by Lin et al. (1986), for temperatures between 1.5 and 300 K. They are shown in fig. 60. It is clear that $\rho(T)$ for USn_3 is different from those we have been discussing above. The room-temperature resistivity is considerably smaller than 100 $\mu\Omega$ cm and $\partial\rho/\partial T$ is positive in the whole temperature range. This is not really surprising because we do not, as should have become clear in the previous sections, consider pure USn_3 as a heavy-electron

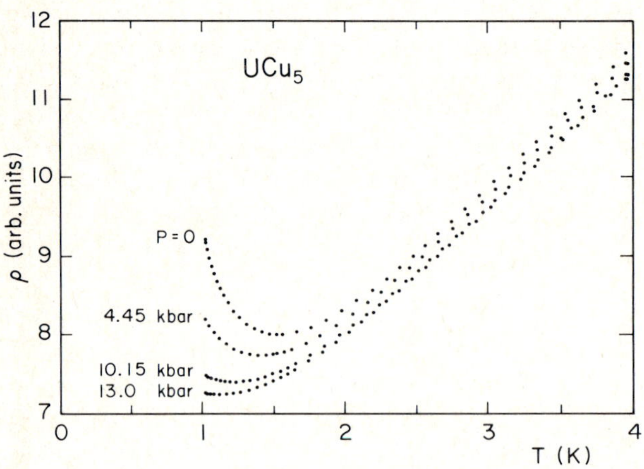

Fig. 59. Influence of external pressure on the low-temperature electrical resistivity of UCu_5 as measured by Thompson et al. (1986).

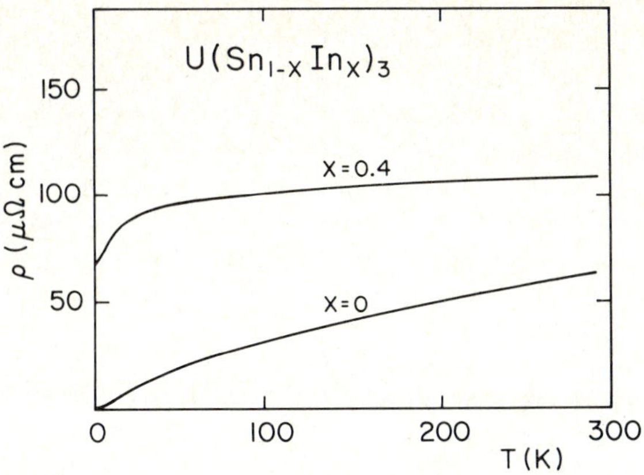

Fig. 60. Comparison of the temperature dependence of the electrical resistivities of USn_3 and $U(In_{0.4}Sn_{0.6})_3$ between 1.5 and 300 K by Lin et al. (1986).

system. Buschow and van Daal (1972) claim a T^2-dependence of ρ below 10 K and a linear increase of ρ with T at temperatures above 40 K. As we demonstrated above, the recent studies by Lin et al. (1986) on $U(Sn_{1-x}In_x)_3$ showed for $x = 0.4$ a strong enhancement of the low-temperature specific heat and magnetic susceptibility. As may be seen from fig. 60 the electrical resistivity $\rho(T)$ for a sample of this composition also approaches the now familiar features of $\rho(T)$ of these materials, indicating a strong enhancement of the low-temperature electrical resistivity as well.

The temperature dependences of the electrical resistivity of these heavy-electron materials clearly share certain common features. Considering that the materials we are discussing are ordered intermetallic compounds, their room-temperature values of ρ are rather high, 100 $\mu\Omega$ cm or more. In some compounds, the resistivity increases even further with decreasing temperature, in others, $\partial\rho/\partial T$ is positive but very small and may turn negative at lower temperatures. For most of the compounds a local minimum in $\rho(T)$ is observed at temperatures above 150 K. For all of the pure binary compounds, $\partial\rho/\partial T$ eventually becomes positive at low enough temperatures requiring, of course, that $\rho(T)$ passes over a maximum. These maximum temperatures may exceed 300 K, as in UPt_3 or be as low as 2.5 K, as in UBe_{13}. The decrease of ρ towards $T=0$ K is not of the same magnitude in all compounds, however, and it has been found to be very sensitive to external influences like magnetic field or external pressures and particularly also to small changes in the chemical composition of the materials.

2.4. *Other transport properties (thermal conductivity, thermoelectric power, Hall effect, ultrasound attenuation)*

Experimental data on transport properties other than electrical resistivity are not abundant for this class of materials. Where available we present them again in subsections devoted to particular substances.

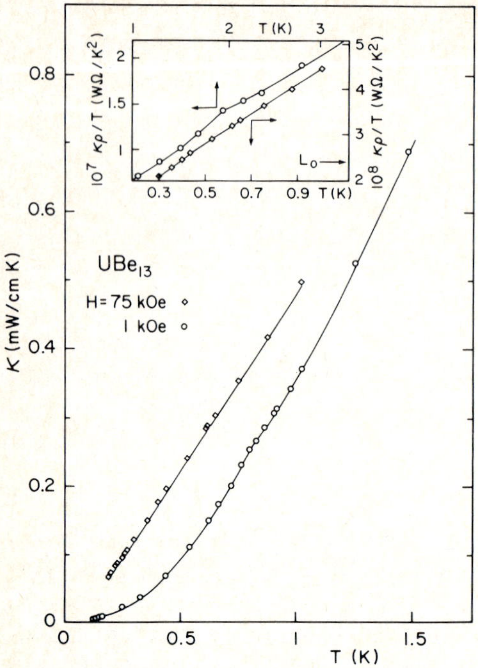

Fig. 61. Temperature dependence of the thermal conductivity of UBe_{13} between 0.15 and 1.5 K in external magnetic fields of 1 and 75 kOe. Data are due to Jaccard et al. (1985a). The inset shows the temperature dependence of the generalized Lorenz number $\kappa\rho/T$.

2.4.1. UBe_{13}

The thermal conductivity κ and the thermoelectric power S of UBe_{13} have been measured by Jaccard et al. (1985a) at temperatures between 0.1 and 3.0 K. Included in this work was a study of the influence of an external magnetic field H on these properties with values of H up to 75 kOe. The main purpose in applying magnetic fields was, of course, to suppress superconductivity and to measure these properties in the normal state to as low temperatures as possible. Considering the sizeable magnetoresistance effects in UBe_{13} shown in section 2.3, it was quite natural to check whether they are somehow reflected in other transport properties as well. In fig. 61 we show their results for κ in magnetic fields of 1 and 75 kOe. Since 1 kOe (this field had to be applied for technical reasons) hardly affects the superconducting state of UBe_{13} at all, these data can be regarded as effective zero-field measurements. The distinct kink in the low-field curve at about 0.8 K indicates the transition to the superconducting state. We shall discuss the $\kappa(T)$ curve below that temperature in section 4, devoted to phase transition phenomena. The high-field curve shows that around 1 K the normal-state thermal conductivity is enhanced by about 50% in a field of 75 kOe. Although on this κ versus T plot the data lie on a straight line, it may be seen that it does not extrapolate through the origin and therefore a simple assignment of a temperature dependence of $\kappa(T)$ is not possible. The inset of fig. 61 shows quite clearly that the resulting Lorenz number $\kappa\rho/T$ from $\kappa(T)$ and $\rho(T)$ measurements in the normal state of UBe_{13} is temperature dependent and mostly exceeds the classical value $L_0 = 2.45 \times 10^{-8}\ \mathrm{W\Omega/K^2}$. This indicates that both $\kappa(T)$

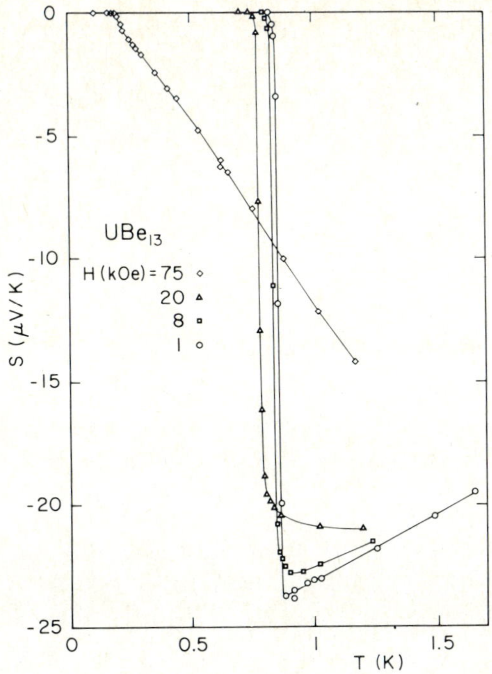

Fig. 62. Temperature dependence of the thermoelectric power S of UBe_{13} between 0.15 and 1.5 K in various external magnetic fields. (Data are due to Jaccard et al. 1985a.)

and $\rho(T)$ cannot be described by simple scattering of quasiparticles in a Fermi system even at these low temperatures.

Figure 62 shows the temperature dependence of the thermoelectric power S between 0.1 and 1.5 K and for various values of an external field. As expected the superconducting transition manifests itself by a sudden drop of S to zero. In the low-temperature normal state of UBe_{13}, S is obviously quite large and negative, and in zero or low magnetic field still increases its absolute value at temperatures around 1 K. For $H = 1$ kOe, S is about $-11\ \mu V/K$ at 4 K. In a field of 8 kOe there is some indication that S passes through a minimum that is located just at about the respective superconducting transition temperature. With still increasing field this minimum is clearly shifted to higher temperatures and for 75 kOe it may even be anticipated that $S(T)$ would change sign at about 0.15 K if the superconducting state would not interfere. It would be of interest to extend this kind of measurements to higher fields and lower temperatures in order to study the ultimate approach of $S(T)$ to $T = 0$ K in the normal state.

Results from Hall-effect measurements on UBe_{13} are available from Alekseevskii et al. (1984) and from Penney et al. (1986). The former authors mainly studied the field dependence of the Hall constant $R_H = V_H/(IH)$ where V_H is the Hall voltage, I is the measuring current and H the external magnetic field, at low temperatures between 1.9 and 100 K and in fields up to 140 kOe. They noted a general increase of R_H in small fields with decreasing temperature which in their case amounts to a factor of 10 between 100 and 4.2 K. Below 4 K they observed an appreciable field

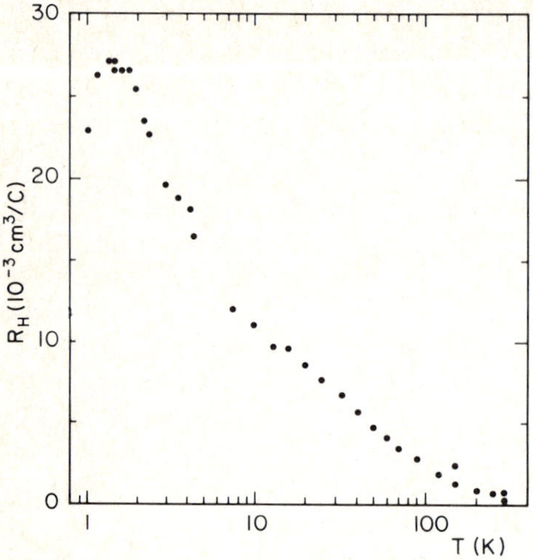

Fig. 63. Temperature dependence of the Hall coefficient R_H of UBe_{13} between 1 and 300 K as measured by Penney et al. (1986).

dependence of the Hall voltage in fields exceeding 60 kOe, resulting in maxima of R_H in fields of the order of 100 kOe and a subsequent decrease of R_H with still increasing field at temperatures below 3 K. Using the classical formula for the Hall coefficient,

$$R_H = \frac{1}{nec}, \tag{9}$$

where n is the charge-carrier density, e is the electron charge and c the velocity of light, they calculate the charge carrier density at 100 K and at 4.2 K, and obtain $4.3 \times 10^{21}\,cm^{-3}$ and $4 \times 10^{20}\,cm^{-3}$, respectively.

Penney et al. (1986) measured R_H in small fields as a function of temperature between the superconducting transition temperature T_c and 300 K. Their result is shown in fig. 63 on a logarithmic temperature scale. The Hall constant is small and positive at room temperature and increases with decreasing temperature by a factor of more than 20 down to liquid-helium temperatures, a much more pronounced increase than is observed in $\rho(T)$ (see section 2.3.1). At 4 K the rate of increase is sharply enhanced and R_H peaks at 1.5 K, distinctly below the temperature where the maximum of $\rho(T)$ is observed and subsequently decreases by about 20% before superconductivity sets in. The maximum value of R_H at 1.5 K corresponds to $n = 2.3 \times 10^{20}\,cm^{-3}$ if interpreted in the same way as discussed above. Within experimental uncertainties, the two sets of data agree reasonably well.

Ultrasound attenuation in UBe_{13} and $U_{1-x}Th_xBe_{13}$ was studied by Golding et al. (1985a,b) and by Batlogg et al. (1985a), mainly in order to obtain information on the superconducting state but also to probe parameters pertinent to the normal heavy-electron state. For a rough analysis the attenuation coefficient in the normal state due to conduction electrons was approximated by (Batlogg et al. 1986)

$$\alpha_n = \frac{4}{5\pi}\left[\frac{(m^* E_1)^2 v}{2\rho v^2 \hbar^3}\right] q l_e. \tag{10}$$

Equation (10) is valid for longitudinal sound waves in the hydrodynamic limit, hence $ql_e \ll 1$. m^* is the effective mass of the itinerant electrons, E_1 the deformation potential, ν the sound frequency, v the sound velocity, ρ the density, q the phonon wave number and l_e the electronic mean free path. Equation (10) implies that the attenuation in the normal state is proportional to l_e or, in other words, the electrical conductivity. Since $q=2\pi\nu/v$, α_n should also be proportional to ν^2. Since at T_c the electrical resistivity of UBe_{13} exceeds $100\,\mu\Omega\,cm$ and has no simple temperature dependence, the electronic attenuation of sound is very small in UBe_{13} and its temperature dependence in the normal state is not well defined. From the measurements by Golding et al. (1985a) in the range between 0.9 and 2.4 GHz, however, the ν^2-dependence of the attenuation is fairly well established as may be seen in fig. 64. The attenuation data can also be used to estimate $(m^* E_1)$, since all other quantities entering eq. (10) are known. Using reasonable values for l_e as obtained from data of the electrical resistivity, the authors find $(m^*/m_0)E_1 \sim 5$–$10\,eV$. Here m_0 is the free-electron mass. This is rather surprising because comparable values are also found for ordinary metals like Cu or Sn. It is suggested that the large enhancement of the electronic mass as evidenced by the low-temperature specific heat is somehow compensated by a reduced lattice coupling.

Another version of eq. (10) has been given by Lüthi (1985), namely

$$\alpha_n = \frac{4}{5}\frac{d_\Gamma^2 N(E_F)}{\rho v\, v_F q l_e} \tag{10'}$$

which is valid for non-interacting electrons and $ql_e \ll 1$. d_Γ is the deformation potential for longitudinal waves and v_F the Fermi velocity. From eq. (10′) it is clear that even for $ql_e \ll 1$, the ultrasound absorption due to electrons may be sizeable only if $N(E_F)$ is large and v_F is small. Both conditions are, of course, fulfilled for heavy-electron materials.

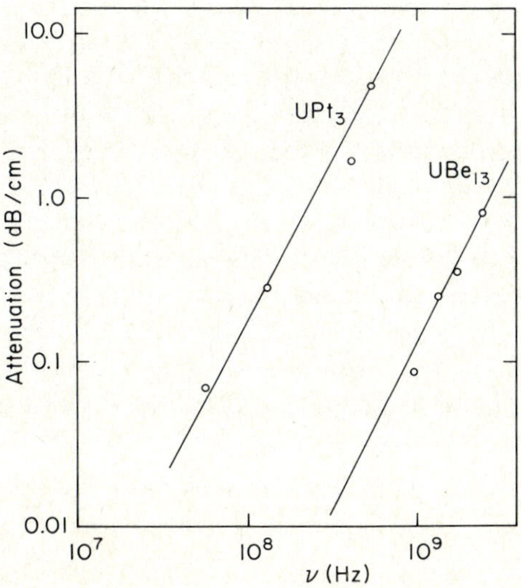

Fig. 64. Frequency dependence of the low-temperature normal-state electronic contribution to the ultrasound attenuation in UPt_3 and UBe_{13}. The solid lines indicate the slope compatible with an ν^2 dependence. (Data are from Golding et al. 1985a.)

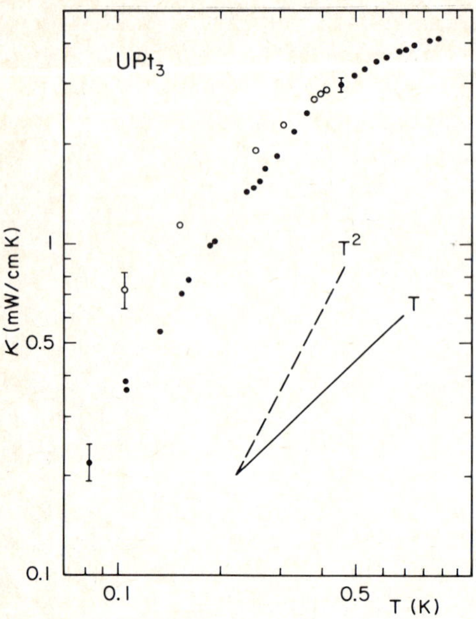

Fig. 65. Temperature dependence of the thermal conductivity of UPt_3 below 1 K. Dots represent data in the superconducting state and open circles represent values obtained in the magnetic-field-induced normal state ($H = 25$ KOe) obtained by Steglich et al. (1985).

2.4.2. *UPt_3*

Experimental data on the thermal conductivity of UPt_3 were reported by Steglich et al. (1985) and by Jaccard et al. (1985b). Both studies involved measurements in the superconducting and magnetic field induced normal state below 1 K and the results appear to be in fairly good qualitative agreement. Franse et al. (1985) also report $\kappa(T)$ data up to 200 K. As an example we show the data given by Steglich et al. (1985) in fig. 65. Here a linear T-dependence of κ is claimed in the normal state below 0.25 K and the resulting Lorenz number calculated by using the value of the residual resistivity is 2.25×10^{-8} $W\Omega/K^2$, only 10% less than the classical Sommerfeld value. In the superconducting state, $\kappa(T)$ obviously decreases more rapidly with temperature. We shall discuss these features, including the data of Jaccard et al. (1985b), which extend to somewhat lower temperatures than those shown in fig. 65, below in section 4. The same authors also investigated the temperature dependence of the thermoelectric power S of UPt_3. In fig. 66 we show the data of Steglich et al. (1985) which cover the temperature range from 2 to 300 K. The positive peak is located at $T \sim 8$ K and with increasing temperature, S changes sign at about 24 K and reaches an almost constant negative value at temperatures exceeding 120 K. Jaccard et al. (1985b) concentrated on measurements below 1 K, a temperature range that was also covered in the work by Steglich et al. (1985). We show the results given by Jaccard et al. (1985b) in fig. 67. Of particular interest is the observation that S in the normal state does not vanish linearly with temperature as T approaches 0 K, a conclusion that was also emphasized by Steglich et al. (1985). Assuming that S can be interpreted by (Barnard 1972)

$$S = \frac{\pi^2 k_B^2}{3|e|} T \left[\frac{\partial \ln N(E)}{\partial E} \right]_{E_F}, \tag{11}$$

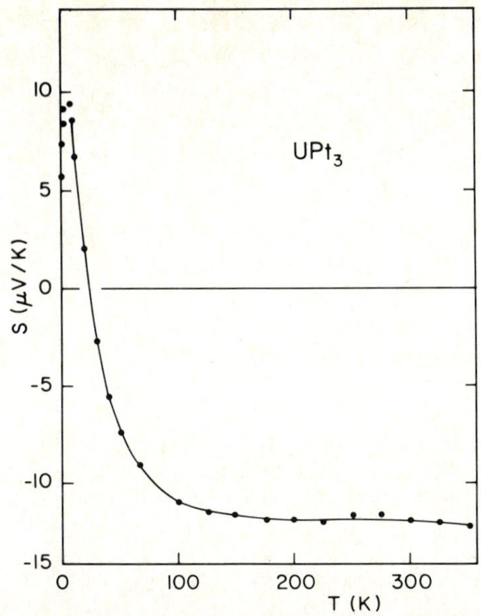

Fig. 66. Temperature dependence of the thermopower of UPt_3 between 2 and 350 K as obtained by Steglich et al. (1985). The line is to guide the eye.

this latter result would imply that the logarithmic energy dependence of the electronic density of states at E_F is still temperature dependent, even for $T \ll 1$ K.

Measurements of the attenuation of ultrasound in single-crystalline UPt_3 in the normal state at temperatures just above the superconducting transition have been described in some detail by Bucher et al. (1985) and by Batlogg et al. (1986). In fig.

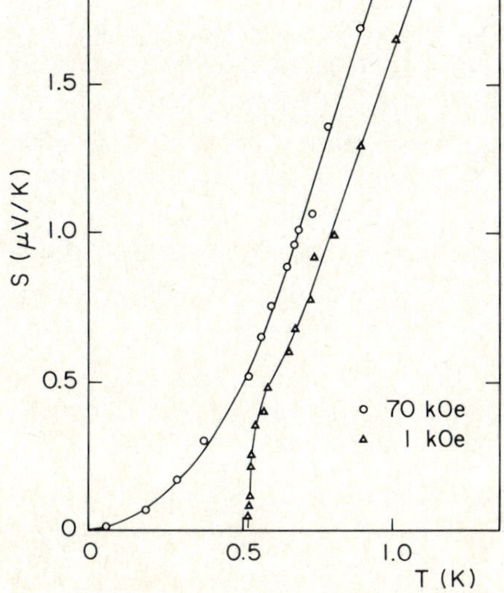

Fig. 67. Temperature dependence of the thermopower of UPt_3 below 1 K as obtained by Jaccard et al. (1985b). The solid lines are to guide the eye. 70 kOe is a supercritical field ($>H_{c2}(0)$).

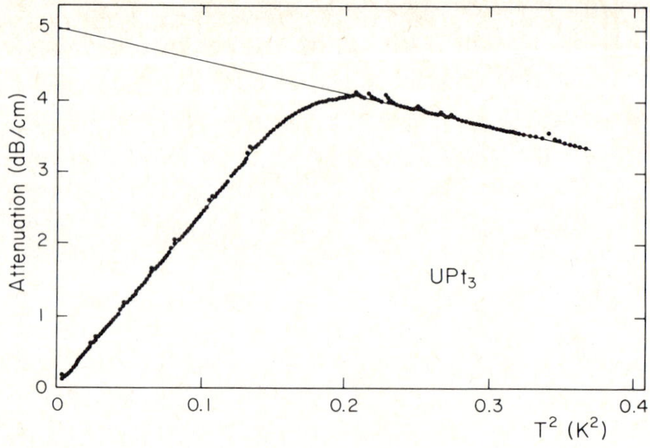

Fig. 68. Temperature dependence of the ultrasound absorption of UPt_3 below 0.6 K as reported by Bishop et al. (1984) and by Batlogg et al. (1986). The solid line is an extrapolation of the normal-state absorption to $T = 0$ K.

68 we show their data for longitudinally polarized waves along the hexagonal axis of UPt_3 at temperatures below 0.6 K where the attenuation is plotted versus T^2. The solid line in fig. 68 is an extrapolation of α_n in the hypothetical normal state. It emphasizes that α_n decreases proportional to T^2 with increasing temperature. This is most likely a result of the l_e-dependence of α_n as expressed in eq. (10). At these low temperatures, the electrical resistivity of UPt_3 can be approximated by eq. (8) and therefore a T^2-dependence of α_n with a negative prefactor, as observed, is anticipated. The frequency dependence as implied by eq. (10) was tested by measurements of α_n at four frequencies between 52 and 508 MHz. We show the results for UPt_3 together with that for UBe_{13} in fig. 64. They obviously confirm the expectation. The comparison in fig. 64 also demonstrates the influence of l_e on α_n. Just as for the electrical conductivity, the attenuation of ultrasound is different by orders of magnitude between UPt_3 and UBe_{13}. To observe the same level of attenuation in UBe_{13} as in UPt_3, much higher sound frequencies have to be applied. An estimate of the microscopic quantity $(m^* E_1)$ for UPt_3 by assuming a range of reasonable values for l_e (200 to 800 Å) as they follow from resistivity values shown in a previous section, again results in 6 to 12 eV for $(m^*/m_0)E_1$, as observed for UBe_{13} but also for quite ordinary metals. With a similar analysis, Yoshizawa et al. (1985) obtain 18 eV for this quantity, using eq. (10′) and assuming that d_Γ is given by the electronic Grüneisen parameter and the f-electron bandwidth.

2.4.3. *UAl_2*

Measurements of the thermal conductivity and the thermoelectric power on UAl_2 have been done by Armbrüster et al. (1979) for temperatures between 0.3 and 300 K. Below 15 K, κ is found to vary linearly with temperature including contributions from both the electrons and from lattice-vibration excitations. Between 20 and 50 K, κ is roughly constant and then increases again linearly with T up to room

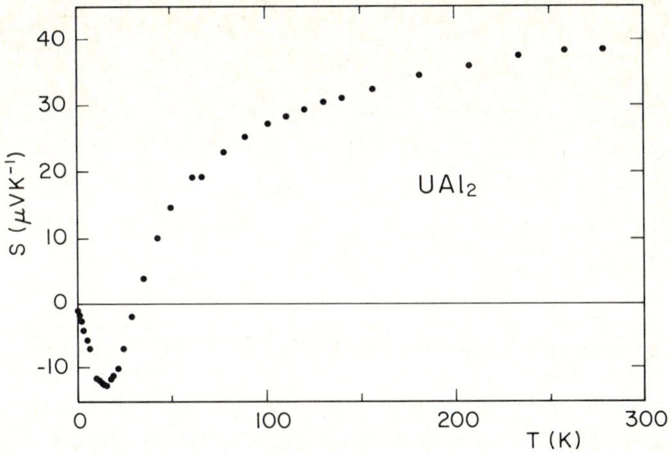

Fig. 69. Temperature dependence of the thermopower of UAl_2 between 0.3 and 300 K as obtained by Armbrüster et al. (1979).

temperature, but with a very much smaller slope. The Lorenz ratio $\kappa\rho/T$ exceeds the classical value at all temperatures, indicating the non-vanishing lattice contributions to κ.

The temperature dependence of S resembles very much that of UPt_3 with opposite sign however (see fig. 69). Below 15 K, S is claimed to vary linearly with T with a slope of $-1.13\,\mu V/K$. At 16 K it reaches a negative maximum of $-12.5\,\mu V/K$, changes sign at 30 K and subsequently increases with a monotonously decreasing slope reaching about $+40\,\mu V/K$ at 300 K.

2.5. *Other thermal properties* (*thermal expansion, magnetostriction, elastic constants*)

Thermal properties that are related to a volume derivative of the free energy F are also expected to be anomalous in heavy-electron materials, because narrow structures in the energy spectrum of electrons are apt to vary considerably with pressure or, in other words, with forced variation of the volume. Obviously such properties are the thermal expansion and the compressibility. These quantities, together with the specific heat, are related via the Grüneisen ratio

$$\gamma_G = \frac{\beta V_m}{\kappa C_V}, \tag{12}$$

where β is the volume expansion coefficient, V_m is the molar volume, κ the compressibility and C_V the molar specific heat at constant volume. γ_G is a measure for the volume dependence of the energy excitation spectrum that determines C_V, β is the contribution to the thermal expansion due to the volume dependence of this same excitation spectrum and κ is the total compressibility. At the low temperatures of interest here, $C_V = C_p$, and all properties are dominated by the heavy-electron subsystem and its coupling to the crystal lattice. The thermal-expansion coefficient is proportional to $\partial^2 F/\partial V\, \partial T$ and the elastic constants that determine the compressibility are proportional to $\partial^2 F/\partial \varepsilon_\Gamma^2$, where ε_Γ is the strain of symmetry Γ connected with a particular elastic constant C_Γ of the same symmetry. In an ordinary electronic system obeying Fermi–Dirac statistics, γ_G is a measure for the volume dependence of the electronic density of states because also

$$\gamma_G = \frac{(\partial S/\partial V)V}{(\partial S/\partial T)} \tag{13}$$

and therefore, because $S = \gamma T$, $\gamma_G = \partial \ln \gamma/\partial \ln V \equiv \gamma_e$, the so-called electronic Grüneisen parameter. Experimentally, γ is clearly temperature dependent for heavy-electron materials in the liquid-helium temperature range and anomalous behaviour of the thermal-expansion coefficient may be expected. The same may be true for elastic constants because they too are influenced by the electronic density of states. In compounds with f-electrons it is also possible that both the thermal expansion and the elastic behaviour of a material are influenced by magnetoelastic coupling between the partially-filled shells of localized f-electrons of the atoms and the crystal lattice. These effects have been demonstrated in numerous metallic rare-earth compounds (Lüthi 1980). In heavy-electron materials, however, such coupling mechanisms may be radically changed in a way similar to suppression of magnetic ordering.

2.5.1. *UBe_{13}*

The low-temperature thermal expansion of single-crystalline UBe_{13} has been measured by Ott (1984) between 0.3 and 10 K. The result for the linear thermal-expansion coefficient α along [100] is shown in fig. 70. It is anomalous in the sense that α increases with decreasing temperature below 7 K, i.e., in the same temperature range where C_p^{el}/T increases with decreasing temperature. The discontinuous change of α below 1 K is associated with the transition to the superconducting state and is dictated by thermodynamics. Around 1 K, γ_{el} is calculated to be about 50, two orders

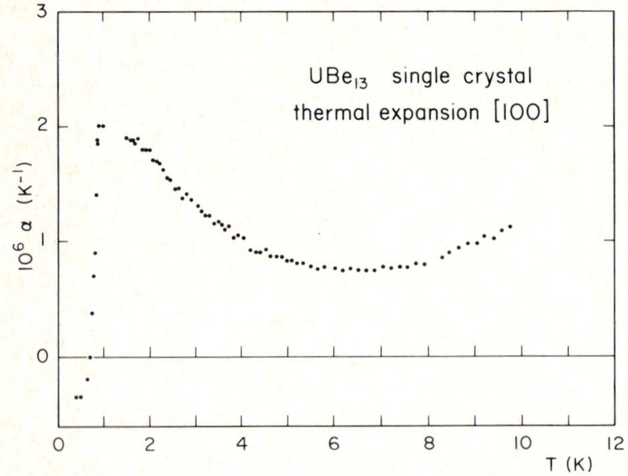

Fig. 70. Temperature dependence of the linear thermal-expansion coefficient of UBe_{13} between 0.3 and 10 K.

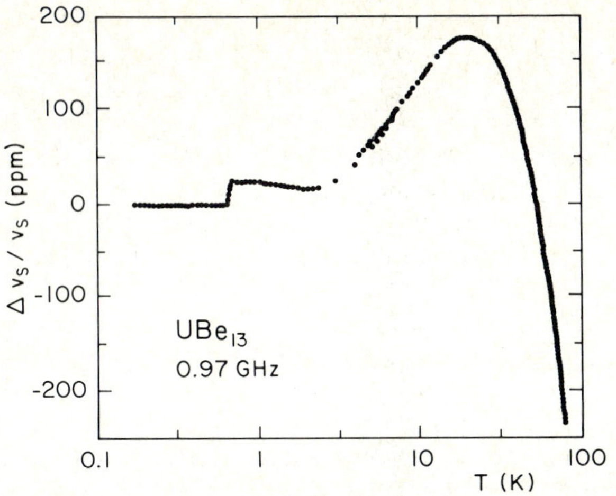

Fig. 71. Temperature-induced shift of the velocity of longitudinal sound waves along [001] in UBe_{13} as measured by Golding et al. (1985a).

of magnitude larger than the free-electron value of 2/3 but comparable to what is found in Ce-based heavy-electron materials like $CeAl_3$ (Andres et al. 1975, Ribault et al. 1979) and $CeCu_6$ (Ott and Fisk 1985). This also means that the electronic density of states at E_F is highly volume dependent. Magnetostriction in UBe_{13} (Ott et al. 1985f) at these temperatures is fairly small in fields up to 15 kOe, at least when compared with compounds where the above mentioned magnetoelastic coupling is effective. This indicates that these coupling effects are not of great importance in UBe_{13} at least at temperatures below 15 K.

Evidence for anomalous behaviour of elastic constants in UBe_3 may be found in the work by Golding et al. (1985a). The temperature induced change of the velocity of longitudinal sound waves along [001] in single-crystalline UBe_{13} at temperatures between 0.2 and 80 K is shown in fig. 71. The discontinuous change just below 1 K is associated with superconducting transition. The sound-velocity increase on cooling from liquid-N_2 temperatures is ascribed to the expected freezing out of anharmonicity. The velocity passes over a maximum at 25 K and through a minimum around 2 K, the temperature at which the specific heat and the electrical resistivity show a maximum. Between 5 and 25 K, a nearly logarithmic temperature dependence of the relative change of the sound velocity is observed. Yoshizawa et al. (1985) measured the longitudinal and the transverse sound velocity of polycrystalline UBe_{13} between 1.5 and 150 K and calculated the corresponding elastic constants $C_L = \rho v_L^2$ and $C_T = \rho v_T^2$. C_L is found to increase with decreasing temperature with a trend to saturation below 50 K. Below 15 K, C_L decreases down to 1.5 K. A monotonous increase with decreasing temperature is found for C_T. The relative changes in this temperature range are small and of the order of 10^{-4} or less. From these measurements the compressibility at 200 K is estimated to be $0.62 \times 10^{12}\,cm^3/erg$. The same data can be used to calculate Poisson's ratio by (Lüthi, 1985)

$$-\frac{\varepsilon_\perp}{\varepsilon_\parallel} = \frac{1}{2}\left(\frac{(v_L/v_T)^2 - 2}{(v_L/v_T)^2 - 1}\right). \tag{14}$$

In the temperature range from 4 to 300 K, $-\varepsilon_{\perp}/\varepsilon_{\parallel}=0.05$ for UBe_{13} (Lüthi 1985) compatible with the very small ratio C_{12}/C_{11} in pure Be.

Robinson et al. (1986) used inelastic-neutron-scattering spectra to evaluate parts of the phonon spectrum of UBe_{13}. From the q-dependence of accoustic phonon energies at 300 and 4.2 K, they calculate the elastic constants C_{11}, C_{12} and C_{44} at these temperatures. As in pure Be, C_{12} is found to be much smaller than C_{11}, in agreement with the results of the ultrasound measurements mentioned above. The room-temperature bulk modulus has also been measured by Mock et al. (1985) by using Brillouin scattering. Compared to other XBe_{13} compounds (where X is a rare-earth or an actinide element) the bulk modulus is not anomalous and the resulting compressibility is $0.8 \times 10^{-12}\ cm^3/erg$ at 300 K.

2.5.2. UPt_3, UAl_2

Thermal expansion data for UPt_3 are available from work by de Visser et al. (1985). Measurements on single crystals revealed a substantial anisotropy for the linear thermal-expansion coefficients along different crystallographic directions between 1.5 and 100 K. The results are shown in fig. 72. Along the c-axis of the hexagonal lattice, α is negative at low temperatures with a negative maximum between 10 and 20 K. It changes sign between 40 and 50 K and subsequently increases with decreasing slope as the temperature is raised. The thermal expansion in the basal plane is isotropic and always positive with a maximum just above 10 K. Also shown in fig. 72 is the expected linear thermal-expansion coefficient of an ideal polycrystalline sample and it agrees fairly well with respective results reported

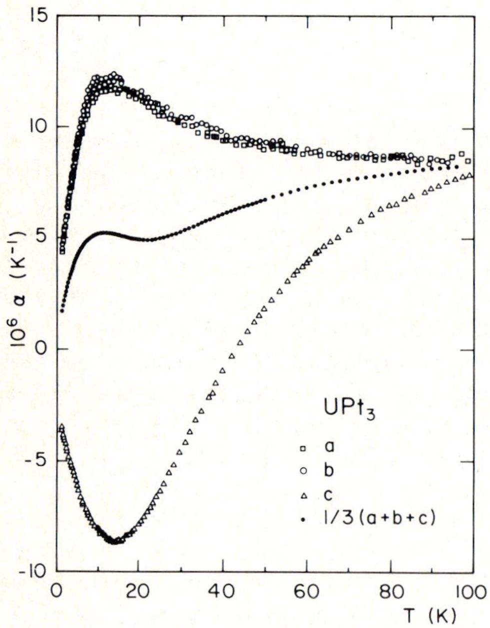

Fig. 72. Temperature dependence of the linear thermal-expansion coefficient along different crystallographic directions of UPt_3 between 1.5 and 100 K, as measured by de Visser et al. (1985). Dots indicate the calculated volume expansion coefficient.

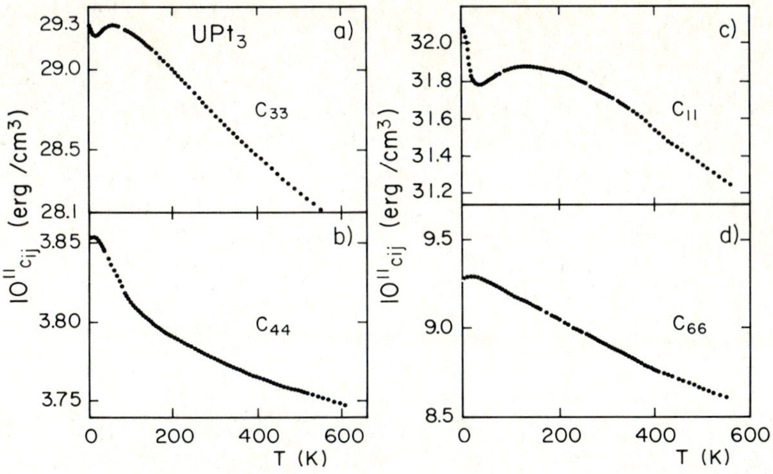

Fig. 73. ($a-d$) symmetry elastic constants C_{ij} of UPt_3 as a function of temperature between 1.5 and 600 K as obtained by Yoshizawa et al. (1985).

previously by the same authors for a polycrystalline sample. Combining the thermal-expansion and the specific-heat results, a very large electronic Grüneisen parameter of 159 is deduced (de Visser et al. 1985). This is clearly at variance with a γ_e value quoted by Yoshizawa et al. (1985) who obtained, from experimental values of the thermal expansion, the specific heat and the compressibility at 1 K, $\gamma_e = 60$. de Visser et al. (1986c) also measured the low-temperature magnetostriction along the c-axis and in the basal plane for external magnetic fields oriented in either direction and with strengths up to 80 kOe. They observe distinct anisotropies and find that the volume magnetostriction is an order of magnitude larger if the field is applied in the basal plane than along the hexagonal axis. When comparisons are possible, the magnetostriction data are in good agreement with expected values calculated from the pressure dependence of the magnetic susceptibility (see above).

Sound velocity measurements on single-crystalline UPt_3 between 1.5 and 600 K were reported by Yoshizawa et al. (1985) who showed their data in the form of the temperature dependence of various symmetry elastic constants (see fig. 73). All modes show distinctly anomalous behaviour at low temperatures, especially prominent for C_{11} and C_{33}. The magnitude of the effect in C_{11} is clearly much larger than what is observed in UBe_{13}. The authors conclude that these anomalies are due to a deformation-type electron–phonon coupling which should be particularly strong for the C_{11} and C_{33} modes. These measurements give a Poisson ratio for UPt_3 between 0.19 and 0.48. A complete set of room-temperature elastic constants for UPt_3 has been determined by de Visser et al. (1985). de Visser et al. also note that the compressibility of UPt_3 is nearly isotropic and they calculate a volume compressibility of $4.8 \times 10^{-13}\,cm^3/erg$ at 300 K. Apparently the corresponding Debye temperature of 217 K is close to that obtained from low-temperature specific heat data. The absolute values of the elastic constants given by

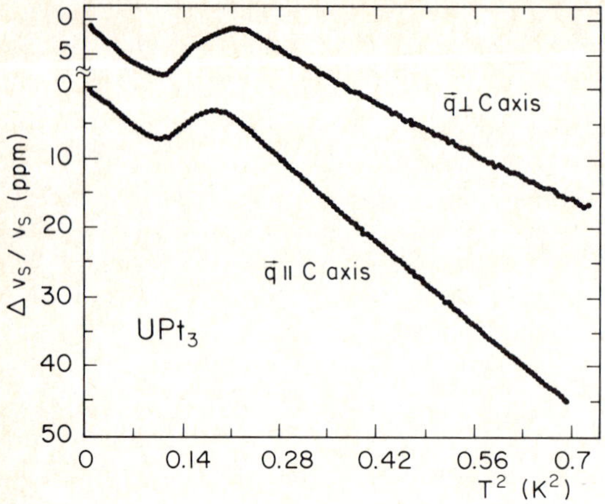

Fig. 74. Temperature-induced shift of the velocity of sound waves in UPt_3 propagating parallel and perpendicular to the hexagonal *c*-axis below 0.8 K. (Data are due to Batlogg et al. 1986.)

de Visser et al. (1985) deviate somewhat from those reported by Yoshizawa et al. (1985).

Results of high-resolution measurements of the ultrasound velocities parallel and perpendicular to the *c*-axis of UPt_3 single crystals at temperatures below 1 K were reported by Batlogg et al. (1986). We show their result in fig. 74 which reveals a T^2-variation of the sound velocity at these low temperatures, both in the normal and in the superconducting state. The same authors also noted a minimum for the sound velocity along the *c*-axis at about 18 K and a large decrease with increasing temperature for the sound velocity in the basal plane with no indication of a minimum up to 20 K. Batlogg et al. (1986) argue that the T^2-dependence at those low temperatures is a result of the very large electronic specific heat varying linearly with temperature.

The low-temperature linear thermal-expansion coefficient of polycrystalline UAl_2 was also measured by de Visser et al. (1985) between 1.5 and 100 K. The result is shown in fig. 75. Again the calculation of γ_e leads to a rather large value of 42.

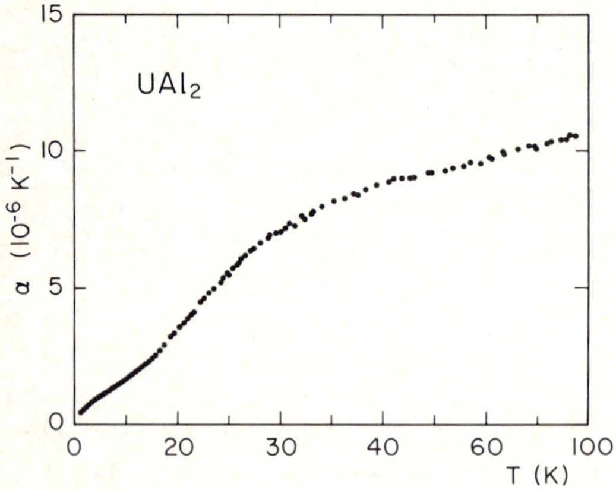

Fig. 75. Temperature dependence of the linear thermal-expansion coefficient of UAl_2 between 1.5 and 100 K, as obtained by de Visser et al. (1985).

2.5.3. *U_2Zn_{17}, UCd_{11}*

The low-temperature thermal expansion of polycrystalline U_2Zn_{17} and UCd_{11} was measured by Ott et al. (1985b) and by Fisk et al. (1984), respectively. The results for the linear thermal-expansion coefficients between 1.5 and 12 K are shown in fig. 76. The distinct anomalies are related to the magnetic phase transitions and are discussed in section 4. The rather small absolute peak values of the anomalies indicate the absence of any appreciable magneto-elastic coupling. The background thermal expansion is compatible with the large electronic specific heat, and the corresponding electronic Grüneisen parameter γ_e is of the order of 2, hence not very much enhanced.

2.6. *Spectroscopic investigations*

Spectroscopic investigations are usually made to answer specific questions concerning microscopic aspects that arise when trying to understand thermal- or transport properties of a material and they often reveal important details related to the dynamics of processes on a microscopic scale. The same is true in the field of heavy-electron materials where these questions are of paramount importance. Below we give an overview of existing data obtained with various different experimental techniques applied on these materials. Examples of such measurements have already been mentioned in section 2.5 dealing with an evaluation of elastic properties using spectroscopic methods.

Direct and *inverse photoemission* are very useful to probe the energy dependence of

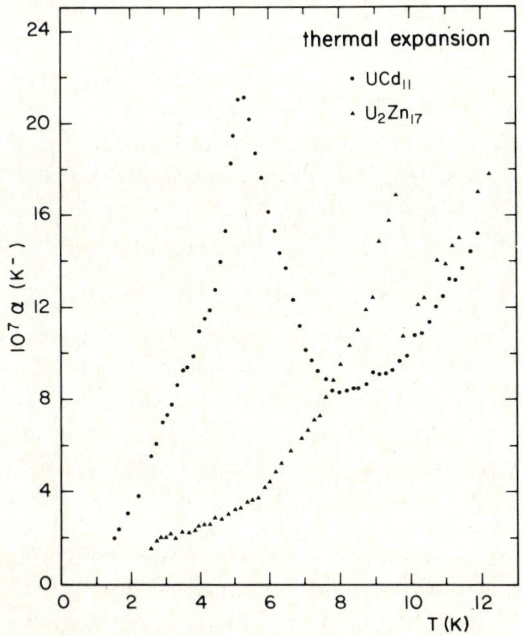

Fig. 76. Temperature dependence of the linear thermal-expansion coefficients of polycrystalline U_2Zn_{17} and UCd_{11} between 1.5 and 12 K. (Data are due to Ott et al. 1985b and Fisk et al. 1984.)

the electronic spectrum. Results of such measurements serve for comparisons with theoretical band-structure calculations or may even provide numerical parameters that enter various model calculations. *Neutron diffraction* is most widely used to establish lattice- or magnetic-structures and, if done as a function of temperature, it can also give information on the temperature dependence of the lattice parameter or the magnetization. *Inelastic neutron scattering* probes lattice vibrations, spin fluctuations, and other possible excitations within the energy spectra of magnons or crystal-field split energy levels of unpaired f electrons. The recent experimental advance of using polarized neutrons gives a more reliable measurement of magnetic scattering, especially if it is weak. *Raman scattering* is yet another method to study certain excitations in a solid. We also should mention the various resonance techniques involving *nuclear magnetic resonance* (NMR), *electron-spin resonance* (ESR) and *muon–spin rotation* (μSR) experiments. They all essentially probe internal magnetic fields and give valuable information on the various relaxation mechanisms which are involved in each specific type of measurement. Measurements of *optical properties* of metallic substances can give information on the energy dependence of the electrical structure close to E_F between 0.1 and 10 eV. They can also be used to evaluate the frequency dependence of the electric conductivity. Finally, we should also like to mention the method of *point-contact spectroscopy*, which is, in principle, suited to provide information on the electron energy spectrum at low energies, and of low-energy scattering processes of the electrons that are injected into the material through the point contact.

All these techniques have been applied to study heavy-electron actinide materials, but so far only on a limited number of compounds, mainly on UBe_{13}, somewhat less on UPt_3 and U_2Zn_{17} and very little on UAl_2 and UCu_5.

2.6.1. UBe_{13}

A fairly complete room-temperature XPS and BIS study of UBe_{13} has been published by Wuilloud et al. (1984). The spectrum of the occupied and empty electron states in the valence band around E_F is shown in fig. 77. The occupied valence-band states of UBe_{13} were also studied by resonant photoemission and from these experiments it was concluded that the photoemission intensity between -4 eV and E_F is essentially due to electrons with 5f symmetry (Landgren et al. 1984, Parks et al. 1984). In this energy range the spectra excited with low-energy (40 eV) and high-energy (1465 eV) photons are almost identical. At higher binding energies the XPS spectrum does not show the sp states of Be because of their very weak scattering cross section at this high photon energy. The valence-band XPS spectrum of UBe_{13} is very similar to that of α-U metal. The BIS spectrum represents almost exclusively the empty 5f states superimposed on an energy-dependent background. The single featureless peak with a maximum at 1.3 eV is distinctly different from analogous BIS spectra obtained for α-U (Baer and Lang 1980, Baer 1984) or other materials of interest here, namely UPt_3 and UAl_2 (see below). Of interest is also the spectrum of the 4f core levels of U in UBe_{13} that was published by Wuilloud et al. (1984), shown in fig. 78. The binding energies of these lines are taken as evidence for a 5f population per U ion that is clearly larger than 2. Furthermore, the asymmetry

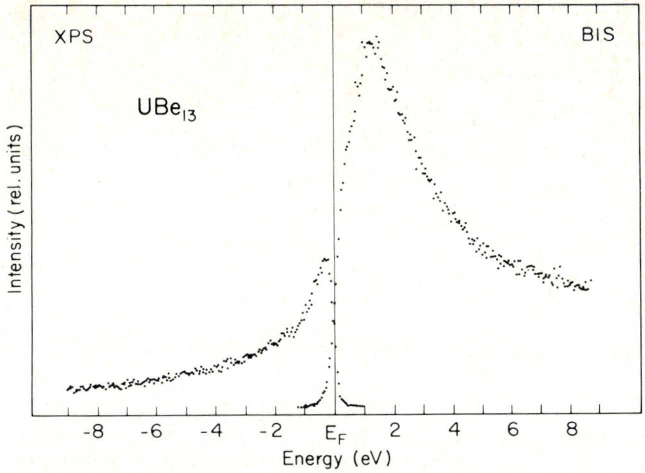

Fig. 77. Combined XPS–BIS spectrum of UBe_{13}. The relative intensities below and above E_F were scaled by a continuity criterion. (For details see Wuilloud et al. 1984.)

of the lines is claimed to reflect a high density of electron states at E_F centered around the U atoms. Wuilloud et al. also demonstrated the absence of such an asymmetry for the spectrum of the Be ls level.

Valence-band spectra excited with low-energy (40 eV) photons but obtained from samples cooled to low temperatures (20 K) and detected with high resolution (0.13 eV) revealed additional structures between -3 eV and E_F. The experimental

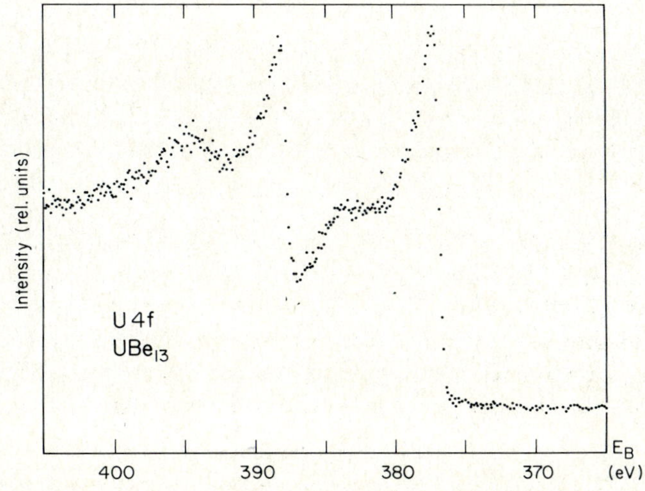

Fig. 78. XPS spectrum of the U 4f levels in UBe_{13} as obtained by Wuilloud et al. (1984).

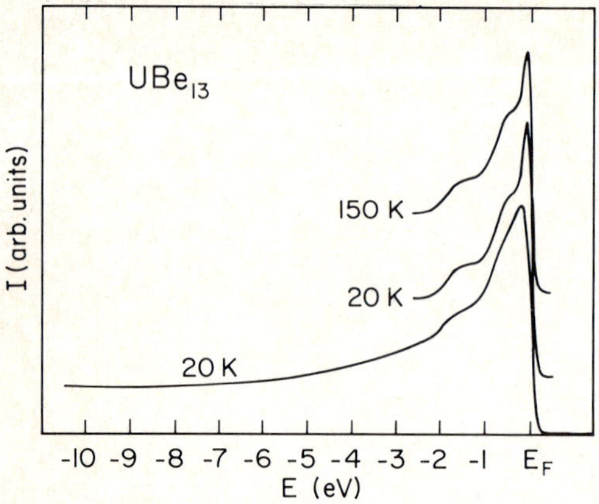

Fig. 79. Valence-band photoemission spectra for UBe_{13} as obtained with 40 eV photon excitation energy by Arko et al. (1984a). The long scan is at 20 K and with 0.3 eV resolution. The short scans are taken with a resolution of 0.13 eV at the indicated temperatures.

results reported by Arko and co-workers (Arko et al. 1984a) are shown in fig. 79. Of special interest is the narrow feature close to E_F with a width of 0.15 eV. Its amplitude, but not its width, was found to vary much more with temperature than would be allowed by a simple broadening effect of the Fermi distribution function. Arko et al. (1984a) claim to have observed the structure that is responsible for the large low-temperature specific heat. The width of the structure close to E_F is, however, still much too large to be seriously considered as the source of the very low renormalized Fermi energies. From constant-initial-rate scans a uranium-derived 6d band intersecting E_F is suggested. The same authors confirm the previously mentioned very small Be s–p contribution to the electronic spectrum at E_F.

A neutron-diffraction study of UBe_{13} and, for reasons of comparison, $ThBe_{13}$ was published by Goldman et al. (1985). With regard to the crystal structure of UBe_{13} it confirmed previous X-ray investigations (Bänziger and Rundle 1949). Goldman et al. (1986a) also reported on inelastic neutron scattering measurements on UBe_{13} at a temperature of 10 K using both polarized and unpolarized neutrons. This study was primarily intended to obtain information on the magnetic scattering in UBe_{13} and therefore experiments using unpolarized neutrons were made on powder samples of both UBe_{13} and $ThBe_{13}$. In spite of much lower signal rates, polarized neutrons were also used, thus eliminating the nuclear background scattering. Unfortunately, this second method also suffers from a much coarser energy resolution of the scattering spectrum. In fig. 80 we show the spectra for UBe_{13} and $ThBe_{13}$ obtained by Goldman et al. (1986a) with unpolarized neutrons. The lower part of fig. 80 represents the difference in intensity of the two spectra which is ascribed to magnetic scattering. The magnetic neutron scattering cross section per magnetic atom can be written as (Marshall and Lovesey 1971, Buyers and Holden 1985)

$$\frac{\partial^2 \sigma}{\partial \Omega \, \partial \omega} = \gamma_0^2 |f_Q|^2 \left[1 - \exp\left(-\frac{\hbar\omega}{kT} \right) \right]^{-1} \frac{\chi''(Q,\omega)}{\pi}, \tag{15}$$

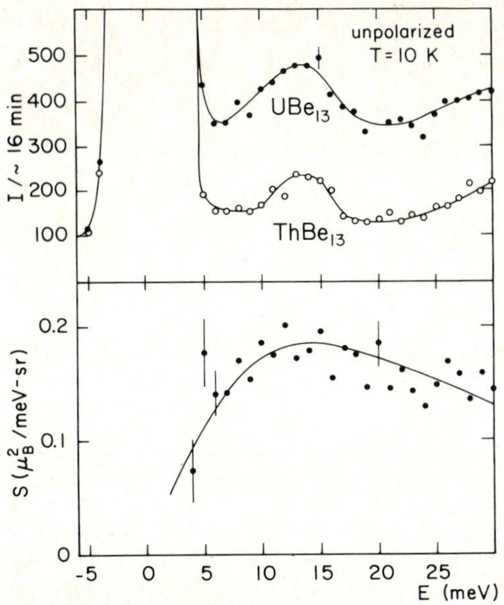

Fig. 80. Top: Inelastic unpolarized-neutron scattering spectra of UBe_{13} and $ThBe_{13}$ at 10 K. Bottom: The difference spectrum of the upper spectra; the solid line is a fit using a modified quasielastic Lorentzian. (Data were obtained by Goldman et al. 1986a.)

where γ_0 is the dipole coupling constant, f_Q is the magnetic form factor and $\chi''(Q,\omega)$ is the imaginary part of the magnetic susceptibility. If all the Q-dependence of the scattering is described by the form factor then

$$\chi''(Q,\omega)=\chi''(\omega)=\frac{A\Gamma\omega}{\Gamma^2+\omega^2}, \tag{16}$$

where A is an amplitude factor and Γ is the halfwidth at half maximum of the response spectrum. An analysis of the results on UBe_{13} based on eqs. (15) and (16) gives $\Gamma=14$ meV and also indicates, by use of the Kramers–Kronig relation combining the imaginary and the real part of χ, that the integrated magnetic intensity accounts for a good part of the bulk value of the magnetic susceptibility measured at 10 K. Although with much poorer statistics, the spectrum obtained with polarized neutrons confirms the result shown in the lower part of fig. 80. This provided a consistency check for the subtraction procedure employed in the unpolarized-neutron measurements and also demonstrated implicitly that the inelastic scattering due to phonons in $ThBe_{13}$ and UBe_{13} is virtually identical at energies above 0.5 meV. This result confirmed an earlier study by Renker et al. (1985) (see below). Stassis et al. (1986) have measured the form factor of the induced magnetization (i.e., in a large applied magnetic field) and found that the spatial extent of the 5f-electrons in these heavy-electron systems is indistinguishable from that in conventional uranium compounds.

Goldman et al. (1986a) also measured the inelastic scattering of unpolarized neutrons in UBe_{13} with very fine energy resolution around $E=0$ above and below the superconducting transition temperature. Their result is shown in fig. 81. They

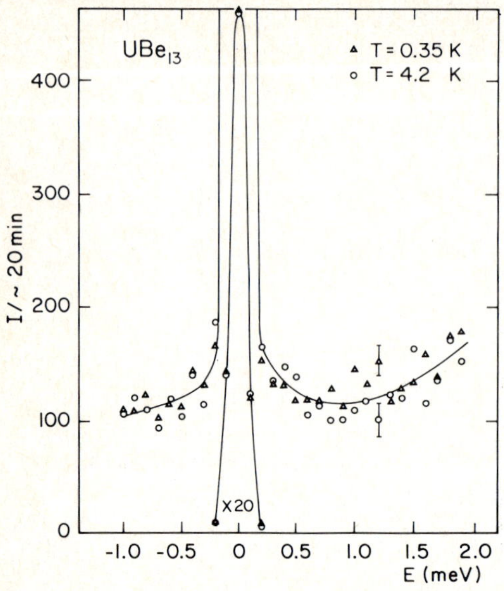

Fig. 81. High-resolution, constant $Q=0.4\,\text{Å}^{-1}$ scan of the small-energy transfer region of the unpolarized-neutron scattering spectrum of UBe_{13} at 0.35 and 4.2 K as obtained by Goldman et al. (1986a).

observed no difference between the two temperatures and claimed two implications: (i) the absence of any narrow quasielastic feature carrying significant spectral weight, and (ii) no change in the scattering rate at energies of the superconducting gap of about 0.3 meV.

Renker et al. (1985) used inelastic neutron scattering to determine the phonon density of states in the heavy-electron compound UBe_{13}, primarily to investigate the possible existence of low-energy phonon modes. The measurements employed a time-of-flight technique and were carried out at temperatures between 8 and 300 K. The spectra could be analyzed down to 1 meV. The generalized phonon density of states consists of two distributions which are separated from each other by a deep valley. At lower energies the intensity peaks around 13 meV while the high-energy part represents a broad spectrum starting at about 25 meV and a cut-off energy of about 90 meV. Maximum intensity is observed around 60 meV. The high-energy part was found to closely resemble that of pure Be metal whereas at low energies some additional intensity was found. This is probably due to a lowering of the accoustic branches as compared to those in Be. From their data Renker et al. (1985) also calculated the low-temperature lattice specific heat and found good agreement with values from direct specific-heat measurements.

A Raman-scattering study on UBe_{13} was recently reported by Cooper et al. (1985). They used a polarized Raman-scattering technique to identify the symmetry and energy of all possible Raman-active optical phonon modes, given the crystal structure of UBe_{13}. In fig. 82 we show their results for two possible configurations of the incident and collected light polarization for one particular wavelength at $T=350$ K. An interesting feature is the increasing scattering intensity with decreasing energy, which is only observed in depolarized spectra [(xy) configuration]. The

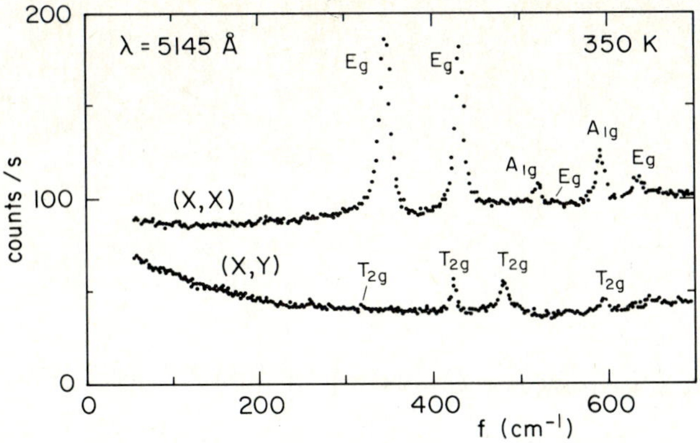

Fig. 82. (x, x) and (x, y) Raman-spectra for UBe_{13} taken by Cooper et al. (1985) at 350 K using a 5145 Å excitation line.

authors conclude that these excitations have T_{1g} symmetry and are therefore indicative of light scattering due to spin fluctuations of itinerant 5f electrons in UBe_{13}. The two low-energy T_{2g} modes stiffen as the temperature is lowered and, in addition, all T_{2g} modes broaden and gain intensity with decreasing temperature. All other modes are virtually unaffected by a temperature reduction and also no new lines appear upon cooling the sample. The low-energy scattering in the (x, y) configuration is also temperature dependent as shown in fig. 83. At 40 K a broad maximum centered around $115\,cm^{-1}$ (14.3 meV) is observed. Cooper et al. (1985) speculate that this structure might be connected with the spin-fluctuation response as observed in neutron scattering experiments, as shown above in the lower part of fig. 80. More recent experiments reveal the absence of such a structure in similar spectra for $ThBe_{13}$ (Cooper et al. 1987).

Measurements of the ^{9}Be nuclear spin-lattice relaxation time T_1 in the normal state of UBe_{13} were reported by Clark et al. (1984) for temperatures between T_c and

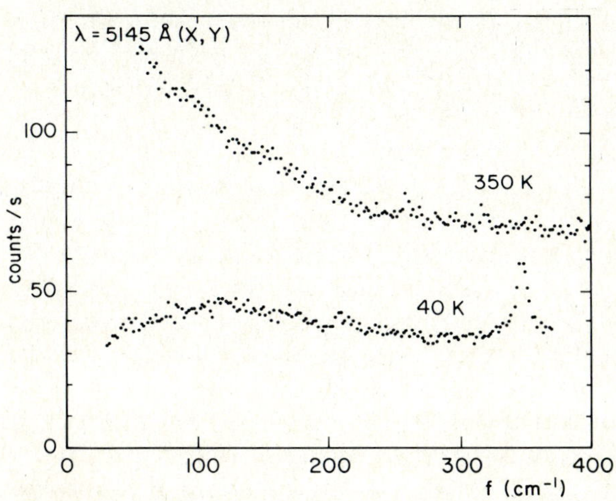

Fig. 83. T_{2g} and T_{1g} spectra of UBe_{13} between 25 and 400 cm at two different temperatures, as obtained by Cooper et al. (1985).

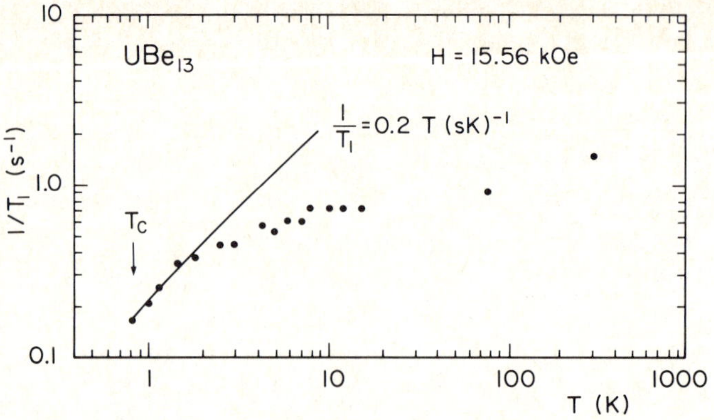

Fig. 84. Temperature dependence of the ^{9}Be nuclear spin-lattice relaxation rate in UBe_{13} between 0.9 and 300 K as obtained by Clark et al. (1984). The solid line indicates the linear temperature dependence of T_1^{-1} below 2.5 K.

room temperature. The data (shown in fig. 84) indicate a linear temperature dependence of T_1^{-1} between T_c and 2 K, the behaviour usually observed in metals. In comparison with Be metal, the relaxation rate is greatly enhanced suggesting the significance of the heavy-electron state. Above 2 K, the relaxation rate tends to a constant value and Lan et al. (1985) claim that a model of nuclear relaxation by conduction electrons in a narrow band width of 10 K gives a good phenomenological fit to the data.

ESR experiments on UBe_{13} doped with dilute local-moment ions like Er, Dy and Gd were reported by Gandra et al. (1985). The measurements were made at temperatures where the C_p^{el}/T ratio is varying rapidly with temperature, i.e., between 1.5 and 12 K. The authors conclude that neither the enhancement of C_p and χ, nor their temperature variation are reflected in a possible anomalous behaviour of ESR data. In fig. 85 we show the result for the inverse relaxation time $1/\gamma T_2$ versus T for polycrystalline UBe_{13} doped with 1% Er. Consistent results were obtained for 1% Er impurities in a single-crystalline specimen and for a polycrystalline sample with 1000 ppm Er concentration. In metallic hosts the spin-relaxation rate is expected to follow the usual Korringa relation (Barnes, 1981),

$$\frac{1}{\gamma T_1} = \frac{\pi}{\hbar}[N(E_F)J]^2 kT, \tag{17}$$

where γ is the gyromagnetic ratio and J is the appropriate exchange integral. The dotted line in fig. 85 indicates the expected temperature dependence of $1/\gamma T_1$ if eq. (17) was applicable in this case and the temperature dependence of $N(E_F)$ is deduced from specific-heat results. In the same temperature interval the g-value deduced for the same sample is constant. The observed g-value is comparable to that of Er local moments in ordinary metals indicating that no large g-shift occurs due to the fairly large magnetic susceptibility of the host material. These experimental facts suggest that local moments substituting on the U sites do not couple significantly to the

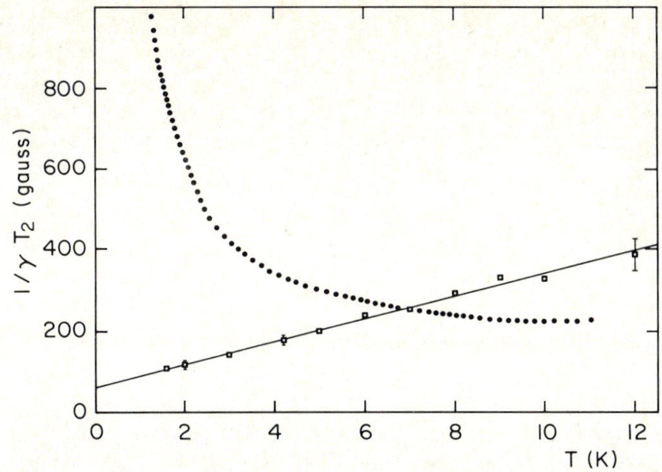

Fig. 85. Temperature dependence of the relaxation rate of ESR in UBe_{13} doped with 1% Er on U sites. The dotted line indicates the expected behaviour if eq. (17) was valid and $N(E_F)$ proportional to the low-temperature specific heat. The data were obtained by Gandra et al. (1985) who assumed that $T_1 = T_2$.

heavy-electron system. This is clearly different from the observations in connection with the nuclear relaxation rate of the Be nuclear spin mentioned above.

Optical experiments on polycrystalline UBe_{13} have been made by Bonn et al. (1985). They measured the reflectance at frequencies between 15 and 27000 cm^{-1} and at various temperatures between 2 and 300 K. In fig. 86 we show their data at low frequencies and low temperatures, which reveal an increasing absorption with decreasing temperature, compatible with the temperature dependence of the d.c.

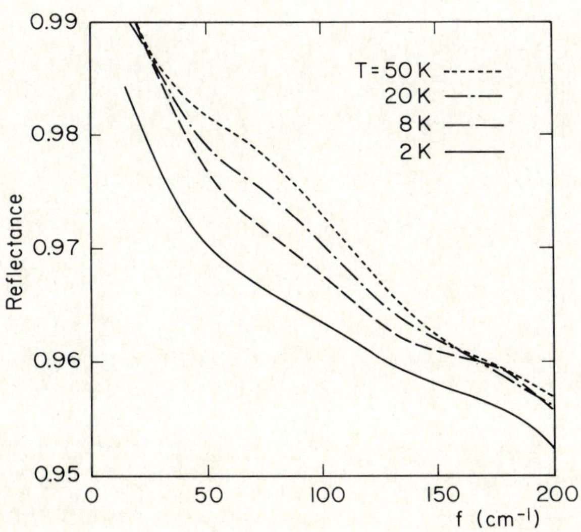

Fig. 86. Frequency dependence of the reflectance of UBe_{13} at low temperatures as obtained by Bonn et al. (1985).

resistivity shown in section 2.3. Anomalous, however, is the non-monotonic behaviour, since at 100 K (not shown) the absorption exceeds that at 50 K. Bonn et al. (1985) also converted their reflectance data to obtain the frequency dependence of the conductivity via a Kramers–Kronig analysis. At 2 K this results in a prominent maximum of conductivity centered around 150 cm^{-1}. With increasing temperature, the low-frequency tail of this peak is rapidly smeared out and hardly observable at 8 K.

2.6.2. UPt_3, UAl_2

From X-ray photoemission experiments on UPt_3, Schneider and Laubschat (1981b) concluded that the position of the f-intensity in the valance band spectra of α-U and UPt_3 was roughly the same, and therefore a band description of the f electrons in UPt_3 seemed appropriate. More recent experiments by Arko et al. (1984a) with low-energy photon excitation (40 eV), higher energy resolution, and samples cooled to low temperatures confirmed the presence of f-symmetry intensity in the valence band spectrum close to E_F. Their result is shown in fig. 87. The very narrow structure at E_F is associated with an f band that is responsible for the heavy-mass electrons. The additional features at higher binding energies are thought to be due to electrons from Pt. As for UBe_{13}, further results obtained by constant-initial-state scans are interpreted as evidence for contributions of U 6d electrons situated at E_F. Allen et al. (1985) combined photoemission using low-energy photon excitation (~200 eV) with BIS measurements recording high-energy photons (~1490 eV). The resonant photoemission data indicating f-electron occupied states very close to E_F confirmed earlier results by Parks et al. (1984). The BIS data revealed some additional structure in the empty-state spectrum that was absent in UBe_{13}. We show these results in fig. 88. The experimental data were compared with relevant band-structure calculations for energies close to E_F. The authors concluded that the width of the experimental spectra was larger than the width resulting from one-electron calculations, probably due to 5f Coulomb interaction.

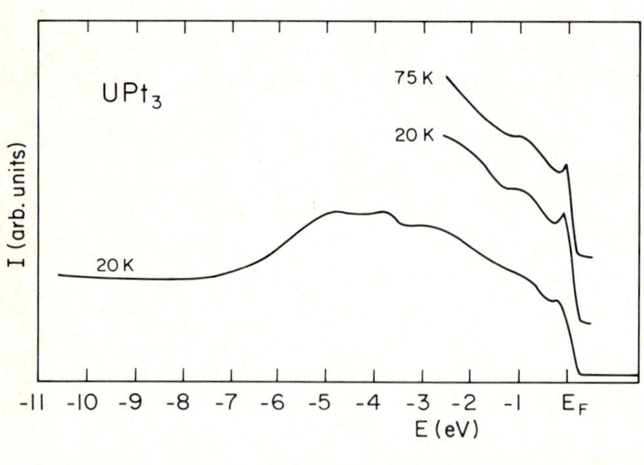

Fig. 87. Valence-band photo-emission spectra for UPt_3 as obtained with 40 eV photon excitation energy by Arko et al. (1984a). The long and short scans were taken with the same conditions as mentioned for fig. 82.

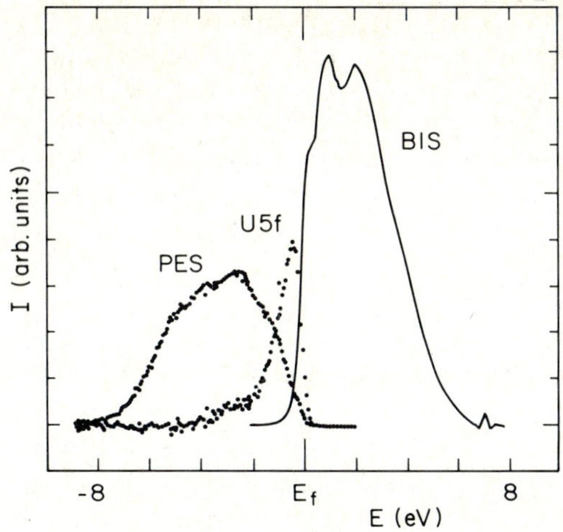

Fig. 88. Photoemission (PES) and Bremsstrahlen-isochromat (BIS) spectra of the valence band region of UPt_3 as obtained by Allen et al. (1985). The spectrum labeled PES was taken with a photon energy of 92 eV. The spectrum labeled U 5f is obtained if curve PES is substracted from the intensity that is obtained with 98 eV photons (resonant 5f emission).

Allen et al. (1985) also show the same kind of spectra for UAl_2 (see fig. 89). Again a comparison with local-density-functional calculations was made. The conclusion regarding these spectra is the same as that for UPt_3. An earlier photoemission study of UAl_2 by Naegele et al. (1980) had shown a considerable intensity increasing towards E_F with the usual cut-off at E_F which was interpreted as being due to uranium 5f and 6d electrons.

Neutron experiments on UPt_3 employing unpolarized and polarized neutrons were carried out by several workers. Aeppli et al. (1985) claim the direct observation of spin fluctuations in polycrystalline UPt_3 from spin-polarized neutron scattering data taken at 1.3 K. A Q-scan at constant energy confirmed the 5f-like form factor of the scattering (Stassis et al. 1986). In fig. 90 we show Aeppli et al.'s result for the energy dependence of a constant-Q spectrum revealing the difference in scattering intensity for neutrons polarized parallel (HF) and perpendicular (VF) to the momentum transfer. The solid line is a fit using eqs. (15) and (16) which results in a linewidth Γ of 9 ± 2 meV. Johnson et al. (1986) also report inelastic neutron scattering on UPt_3 in which they find no evidence for magnetic correlations.

Neutron work on UAl_2 was reported by Löwenhaupt et al. (1979). Distinct inelastic excitations were not observed, but magnetic scattering intensity in the

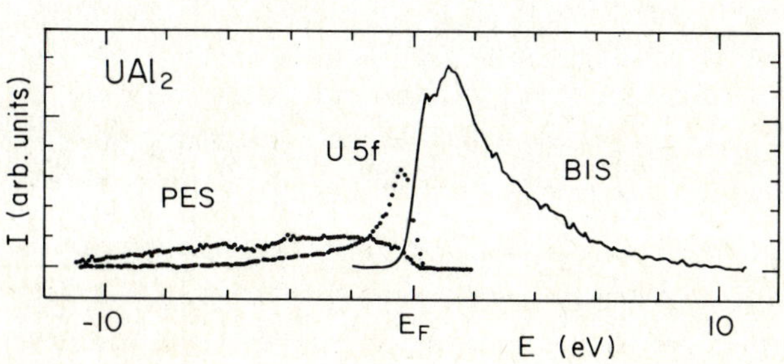

Fig. 89. Photoemission (PES) and Bremsstrahlen-isochromat (BIS) spectra of the valence-band region of UAl_2 as obtained by Allen et al. (1985). The labelling of the curves corresponds to that in fig. 88.

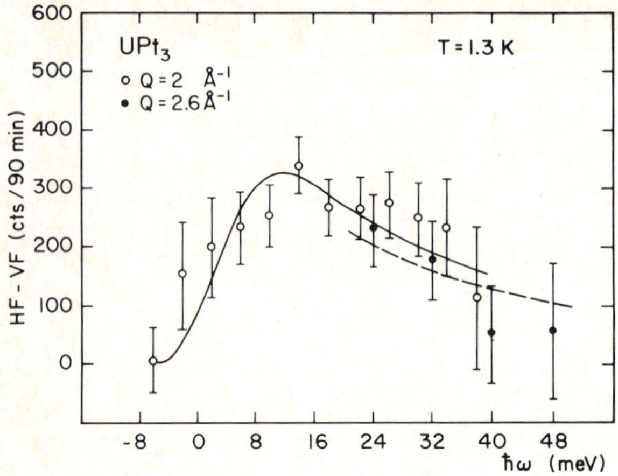

Fig. 90. Constant-Q inelastic polarized-neutron scattering spectra for UPt_3 as obtained by Aeppli et al. (1985). The solid line is a fit as mentioned in the text. For the meaning of the broken line see the work by Aeppli et al. (1985).

energy spectrum with a sizeable, temperature-dependent linewidth Γ was detected. In character the magnetic scattering is very similar to that observed in UBe_{13} and UPt_3 but the low-temperature value of the linewidth Γ is 25 meV (HWHM). Recent measurements with pulsed neutrons, which can access a wider energy range, by Loong et al. (1986) have confirmed the large value of Γ, which is assumed to arise from spin fluctuations (see also Buyers and Holden 1985, and Shapiro 1986).

Optical properties of UPt_3 were reported by Schoenes and Franse (1985). From room-temperature reflectivity measurements on single crystals of UPt_3 they deduced a pronounced anisotropy of the optical properties between light polarized in the basal plane and along the hexagonal axis. At low-energies below 0.3 eV the reflectivity reveals the same anisotropy as the room-temperature electrical resistivity. Part of the structure in the spectrum was associated with itinerant f electrons.

UPt_3 single crystals were also investigated by means of point-contact spectroscopy with both parts of the contact consisting of UPt_3 (Moser et al. 1986). In fig. 91 we show the resulting $\partial U/\partial I$ and $\partial^2 U/\partial I^2$ spectra obtained with the sample cooled to 4.2 K. The $\partial U/\partial I$ curve is interpreted as revealing a narrow-peak structure in the electronic density of states at E_F with a width of about 4 meV.

2.6.3. *U_2Zn_{17}, UCu_5*

Preliminary experiments employing the resonant-photoemission technique (Reihl 1984) revealed the presence of 5f-derived intensity in the valence-band spectrum of U_2Zn_{17} close to E_F. Neutron diffraction experiments on U_2Zn_{17} (Cox et al. 1986) confirmed the previously reported crystal structure. No inelastic scattering data are yet available, but low-temperature diffraction experiments confirmed the previously established magnetic ordering below 10 K. These experiments will be discussed in section 4.

For UCu_5, results from XPS- and resonant-photoemission experiments are available. From an early XPS study, Grohs et al. (1980) concluded that in UCu_5, U is in an intermediate-valent state between U^{3+} and U^{4+}, while Schneider and

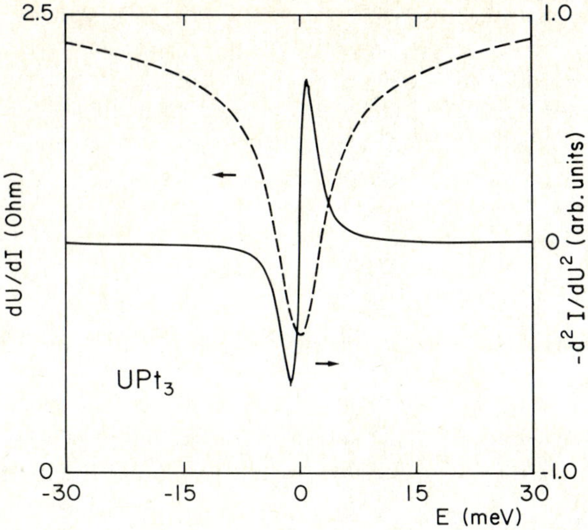

Fig. 91. $\partial U/\partial I$ (broken line) and $\partial^2 I/\partial U^2$ (solid line) of a point contact of two pieces of UPt_3, taken by Moser et al. (1986) at 4.2 K.

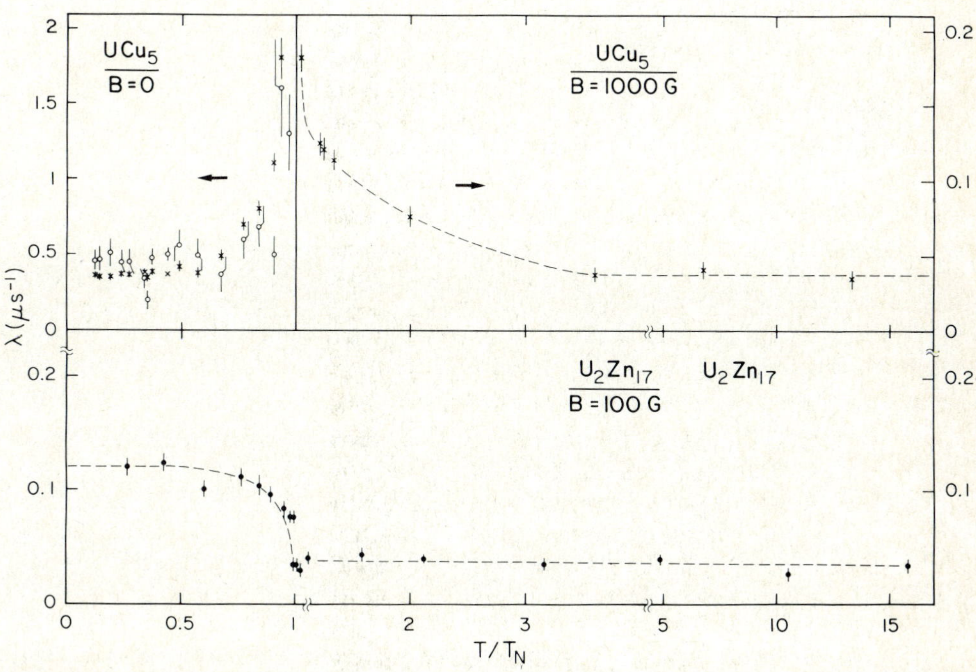

Fig. 92. Transverse-field μSR relaxation rates of UCu_5 (top) and U_2Zn_{17} (bottom) as a function of renormalized temperature, as obtained by Barth et al. (1986a).

Laubschat (1981a) claimed that none of the features of their own XPS data on UCu_5 allowed for such a conclusion. Following this controversy, Schneider et al. (1982) used the resonant-photoemission technique to unravel the f-electron contributions to the valence-band spectrum with high energy resolution. Again these experiments gave no evidence for two different final-state 5f multiplets in the valence bands of UCu_5, a prerequisite for an intermediate-valent state. Instead it was suggested that the 5f electrons have an itinerant character, partly due to hybridization with 3d electrons.

μSR experiments have been made on polycrystalline samples of U_2Zn_{17} and UCu_5 between 1.5 and 300 K and in external magnetic fields up to 7.5 kOe. In fig. 92 we show low-field data by Barth et al. (1986a) for the relaxation rate λ of the time-dependent amplitude of the muon-spin precession. Results for both the magnetically-ordered and the paramagnetic state are shown here. For UCu_5 the relaxation rate increases with decreasing temperature in the paramagnetic state, reaching a λ-type maximum at the transition temperature, a feature that is often observed in magnetically ordering materials and indicating the enhancement of the inhomogeneity of the internal-field distribution around the critical temperature. Quite in contrast is the behaviour of λ in U_2Zn_{17} where up to values for T/T_N of 15 virtually no temperature-dependence is observed, indicating that the internal field inhomogeneities at the muon site do not depend on temperature. This is quite a remarkable observation when considering the temperature dependence of the magnetic susceptibility (see fig. 25). Also quite different in the two compounds is the temperature dependence of λ in the magnetically-ordered state. Those features will be briefly discussed in section 4.

3. Physical models of heavy-electron behaviour

3.1. *Introduction*

Detailed theoretical understanding of the behaviour of heavy-electron systems is only in its infancy. There is a general hope that solutions of the Anderson-lattice Hamiltonian may reproduce much of the physics of heavy-electron systems: one can appreciate the difficulties here by recalling the tortured history of the single-impurity Anderson model. Taking another extreme starting position, one can hope that some kind of inclusion of many-body effects into the one-electron theory represented by band-structure calculations will also contain the heavy-electron physics.

Our purpose in section 3 is to indicate the theoretical ideas which we have found useful in our thinking about heavy-electron systems, both in understanding the results of experiments and in suggesting further experiments. We suggest that the reader consults two excellent sources: (i) 'Proceedings of the 8th Taniguchi Symposium on Heavy Fermions and Valence Fluctuations' (Kasuya and Saso 1985), and (ii) the critical review on 'Theories of Heavy-Electron Systems' by Lee et al. (1986).

3.2. *Phenomenology*

(a) The extremely large values of the electronic specific-heat coefficient γ serving to distinguish the heavy electron materials immediately suggest the presence of a very narrow 'band' or resonance at the Fermi level. It is the first order of business for theory to provide a way to think about this. The many-body theory of the Anderson model does provide a description for the impurity-atom situation, and the extension to the ordered lattice is a very active current research area. Band theory gives a completely different description of the situation.

The observation that this γ has a large temperature dependence has two immediate possible explanations. One is that the narrow 'band' has an effective Fermi temperature T_F which is of order 10 K, so that T/T_F is already large in the range of temperatures where one usually extracts γ from specific heat data of ordinary metals. One then has a non-degenerate electron gas above this temperature, $T > T_F$. It seems possible that the electron correlations might be such as to make it reasonable to continue to regard the electrons as occupying this band, rather than spilling over into lower-effective-mass states near the Fermi level.

The other immediate explanation is to regard the high γ as arising from a 'Kondo' or 'Abrikosov–Suhl' resonance at the Fermi level. This resonance then has itself an intrinsic temperature dependence such that it no longer exists for $T > T_0$, the characteristic temperature.

(b) A primary question in studies of the magnetism of actinide (and transition-metal) systems has been that of whether or not the magnetic order observed in many compounds derives from itinerant or localized electrons. A new variant of this question arises with the heavy-electron systems: the electronic specific heat γ is generally taken to be a property of itinerant electrons, yet the magnetically ordered heavy-electron systems seem to possess magnetic moments localized at lattice sites. Furthermore, very large entropies are developed already at low temperature in the electronic system, entropies of the size expected from the releasing of local internal degrees of freedom.

Our attention is turned by these observations to considerations of how magnetic moments can be lost in metals. One way, of course, is by hybridization which, in principle, is included in the Anderson model. The extensive results on the Anderson model and its limiting case, the Kondo-impurity model, provide a very different way to lose magnetic moment in which conduction electrons effectively form a singlet with the local spin moment. The usefulness of this way of thinking is partially obscured by the lack of definite evidence that it is relevant to uranium systems. For cerium systems the evidence is clearer.

(c) There appears to be two clearly distinct temperature regimes relevant to the heavy-electron systems. The low-temperature regime is characterized by Fermi-liquid behaviour or an approach to this; the high-temperature regime is characterized by behaviour reminiscent of a collection of independent magnetic impurities. The transition from high- to low-temperature regimes is now generally referred to as the onset of coherence, meaning that some kind of phase coherence is being established in the electronic scattering at different lattice sites. The idea that a highly correlated Fermi liquid exists in these materials at sufficiently low temperatures is not itself surprising, but at the same time not obvious. What is difficult is trying to understand how the parameters characterizing the high-temperature regime evolve into those characterizing the low-temperature one. We also expect that a correct Fermi-liquid description will involve some attempt to incorporate the effects of spin-orbit coupling, known to be important in actinide systems, and real band-structure effects, already evident in the anisotropies of macroscopic properties such as magnetic susceptibilities and electrical resistivities (Brooks et al. 1984).

3.3. *Theoretical models*

3.3.1. *Anderson model*

Understanding the occurrence of magnetic moments in solids is a long standing problem. One approach is by consideration of how one impurity atom in an otherwise non-magnetic host can develop a moment. In metals much of the thinking along these lines has been guided by results obtained for the Anderson Hamiltonian (Anderson 1961):

$$H=\sum_{k\sigma} E_k n_{k\sigma} + E_{\mathrm{f}} \sum_{\sigma} n_{\mathrm{f}\sigma} + U n_{\mathrm{f}\sigma} n_{\mathrm{f}-\sigma} + \sum_{k\sigma} (V_{k\mathrm{f}} C^{+}_{k\sigma} C_{\mathrm{f}\sigma} + V_{\mathrm{f}k} C^{+}_{\mathrm{f}\sigma} C_{k\sigma}). \tag{18}$$

The host band states labelled by $k\sigma$ have energy E_k, occupation $n_{k\sigma}$; $C^{+}_{k\sigma}$ is the band-state creation operator. The impurity has its unperturbed f-level at energy E_{f} and this level will be taken to have a degeneracy N_{f}, U is the Coulomb correlation energy which restricts double-f occupancy of the impurity. $V_{k\mathrm{f}}$ is the f-conduction electron hybridization: this gives the f-level a width $\Delta = \pi N(0)|V_{k\mathrm{f}}|^2$, $N(0)$ being the density of states at the host Fermi level.

The Hartree–Fock solution to the orbitally non-degenerate case gives a magnetic moment for $(E_{\mathrm{F}} - E_{\mathrm{f}})/U$ and Δ/U, both sufficiently small. A good discussion of this can be found in work by Blandin (1973) and by Kittel (1963). The remarks made so far concerning the Anderson Hamiltonian apply only to $T=0\,\mathrm{K}$. For $k_{\mathrm{B}}T$ much less than the energies involved in the problem we expect the conclusions to remain valid. It is a curious fact, and one very important to our topic, that at low temperature many-body effects not envisioned in the Hartree–Fock solution mentioned above can cause the conduction electrons to screen out the impurity moment whose development is seen in the Hartree–Fock solution.

This disappearance of the local moment in some situations at low temperature was first understood within the Kondo model. Here we have an interaction Hamiltonian (Kasuya, 1956),

$$H_{\mathrm{K}} = 2J\boldsymbol{S}\cdot s, \tag{19}$$

which involves the contact interaction between a localized impurity spin $\boldsymbol{S}$ and a conduction-electron spin s. The singular behaviour occurs for $J<0$ and leads to the gradual formation of the Kondo singlet state below a characteristic temperature

$$T_0 \sim \exp\left(-\frac{1}{JN(0)}\right). \tag{20}$$

Associated with the development of the singlet is a logarithmic increase of resistivity on cooling towards $T = 0$ K, which varies as $\ln(T_0/T)$ and then saturates, and a magnetic susceptibility which approaches a $T = 0$ K limiting value of $\chi = C/T_0$, where C is the Curie constant for the local moment. The impurity-spin entropy is also lost over a temperature range of T_0.

It is known that the Kondo Hamiltonian can be derived from the Anderson one, and as there are now extensive Bethe-Ansatz solutions to the Anderson Hamiltonian including orbital degeneracy (which is more difficult within the Kondo model) and, in some cases crystal-field splittings, one can consider these results, within the framework of the terminology developed for the Kondo problem (Kondo 1964).

The first result of interest to us is that of Gunnarsson and Schönhammer (1983) who developed a large-N_f variational solution to the Anderson model specifically designed to explain the XPS/BIS spectra of Ce compounds. They were quite successful: they found an effective density of states with a broadened peak near E_F, an empty peak near $E_F + U$ and a resonance just above the Fermi level at T_0 with width $\sim \pi T_0/N_f$. In addition to explaining a great deal of the spectroscopic data even on concentrated Ce compounds, their solution showed that the limiting $T = 0$ K values of the magnetic susceptibility and specific heat would both be proportional to $1/T_0$. Many further calculations involving the Bethe-Ansatz approach as well as the $1/N_f$ method can be studied in the aforementioned Proceedings of the 8th Taniguchi Symposium (Kasuya and Saso 1985).

There are several important caveats to be raised here regarding the applicability of these results to uranium systems. The first is that the observed XPS/BIS spectra for heavy-electron uranium compounds show in general a very broad response peaked somewhat above the Fermi level, very different from the cerium compound data where there is good evidence for the f-level being about 2 eV below the Fermi energy. A further puzzle is the lack of more evidence in inelastic neutron scattering for 5f crystal-field splittings in heavy-electron uranium compounds, and which are usually seen in cerium compounds. Observation of such splittings for uranium compounds would give more confidence that the underlying physics of the cerium and uranium heavy-electron compounds is similar (see Buyers and Holden 1985).

A further problem mentioned by Lee et al. (1986) is that the uranium problem is intrinsically more complicated. This is because two magnetic configurations, $5f^2$ and $5f^3$ are involved, and it is not clear exactly how to incorporate the term involving U in the Hamiltonian.

In spite of this, the Anderson impurity-model results provide a very useful framework for thinking about the heavy-electron systems of uranium. For example, the ratio of χ/γ approaches a value at low temperature which is generally within a small numerical factor of the free-electron value, indicating the irrelevance of T_0 for the ratio, as expected. Exactly, what this ratio should be depends on both the angular momentum J of the f-configuration and also on the crystal-field ground state, about which we have no distinct information.

The applicability of impurity results to the ordered-compounds situation appears, at first site, very questionable. However, for cerium compounds, e.g., in the alloys $La_{1-x}Ce_xAl_2$, it has been shown that the electrical resistivity per Ce is independent of x for $T > 20$ K (Onuki et al. 1984). (We know of no uranium compound data showing a similar independence of x.) It is possible that until coherence develops, the f-atoms act as impurities and that intersite interactions do not affect T_0. There is no convincing theory as yet which shows how the coherence temperature, which for this discussion we may take as the position of the peak of the resistivity for those compounds with resistance peaks, depends on T_0 or the physical parameters of the system. We note in this connection that Batlogg et al. (1987) have analyzed their magnetoresistance data on UBe_{13} using Schlottmann's calculations (Schlottmann 1983) and find T_0 (T^* in section 2) to be strongly decreasing with temperature below 10 K, but weakly temperature dependent above 10 K (see fig. 43). The peak in resistivity in UBe_{13} is near 2 K, there being a shoulder near 10 K.

Experimentally it seems very easy to upset the coherence with small amounts of added impurities, at the 1–2% level. Data are available for U_2Zn_{17} and UBe_{13}, for example. One finds some effect on the specific heat and magnetic susceptibility – the lattice-parameter change seems of dominant importance to these properties. The more striking effects are seen, however, in the electrical resistivity and Hall effect. For the latter, there seems to be in several cases a change in sign of the Hall constant when the coherence sets in, the high-temperature Hall constant being accounted for by the skew-scattering impurity result of Coleman et al. (1985).

In the electrical resistivity, the drop to low values below the coherence temperature is rapidly lost with impurity addition, the incremental resistivities per impurity being often tens of μΩ-cm, seemingly well beyond the unitarity limit. In addition, there is often a rise in resistivity at lowest temperature instead of a tendency to a constant value. A further question, and possible complication, to the analysis of resistivity data is the relevance of the phenomenon of resistance saturation so often seen in transition-metal systems (Fisk and Webb 1976). A number of heavy-electron uranium compounds have monotonically increasing (with temperature) electrical resistivities (e.g. UPt_3, $UAuPt_4$) which are very similar to the electrical resistivities seen in A15 transition-metal compounds such as Nb_3Sn. Is the coherence temperature for these at some high temperature, or possibly is the coherence temperature renormalized with T as in the Yu–Anderson theory (Yu and Anderson 1984) of A15 high-T_c superconductors?

3.3.2. *Anderson-lattice calculations*

The interesting low-temperature properties of heavy-electron systems only occur for ordered lattices of f-atoms or alloys close to this. It is therefore essential for theory to come to grips with the full interacting lattice. It is not yet clear in exactly which way the two-impurity-site problem differs from the single-site one. There are, of course, the RKKY interaction, the strength of which (k_BT_{RKKY}) compared with k_BT_0 is naturally important, and higher-order terms in the RKKY (Abrahams and Varma 1985, Suhl 1985), the singular character of which has been a contested issue (Wilkins 1985). Suffice it to say here that formation of the heavy-electron state would seem to require that T_{RKKY} be smaller than temperatures such as T_{coh} and T_0.

An immediate new feature of the lattice problem not present in the single-site one is that the Fermi level has to be self-consistently found, since it is no longer determined by the 'host' of non-f electrons. A further question concerning the lattice problem (Nozières 1985) is whether there are enough conduction electrons present to screen each lattice site completely. Nozières suggests that the screening is actually done by a mobile screening cloud, with this cloud itself being responsible for the heavy-electron phenomena.

One general line of attack on the Anderson lattice is to derive from it an effective one-particle Hamiltonian which will give the quasi-particle energy spectrum near the Fermi energy. A number of techniques have been used for this. The work of Read et al. (1984) using a functional-integral technique results in an effective Hamiltonian with a reduction of the hybridization term V_{kf} by the factor $\sqrt{1-n_f}$ for the single f-electron or f-hole cases they considered, and a shift of E_F.

A quite different approach is that by Rice and Ueda (1985). They use the Gutzwiller approximation (Gutzwiller 1964) with n_f fixed to derive the effective Hamiltonian. In this case, V is renormalized by a factor

$$\left(\frac{1-n_f}{1-n_{1s}}\right)^{1/2} \tag{21}$$

where n_{1s} is the occupancy of electrons with quantum numbers 1s and n_f is the number of f-electrons per site. In the large N_f limit, $n_{1s}\to 0$ and this renormalization factor is equal to that found by Read et al. (1984). In the approach here, E_F is shifted so as to keep n_f fixed. The condensation energy per site in the coherent state for this Gutzwiller approach has a different form than that found by Read et al. In the Gutzwiller model the f-electrons hop from lattice site to lattice site, making the quasi-particle bands f-like.

A third approach to finding an effective Hamiltonian uses Coleman's slave-boson technique for restricting $n_f \leqslant 1$, a method previously applied in the quite successful large-N_f expansion for the single-site Anderson model (Coleman 1984). Read and Newns (1984) extended the method to the lattice problem. For large N_f, the results are essentially the same as in the Gutzwiller method. There is a spurious phase transition which occurs for finite N_f due to the broken symmetry for N_f infinite. This is discussed by Read (1985) and by Coleman (1985).

We refer the reader also to the early work by Yoshimori and Kasai (1983) on an extension of the single-site Anderson model for Ce-compounds, Brandow's diagrammatic technique (Brandow 1986) giving results similar to those of the Gutzwiller approach and Kuramoto's Green's-function method (Kuramoto 1985).

A quite different approach to the lattice problem is that of Razafimandimby et al. (1984), and further developments of it by d'Ambrumenil and Fulde (1985). They treat the special case of $CeCu_2Si_2$ with an eye towards an electron–phonon induced superconductivity. They deal with the compound as a Kondo lattice and put the Kondo scattering phase shift in by hand via a resonance width T_K, and then perform a Korringa–Kohn–Rostoker (KKR) calculation. A feature of their results is the presence of both light and heavy electrons at the Fermi surface.

The kind of result one is looking for in all these calculations, we reiterate, is a description of how the coherent state develops from the high-temperature local-moment regime. We want to know how the coherence temperature T_{coh} is related to the single-site characteristic temperature T_0. This is likely to be complicated. For example, in UBe_{13}, T_{coh} must be somewhat less than T_0, judging by the position of the resistance maximum near 2 K and Batlogg et al.'s magneto-resistance estimates of T_0 (Batlogg et al. 1987). UPt_3, on the other hand, would seem to have these temperatures closer together: the magnetic susceptibility has something like Curie–Weiss behaviour below temperatures for which one sees the large variations in the electrical resistivity and Hall constant (Schoenes and Franse 1986, Hadzic–Leroux et al. 1986). An entropy integregation could also be used to suggest the magnitude of T_0.

3.3.3. *Band-structure calculations*

Although band-structure calculations exist for a number of uranium intermetallics (Brooks et al. 1984, Arko et al. 1984b), the only heavy-electron compounds for which we have results are UBe_{13} (Takegahara et al. 1985), UPt_3 (Albers et al. 1985, Albers 1985, Boring et al. 1985b, Oguchi and Freeman 1985a, Strange and Gyorffy 1985, Sticht and Kübler 1985) and UAl_2 (Boring et al. 1985a). These have all been done self-consistently within the local-density approximation with fully relativistic codes.

The first point to make is that these results are clearly $T=0$ K results and cannot be expected to reproduce the transition to the high-temperature regime. As might be expected, the f-bands for all these compounds are narrow and pinned at the Fermi level. Due to the large degeneracy of the f-shell, the Fermi level is forced towards the bottom of the f-band. The f-band widths are found to be of order 1 eV. The immediate question is: can the large masses be accounted for? The answer, so far, is no. While the results give band masses of order $10\,m_e$, considerable further enhancement is needed to achieve the values estimated from experiment. But we must not overlook the important contribution these calculations make, even if the 'heaviness' of the electrons is not predicted. It must be the case that the f-neighboring ligand interactions are the most important for developing heavy-electron behaviour (Koelling et al. 1985). This is clearly a question well addressed by these calculations, and attempts to deconvolute such effects in UPt_3 have been made (Albers 1985). The very detailed calculations by Oguchi and Freeman (1985a) have also been used to directly address the superconductivity of UPt_3. A problem here is the apparent instability against magnetic order. One hopes, in fact, that some criteria for when heavy-electron behaviour will occur might be derivable from band calculations. Attempts to include realistically the effects of the crystal structure on heavy-electron behaviour can only be soundly based on such calculations.

We finally mention results on $CeSn_3$, the one intermediate-value γ material for which de Haas–van Alphen results exist (Johanson et al. 1983). These results, in fact, caused quite a surprise as they gave for the first time real confidence in the idea that an intermediate-valence compound might have a well-defined Fermi surface at low temperatures. Band-structure calculations (Koelling 1982, Yanase 1983) matched the data quite well over large parts of the Fermi surface. Critics have pointed out that this matching actually does not provide a stringent test of the calculations. Also of interest is the calculation by Hewson et al. (1985) for $CeSn_3$, initially based on an effective one-body Hamiltonian derived from the Anderson-lattice model. It proved necessary, due to the difficulty of the calculation, for them to use the hybridization renormalization parameter $\sqrt{1-n_f}$ as an adjustable parameter not determined self-consistently. Here also good agreement with experimentally determined external Fermi surface areas was obtained.

3.3.4. *Fermi-liquid theory*

We have mentioned before that it is generally believed that the low-temperature coherent state of heavy-electron systems should be described in Fermi-liquid terms. There is no theorem that says this must be so, but the linear specific heats at low temperature, the Pauli-like susceptibilities and the observation of T^2-variations of the electrical resistivity all suggest Fermi-liquid behaviour. At the same time it is to be emphasized that crystalline anisotropies of the electronic properties and spin–orbit effects have not been included in any of the Fermi-liquid approaches to heavy-electron systems. For example, the anisotropy of the resistivity and magnetic susceptibility of UPt_3 is very large – almost a factor of 2 between basal plane and c-axis.

The great success of Fermi-liquid theory has been in its application to ^{3}He, for which Landau (1956) invented it. The application to heavy-electron materials has been partly motivated by the analogy of these materials to ^{3}He as both being nearly localized. All the applications to date have only involved the Landau interaction parameters F_0^s and F_0^a. For a discussion of Landau's theory see the work by Leggett (1975). It might be expected that $l=3$ scattering would be very important in 4f and 5f materials, hence making the corresponding parameters in the Legendre expansion of the Landau interaction parameters important. It seems to be the hope of the Fermi-liquid practitioners that the f-electrons only serve to make the conduction electrons heavy and that nothing is really happening in the $l=3$ phase shift.

All authors agree that spin fluctuations are strong and that charge fluctuations can be neglected. F_0^s is universally agreed to be large and positive. The magnetic susceptibility is enhanced over that of free electrons by the factor $S=(m^*/m)/(1+F_0^a)$. The 'Wilson ratio' based on experimental results is here given by

$$R=\frac{\pi^2 k_B^2 \chi(T=0)}{g^2 J(J+1)\mu_B^2 \gamma(T=0)}=\frac{1}{1+F_0^a}. \tag{22}$$

There is uncertainty as to what J is appropriate for uranium, and even knowing J, we still have to decide about what the ground state of the crystal-field-split J multiplet is. We also note here the discussion by Brown et al. (1985) about the effects of hybridization in lowering the f-moment and thus changing R. Their argument was directed to suggesting that small R implies good hybridization which, they believed, favored superconductivity.

A goal of the Fermi-liquid approach has been to use the parameters determined from the normal-state properties to say something about the superconducting properties. Pethick et al. (1986) have done this for the case of UPt_3 in a finite wave-vector extension of the theory, being able to correctly give the pressure dependence of T_c with a triplet-pairing model, given the pressure dependence of the electronic specific heat. Rodriguez (1985) has computed the electronic viscosity tensor for evaluating the ultrasonic attenuation in the normal and superconducting states. A point of controversy with this calculation is whether or not the scattering time is energy dependent.

Pethick et al. (1986) have also been able to get $\chi(q, \omega)$ from their approach and find agreement with neutron-scattering results on the spin-fluctuation parameters from the T^3 and $T^3 \ln T$ terms in the specific heat. They conclude that there are contributions to the T^3 term in UPt_3 not coming from phonons or spin fluctuations.

A $T^3 \ln T$ term in the specific heat is often taken as a marker of spin fluctuations but it is also predicted by Fermi-liquid theory. For large values of $1/(1+F_0^a)$, the spin-fluctuations term appears as the dominant contribution to the $T^3 \ln T$ term. For UPt_3, Coffey and Pethick (1986) estimate $1/(1+F_0^a) \sim 5$. There are contributions to the $T^3 \ln T$ term other than from F_0^a and these are known to be important for ^{3}He. We also note that UBe_{13} has no clear $T^3 \ln T$ feature in its low-temperature specific heat and, in fact, applicability of the Fermi-liquid theory is not obvious since superconductivity sets in before one has established the usual Fermi liquid properties such as T^2 electrical resistivity. Possibly high pressure and/or high magnetic fields will make the Fermi-liquid regime accessible. The T^2 term in the electrical resistivity follows from general phase–space arguments within Fermi-liquid theory. The magnitude of the T^2 coefficient, however, is more difficult to obtain. Spin-fluctuation theory gives a T^2 term in ρ, as does the Baber electron–electron scattering (Baber 1937). In addition, there is a T^2 variation expected in χ within Fermi-liquid theory, and a possible T^2 variation of χ which follows from the second term in a T/T_F expansion of the Pauli susceptibility.

UBe_{13}, however, has been approached in a different way and in analogy to ^{3}He, by Rice et al. (1985). They use an extension of the Brinkman–Rice theory (Brinkman and Rice 1970). Using the Gutzwiller approximation applied to the Anderson model in the Kondo limit, they calculate the mass renormalization (to the band mass) and F_0^a. The latter they find to depend on the f-orbital degeneracy. An interesting conclusion is that the paramagnetic ground state will only be stable for high degeneracy. This would make an interesting distinction between the superconducting and magnetic systems if it could be experimentally verified. In this connection we note that, as presented in the section on systematics, the magnetically ordering heavy-electron systems have smaller γ's in the paramagnetic state just above T_N than the superconducting ones, at least consistent with the notion of a higher degeneracy f-state underlying the heavy-electron system of the superconductors. Rice et al. (1985) also integrate the temperature dependence of the entropy in their model. Fitting this to the experimental data gives a Fermi energy in UBe_{13} of 520 K and an effective mass $m^* = 25\, m_b$, where m_b is the band effective mass.

Fermi-liquid ideas are naturally limited in their applicability to temperatures small compared with the effective Fermi temperature. In this range, however, band theory should be able to produce the appropriate Landau parameters. There are, in addition, other ways in which the Fermi-liquid approach has been useful. One is Nozières' realization that the low-temperature limit of the single-impurity Kondo problem is a local Fermi liquid, in which the scattering of electrons by the impurity is treated by a phase shift (Nozières 1974). This approach is not easily generalized to the Anderson lattice: general arguments do not serve to determine the Landau parameters. Above we remarked that some of this thinking went into the periodic-lattice models of Razafimandimby et al. (1984) and d'Ambrumenil and Fulde (1985). Another interesting application of Fermi-liquid ideas has been that of Martin and co-workers: the Luttinger theorem has been applied to cerium f-resonances (Martin 1982). In one application, the peak observed in the low-temperature electronic specific heat of $CeAl_3$, for example, is seen as the formation of a pseudo-gap in the resonant density of states at the Fermi level, due to coherence (Bredl et al. 1984).

Finally, we point to a problem that has been emphasized by Varma (1985a), namely the experimental fact that many of the properties observed in heavy-electron systems have 'normal' values. For example, Korringa relaxation rates in EPR (Gandra et al. 1985) and electron–phonon couplings (Batlogg et al. 1987) have values typical for metals. Varma argues that some properties scale with the effective mass, the specific heat, for example, some do not. Essentially, m^* appears in both the numerator and denominator of certain expressions and cancels out. A related problem is how to recognize whether there are any light electrons at the Fermi level in the heavy-electron systems. Other explanations than these vertex corrections seem possible, partly because the measurements are not always measuring exactly what one wants. The NMR relaxation rates of ^{9}Be in UBe_{13}, for instance, can be argued to be very fast, whereas the EPR rates measured on atoms substituted for U are normal: we can claim that the EPR does not see the heavy-electron system because the EPR-active atoms are not hybridized in the same way the uranium atoms are.

4. Phase transitions

The high current interest in heavy-electron materials is largely due to the occurrence of low-temperature phase transitions in these substances and, in particular, the observation of superconductivity involving the heavy quasi-particles. Since superconductivity was so unexpected in this kind of materials, it did not take long for arguments to be brought forward suggesting that unconventional mechanisms (not the usual electron–photon interaction) are responsible for this superconductivity. In this case the resulting superconducting state is characterized by non-spherical symmetry, opposite to what is accepted for ordinary superconductors (Varma 1984b, Anderson 1984a). If one seriously considers that spin-fluctuation interactions might be responsible for superconductivity in these materials it seems natural also to expect magnetic-ordering phenomena, possibly again involving the heavy but itinerant electrons. As examples for this behaviour we discuss below some aspects of phase transitions that are observed in these materials. According to the above remarks, it seems natural to look for similarities between the two different types of phase transitions and we attempt to point them out where they exist.

4.1. *Experiment*

4.1.1. *Superconductivity*

In the class of substances under consideration, two compounds are superconducting below critical temperatures that are less than 1 K, namely UBe_{13} and UPt_3.

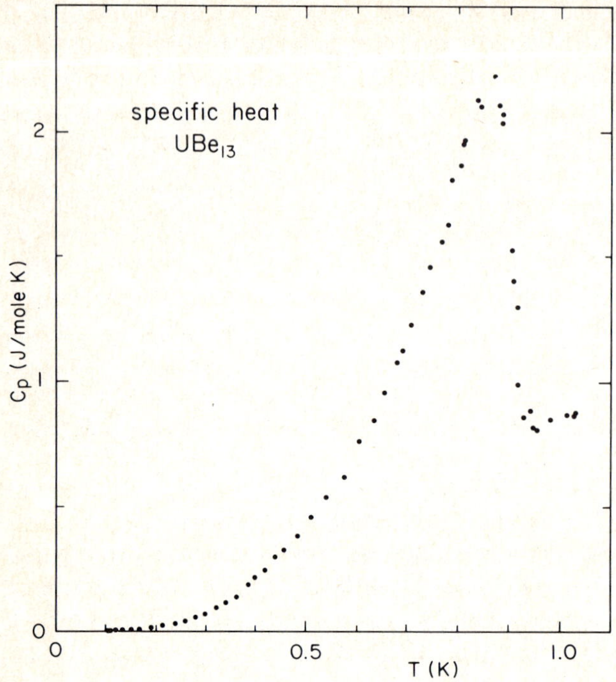

Fig. 93. Temperature dependence of the specific heat of UBe_{13} between 0.12 and 1 K, as obtained by Ott et al. (1984c).

Traces of superconductivity in UBe_{13} were first observed by Bucher et al. (1975) by measurements of the magnetic susceptibility, but interpreted as being due to precipitated U filaments in the samples. The demonstration of bulk superconductivity of heavy electrons in UBe_{13} below 0.9 K by specific-heat measurements by

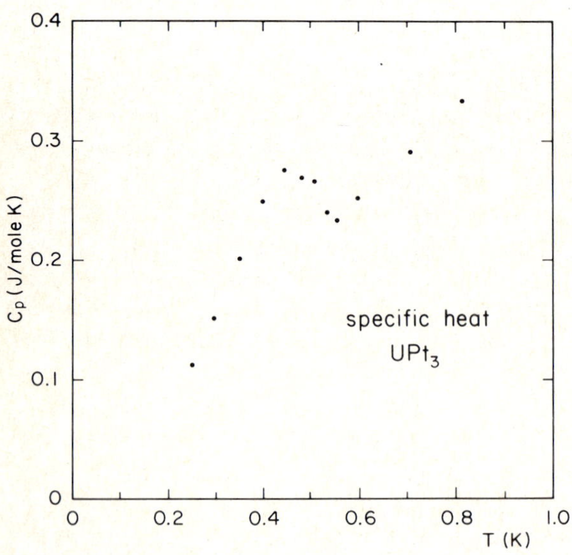

Fig. 94. Temperature dependence of the specific heat of UPt_3 between 0.25 and 0.8 K as obtained by Stewart et al. (1984c).

Ott et al. (1983) was important because it showed that the often debated superconductivity involving quasi-particles with large effective mass in $CeCu_2Si_2$ as reported by Steglich et al. (1979) also occurred in other substances and, in addition, was also possible in materials with 5f electrons. The subsequent discovery of superconductivity of UPt_3 by Stewart et al. (1984c) gave additional impetus to the field.

In figs. 93 and 94 we show the temperature dependence of the specific heats of UBe_{13} and UPt_3 just above and below the respective superconducting phase transitions. The discontinuity of C_p at T_c is very large for UBe_{13} but consistent with the large electronic specific heat in the normal state just above T_c. Different workers report essentially the same curves, depending only on somewhat different T_c's of different samples (see also table 1). The anomaly of C_p at T_c is distinctly smaller for UPt_3 and is therefore usually shown as a C_p/T plot to elucidate it. Different workers using different samples of UPt_3 report somewhat different temperature dependences of C_p, especially in the superconducting state. A distinct difference between UBe_{13} and UPt_3 is also found for $C_p(T)$ at temperatures well below T_c. While it is immediately recognized that for UPt_3 a considerable contribution to C_p varies linearly with temperature in the superconducting state (Steglich et al. 1985, Sulpice et al. 1986) the same is apparent for UBe_{13} only at temperatures $T \ll T_c$, i.e., below 0.15 K. Very recent and as yet unpublished experiments below 0.1 K indicate that also for UBe_{13} a small linear-in-T term of the specific heat must be considered (Flouquet 1986, Ott et al. 1986d). The significance of these findings are of utmost importance to theoretical concepts and are in the course of being clarified (Hirschfeld et al. 1986). While simple BCS superconductors (Bardeen et al. 1957) exhibit an exponential decay of C_p at $T \ll T_c$, alternative superconducting states would manifest themselves by other characteristic temperature dependences of C_p in this temperature range. We should like to point out here that in all published work on the specific heat in the superconducting state of UBe_{13} the entropy balance between the hypothetical normal and the superconducting state is never quite zero. A slightly positive $\Delta S = S_s - S_n$ between 4% and 10%, depending on the sample measured, is reported in all cases. It is not clear whether this imbalance is due to an as yet unknown feature of the hypothetical normal state as $T \to 0$ K. Measurements below 0.1 K and in supercritical external magnetic fields [i.e., $H > H_{c2}(0)$] will have to clarify this.

In fig. 95 we show the ratio of the specific heat $C_p(T)$ of UBe_{13} in the superconducting state with the specific heat in the normal state at T_c in the form $\log[C_p^s(T)/C_p^n(T_c)]$ versus T_c/T. A comparison with the universal BCS curves shows considerable deviations both at $T \sim T_c$ and $T \ll T_c$. The temperature variation of C_p is fairly well approximated by a power law T^n, where $n = 2.85 - 2.90$ (Ott et al. 1986a, Mayer et al. 1986), hence close to 3. Also shown in fig. 95 are two curves that were calculated by assuming that the superconducting energy gap is anisotropic and has nodes at certain points of the Fermi surface, as is possible for non-singlet-type pairing of electrons in unconventional superconductors. More details about the calculations can be found in the article by Ott et al. (1984c), but we note that the salient features of the experimental data are fairly well reproduced

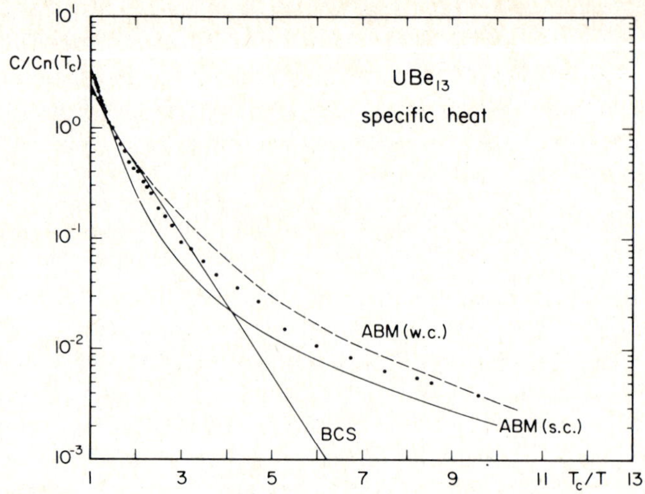

Fig. 95. $C_s/C_n(T_c)$ for superconducting UBe_{13}. Broken line: weak-coupling ABM state; solid lines: BCS and strong coupling ABM state. (Data and calculations are from Ott et al. 1984c.)

by assuming a strong-coupling axial or Anderson–Brinkman–Morel (ABM) state for the superconducting state of UBe_{13}.

The temperature dependence of C_p in the superconducting state of UPt_3 has been investigated in detail by at least two groups (Steglich et al. 1985, Sulpice et al. 1986) with slightly different results. Both claim two distinct contributions proportional to T and T^2, respectively, but there is some disagreement on the magnitude of the two terms. In fig. 96 we show the results of Sulpice et al. (1986) where C_p/T is plotted versus T. The existence of the linear term in T is taken as evidence that not all electrons participate in the formation of the superconducting state. Also in UPt_3, the

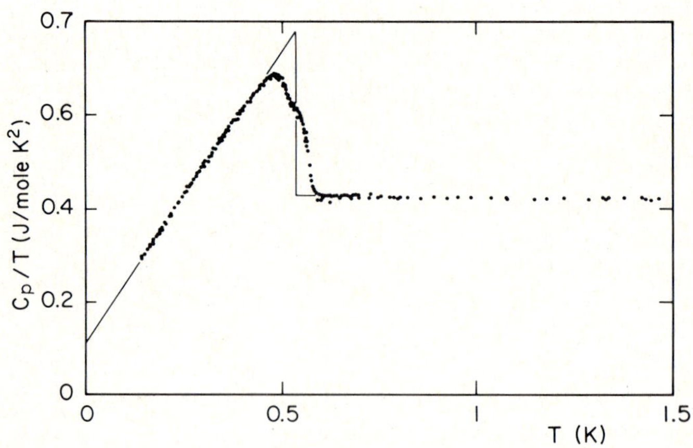

Fig. 96. C_p/T versus T of UPt_3 between 0.15 and 1.5 K as obtained by Sulpice et al. (1986). The solid line represents linear extrapolations from the normal and superconducting state towards $T = T_c$ and $T = 0$ K.

problem of the zero-entropy balance is unsolved mainly because the reduced temperatures T/T_c that can be reached experimentally are not very low. Some of the estimates of the normal-state contributions below T_c that are mentioned above are based on the assumption that $\Delta S=0$ for the superconducting transition.

An important quantity of every superconductor is the temperature dependence of its critical magnetic field. Since both UBe_{13} and UPt_3 are type-II superconductors, the superconducting phases are limited by the upper critical field H_{c2}. For both compounds, $H_{c2}(T)$ shows various remarkable features. In fig. 97 we show results by Chen et al. (1985) on $H_{c2}(T)$ of UBe_{13} from resistivity measurements on a single-crystalline and a polycrystalline sample, respectively. First we note the extremely steep slope of $\partial H_c/\partial T$ at T_c which is at least of the order of 0.5 MOe/K but also a vertical or even positive slope very close to T_c cannot be excluded from the available data (Maple et al. 1985, Rauchschwalbe et al. 1985). For the single-crystalline sample, where the external field was oriented parallel to the cubic crystal axis, the temperature derivative decreases rapidly with decreasing temperature and reaches a constant value of about -90 kOe/K at 0.6 K and stays constant down to the lowest temperature of 50 mK. Even more intriguing is $H_{c2}(T)$ for the polycrystalline sample where below about 0.5 K the negative slope $(\partial H_{c2}/\partial T)$ increases again with decreasing temperature without any sign of saturation down to 0.05 K (Chen et al. 1985, Rauchschwalbe et al. 1985). The features of the two results shown in fig. 97 might be due to a possible anisotropy of H_{c2} of UBe_{13} at very low temperatures. In a recent publication, Alekseevskii et al. (1985) claim the observation of such an anisotropy from resistivity measurements on a single crystal with the external magnetic field oriented along different crystallographic directions. According to their results the anisotropy increases with decreasing temperature.

The general shape of the critical-field curve of UBe_{13} has been confirmed by measurements of the thermoelectric power S (Jaccard et al. 1985a), and of the

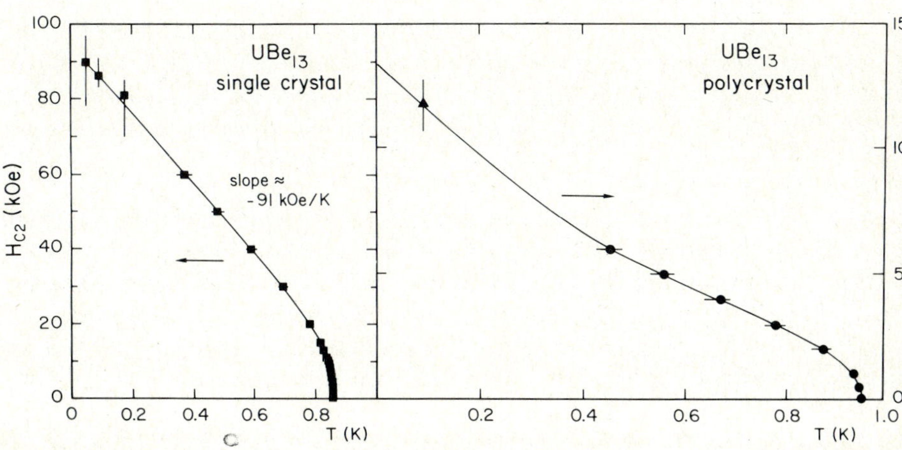

Fig. 97. Temperature dependences of the upper critical field H_{c2} for superconducting UBe_{13} in single-crystalline and polycrystalline form, as obtained by Chen et al. (1985).

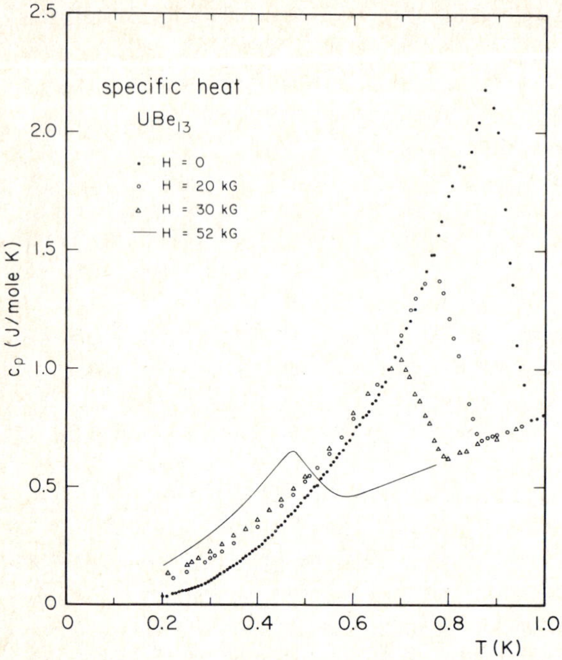

Fig. 98. Influence of external magnetic fields on the anomaly of the specific heat at the superconducting transition of UBe_{13} as obtained by Ott et al. (1986a).

specific heat (Ott et al. 1986a, Mayer et al. 1986), in external magnetic fields. Since below 1 K, the normal-state thermopower of UBe_{13} is quite large and negative, the sudden drop to zero at $T_c(H)$ gives a good indication for the superconducting transition. Also the specific-heat anomalies stay surprisingly sharp even in high magnetic fields, a feature that was also observed for the resistive transitions in even higher fields. In fig. 98 we show the influence of magnetic fields on the specific-heat anomaly at the superconducting transition of UBe_{13} (Ott et al. 1986a). The critical-field curve obtained with these bulk measurements is in very good agreement with that determined by resistivity measurements.

The main interest in the critical-field curve of UPt_3 is focussed on its anisotropy. The first results reported by Chen et al. (1984a), which are shown in fig. 99, indicated that the anisotropy of $H_{c2}(T)$ is largest at T_c. The same data also show a positive curvature of the $H_{c2}(T)$ curves near T_c but this effect is much more pronounced for the configuration when the electrical current of the resistivity measurement flows along the hexagonal c-axis and the magnetic-field direction is in the basal plane ($H_{c2}^{\perp}$). More recent measurements by Rauchschwalbe et al. (1985) confirm the general behaviour shown in fig. 99. They also note a distinctly smaller initial slope ($\partial H_{c2}/\partial T$) close to T_c for $H_{c2}^{\perp}(T)$ compared to $H_{c2}^{\parallel}$, but the change in slope for $H_{c2}^{\perp}(T)$ at lower temperatures appears to be more discontinuous than what was observed by Chen et al. (1984a).

The observation of the non-exponential decrease of the specific heat in the superconducting state triggered additional experiments to investigate the temperature dependence of various other physical properties below T_c. Non-exponential

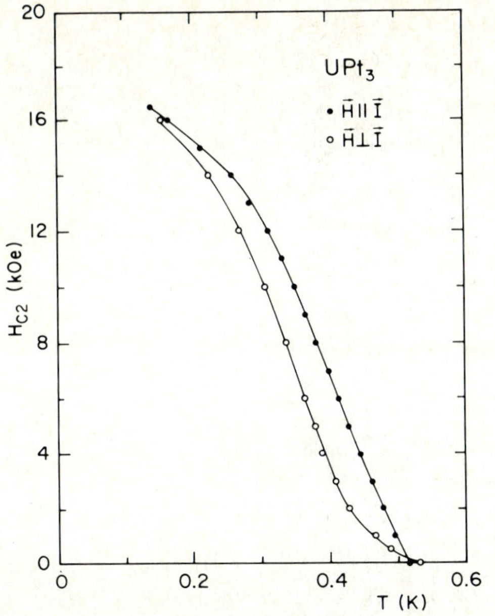

Fig. 99. Anisotropy of the upper critical field $H_{c2}(T)$ of UPt_3 as found by Chen et al. (1984a). The current I was parallel to the hexagonal c axis.

temperature variations were found for the inverse spin-lattice relaxation time T_1 of ^9Be nuclei in UBe_{13} (MacLaughlin et al. 1984), for the ultrasound absorption in UBe_{13} (Golding et al. 1985b) and UPt_3 (Bishop et al. 1984), and a power-law-type temperature dependence was reported for the thermal conductivity in the superconducting state of both UBe_{13} (Jaccard et al. 1985a) and UPt_3 (Jaccard et al. 1985b, Steglich et al. 1985). Examples of these findings are shown in figs. 61, 65, 68, 100, 101 and 103. Below T_c, T_1^{-1} decreases proportional to T^3 at least down to 0.2 K where a tendency towards saturation is indicated by the presently available data which are shown in fig. 100. A similar tendency was noted in analogous measurements on $CeCu_2Si_2$ (Kitaoka et al. 1984). In fig. 101 we show the temperature dependence of the ultrasound absorption coefficient of UBe_{13} well below T_c as measured by Golding et al. (1985b) demonstrating the T^2 variation. The same temperature dependence was reported by Bishop et al. (1984) for the attenuation in superconducting UPt_3 at very low temperatures as may be seen from fig. 68. Conflicting results are due to very recent experiments by Müller et al. (1986a) who reported a T^3 variation of the ultrasound absorption in the superconducting state of UPt_3 (see fig. 103).

The T^2 dependence of the thermal conductivity κ in the superconducting state of UBe_{13} measured by Jaccard et al. (1985a) is shown in fig. 61. A T^2 term was also found for κ in the superconducting state of UPt_3 by Steglich et al. (1985) and by Jaccard et al. (1985b). However, as in the specific heat, a considerable contribution linear T is found, again indicating the possible presence of normal electrons even well below T_c. Relevant data are shown in fig. 65.

Clearly, all these results indicate that for both UBe_{13} and UPt_3 the superconduct-

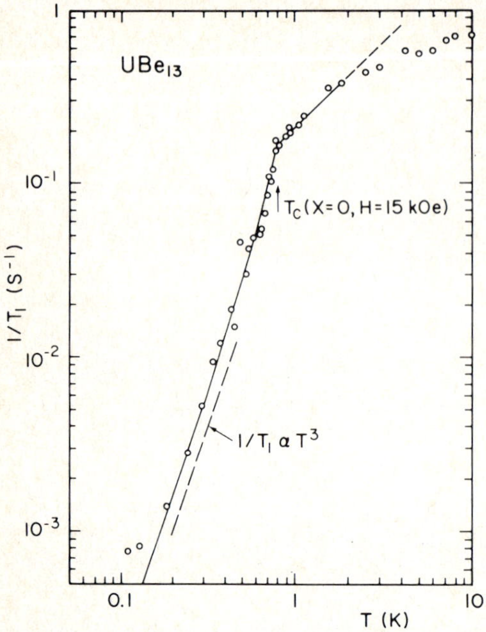

Fig. 100. Temperature dependence of the [9]Be NMR spin-lattice relaxation rate in UBe_{13} between 0.1 and 10 K. (Data are due to MacLaughlin et al. 1984.)

ing energy gap is not non-zero over the entire Fermi surface. This gap anisotropy, including gap zeroes, is particularly intriguing for the cubic compound UBe_{13}. Also for UPt_3 any crystal-structure induced anisotropy should not necessarily lead to gap zeroes. These gap anisotropies are used to argue that the superconducting states of UBe_{13} and UPt_3 are characterized by $l \neq 0$ pairing of the electrons which would then imply that the essential pairing mechanism is not due to the usual electron–phonon interaction. Although the arguments on the validity of this conclusion are still controversial, the mere possibility of the fact has induced theoretical

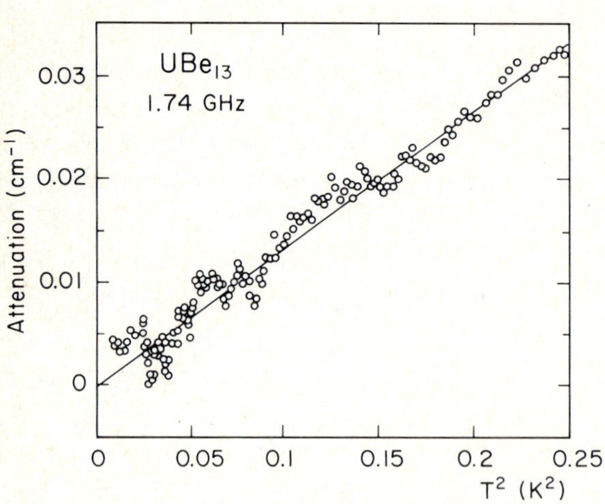

Fig. 101. Temperature dependence of the attenuation of ultrasound in superconducting UBe_{13} at temperatures well below T_c, as found by Golding et al. (1985b). The T^2 dependence is emphasized by the solid line.

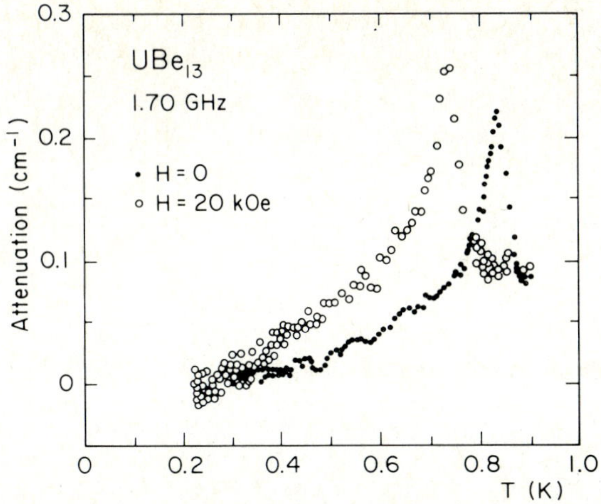

Fig. 102. Attenuation of ultrasound in UBe_{13} below 1 K in zero and 20 kOe external magnetic field. (Data are due to Golding et al. 1985a.)

work that is needed for a thorough interpretation of the experimental results. That new aspects have to be considered for the description of superconductivity in these compounds is also clear from various other experimental facts. Recently Golding et al. (1985b) reported the observation of a pronounced peak in the ultrasound absorption just below the superconducting transition of pure UBe_{13} (fig. 102), a feature that has only once been observed before in any superconductor, namely in V_3Si (Fukase et al. 1976). The magnetic-field induced shift of the peak demonstrates that it is related to the transition to the superconducting state. The temperature width of the peak seems to exclude explanations due to excitations across the opening gap just below T_c. The energy compatible with the ultrasonic frequency would only allow for excitations in a narrow range of milli-Kelvin width just below

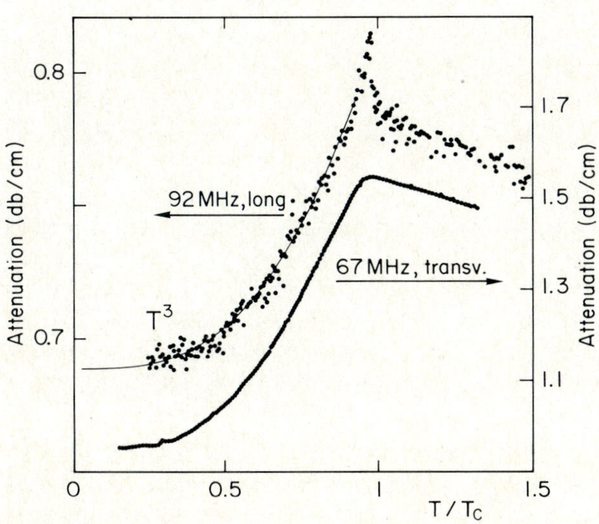

Fig. 103. Attenuation of ultrasound in UPt_3 below 1 K. Propagation of ultrasound is along the hexagonal c-axis. (Data are due to Müller et al. 1986a.)

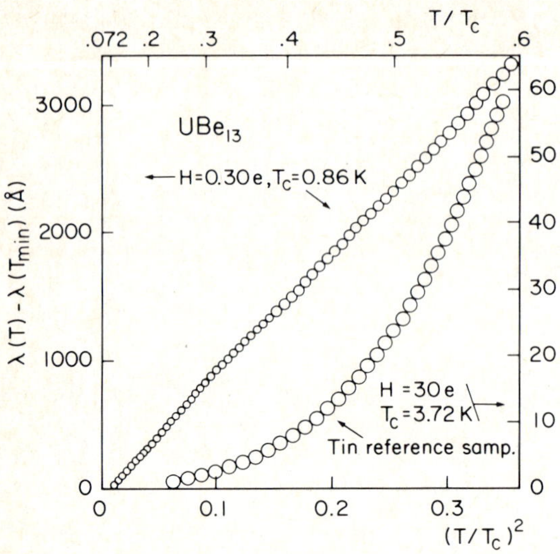

Fig. 104. Incremental magnetic-field penetration depth $\lambda(T)-\lambda(T_{min})$ of UBe_{13} versus T^2 (left scale). For comparison, the same quantity as measured for tin is shown in the same reduced-temperature range (right scale). (Data are due to Einzel et al. 1986.)

T_c. The attenuation peak observed in V_3Si by Fukase et al. (1976) is indeed much more narrow in temperature and clearly much smaller relative to the overall attenuation. What is not understood is the fact that the anomalous attenuation peak in UBe_{13} gains intensity in an external magnetic field. Its appearance was tentatively ascribed to the excitation of collective modes in an anisotropic superconducting state (Golding et al. 1985b). Müller et al. (1986) reported the observation of a much smaller but also distinct peak in the ultrasound absorption coefficient for longitudinal sound waves along the hexagonal axis of UPt_3 as shown in fig. 103. Its magnitude is strongly frequency dependent and the feature is not observed for transversely polarized sound waves along the same direction.

Non-classical behaviour was also found for the temperature dependence of the magnetic-field penetration depth λ of UBe_{13}. From BCS theory the penetration depth $\lambda(T)$ is expected to vary as $[1-(T/T_c)^4]^{-1/2}$. In a recent investigation, Gross et al. (1986) found that $\lambda(T)$ of UBe_{13} is rather proportional to $(T/T_c)^2$. In fig. 104 we show a comparison of the temperature variations of λ for UBe_{13} and Sn in the same interval of renormalized temperatures T/T_c (see also Einzel et al. 1986).

In section 2 we pointed out that small amounts of impurities often drastically change the low-temperature physical properties of these compounds. The same is true for the superconducting state of UBe_{13} and UPt_3. The superconductivity of UPt_3 is extremely sensitive to impurities and defects. Powdered UPt_3 obtained by grinding superconducting single crystals is reported to have lost the superconducting properties down to temperatures of a few milli-Kelvin (Stewart et al. 1984c). Small amounts of Pd substituted for Pt have the same devastating effect (de Visser et al. 1986a). More details are known on the influence of impurities on the superconducting state of UBe_{13}. A very strong suppression of T_c was reported when Be is replaced by Cu, Ga or B (Giorgi et al. 1984). Replacing U by Th results in a very unusual change of T_c with increasing Th content x as shown in fig. 105. What is

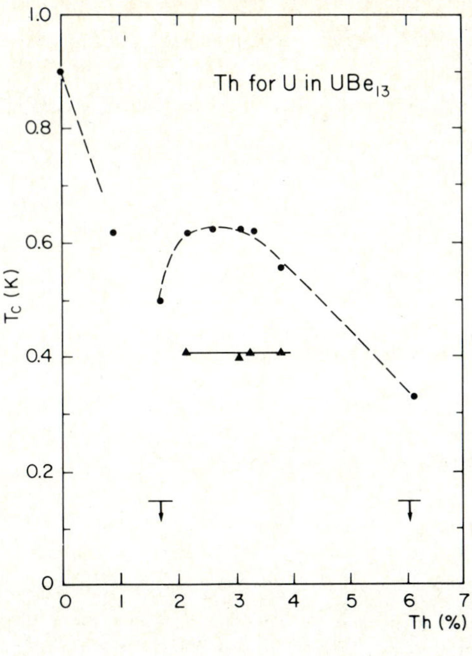

Fig. 105. Superconducting phase diagram for $U_{1-x}Th_xBe_{13}$ as obtained by specific-heat measurements by Ott et al. (1986b). The dots denote the usual superconducting transition, the triangles mark the lower transition at T_{c2}. No second transitions above 0.2 K were observed for $x=0.017$ and 0.06.

even more intriguing is the appearance of a second phase transition in the superconducting state as evidenced by measurements of the specific heat (Ott et al. 1985a). The nature of this transition is not definitely established but it clearly does not destroy superconductivity. In fig. 106 we show the specific heat for $U_{0.967}Th_{0.033}Be_{13}$, where a second anomaly at about 0.4 K indicates this unexpected phase transition. Additional measurements of the thermal expansion

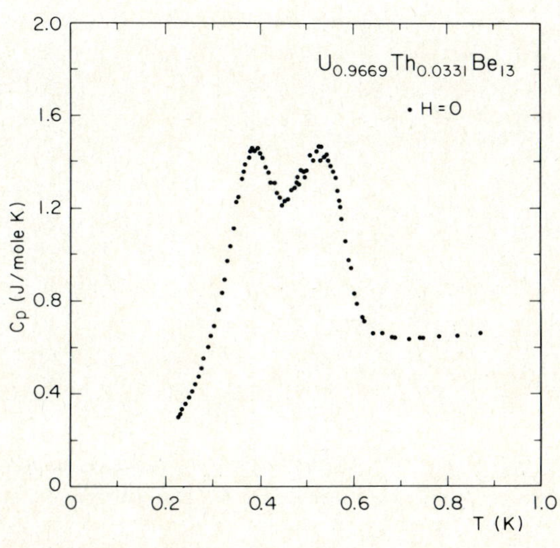

Fig. 106. Example for two phase transitions as evidenced by the temperature dependence of the specific heat in $U_{1-x}Th_xBe_{13}$ compounds. (Data are from Ott et al. 1985a.)

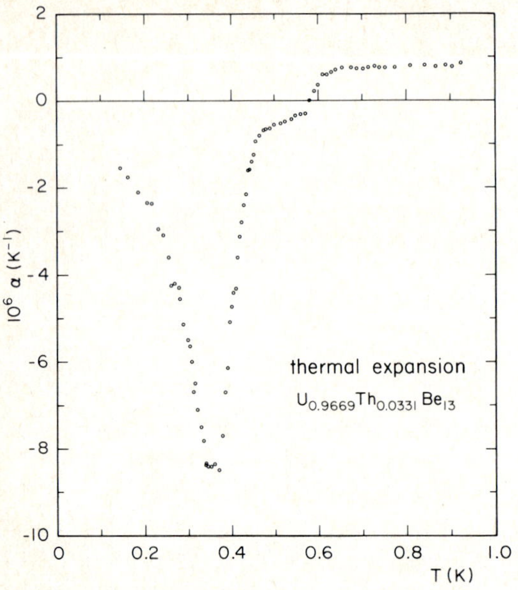

Fig. 107. Temperature dependence of the linear thermal-expansion coefficient of $U_{0.9669}Th_{0.0331}Be_{13}$ below 1 K. (Data are from Ott et al. 1985a.)

(Ott et al. 1986a) (fig. 107) and the ultrasonic attenuation (Batlogg et al. 1985a) (fig. 108) reveal that this second transition at T_{c2} couples quite strongly to the lattice. Various interpretations for this transition have been proposed. A very original interpretation based on the idea of a superconducting-glass formation was proposed by Volovik and Khmel'nitzkii (1984). Joynt et al. (1986) argue that the additional attenuation might be due to domain walls between anisotropic superconducting phases with varying orientation of the order parameter (see also section 4.2.1).

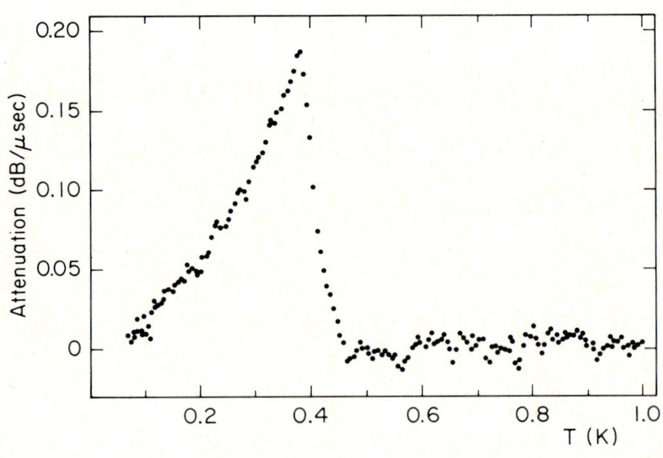

Fig. 108. Attenuation of ultrasound in Th-doped UBe_{13}. The large anomaly is associated with the second transition at T_{c2}. On that level of sensitivity, the anomaly at T_{c1} cannot be detected. (The results are from Batlogg et al. 1985a.)

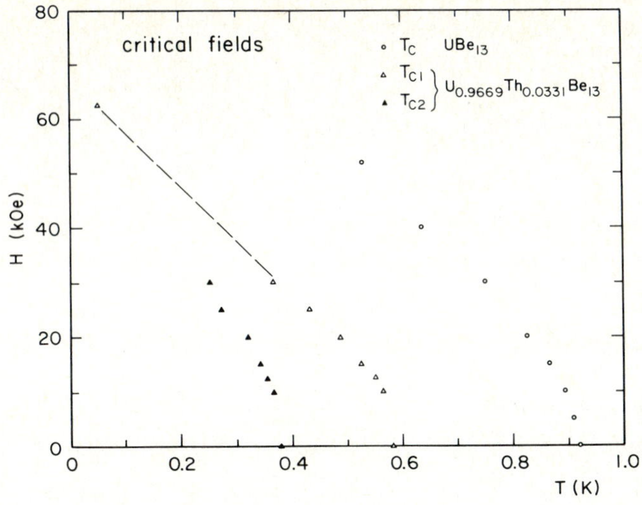

Fig. 109. Temperature dependence of critical fields of UBe_{13} and $U_{0.9669}Th_{0.0331}Be_{13}$ as obtained from specific-heat measurements. (Ott et al. 1986a.)

Another explanation assuming some kind of magnetic ordering (Batlogg et al. 1985a) cannot be ruled out, but it has little support from measurements of the nuclear magnetic resonance of Be nuclei (MacLaughlin et al. 1984). Very recent μSR relaxation-rate measurements (Heffner et al. 1986), which put an upper limit of any ordered moment below T_{c2} to $10^{-4}\,\mu_B$, indicate that the occurrence of magnetic order is very unlikely. The possibility of a structural phase transition has to be kept in mind, but there is no real evidence for it considering the temperature dependence of the ultrasound velocity which shows no sign of a precursor softening (Batlogg et al. 1985a). These data are, however, incomplete and should also be obtained for other symmetry modes. The magnetic-field dependence of this lower transition is similar to the upper critical field of the superconducting transition as shown in fig. 109 (Ott et al. 1986a).

Other non-magnetic impurities also drastically change the anomaly of the specific heat at the superconducting transition of UBe_{13}. This is demonstrated in fig. 110 where we show $C_p(T)$ for equal amounts of Lu and Zr replacing U. The superconducting transition of the Lu-doped sample is hardly recognizable, and the anomaly and specific heat in the superconducting state of the Zr-doped specimen (Ott et al. 1986b) are reminiscent of the data that were obtained for pure UPt_3. A compilation of how T_c of UBe_{13} is affected by various different impurities is shown in fig. 111 (Smith et al. 1985). All these results give evidence that the superconducting state of UBe_{13} is much more affected by small amounts of non-magnetic impurities than is usually the case for ordinary superconductors, and therefore give additional support for some unconventional superconductivity in this material. Although the amount of this kind of data is much smaller for UPt_3, the same conclusion seems to apply.

4.1.2. *Magnetic order*

As was mentioned in the Introduction it is often argued that the large specific

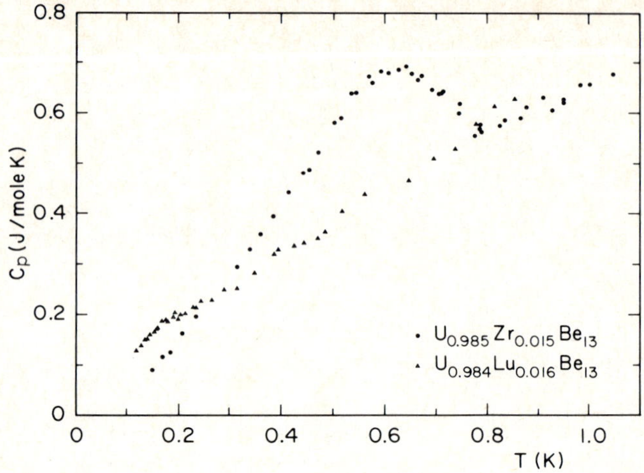

Fig. 110. Examples of the influence of small amounts of non-magnetic impurities on the specific-heat anomaly at the superconducting transition of doped UBe_{13}. (Data were obtained by Ott et al. 1986b.)

heat at very low temperatures in this class of compounds is due to the absence of a cooperative magnetic phase transition (Varma 1985a). This explanation seems particularly plausible in the case of the standard material $CeAl_3$, where the entropy of the doublet ground state of the crystal-field split Hund's rule ground-state multiplet is recovered between 0 and about 12 K. This implies a large value for

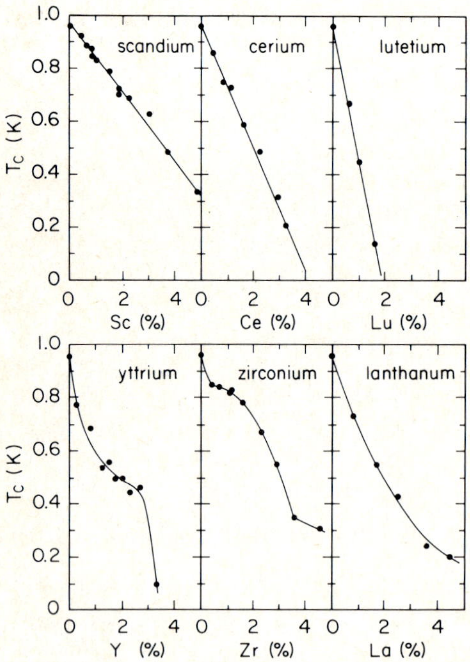

Fig. 111. $T_c(x)$ for low concentrations x of various impurities in superconducting UBe_{13} as obtained by Smith et al. (1985).

$\partial S/\partial T$ at low temperatures and necessarily also a large value for γ, defined as the ratio C_p^{el}/T. Some doubts that this view is generally adaptable to all heavy-electron materials arise when considering, e.g., the properties of UCu_5, where we argued above that the heavy-electron state is formed at temperatures at which the material has already entered a magnetically-ordered state (Ott et al. 1985c). Below we discuss some examples where magnetic ordering is observed although the heavy-electron state is already formed. So far only antiferromagnetic ordering has been found for all magnetically ordering heavy-electron materials. For completion we also mention the new low-temperature phase transition observed in UCu_5, although its character has not yet been established. Where data exist we also point to the fact that these magnetic phase transitions are very sensitive to impurities on the level of a few atomic percent. This is certainly different from the behaviour of conventional magnets with localized moments, and rather reminiscent of the influence of impurities on the superconducting state as mentioned above.

U_2Zn_{17}. The magnetic phase transition of U_2Zn_{17} (Ott et al. 1984b) is, at least from a comparison of the specific-heat features, very similar to that of $NpSn_3$ (Trainor et al. 1976), which will also be discussed below. As we have shown in section 2, the most obvious indication for the phase transition of U_2Zn_{17} is the observed specific heat anomaly (see fig. 11). What is especially noteworthy in this case, and also in the other heavy-electron magnets, is the remanence of a large contribution to the specific heat that varies linearly with T for $T \to 0$ K a feature that is usually not observed in ordinary antiferromagnets. It suggests that a considerable portion of the Fermi surface with high electronic density of states survives the phase transition. Amazingly, the percentage of this portion seems to be roughly the same in all cases that have been investigated so far and is of the order of 30%. The magnetic specific heat in the ordered state of U_2Zn_{17} is also characterized by a fairly large contribution that is proportional to T^3 below 5 K, indicating the presence of spin–wave-type excitations. The cut-off temperature of this excitation spectrum is estimated to be of the order of 100 K.

An interesting aspect of this type of phase transition, especially also with respect to possible analogies with the corresponding superconducting transitions, is the question of the entropy change involved in the transition. In a conventional rare-earth magnet, the entropy change is determined by the change of the excitation spectrum associated with the localized f-electrons, which usually consists of a lifting of degeneracies of electronic states due to the lowering of the symmetry. If instead antiferromagnetism of itinerant electrons is envisaged, we recall the work by Fedders and Martin (1966) who demonstrated analogies between this phenomenon and superconductivity, claiming that thermodynamic aspects of the magnetic phase transition like the anomaly of the specific heat should be described by the BCS-theory of superconductivity (Bardeen et al. 1957). One essential requirement is then that the entropy balance through the transition is zero and this is certainly not the case for U_2Zn_{17}. The data shown in section 2 reveal that the entropy associated with the electronic specific heat γ_0 below T_N is 3.84 J/mol K and the remaining part of the magnetic specific heat involves 3.905 J/mol K up to the ordering temperature. The sum of these two contributions is, however, considerably less than the entropy

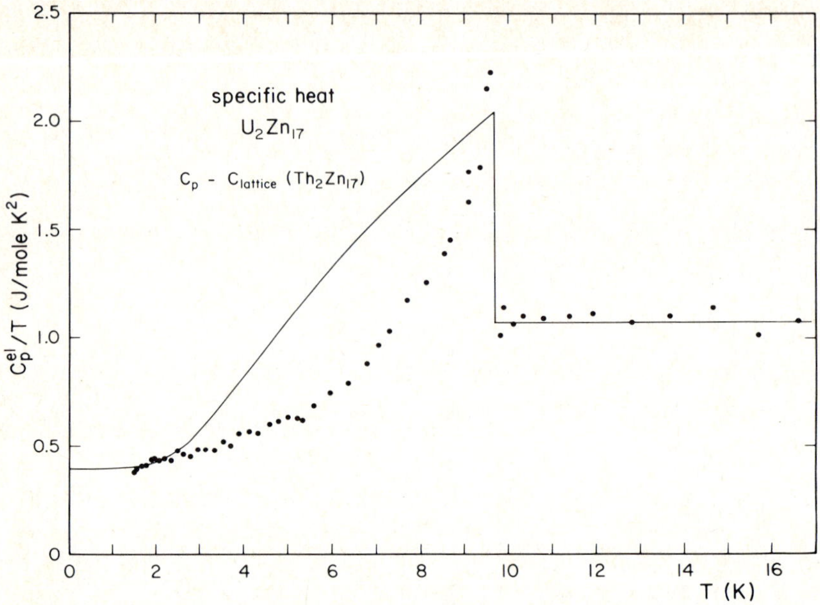

Fig. 112. Comparison of the temperature dependence of the specific heat of magnetically-ordering U_2Zn_{17} with a BCS-type calculation, considering a finite linear-in-T contribution to C_p as $T \to 0$ K. (Experimental data are due to Ott et al. 1984b.)

of about 10 J/mol K associated with the electron system, which is anticipated for a paramagnetic state between 0 K and T_N. We also note that the magnetic entropy is much less than $2R\ln 2 = 11.52$ J/mol K, the entropy that would be associated with a doubly degenerate ground state per U atom. In fig. 112 we demonstrate that a general BCS curve for $C_p(T)$, taking into account the remaining electronic specific heat below T_N, does not describe the experimental data.

Microscopic studies of the magnetic phase transition of U_2Zn_{17} have been made using neutron diffraction (Cox et al. 1986) and the muon-spin-rotation (Barth et al. 1986a) technique. The neutron study revealed a relatively simple magnetic structure of U_2Zn_{17} below T_N, confirming the antiferromagnetic nature of the ordering as indicated by the magnetic susceptibility (see fig. 25). The structure is shown schematically in fig. 113. Because the primitive unit cell contains two U atoms, the magnetic and the chemical cell have the same size. The saturated ordered moment in zero magnetic field is 0.8 μ_B/U ion and it is oriented within the basal plane of the rhombohedral crystal lattice. Considering the effective moment of 3.3 μ_B/U ion as determined from χ-measurements on single-crystalline U_2Zn_{17} (see section 2), the ordered moment appears considerably reduced. Somewhat less quantitative information was obtained from presently available μSR data. In fig. 92, section 2, we show a graph where the muon-spin relaxation rate λ is plotted versus temperature. For both compounds shown, $\lambda(T)$ clearly reflects the transition but in an obviously different way. For UCu_5, $\lambda(T)$ also suggests some kind of critical behaviour below

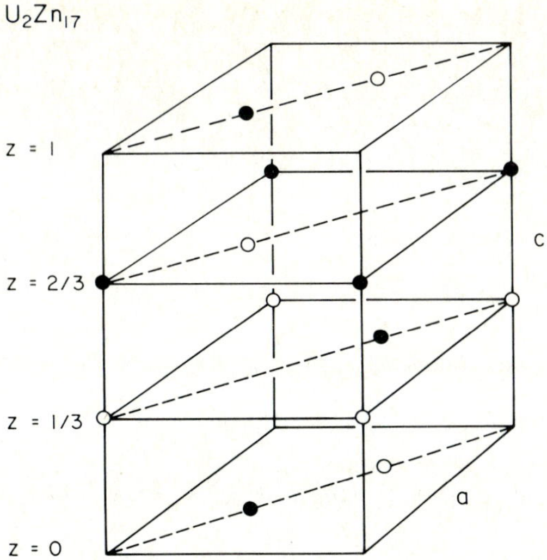

Fig. 113. Proposed magnetically-ordered structure of U_2Zn_{17} on the basis of neutron-diffraction results by Cox et al. (1986). Open and closed circles denote atoms with oppositely oriented moments. The moment directions are within the basal plane.

T_N, where the relaxation rate decreases rapidly with decreasing temperature and reaches a constant value at $T/T_N \sim 0.7$. The relaxation rate well below T_N is roughly an order of magnitude larger than that well above T_N. In U_2Zn_{17}, $\lambda(T)$ rises suddenly with decreasing temperature at T_N with λ being temperature independent in the paramagnetic state. The saturation value for $T \rightarrow 0$ K is $0.12\,\mu s^{-1}$ as compared to $0.035\,\mu s^{-1}$ above T_N.

Next we should like to point out that the phase transition of U_2Zn_{17} is drastically influenced by impurities replacing Zn on the anion sites (Willis et al. 1986). The relevant figures demonstrating the effects on the anomalies of the specific heat, the magnetic susceptibility and the electrical resistivity are contained in section 2. Available data with Mn and Cu as substitutions for Zn show that impurities on the few-at% scale are sufficient to push T_N from 9.7 K to below 1 K. Unpublished data of χ-measurements down to 20 mK indicate that 2% Cu on the Zn sites suppress magnetic order above this very low temperature. The magnetic ordering temperature can also be lowered by replacing U with Th but the effect of this kind of substitution is apparently less dramatic, 10% Th for U shift T_N to 7.5 K (Willis et al. 1986). The large low-temperature specific heat observed in the paramagnetic state of U_2Zn_{17} is hardly affected at all by all these mentioned substitutions on either the cation or the anion sites (see fig. 12, section 2) and also the magnitudes of the magnetic susceptibility and the electrical resistivity of the paramagnetic state do not change significantly. It is clearly those properties that are related with the cooperative phase transition which are altered very effectively by impurities.

Information on how the phase transition is affected by external magnetic fields is contained in the $\rho(H)$ data that are shown in fig. 53, section 2. We note a surprisingly small shift of T_N with increasing field and an unexpected positive magnetoresistance in the magnetically ordered phase. Variation of another external parameter, namely increasing hydrostatic pressure enhances the magnetic-ordering temperature of U_2Zn_{17}, in qualitative agreement with the sign of the thermal-expansion anomaly shown in fig. 76, section 2. Resistivity measurements up to 17 kbar (Thompson et al. 1986) indicate that at still higher pressures a negative shift of T_N with increasing pressure is to be expected. The pressure derivative of T_N, $\partial T_N/\partial p$, extrapolated to $p=0$, is comparable in magnitude to pressure-induced shifts of the critical temperature T_c of superconductors.

UCd_{11}, $NpSn_3$. There is no doubt that the specific-heat anomaly at 5.0 K observed in UCd_{11} (see fig. 13, section 2) is related with a magnetic phase transition, as has recently been confirmed by spectroscopic measurements using the muon-spin-rotation technique (Barth et al. 1986b). Interesting features concerning this phase transition were obtained by electrical-resistivity measurements under external pressure. As observed in U_2Zn_{17} and also to be expected from the thermal expansion results (see fig. 76, section 2), T_N of UCd_{11} also increases with increasing pressure. Additional anomalies in the $\rho(T)$ curves under hydrostatic pressure at temperatures below T_N ($p=0$) led Thompson et al. (1986) to suggest additional lines in the $[p, T]$-diagram of UCd_{11}.

The magnetic character of the phase transition of $NpSn_3$ was unambiguously demonstrated by Mössbauer spectroscopy using the ^{119}Sn resonance (Trainor et al. 1976). The shape of the Mössbauer spectrum indicated two magnetically inequivalent sites, typical for certain antiferromagnetically ordered structures. In fig. 114 we show that also for $NpSn_3$, a generalized BCS-type specific-heat curve does not describe the experimental data below T_N, contrary to what was claimed in the original work by Trainor et al. (1976). As in U_2Zn_{17} the γ-value extrapolated to $T=0$ K is between 30 and 40% of the value deduced from data above T_N.

UCu_5. UCu_5 was, as mentioned in section 2, identified as ordering antiferro-

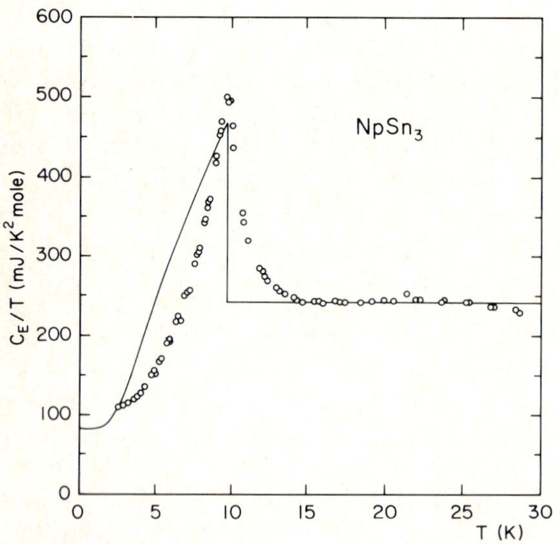

Fig. 114. Comparison of the temperature dependence of the specific heat of magnetically ordering $NpSn_3$ with a BCS-type calculation, considering a finite linear-in-T contribution to C_p as $T\to 0$ K. (Experimental data are due to Trainor et al. 1976.)

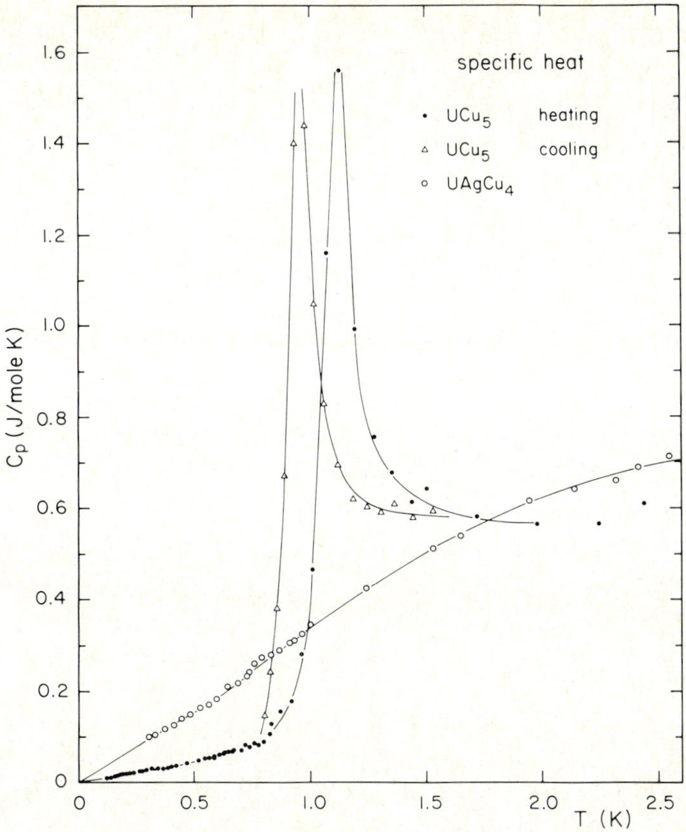

Fig. 115. $C_p(T)$ for UCu_5 and $UAgCu_4$ between 0.15 and 2.5 K as found by Ott et al. (1985c).

magnetically at 15 K by measurements of the magnetic susceptibility (Misiuk et al. 1973) and by neutron diffraction experiments (Murasik et al. 1974). As we also pointed out in section 2, the special feature of UCu_5 is that the enhancement of the specific heat with decreasing temperature occurs in the magnetically-ordered state well below T_N (Ott et al. 1985c). As was shown by specific-heat and electrical-resistivity measurements this low-temperature state with enhanced specific heat is again unstable and undergoes another phase transition of, as yet, unidentified character. In fig. 115 we show the corresponding anomaly of the specific heat which occurs at slightly different temperatures, depending on whether C_p is measured upon cooling or warming the sample. Although the transition is hysteretic in temperature, no latent heat was observed. In fig. 116 it is shown that the transition is accompanied by a considerable increase in $\rho(T)$ with decreasing temperature. In contrast to the generally observed trend towards a low-resistive state in heavy-electron materials as $T \to 0$ K, $\rho(T)$ is still very high at $T \sim 20$ mK (Willis et al. 1985c). This lower transition in UCu_5 can be suppressed by replacing one Cu atom per formula

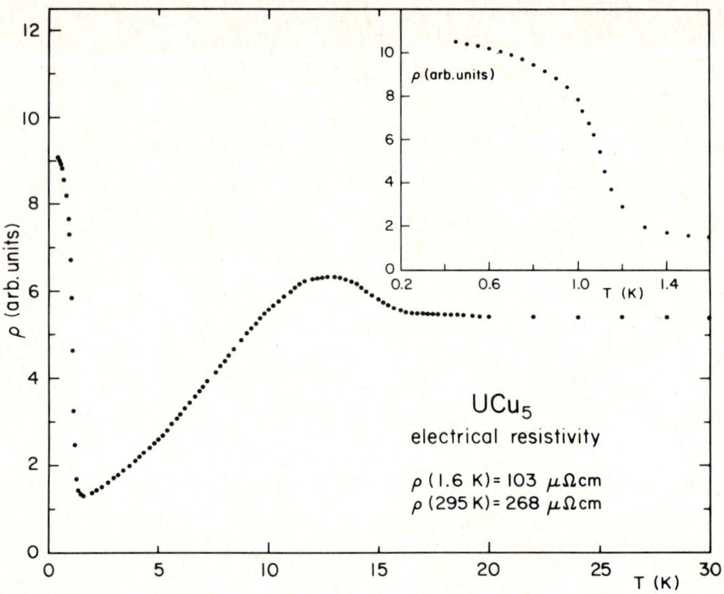

Fig. 116. Temperature dependence of the electrical resistivity of UCu_5 between 0.4 and 30 K as measured by Ott et al. (1985c).

unit by Ag. We note, however, that this substitution does not suppress magnetic order at higher temperatures but rather enhances it (see fig. 19, section 2). Figure 115 also shows the linear temperature dependence of C_p of $UAgCu_4$ as T approaches zero with a prefactor that is roughly four times larger than that of UCu_5. Thus, the loss of Fermi surface induced by the transition is again of the same magnitude as for the magnetic cases mentioned above. An increasing external magnetic field leads to an increase of the transition temperature. Specific-heat data (Ott et al. 1985g) in a field of 50 kOe give peaks that are 0.32 K higher in temperature than those obtained in zero field. The separation of the two C_p peaks is about 0.15 K in both cases. From measurements of the electrical resistivity in fields up to 200 kOe, (Batlogg et al. 1985c and Batlogg 1986) the field dependence of the transition temperature was determined as shown in fig. 117. From analogous data obtained with $U_{0.973}Th_{0.027}Cu_5$ it was found that the vertical phase boundary is bent in such a way that the low-temperature phase is no longer attained for $H = 0$, but only above a limiting field of $H = 115$ kOe, as shown also in fig. 117. The effect of external pressure on the phase transition is shown in fig. 59, section 2. The resistivity increase induced by the transition is obviously shifted to lower temperatures indicating a suppression of the transition by increasing external pressure. Unfortunately, no data below 1 K are yet available to check whether the transition is simply shifted in temperature or suppressed altogether.

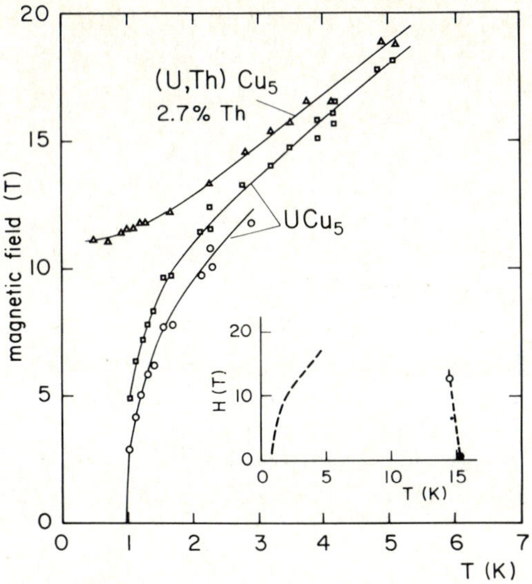

Fig. 117. Magnetic phase diagram for UCu_5 and (U,Th) Cu_5 as obtained from magnetoresistance experiments by Batlogg (1986).

4.2. *Theory*

4.2.1. *Superconductivity*

As we mentioned in the introduction to section 4, the occurrence of superconductivity in heavy-electron materials attracted considerable interest, especially also among theorists, because various experimental facts mentioned above make it quite possible that these superconductors are the first where the pairing of the electrons is not due to phonon exchange, but rather due to electron–electron interactions. Heuristic arguments supporting this possibility are:

(1) the characteristic temperature of the heavy-electron subsystem (the renormalized Fermi temperature) is equal or smaller than the Debye temperature which characterizes the lattice-vibration spectrum,

(2) magnetic interactions (possibly spin fluctuations) dominate the low-temperature physical properties as, e.g., the electrical resistivity or the specific heat. Such interactions are known to be harmful to ordinary superconductivity.

In addition, and in a more general sense, also the analogy with liquid ^{3}He, itself a strongly-interacting Fermi liquid with large effective quasi-particle mass, a correspondingly large magnetic susceptibility and adopting a superfluid state with spin-triplet, $l=1$ pairing, because of the strong repulsion between ^{3}He atoms, is often used in favor of unconventional superconductivity in these compounds. These claims are, however, disputed and arguments favoring electron–phonon induced superconductivity have been proposed. They rest on ideas that the electron–phonon interaction is drastically altered by the abnormal electronic spectrum, leading to an enhancement of this interaction at low temperatures, which might then lead to a transition to a superconducting state of conventional character. The most recent reviews of pertinent theoretical work, with regard to superconductivity in heavy-electron materials can be found in articles by Varma (1985b) and by Lee et al. (1986). Here we do not intend to present a complete overview of existing relevant theoretical work, but rather mention a few points that have been investigated recently. In all formula that follow below we use the nomenclature of the original work that is cited.

Actual calculations demonstrating a superconducting transition induced by electron–electron interactions in these materials is, of course, a very formidable task, if one considers that such a calculation would also have to predict the transition temperature. Predictions of transition temperatures are known to be quite difficult even in conventional superconductors and also for superfluid ^{3}He. In our case, additional complications due to the peculiarities of the electronic energy spectrum and strong spin–orbit interactions make calculations using known techniques and approximations even less meaningful.

Calculations based on the Fermi-liquid concept used for ^{3}He are due to Valls and Tesanovic (1984) and to Pethick et al. (1986). Valls and Tesanovic use a BCS-type formula for the critical temperature,

$$T_c = 2T^* \exp(-1/\lambda), \tag{23}$$

where T^* represents a cut-off energy which, in heavy-electron materials is given by the width of the narrow feature in the density of electronic states that gives rise to the large electronic specific heat, and λ is given by

$$\lambda = -\sum_{l=0}^{1} (-1)^l (A_l^s + A_l^a)/12. \tag{24}$$

The parameters $A_l^{s,a}$ are evaluated from Landau parameters inherent in the Fermi-liquid theory. In their work λ represents an attractive interaction for $l=1$ pairing while it is repulsive for an $l=0$ or s-type pairing. Unfortunately, a straightforward prediction of this model, an increase of T_c with decreasing lattice parameters, is not corroborated by experiment, neither in UBe_{13} (Chen et al. 1984b) nor in UPt_3 (Brodale et al. 1986).

In another very similar discussion, based on the induced-interaction model for ^{3}He (Babu and Brown 1973), of how to calculate interaction parameters that would enter a formula of the type of eq. (23), Bedell and Quader (1985) discuss the strength of the electron–phonon interaction in such systems. They conclude that it would scale like $(T_F/\theta_D)^2$ and because $T_F < \theta_D$, the pairing interaction in the s-channel would be weak and hence p-type pairing finally should dominate. They also state that the additional complications due to lack of translational invariance, band-structure effects or spin–orbit interactions would not change their result qualitatively.

In the model calculation by Pethick et al. (1986), which assume that the critical temperature is again given by an equation of the same form as eq. (23), λ is given by

$$\lambda = -(1 + A_0^a)/6, \tag{25}$$

where A_0^a is large and negative. Here the Fermi-liquid parameters are evaluated from comparison with experimental data on UPt_3. The cut-off energy is assumed to be about one hundredth of T_F and T_F is calculated from the prefactor of the $T^3 \ln T$ term of the low-temperature specific heat. This model seems to be able to reproduce the volume dependence of the low-temperature specific heat and of the critical temperature T_c of UPt_3 (Brodale et al. 1986).

Razafimandimby et al. (1984) argue that the heavy-electron systems can be interpreted as Kondo lattices (atoms providing Kondo scattering are homogeneously distributed on regular sites of the crystal lattice) and that the electron–phonon interaction in such systems has two additional coupling mechanisms that are due to the presence of Kondo ions. One contribution originates from an additional phase shift of the scattered electrons caused by the Kondo ions, and the second is connected with the volume dependence of the so-called Kondo temperature T_K. The role of this volume- or pressure-dependence is similar to that of a deformation potential. Estimates of T_c using this model are also based on an equation for T_c with the same form as eq. (23), and the cut-off energy is now proportional to T_K. The electron–phonon interaction parameter has the same mass- and frequency-dependence as that for ordinary superconductors but the coupling constant is now proportional to η^2, where $\eta = \delta \ln T_K / \delta \ln V$, a factor that appears to be fairly large in these systems. The authors also argue that the weakening of the pairing interaction due to electron repulsion is not substantial, another special feature of Kondo lattice systems (Razafimandimby et al. 1984). Inserting actual numbers in the formula for T_c results in values of roughly 1 K for this quantity. We should like to note here that the choice of values for some of the necessary input parameters seems problematic, to say the least, and we do not consider these estimates to be more reliable than those mentioned above.

Various other calculations of T_c assuming that superconductivity is triggered by electron–phonon interaction are due to three Japanese groups (Ohkawa and Fukuyama 1984, Miyake et al. 1984, Tachiki and Maekawa 1984). They all have in common that the relevant electron spectrum can be viewed as a narrow band of width T_0 and that the pairing occurs within this band of heavy quasiparticles. Different in these calculations are the models for the electron–phonon interaction. While Miyake et al. (1984) also consider the volume dependence of T_K as the main source for an enhanced electron–phonon coupling, Ohkawa and Fukuyama (1984) apply a deformation-potential model whereby the resulting induced electron–electron interaction (attractive or repulsive) depends on the symmetry of the lattice structure. The calculations confirm that in most cases, s-type pairing is the most favorable. A very interesting conclusion is due to Ohkawa and Fukuyama (1984) who find that the energy gap for $l=0$ pairing also may have lines of nodes under particular circumstances.

In view of the possibility that the superconducting state of heavy-electron materials is characterized by unconventional pairing, various recent theoretical papers dealt with the question as to what particular symmetries for order parameters might be possible. The first analyses of possible forms are due to Anderson (1984b) and to Volovik and Gor'kov (1984). It was also Anderson (1984a) who pointed out that, strictly speaking, the terms 's-' or 'p-wave' or 'singlet' and 'triplet' superconductivity are not quite correct and that the classification of 'even-' and 'odd-parity' superconductivity would be more appropriate. The most extensive investigation of this question applying group-theoretical methods was published by Blount (1985). The detailed form of the order parameter clearly depends very much on the actual crystal structure of the system under investigation. For cubic materials, like UBe_{13}, four different forms seem possible. Each form is more or less stable than the others and, therefore, four different critical temperatures have to be considered. Possible forms for the order parameters in the odd-parity case for cubic, tetragonal and hexagonal crystal symmetry were investigated by Blount (1985), by Ueda and Rice (1985) and by Volovik and Gor'kov (1985). The latter authors also worked out the even-parity case. While lines of nodes for the energy-gap function on the Fermi surface seem possible in the even-parity case, symmetry arguments prohibit this to happen in the odd-parity case where only discrete points of zeroes are allowed. Volovik and Gor'kov (1985) also point out that in certain cases of the odd-parity state, finite magnetic moments could occur and lead to domain formations. Joynt and Rice (1985) point to the fact that anisotropic superconducting states might lead to crystal-lattice distortions and deformations of cubic crystals to tetragonal or rhombohedral structures.

Two interesting theoretical papers resulted from the observation of the unusual $T_c(x)$ diagram for $U_{1-x}Th_xBe_{13}$ shown in fig. 105. Volovik and Khmel'nitskii (1984) investigate the possibility of the formation of a superconducting glass for $0.02 < x < 0.05$ in the sense that a non-zero order parameter with random orientation characterizes the state at T_{c1} with a subsequent transition to a fully-ordered superconducting state at T_{c2}. Joynt et al. (1986) follow the idea of domain formations and also argue for various different superconducting phases in the diagram of fig. 105. Experimental observations of anomalies of the thermal expansion and the ultrasonic attenuation, both at T_{c2} (see section 4.1.1), are interpreted as being connected with domain walls between differently-oriented superconducting domains.

In a special attempt to calculate the spin-fluctuation contribution to the pairing interaction in UPt_3, Fay and Appel (1985) analyzed the normal-state specific heat and magnetic susceptibility of this compound in a similarly phenomenological way as Boring et al. (1985a) did for UAl_2 and they used the paramagnon theory to calculate the p-wave pairing parameter λ^{p}_{SF}. They conclude that a one-band model favors p-wave pairing. However, a two-band model seems to preferrentially allow conventional s-wave superconductivity, because the authors find little suppression of s-wave pairing in a d band through spin fluctuations within the narrow f band. This gives some support for a T_c calculation on the basis of band-structure calculations, and an electron–phonon induced superconductivity for UPt_3 as proposed by Oguchi and Freeman (1985b).

Various results on physical properties in the superconducting state of UBe_{13} and UPt_3 have been used to argue for the existence of anisotropic energy gaps in the superconducting state. They mainly consist of the observation of power-law temperature dependences of quantities like the specific heat, the ultrasonic attenuation or the thermal conductivity, rather than the usual activation-type exponential behavior expected for gaps that are non-zero over the entire Fermi surface (see also sections 2 and 4.1).

The available theoretical literature aiming at rigorous calculations that would give a firm basis with which these experimental observations could be compared and thus lead to unequivocal interpretations is almost non-existent. Up to now, the situation is rather that some papers describing experimental data also contain a theoretical section presenting approximate calculations, the results of which are used for the interpretation of the measurements. Examples of this sort are the following:

(1) Ott et al.'s (1984c) interpretation of the temperature dependence of the specific heat in the superconducting state of UBe_{13}, shown in fig. 95. These calculations explicity assume an anisotropic gap function of the form of that of the ABM state (A phase) in superfluid ^{3}He with points of zeroes on the Fermi surface. Strong coupling corrections are incorporated in a similar way as was done by Brinkman et al. (1974), again for superfluid ^{3}He. The calculation of $C_p(T)$ below T_c then requires two parameters where one of them is determined by the magnitude of the specific-heat discontinuity at T_c, and the other is an interpolation parameter for intermediate temperatures. At very low temperatures, C_p varies proportional to T^3 in this case. The whole procedure has obvious weaknesses. First, the treatment by Brinkman et al. (1974) is only approximate even for ^{3}He and second, as pointed out by Anderson (1984b), a straightforward analogy of these materials with ^{3}He is not allowed. The inherent symmetries of crystal lattices and the fairly strong spin–orbit interactions make arguments that are applicable in the case of ^{3}He invalid for heavy-electron materials. A discussion of analogies and differences between heavy-electron materials and ^{3}He is given by Rice et al. (1986).

(2) In another comparison between experiment and theory, Bishop et al. (1984) calculate the temperature dependence of the ultrasonic-attenuation coefficient α_s in the superconducting state of UPt_3. In their expression for α_s the gap function for the superconducting state is again the decisive element. In order to obtain agreement with experiment, this gap function has to be assumed to be compatible with lines of zeros on the Fermi surface. A comparison of this calculation with experiment is shown in fig. 118. The problem here is that an independent calculation of α_s by Rodriguez (1985) leads to the conclusion that the observed T^2 dependence of α_s in UPt_3 at temperatures distinctly below T_c implies a gap function with points of zeroes

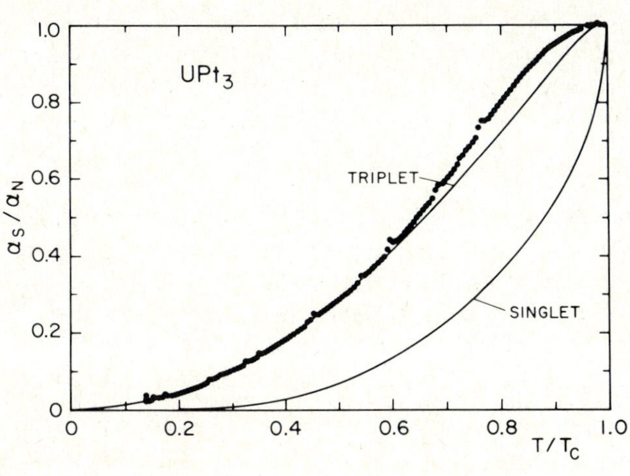

Fig. 118. Comparison of experiment and model calculation for the ultrasound attenuation in the superconducting state of UPt_3 as reported by Bishop et al. (1984).

on the Fermi surface. This result is obtained by solving a scalar quasi-particle Boltzmann equation in the relaxation-time approximation, and under the conditions of the hydrodynamic limit as given by the actual experimental situation (see also section 2). This solution has been questioned by others because the quasi-particle mean free path is assumed to be constant. Lee et al. (1986) argue that in the superconducting state the scattering rate, and hence the mean free path, might well be energy dependent and this would lead to distinct alterations of Rodriguez' result.

(3) The most extensive calculations in direct combination with an experiment are due to Einzel et al. (1986) who investigated the temperature dependence of the magnetic-field penetration depth λ_L in superconducting UBe_{13} (see also section 4.1). Because of the unusual features of λ_L for UBe_{13} (see fig. 104) they generalized the BCS theory for the electromagnetic response of superconductors to systems with anisotropic energy gaps. In their calculations they consider odd-parity states with uniaxal gap anisotropies of the form

$$\Delta(\boldsymbol{k}) = \Delta_0 f(\boldsymbol{k} \cdot \boldsymbol{l}), \tag{26}$$

where $\boldsymbol{l}$ denotes the symmetry axis, anisotropic states with 's-wave-like' gaps, and the gapless case of conventional superconductivity. A comparison of their results with experimental data does not lead to an unambiguous conclusion, because the technique used in these measurements gives the temperature dependence of λ_L only but not its absolute value. However $\lambda_L(T)$ may be explained very well by assuming a gap with point nodes on the Fermi surface, provided the Landau parameter F_1^s, which enters the calculation, is not large. Given these assumptions, the zero-temperature value of the penetration depth is found to be about 5000 Å. We show a comparison of possible fits in fig. 119.

A critical quantity of a superconductor is not only its transition temperature but also the temperature dependence of its critical field. Varma (1984a) pointed out that the anisotropy of the upper critical field H_{c2} of UPt_3 provided convincing evidence

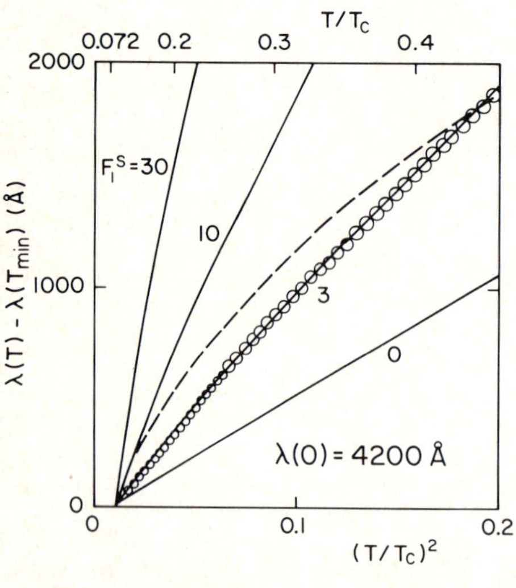

Fig. 119. Possible fits to the experimental results for the temperature dependence of the magnetic-field penetration depth of superconducting UBe_{13}. The solid lines are based on the assumption of point zeroes of the gap. More details can be found in work by Einzel et al. (1986) and by Gross et al. (1986).

for a superconducting state with odd-parity pairing, because the maximum anisotropy is observed close to T_c. Varma argued that this was impossible for an $l=0$ paired state. Calculations by Scharnberg and Klemm (1985) and by Klemm and Scharnberg (1986) were also intended to determine the temperature dependence H_{c2} for various possible, differently-paired states. Whereas their results can be used to estimate the relative stability of these different states, they seem less suited to establish such a state by comparison of theory and experiment. Balatskii et al. (1986) provided fairly elaborate calculations investigating the anisotropy of H_{c2} for various different symmetries of the order parameter and showed that for some odd-parity states an anisotropy of H_{c2} may even be expected in a system with a cubic crystal lattice, certainly a possibility of high interest with regard to comparison with experiment. We note here that an anisotropy of H_{c2} was recently reported for the cubic material UBe_{13} by Alekseevskii et al. (1985). Based on a Ginzburg–Landau-type calculation, Machida et al. (1985) derived the angular dependence of H_{c2} for p- and d-wave superconductors with cubic and tetragonal crystal symmetry. As others they find different anisotropies for different configurations of the order parameter. Calculations of $H_{c2}(T)$ in connection with experimental results on the heavy-electron superconductors UBe_{13}, UPt_3 and $CeCu_2Si_2$ are contained in the work by Rauchschwalbe et al. (1985). In this analysis it is pointed out that Pauli-limiting effects are observed in $H_{c2}(T)$ for $CeCu_2Si_2$ and UBe_{13}, and are taken as evidence for conventional or $l=0$ paired superconductivity. In another attempt to interpret the experimental $H_{c2}(T)$ curve of UBe_{13}. Fenton and Aers (1985) considered both singlet- and triplet-pairing for the superconducting state and arbitrary spin–orbit scattering. Their conclusions imply that the superconducting state of UBe_{13} is either an odd-parity state or, if singlet-pairing occurs, the $H_{c2}(T)$ curves can only be explained by assuming a strong spin–orbit-scattering enhancement with increasing magnetic field.

Since it was first believed that Josephson-tunneling experiments would unambigously decide the question of unconventional superconductivity in these materials, namely by the absence of any Josephson-type supercurrent between an established conventional superconductor and a new material under question, some effort went into theoretical calculations dealing with this problem. Fenton (1985) showed that spin–orbit scattering in the tunneling region may allow a reduced Josephson current even between superconductors whose electron pairing is of different symmetry. More recent work by Sauls et al. (1986) goes even further and claims that under certain conditions a regular Josephson current may be expected for this configuration. Similar conclusions were drawn in the work by Geshkenbein and Larkin (1986).

4.2.2. *Magnetic order*

To our knowledge, no special theoretical attempts to investigate the magnetically-ordered ground state of heavy-electron materials have been made so far. This may be due to the fact that much less is known experimentally of these systems, compared to those that are superconducting, but it may just as well be due to the notorious difficulties of describing magnetism. These difficulties mainly arise if either the concept of localized ordered moments is not adequate to describe

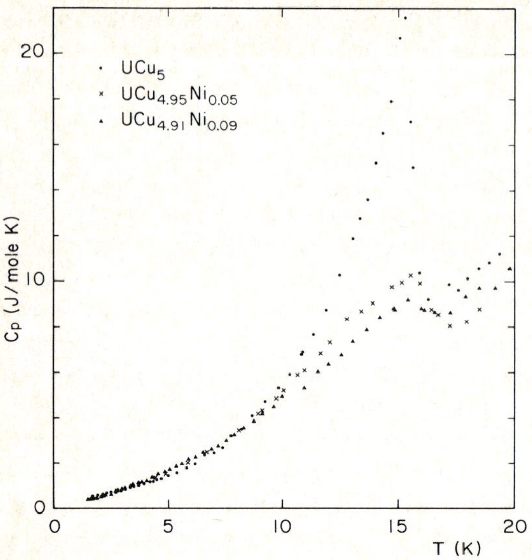

Fig. 120. Influence of small amounts of Ni impurities on the specific heat anomaly at the magnetic transition of UCu_5 as found by Ott et al. (1986c).

experimentally-observed features of the magnetically-ordering systems, but also if models that ascribe the magnetism entirely to itinerant electrons are not satisfactory. This is quite clearly the case here. In figs. 112 and 114 we have shown that the model of itinerant antiferromagnetism by Fedders and Martin (1966) does not describe the specific-heat anomalies of U_2Zn_{17} and $NpSn_3$, respectively, and it is clear that also the C_p anomalies of UCd_{11}, UCu_5 or $NpBe_{13}$ cannot be described in this way. Alternatively, entropy considerations indicate that a straightforward interpretation of ascribing the magnetic moments to 5f electrons of U ions with integral occupany of their unfilled 5f-electron shell leads to comparable difficulties. Furthermore, the remarkable suppression of magnetic order with rather small amounts of impurities as evidenced in fig. 12 for U_2Zn_{17} does not seem to be compatible with conventional ordering of localized magnetic moments. A similar trend is observed for the magnetic transition in UCu_5 if Cu is replaced by small amounts of Ni. Although, as mentioned in section 2, the magnetic ordering of UCu_5 at 15 K seems quite conventional in some ways, the drastic reduction of the specific-heat anomaly with 1 or 2% Ni for Cu in UCu_5 (Ott et al. 1986c) as shown in fig. 120 points to complications that are not really understood. One very important fact in connection with the magnetically-ordered ground state of heavy-electron materials is the remanence of a still considerable electronic contribution to the specific heat as $T \to 0$ K (see sections 2 and 5). This indicates that only parts of the Fermi surface seem to be affected by the ordering process and it is certainly an attractive project for theorists to investigate whether instabilities of Fermi surfaces may have local character, in the sense that induced gaps are very anisotropic. This question seems particularly intriguing in the case of U_2Zn_{17} where neutron-diffraction work (Cox et al. 1986) indicates that no doubling of the unit cell occurs upon antiferromagnetic ordering.

Since we do not wish to mention the whole theoretical literature dealing with anomalies of magnetism in more conventional materials and to suggest what might be appropriate in our cases, we conclude this section by expressing our belief that future experimental results will probably also initiate more theoretical work in this field.

5. Systematics

A convenient place to start in considering the systematics of the occurrence of heavy electron materials is with the Hill plots (Hill 1970). Hill's idea with these was simple and attractive: he suggested that there is a clean dichotomy between magnetic, non-bonding electronic states and non-magnetic, bonding ones. f-orbitals bond most easily with other f-orbitals, hence a plot of magnetic behaviour (such as magnetic-ordering temperature) versus f-atom interatomic separations should give a clear border between the magnetic and non-magnetic intermetallic compounds, the so-called Hill limit (~ 3.4 Å). Updated versions of the Hill plots for U, Np and Pu are shown in figs. 121–123.

We can see that the Hill limit gives a necessary (except in the case of UNi_2), but not sufficient, condition for the minimum f–f spacing above which magnetic ordering can occur. Above this spacing, however, there are known many non-ordering

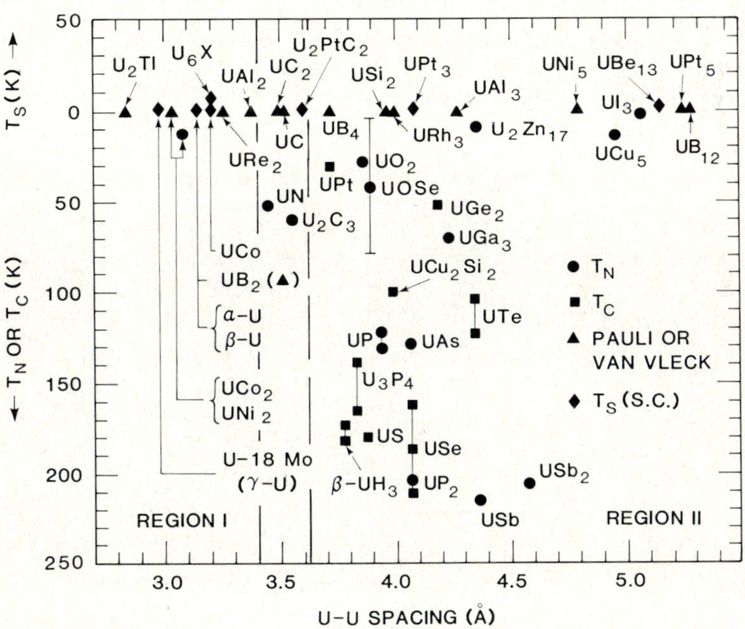

Fig. 121. Hill plot for U-compounds of magnetic versus non-magnetic behaviour according to average spacing between U-ions. (After Hill 1970.)

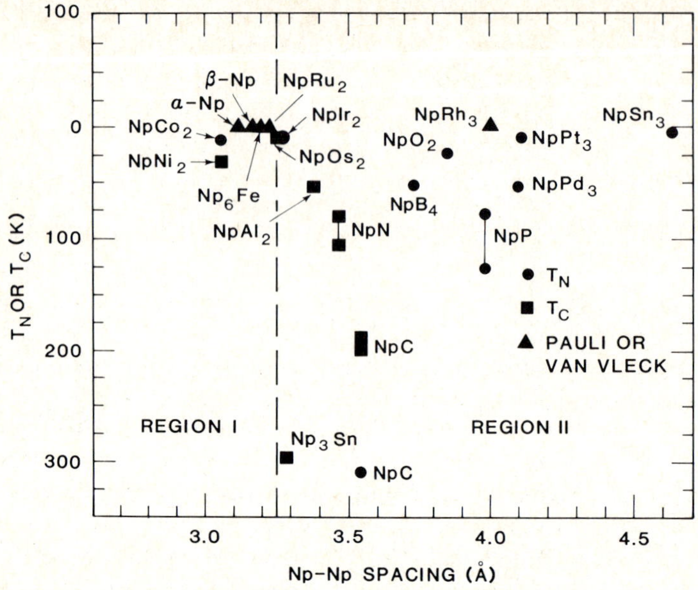

Fig. 122. Hill plot for Np-compounds. (After Hill 1970.)

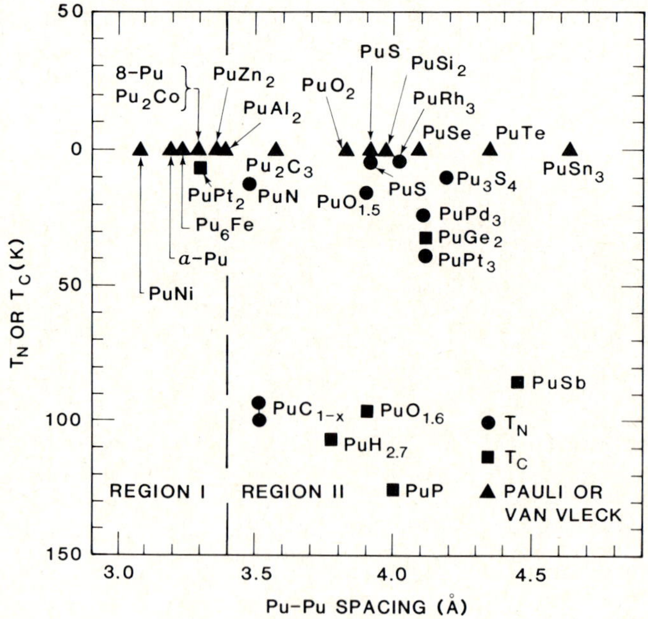

Fig. 123. Hill plot for Pu-compounds. (After Hill 1970.)

compounds. Considering U examples, the most thoroughly studied are the Cu_3Au structure compounds USi_3, UGe_3, USn_3, URh_3 and UIr_3.

There is good experimental and theoretical evidence (Arko et al. 1983, 1984b, Brooks et al. 1984) for these compounds that the 5f electrons are strongly hybridized with p-electrons (Si, Ge, Sn cases) or d-electrons (Rh and Ir). This indicates that fine details concerning the energy-level structure of the ligands can result in demagnetizing the 5f-shell via hybridization for spacings greater than the Hill limit. There are at present no general rules for deciding a priori when this will occur.

There are, nevertheless, several rules of thumb which have proved quite useful to the experimentalist, if not the theorist. The first is that the f-atoms in heavy-electron compounds have no f-atom-like neighbors in the first coordination sphere. This seems to be a requirement even when the f-atom separation is around 4 Å. This rule is consistent with the thinking which believes that the heavy-electron state is developed by the f-ligand interaction.

The second rule of thumb derives from the idea of long standing in Ce research that the metallic radius of Ce in a compound is an important indicator of the f-electron situation. The smooth contraction in metallic radius of the trivalent rare-earth elements between La and Lu, due to the only partial screening of the atomic core by the inner 4f-electrons, is also clearly evident in analogous sequences of isostructural rare-earth compounds. Deviations from the smooth variation are indicative of departures from strictly trivalent behaviour of the rare earth.

In fig. 124 we show the variation of the metallic radius across the rare-earth and actinide sequences (see also Brooks et al. 1984). We can see that there is an actinide contraction in the later actinides which is quite similar to the lanthanide contraction, but the elements Pu and earlier deviate strongly from the 'expected' line. Also note that the radii of the later actinides are quite close to those of the rare earths with the same occupation of the f-shell. This last remark also appears to be approximately correct for isostructural actinide and rare-earth compounds, and we illustrate this in fig. 125.

The variable size of elemental Ce finds its counterpart in the actinide elements Np and Pu. In intermetallic compounds, the actinide counterparts of Ce are U, Np and Pu (little data exists for the short half-life Pa). U, for example, varies in size in various intermetallic compounds between the sizes of Gd and Er in similar compounds structures. The observation in heavy-electron compounds is that U is large, close to Gd in size, with a metallic radius $r_M = 1.80$ Å. By comparing the lattice parameters of isostructural rare-earth and actinide intermetallics, one can eliminate the size changes which are due to compound formation and have nothing to do with f-anomalies, and will be the same for all the trivalent rare earths and actinides.

Of course, it is not always possible to apply this size criterion since the analogous compound sequences are not always found. In the case of UPt_3, the analogous second-half rare-earth compounds do not crystallize in the hexagonal DO_{19} structure, but rather the cubic $L1_2$ structure. DO_{19} is, in fact, the hexagonal stacking of $L1_2$, and a study by van Vucht and Buschow (1965) has shown that DO_{19} is favored over $L1_2$ stacking for larger A to B atom-radius ratios in AB_3 compounds. So this

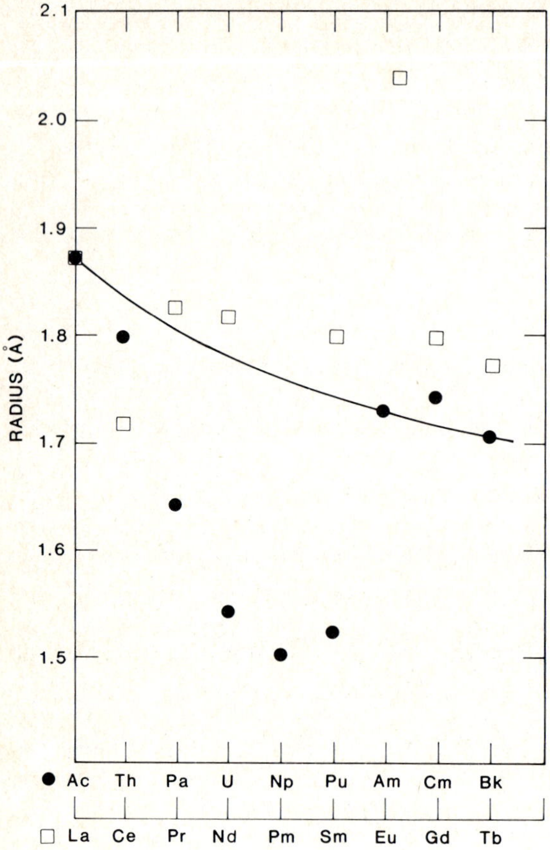

Fig. 124. Variation of metallic radius across the rare-earth and actinide sequences. See Smith et al. (1983) and Brooks et al. (1984) for further discussion.

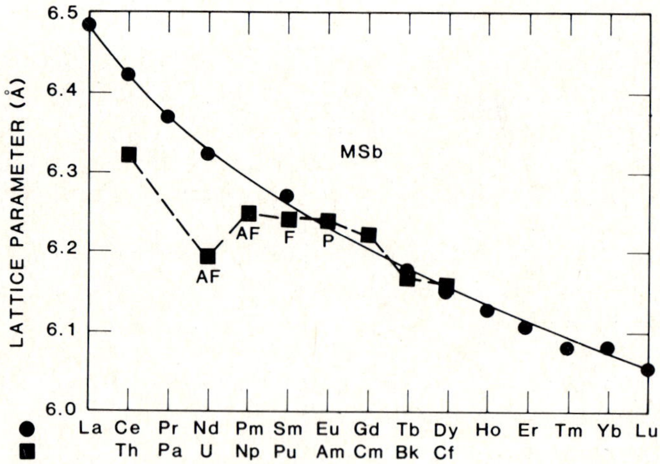

Fig. 125. Variation of lattice parameter of the cubic mono-antimonides across the rare-earth and actinide sequences, as discussed by Smith et al. (1983).

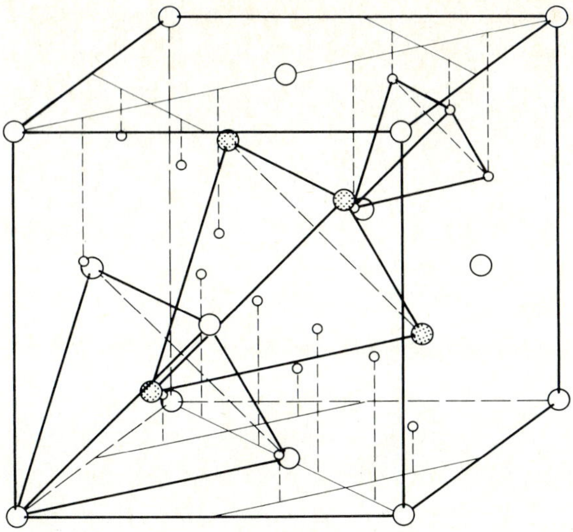

Fig. 126. The $AuBe_5$ structure (Misch 1935). The large open circles denote the Au sites. The lines are drawn to visualize the individual positions.

implies, albeit weakly, that U in UPt_3 is larger than the second-half rare earths. It is, finally, important to emphasize that these size considerations tell one the same thing as the temperature dependent magnetic susceptibilities, namely whether or not the actinide will act as though it has a local moment at high temperature. The utility of the size criterion is that it allows prospective compounds to be selected from tables of crystallographic data.

We have emphasized above the importance of the local environment of the f-atom for the development of the heavy-electron state. One need only make the comparison between URh_3, UIr_3, UPd_3 and UPt_3 to see this. At present there is no clear indication of the role played by local-site symmetry. The non-f near-neighbors are seen to play a crucial role. A particularly interesting example involves UPt_5, a cubic $C15_b$ compound (see fig. 126). An instructive way to write the formula is $U_{0.5}Pt_{0.5}Pt_2$, which emphasizes the relationship to the C15 Laves-phase compounds of which UAl_2 is an example. In UAl_2, for example, the U-atom sites form a diamond lattice, namely two fcc lattices displayed by (1/4, 1/4, 1/4). In UPt_5, one of these fcc lattices is occupied by Pt, the tetrahedron of nearby U neighbors to a given U in UAl_2 is now replaced by Pt neighbors in UPt_5. UPt_5 has a γ of 93 mJ/mol K^2 (Frings 1984), only slightly enhanced above the sum of the elemental values. Substitution of Au for one Pt to make $UAuPt_4$ has a spectacular effect, raising the γ above 700 mJ/mol K^2 (Ott et al. 1987a). It is suspected that the Au atoms occupy the special fcc lattice, although the X-ray data is unable to confirm this, because of the similar scattering powers of Pt and Au. One sees clearly, independent of the ordered or disordered nature of the substitution, the extreme importance of the local chemistry.

It is an important question as to how to unambiguously identify a heavy-electron system when the system undergoes a low-temperature cooperative phase transition. Such transitions involve the opening of a gap across essentially all (superconducting

TABLE 4
Loss of electronic specific heat upon magnetic ordering.

Compound	$\gamma(T=0)/\gamma(T=T_N)$
UCu_5 [a]	0.26 (T_{N2})
U_2Zn_{17}	0.37
UCd_{11}	0.30
$NpSn_3$	0.33

[a] Rough estimate due to uncertainty involved in estimating γ above T_{N2}.

case) or part (spin-density and charge-density wave cases) of the Fermi surface. In the superconducting case, one expects the normalized jump at T_c to be of order $\Delta C/\gamma T_c \sim 1$. In the case of UPt_3, this jump is about half this value for reasons which are not understood.

In the magnetic (spin-density wave) case, there is a curious systematic which we will refer to as the (approximate) rule of 1/3: roughly 2/3 of the Fermi surface is lost on going through the transition, resulting in a γ below the ordering of 1/3 the value above the transition. In table 4 we give data to support this. A number of compounds that order magnetically at relatively high temperatures can behave as heavy-electron materials at low temperatures, UCu_5 (Ott et al. 1985c) and $NpOs_5$ (Brodsky and Trainor 1977) being examples of this. Integrating the specific heat divided by T up through a magnetic transition to obtain the entropy involved, will not necessarily allow a decision whether or not the transition involves a localized spin or itinerant electrons. This is because, thinking in Kondo terms, the compensation of a localized spin will itself involve an entropy $k\ln 2$ per spin for spin $\frac{1}{2}$. If the 'effective' Kondo temperature is of order of the magnetic transition temperature, an entropy of this order might be expected to be involved in the transition.

Some experiments, notably on U_2Zn_{17} (Willis et al. 1986), point to an extreme sensitivity of the cooperative magnetic transition to impurities. This sensitivity can be utilized in such cases to suppress the transition, and hence measure directly the γ as $T\to 0$ K. The data for this in U_2Zn_{17} is shown in fig. 12. A related situation holds between the compounds UCu_5 and $UAgCu_4$ (Ott et al. 1985c).

A striking fact about the heavy-electron systems is that $\chi(T\to 0)$ nearly scales with $\gamma(T\to 0)$. This is to say that the free-electron relation between χ and γ approximately holds. This is somewhat surprising considering the underlying angular momentum of the appropriate f-states. In fig. 127 we present the data for uranium compounds and some other heavy-electron materials. It is interesting that no compounds fall above the 'free electron' line drawn in the figure. In cases other than spin $\frac{1}{2}$, the ratio is referred to as the Wilson ratio:

$$\frac{\chi}{\gamma} = \frac{Ng^2\mu_B^2 J(J+1)}{\gamma}. \tag{27}$$

If the underlying cause of the heavy-electron state is an f-level with angular

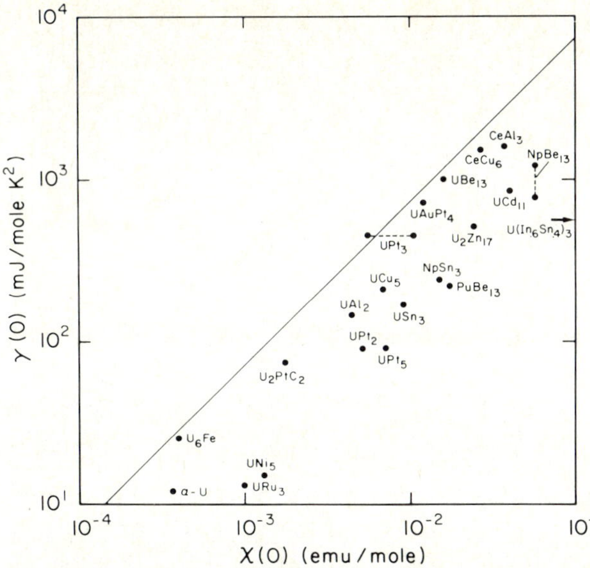

Fig. 127. Low-temperature γ versus χ values for U compounds and some other heavy-electron materials.

momentum J, one expects some lifting of the $(2J+1)$-fold ground-state degeneracy by the crystalline environment, and some appropriately modified Wilson ratio relevant to the crystal-field ground state should be used.

A hallmark of the magnetically ordered heavy-electron state, although admittedly based on a relatively small amount of data, is the very weak dependence of T_N on magnetic field. In rough terms, a field of 10 kOe corresponds to a temperature of 1 K for spin $\frac{1}{2}$. One can expect, therefore, an effect of this approximate magnitude on T_N. This does not appear to be the case. For U_2Zn_{17}, an approximate value of $dT_N/dH = 5 \times 10^{-6}$ K/Oe is observed, for example (see fig. 53). It is to be noted here that U_2Zn_{17} is anisotropic, and the anisotropy of this derivative is not known. Also in the case of UCu_5 dT_N/dH is not particularly large (see fig. 117).

We have alluded above to the difficulties involved in making generalizations about heavy-electron systems due to the relatively small number of such systems known. With this caution in mind, we introduce one further correlation: γ per unit volume, γ_V's along with their low-temperature ground states. In table 5 we give an ordered list of uranium compound γ_V's along with their low-temperature ground states. What we see is, starting from low γ_V, spin-fluctuation behaviour progressing to magnetic order to, finally, superconductivity (Fisk et al. 1985). A large value of γ_V, correlates with superconductivity rather than with magnetic order. The motivation for considering γ_V, is, of course, free electron ideas: so many properties of metals are correlated with electron concentration, suggesting γ per unit volume as perhaps an important parameter.

The above γ_V correlation exists between the low-temperature state and γ_V, saying nothing about how large a γ value is likely to be. An attempt to address the development of large γ as being dependent of f–f separation has been given by Meisner et al.

TABLE 5
Electronic specific heat per mol U and per unit volume for various U compounds.

Compound	γ(mJ/mol K^2)	γ_V (mJ/cm^3 K^2)	Ordering[a]	Ref.[b]
α-U	9.6	0.77	s(?)	[1]
UPt_5	92	1.49	n	[2]
URu_2Si_2	75	1.52	m, s	[3]
U_2PtC_2	75	1.59	s	[4]
UIr_2	52	1.63	n	Quoted in [4]
URu_4B_4	170	1.63	n	[5]
U_6Fe	25	1.79	s	[6]
$U_2Co_3Si_5$	115	2.55	n	[7]
USn_3	169	2.84	n	[8]
UCu_5	210	4.03	m	[9]
UAl_2	150	4.25	n	[10]
U_2Zn_{17}	500	5.08	m	[11]
UCd_{11}	840	5.21	m	[12]
UBe_{13}	825	9.47	s	[13]
UPt_3	450	10.59	n	[14]
$UAuPt_4$	725	11.7	s	[15]

[a] n = no ordering, m = magnetic ordering, s = superconducting.
[b] References:
[1] Bader et al. (1975).
[2] Frings (1984).
[3] Schlabitz et al. (1984).
[4] Meisner et al. (1984).
[5] Batlogg (1986).
[6] DeLong et al. (1983).
[7] Willis (1985).
[8] Buschow and van Daal (1972).
[9] Ott et al. (1985c).
[10] Trainor et al. (1975b).
[11] Ott et al. (1984b).
[12] Fisk et al. (1984).
[13] Ott et al. (1984d).
[14] Stewart et al. (1984c).
[15] Ott et al. (1987a).

(1984). They suggest that there is a smooth upper bound to γ which is a function of f–f separation. We present a version of their plot for cerium, uranium and neptunium compounds in fig. 128.

Very little has been said in this section about neptunium and plutonium compounds, principally because the available data is fairly sparse. We refer the reader to the articles by Brodsky (1978), Fournier and Troć (1985) and papers by Burlet et al. (1984) and Mattenberger et al. (1986). There is clearly much further work to be done here. One of the curious features seen in the comparison between compounds of the light actinide elements is that the neptunium compounds appear often to be the 'most' magnetic. If one supposes that the f-electrons are becoming more firmly bound as Z increases, then plutonium compounds would be expected to be the most localized and, hence, most magnetic. The fact that this is not observed does not necessarily mean that the expected trend with increasing Z is wrong. It is possible, for example, that plutonium wants to adopt a $5f^5$ configuration like Sm in the 4f-electron series, which has a number of peculiarities due to other J-multiplets close to the ground state. The stabilization of intermediate valency between fairly well localized ground state configurations is also a possibility, as is seen in SmB_6 (Kasuya et al. 1977) and SmS (Jayaraman et al. 1975).

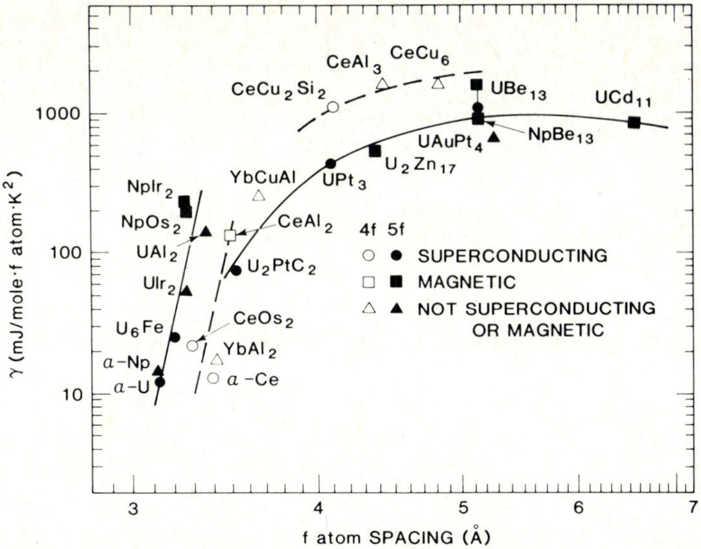

Fig. 128. Envelope of γ versus f–f separation for large -γ compounds (Meisner et al. 1984.)

6. Concluding remarks

The qualitative picture we have of heavy-electron materials is quite simple. At high temperatures these compounds show behaviour usually associated with Kondo-impurity systems. This single-site viewpoint becomes inadequate in a smooth manner on cooling to low temperature, where these systems approach a ground state possessing some of the characteristics of a Fermi liquid. The temperature regime where this change-over occurs is generally referred to as the 'coherence' temperature and is most clearly visible in resistivity- and Hall-effect measurements. Distinct markers are also apparent in inelastic-neutron-scattering results. The Fermi-liquid state in the sense of Landau is, however, only established at temperatures that are two orders of magnitude lower. This is particularly clear from investigations on Ce-based heavy-electrons materials. As we pointed out, this low-temperature state is adopted without a distinct phase transition, but it may itself be unstable against phase transitions that often occur before the Landau Fermi-liquid state is fully developed.

But these high- and low-temperature viewpoints ride above an enormous ignorance of the microscopic aspects of the problem. We know next to nothing about how to describe the development of coherence and very little about how parameters describing the high-temperature regime can be used to describe the low-temperature behaviour. So far, only model calculations have been able to reproduce the salient features of experimental quantities. Finally, we have little predictive capabilities as to which materials will develop heavy-electron behaviour and even less success with predicting possible phase transitions.

On the experimental side, new heavy-electron materials are being sought involving other rare-earths, notably Yb, Tm and Eu, and further investigation of Np and Pu compounds proceeds. Further detailed impurity work involving Yb and U impurities in various hosts, of the quality of earlier Ce-impurity work, is clearly needed. Many low-temperature experiments are being refined. Higher-resolution photo-emission, high-frequency dynamical-response and de Haas–van Alphen-type measurements should elucidate general properties of the low-temperature state. Specific-heat, ultrasound, and thermal-conductivity experiments at extremely low temperatures will certainly be valuable in connection with superconductivity and magnetic order in these systems.

A very natural question presents itself as to the relevance of heavy-electron physics to d-electron transition-metal physics. The generally held opinion is that only spin fluctuations are important in the heavy-electron case, while both charge- and spin fluctuations are important in the d-transition-metal situation. A different way of stating this is to say that heavy-electron effects play a minor role with respect to the overall free energy of heavy-electron compounds, although they become dominant, as demonstrated, at low temperature. In the d transition metals, the electrons which could give rise to heavy-electron effects are also intimately involved in bonding. It seems likely, nevertheless, that important heavy-electron effects will also be found in a number of transition-metal systems.

Various interesting problems are common to both types of materials. We can illustrate this with the example of the temperature dependence of the electrical resistivity. Heavy-electron compounds have resistivities which are nearly temperature independent at room temperature, just as do many transition-metal compounds, high-T_c superconductors for example. It is believed that this situation results because the electronic mean free path is of order of the interatomic spacing. In heavy-electron compounds, the resistivity decreases on cooling in some cases but increases in others. If, however, the mean free path is already of order of the interatomic spacing, it seems puzzling that the resistivity should increase even further. A similar problem arises with the d-transition compounds and alloys. Only in this case the resistivity increases with decreasing temperature only if the material is largely disordered, or contains a large amount of impurities. Consequently one observes a very large residual resistivity at $T=0$ K, contrary to the coherence effects that are characteristic for pure and well-ordered heavy-electron materials. The fact remains that very similar, and so far not understood, problems exist regarding the temperature dependence of the electrical resistivity in both heavy-electron and d-transition-metal compounds.

It is also fair to claim that heavy-electron physics will lead to a better understanding of possible states and processes in solids in general. For example, the development of theoretical techniques of how to treat many-body effects among electrons in metals, and the experimental verification of these results, will be major topics in the future.

Acknowledgements

We should like to thank Mrs. C. Cole and Mrs. M.L. van der Mark for their patience and dedication in typing this manuscript. We are also indebted to Mrs. G. Kägi for her careful and efficient drawing of most of the figures. H.R. Ott would like to thank the Center for Materials Science at Los Alamos National Laboratory for its hospitality when writing parts of the manuscript.

Recent developments

After the writing of this article had been completed and the manuscript was undergoing editorial treatment, various important new results concerning the physics of heavy electrons in metals appeared in the literature or were brought to our attention. In this section we should like at least to mention those that seem relevant with respect to the topics that are treated in the main text. We have chosen to arrange these citations in the same sequence as we organized the review. We therefore first address to physical properties in the normal state as in section 2, then mention some new work concerning theoretical aspects that were treated in section 3, and finally cite some recent work that deals with phase-transition aspects in connection with superconducting or magnetically-ordered ground states of heavy-electron actinide compounds as it is the topic of section 4 of this chapter.

Section 2. One of the really important contributions in connection with general questions concerning heavy electrons in metals is the experimental observation of de Haas–van Alphen (dHvA) oscillations in such materials, as we pointed out in section 6. Results of such experiments were now indeed reported by Reinders et al. (1986) for $CeCu_6$ and by Taillefer et al. (1987) for UPt_3. The success of these experiments depends mainly on the availability of very-good samples and the possibility of performing dHvA measurements at very low temperatures ($T \leqslant 100$ mK) and in high magnetic fields of order 100 kOe. A rather surprising but clarifying result of Taillefer et al. is the fact that all the orbitals that contribute to dHvA oscillations are connected with heavy-mass electrons with $m^*/m_e \geqslant 25$. The so-far reported results were obtained for orbitals that are perpendicular to the a-axis of the hexagonal crystal structure and the largest effective mass observed was $m^* = 90\, m_e$. According to preliminary data (Lonzarich and Taillefer 1987) in configurations where the magnetic-field direction is tilted out of the basal plane, m^*/m_e increases even further. This observation seems to rule out all the models that associate the superconductivity of UPt_3 with 'light' electrons. At first sight also surprising is the fact that the band-structure calculations mentioned in section 3 seem quite successful in predicting the size of different parts of the Fermi surface of UPt_3, but the agreement indicates that the band structure itself is not greatly altered by the interactions that provide the large effective mass. These calculations, however, miss by far in predicting the observed effective-mass values. These experiments demonstrate unambiguously, however, that the low-temperature state of these materials can indeed be visualized as a Fermi liquid of heavy-mass quasiparticles.

Another experiment that should be kept in mind when discussing normal-state properties of UPt_3 is the measurement of an ultrasound attenuation peak for longitudinal sound waves at about 12 K (Müller et al. 1986b). The intensity of this attenuation increases with v^2 and it is noted that an analogous feature is missing for transversely excited sound waves. Müller and co-workers interpret this observation as a clear indication for a narrow resonance in the quasiparticle excitations spectrum near E_F.

Additional optical data for UBe_{13} have been reported by Eklund et al. (1987). Unfortunately these measurements of the reflectance of photons with energies between 0.1 and 30 eV have only been made at room temperature. The usual analysis gives an optical effective mass of order 1 m_e which is compatible with a rather wide conduction band at E_F (see, e.g., section 2.6.1). The choice of this mass and the experimental value of the plasma frequency ω_p are consistent with about 3 free charge carriers per formula unit of UBe_{13} (see also Wuilloud et al. 1984).

Section 3. Regarding theoretical work on the general problem of how to describe heavy electrons in metals we should like to mention a recent review by Fulde et al. (1987) which covers a wide range of aspects including the description of quasiparticles and their interaction, using different approaches and methods and also contains a discussion on the superconductivity of these systems. Another review which discusses extensively the Fermi liquid aspect of heavy electrons is due to Pethick and Pines (1987).

Rice and Ueda (1986) presented a detailed discussion of the application of the Gutzwiller method to describe the heavy-electron state by considering the Kondo limit of the periodic Anderson model. They conclude that the heavy electrons are simply strongly correlated and almost localized f electrons. They also consider the effect of disorder on the heavy-electron Fermi liquid which might, e.g., drastically change the transport properties in comparison with the ideal case. Kotliar and Ruckenstein (1986) investigated the relationship of the Gutzwiller approximation with other methods that are currently used in dealing with these problems and which are, at least partially, mentioned in section 3.

A different approach to solve the 'Kondo-lattice' problem was recently discussed by Jones and Varma (1987) by considering the coupling of two impurities. They find that even when the interaction energy between the moments is much smaller than the Kondo energy, there results a low-temperature behaviour that can be ascribed to a 'correlated' Kondo effect. Their study seems to indicate that the lattice situation can be treated by simply considering a sum of pair Hamiltonians which gives an adequate description of the heavy-electron state at low temperatures.

Finally we cite the work of Zou and Anderson (1986) who give a discussion of the heavy-electron bands on the basis of the relativistic Korringa–Kohn–Rostoker (KKR) equation. Because only one particular local f-electron state out of a multiplet can hybridize with conduction states, they conclude that the magnetic moments connected with the heavy electrons are naturally quenched and therefore a reduced Wilson ratio χ/γ (see also section 5) is expected. Various authors (Zhang and Lee 1987, Aeppli and Varma 1987 and Cox 1987) critizised this work because an important contribution to χ (Van Vleck term) had not properly been taken into account in the above mentioned discussion.

Section 4. Some new experiments probing the superconducting state of UBe_{13} and $U_{1-x}Th_xBe_{13}$ have been reported. First, Ravex et al. (1987) reported new results on

$C_p(T)$ of superconducting UBe_{13} from which they inferred that in the limit of $T=0$ K, a considerable contribution to C_p varies linearly with T. They ascribed this non-zero γ value of about 100 mJ/mol K^2 to the presence of still normal-conducting electrons even when $T \ll T_c$. Subsequent measurements by Ott et al. (1987b) of $C_p(T)$ of superconducting UBe_{13} down to 65 mK did not confirm the above mentioned result. These new data below 180 mK could, in a conventional way, be described by a sum of two electronic contributions of the form $\gamma_0 T + \beta_0 T^3$, where γ_0 is only 22 mJ/mol K^2. A more sophisticated theoretical treatment of $C_p(T)$ well below T_c led to the conclusion that the existing data below 0.4 K could rather well be fitted by considering the influence of resonant impurity scattering on the specific heat well within the superconducting state, assuming an anisotropic state of axial symmetry (see also sections 3 and 4). The treatment of Ott et al. was based on the work of Hirschfeld et al. (1986) who have made a similar analysis for the case of UPt_3, where a polar state appears to be established. Additional work dealing with theoretical aspects of this problem is due to Schmitt-Rink et al. (1986). Very recent experiments by Ott et al. (1987c) indicate that also in very pure UPt_3, the linear-in-T contribution to C_p well below T_c is at least an order of magnitude smaller than observed before (see section 4.1.1). In any case, these new experimental data provide further support to the fact that no exponential temperature dependence of C_p is observed in these superconductors even when $T_c/T \geqslant 10$.

Another result which appears to support the idea of unconventional superconductivity in UBe_{13} is the report of a negative s-wave proximity effect in superconducting UBe_{13} (Han et al. 1986). What is observed is a decrease of the critical Josephson current between a Ta wire and a proximity-induced $l=0$ paired superconducting state in UBe_{13} when the temperature is lowered to below T_c of UBe_{13}. Since the interpretation of this experimental observation is certainly problematic, because both a strong coupling to assure a proximity effect and a weak link for the observation of Josephson tunneling is required simultaneously, we do not wish to make any further comments here.

Application of external pressure on samples of Th-doped UBe_{13} and its influence on T_c (Lambert et al. 1986) gave additional support to the conjecture of different superconducting states separated by the phase boundaries shown in fig. 105. The pressure derivative $\partial T_c/\partial p$ for pure UBe_{13} is markedly different from that measured for, e.g., $U_{0.97}Th_{0.03}Be_{13}$ and moderate pressures of order 10 kbar are sufficient to shift the minimum at $x \simeq 0.017$ for $p = 0$ to zero temperature. Superconductivity is, however, again observed at these and higher pressures when x exceeds 0.04. From this the authors suggest the existence of at least two different superconducting phases in the phase diagram of fig. 105.

A somewhat different conclusion was drawn from recent measurements of the lower critical field H_{c1} of $U_{0.97}Th_{0.03}Be_{13}$ by Rauchschwalbe et al (1987a). A distinct increase of the slope $\partial H_{c1}/\partial T$ at T_{c2} lead these authors to suggest that at T_{c2} a new superconducting transition occurs, below which a new order parameter adds to the one which exists already below T_{c1}. This is, of course, a different situation than that where one superconducting state is replaced by another. It is further suggested that this may happen in this way if at T_{c1} and T_{c2} two different parts of the Fermi surface become unstable with respect to a superconducting transition. This is clearly a fairly unconventional concept as well. These results are then used to propose a new interpretation (Rauchschwalbe et al. 1987b) of the phase diagram shown in fig. 105.

Various recent studies dealt with theoretical aspects of superconductivity that is observed in UBe_{13} and UPt_3. We already mentioned that by Schmitt-Rink et al. (1986) who offer explanations for the observed transport and thermal properties of the superconducting states in terms of an anisotropic order parameter which results in gap zeroes on the Fermi surface. In addition they argue for an effective mean free path that is essentially temperature independent except for $T \ll T_c$ and which is similar in magnitude as that of the normal state. This statement follows from a treatment of impurity scattering in the unitarity limit. As a major favourable result one may consider that these calculations reproduce the anisotropy of ultrasound attenuation in UPt_3, which has been shown experimentally to depend on the polarization of the sound waves in the work of Shivaram et al. (1986).

A possible explanation for the attenuation peaks at T_c of UBe_{13} and UPt_3 (see figs. 102 and 103) was offered by Miyake and Varma (1986). They suggest that this anomalous ultrasound attenuation close to T_c is due to a relaxing of the order-parameter amplitude via the Landau–Khalatnikov mechanism. At lower temperature the same mechanism would give a power-law temperature dependence of the attenuation.

Varma et al. (1986) analyzed the influence of Fermi-liquid corrections on the behaviour of the London penetration depth λ_L. They concluded that the effective dynamic mass is different from the bare mass and that the heavy mass appears in λ_L at all temperatures, a previously disputed issue. They agree with Einzel et al. (1986) that the temperature dependence of λ_L of UBe_{13} (see figs. 104 and 119) indicates anisotropic superconductivity but they suggest another symmetry for the adopted state (polar instead of axial).

Sigrist et al. (1987) studied the possibility of additional phases of anisotropic superconductors when exposed to uniaxial stress. They found that in certain cases the superconducting transition may split under uniaxial stress and conclude that this might provide an experimental means to actually probe the form of the superconducting order parameter, a very attractive possibility, of course. Finally we mention the work of Miyake et al. (1986) who conclude that antiferromagnetic spin-fluctuations, when considered as the pairing mechanism, would favour even-parity superconductivity. This conclusion was also obtained by various other authors, among them Lavagna et al. (1987).

A very interesting material that was only mentioned marginally in the main text is URu_2Si_2. It apparently undergoes a distinct phase transition at 17 K below which only a tiny ordered moment is observed (Broholm et al. 1987b). At still lower temperatures, i.e., around 1 K, a superconducting transition is observed. Palstra et al. (1985) and Maple et al. (1986) studied this low-temperature behaviour in detail. The magnitude of the electronic specific heat just barely qualifies URu_2Si_2 as a heavy-electron material. The observation of only a small moment of less than $0.1\,\mu_B$ may indicate that the primary order parameter of the transition at 17 K has not been established yet. It is certainly also worth mentioning that the temperature derivative of the upper critical field $\partial H_{c2}/\partial T$ at T_c is unusually large but comparable to that of the other heavy-electron superconductors.

As may be seen in fig. 8 of the main text, small amounts of Pd substitutions on the Pt sites of UPt_3 may induce a distinct phase transition at about 6 K. The same seems possible by similar amounts of Au on the Pt sites or Th on the U sites. These findings are discussed in some detail by de Visser et al. (1986a) and by Batlogg et al. (1987). Recently it was shown for $U_{0.95}Th_{0.05}Pt_3$ that this phase transition is of antiferromagnetic nature, namely by neutron-diffraction experiments (Goldman et al. 1986b). The ordered moment below T_N is estimated to be $0.65\,\mu_B/U$.

An interesting study regarding the magnetic phase transition of U_2Zn_{17} has been made by Broholm et al. (1987a) using inelastic neutron scattering. Instead of the usual spin-wave excitation spectrum with two accoustic branches in the (ω, q) plane, they observed a broad ridge of magnetic scattering intensity peaking at $\omega = 0$ and at a commensurate q value. This result, which is not due to inadequate resolution, bears some similarities with earlier observations in Cr. The most interesting feature, however, is that the phase transition is apparently not driven by a divergence of the magnetic susceptibility as usual, but by a strong temperature dependence of the interaction between the moments. A temperature-induced renormalization of this interaction is also the result of a theoretical study (Abrahams and Varma 1985) investigating the competition of RKKY and Kondo interactions in heavy-electron metals.

References

Abrahams, E., and C.M. Varma, 1985, preprint.

Aeppli, G., and C.M. Varma, 1987, Phys. Rev. Lett. **58**, 2729.

Aeppli, G., E. Bucher and G. Shirane, 1985, Phys. Rev. B **32**, 7579.

Albers, R.C., 1985, Phys. Rev. B **32**, 7646.

Albers, R.C., A.M. Boring, P. Weinberger and N.E. Christensen, 1985, Phys. Rev. B **32**, 7571.

Aldred, A.T., and D.J. Lam, 1976, quoted in: Trainor et al. (1976).

Alekseevskii, N.E., V.N. Narozhnyi, V.I. Nizhankovskii, E.G. Nikolaev and E.P. Khlybov, 1984, Pis'ma v Zh. Eksp. & Teor. Fiz. **40**, 421 [JETP Lett. **40**, 1241].

Alekseevskii, N.E., A.V. Mitin, V.I. Nizhankovskii, V.I. Firsov and E.P. Khlybov, 1985, Pis'ma v Zh. Eksp. & Teor. Fiz. **41**, 335 [JETP Lett. **41**, 410].

Allen, J.W., and R.W. Martin, 1982, Phys. Rev. Lett. **49**, 1106.

Allen, J.W., S.J. Oh, L.E. Cox, W.P. Ellis, M.S. Wire, Z. Fisk, J.L. Smith, B.B. Pate, I. Lindau and A.J. Arko, 1985, Phys. Rev. Lett. **54**, 2635.

Anderson, P.W., 1961, Phys. Rev. **124**, 41.

Anderson, P.W., 1984a, Phys. Rev. B **30**, 1549.

Anderson, P.W., 1984b, Phys. Rev. B **30**, 4000.

Andres, K., J.E. Graebner and H.R. Ott, 1975, Phys. Rev. Lett. **35**, 1779.

Arko, A.J., M.B. Brodsky and W.J. Nellis, 1972, Phys. Rev. B **5**, 4564.

Arko, A.J., D.D. Koelling and B. Reihl, 1983, Phys. Rev. B **27**, 3955.

Arko, A.J., C.G. Olson, D.M. Wieliczka, Z. Fisk and J.L. Smith, 1984a, Phys. Rev. Lett. **53**, 2050.

Arko, A.J., D.D. Koelling and J. Schirber, 1984b, in: Handbook on the Physics and Chemistry of the Actinides, Vol. 2, eds A.J. Freeman and G.H. Lander (North-Holland, Amsterdam) p. 175.

Armbrüster, H., W. Franz, W. Schlabitz and F. Steglich, 1979, J. Phys. Colloq. **40**(C4), 150.

Baber, W.G., 1937, Proc. R. Soc. London Ser. A **158**, 383.

Babu, S., and G.E. Brown, 1973, Ann. Phys. **78**, 1.

Bader, S.D., N.E. Phillips and E.S. Fisher, 1975, Phys. Rev. B **12**, 4929.

Baer, Y., 1984, in: Handbook on the Physics and Chemistry of the Actinides, Vol. 1, eds A.J. Freeman and G.H. Lander (North-Holland, Amsterdam) p. 271.

Baer, Y., and J.K. Lang, 1980, Phys. Rev. B **21**, 2060.

Balatskii, A.V., L.I. Burlachkov and L.P. Gor'kov, 1986, Zh. Eksp. & Teor. Fiz. **90**, 1478 [Sov. Phys.-JETP **63**, 866].

Bänziger, N.C., and R.E. Rundle, 1949, Acta Crystallogr. **2**, 258.

Bardeen, J., L.N. Cooper and R.J. Schrieffer, 1957, Phys. Rev. **108**, 1175.

Barnard, R.D., 1972, Thermoelectricity in Metals and Alloys (Taylor & Francis, London).

Barnes, S.E., 1981, Adv. Phys. **30**, 6801.

Barth, S., H.R. Ott, F.N. Gygax, A. Schenck, T.M. Rice and Z. Fisk, 1986a, Hyperfine Interactions **31**, 397.

Barth, S., H.R. Ott, F. Hulliger, F.N. Gygax, A. Schenck and T.M. Rice, 1986b, Hyperfine Interactions **31**, 403.

Batlogg, B., 1986, private communication.

Batlogg, B., D.J. Bishop, B. Golding, C.M. Varma, Z. Fisk, J.L. Smith and H.R. Ott, 1985a, Phys. Rev. Lett. **55**, 1319.

Batlogg, B., Z. Fisk, J.L. Smith and H.R. Ott, 1985b, unpublished.

Batlogg, B., Z. Fisk and H.R. Ott, 1985c, unpublished.

Batlogg, B., D.J. Bishop, B. Golding, E. Bucher, J. Hufnagl, Z. Fisk, J.L. Smith and H.R. Ott, 1986, Phys. Rev. B **33**, 5906.

Batlogg, B., D.J. Bishop, E. Bucher, B. Golding, A.P. Ramirez, Z. Fisk, J.L. Smith and H.R. Ott, 1987, J. Magn. & Magn. Mater. **63/64**, 441.

Bedell, K.S., and K.F. Quader, 1985, Phys. Rev. B **32**, 3296.

Benoit, A., A. Berton, J. Chaussy, J. Flouquet, J.C. Lasjaunias, J. Odin, J. Palleau, J. Peyrard and M. Ribault, 1981, in: Valence Fluctuations in Solids, eds L.M. Falicov, W. Hanke and M.B. Maple (North-Holland, Amsterdam) p. 283.

Bishop, D.J., C.M. Varma, B. Batlogg, E. Bucher, Z. Fisk and J.L. Smith, 1984, Phys. Rev. Lett. **53**, 1009.

Blandin, A., 1973, Magnetism V, ed. H. Suhl (Academic, New York) p. 58.

Blount, E.I., 1985, Phys. Rev. B **32**, 2935.

Bonn, D.A., R.J. Klassen, T. Timusk, J.L. Smith and Z. Fisk, 1985, unpublished.

Boring, A.M., R.C. Albers, G.R. Stewart and D.D. Koelling, 1985a, Phys. Rev. B **31**, 3251.

Boring, A.M., R.C. Albers, F.M. Mueller and D.D. Koelling, 1985b, Physica B **130**, 1711.

Brandow, B.H., 1986, Phys. Rev. B **33**, 215.

Bredl, C.D., S. Horn, F. Steglich, B. Lüthi and R.M. Martin, 1984, Phys. Rev. Lett. **52**, 1982.

Brinkman, W.F., and T.M. Rice, 1970, Phys. Rev. B **2**, 4302.

Brinkman, W.F., J.W. Serene and P.W. Anderson, 1974, Phys. Rev. A **10**, 2386.

Brodale, G.E., R.E. Fisher, N.E. Phillips, G.R. Stewart and A.L. Giorgi, 1986, Phys. Rev. Lett. **57**, 234.

Brodsky, M.B., 1978, Rep. Prog. Phys. **41**, 1547.

Brodsky, M.B., 1979, J. Phys. Colloq. **40**(C4), 147.

Brodsky, M.B., and N.J. Bridger, 1973, in: Magnetism and Magnetic Materials, AIP Conf. Proc., Vol. 18, eds C.D. Graham and J.J. Rhyne (AIP, New York) p. 357.

Brodsky, M.B., and R.J. Friddle, 1975, in: Magnetism and Magnetic Materials, AIP Conf. Proc., Vol. 24, eds C.D. Graham Jr, G.H. Lander and J.J. Rhyne (AIP, New York) p. 353.

Brodsky, M.B., and R.J. Trainor, 1977, Physica B **91**, 271.

Broholm, C., J.K. Kjems, G. Aeppli, Z. Fisk, J.L. Smith, S.M. Shapiro, G. Shirane and H.R. Ott, 1987a, Phys. Rev. Lett. **58**, 917.

Broholm, C., J.K. Kjems, W.J.L. Buyers, P. Matthews, T.T.M. Palstra, A.A. Menovsky and J.A. Mydosh, 1987b, Phys. Rev. Lett. **58**, 1467.

Brooks, M.S.S., B. Johansson and H. Skriver, 1984, in: Handbook on the Physics and Chemistry of the Actinides, Vol. 1, eds A.J. Freeman and G.H. Lander (North-Holland, Amsterdam) p. 153.

Brown, G.E., K.S. Bedell and K.F. Quader, 1985, preprint.

Bucher, E., J.P. Maita, G.W. Hull, R.C. Fulton and A.S. Cooper, 1975, Phys. Rev. B **11**, 440.

Bucher, E., B. Batlogg, D.J. Bishop, C.M. Varma, Z. Fisk and J.L. Smith, 1985, J. Appl. Phys. **57**, 3060.

Burlet, P., S. Quezel, J. Rossat-Mignod, J.C. Spirlet, J. Rebizant, W. Müller and O. Vogt, 1984, Phys. Rev. B **30**, 6660.

Buschow, K.H.J., and J.F. Fast, 1966, Z. Phys. Chem. **50**, 1.

Buschow, K.H.J., and H.J. van Daal, 1972, in: Magnetism and Magnetic Materials, AIP Conf. Proc., Vol. 5, eds C.D. Graham and J.J. Rhyne (AIP, New York) p. 1464.

Buschow, K.H.J., H.J. van Daal, F.E. Maranzana and P.B. van Aken, 1971, Phys. Rev. B **3**, 1662.

Buyers, W.J.L., and T.M. Holden, 1985, in: Handbook on the Physics and Chemistry of the Actinides, Vol. 1, eds A.J. Freeman and G.H. Lander (North-Holland, Amsterdam) p. 239.

Cafasso, F.A., H.M. Feder and D.M. Gruen, 1963, J. Chem. Phys. **38**, 1256.

Chen, J.W., S.E. Lambert, M.B. Maple, Z. Fisk, J.L. Smith, G.R. Stewart and J.O. Willis, 1984a, Phys. Rev. B **30**, 1583.

Chen, J.W., S.E. Lambert, M.B. Maple, Z. Fisk, J.L. Smith and H.R. Ott, 1984b, in: Proc. 17th Int. Conf. on Low Temperature Physics, eds U. Eckern, A. Schmid, W. Weber and H. Wühl (North-Holland, Amsterdam) p. 325.

Chen, J.W., S.E. Lambert, M.B. Maple, M.J. Naughton, J.S. Brooks, Z. Fisk, J.L. Smith and H.R. Ott, 1985, J. Appl. Phys. **57**, 3076.

Clark, W.G., Z. Fisk, K. GLover, M.D. Lan, D.E. MacLaughlin, J.L. Smith and C. Tien, 1984, in: Proc. 17th Int. Conf. on Low Temperature Physics, eds U. Eckern, A. Schmid, W.

Weber and H. Wühl (North-Holland, Amsterdam) p. 227.

Coffey, D., and C.J. Pethick, 1986, Phys. Rev. B **33**, 7508.

Coleman, P., 1984, Phys. Rev. B **29**, 3035.

Coleman, P., 1985, in: Theory of Heavy Fermions and Valence Fluctuations, eds T. Kasuya and T. Saso (Springer, Berlin) p. 163.

Coleman, P., P.W. Anderson and T.V. Ramakrishnan, 1985, Phys. Rev. Lett. **55**, 414.

Cooper, S.L., R.T. Demers, M.V. Klein, Z. Fisk and J.L. Smith, 1985, Physica B **135**, 49.

Cooper, S.L., M.V. Klein, Z. Fisk, J.L. Smith and H.R. Ott, 1987, Phys. Rev. B **35**, 2615.

Cox, D.E., G. Shirane, S.M. Shapiro, G. Aeppli, Z. Fisk, J.L. Smith, J.K. Kjems and H.R. Ott, 1986, Phys. Rev. B **33**, 3614.

Cox, D.L., 1987, Phys. Rev. Lett. **58**, 2730.

d'Ambrumenil, N., and P. Fulde, 1985, J. Magn. & Magn. Mater. **47/48**, 1.

de Visser, A., J.J.M. Franse, A. Menovsky and T.T.M. Palstra, 1984a, Physica B+C **127**, 442.

de Visser, A., J.J.M. Franse and A. Menovsky, 1984b, J. Magn. & Magn. Mater. **43**, 43.

de Visser, A., J.J.M. Franse and A. Menovsky, 1985, J. Phys. F **15**, L53.

de Visser, A., J.C.P. Klaasse, M. van Sprang, J.J.M. Franse, A. Menovsky and T.T.M. Palstra, 1986a, J. Magn. & Magn. Mater. **54–57**, 375.

de Visser, A., R. Gersdorf, J.J.M. Franse and A. Menovsky, 1986b, J. Magn. & Magn. Mater. **54–57**, 383.

de Visser, A., E. Louis, J.J.M. Franse and A. Menovsky, 1986c, J. Magn. & Magn. Mater. **54–57**, 387.

DeLong, L.D., J.G. Huber, K.N. Yang and M.B. Maple, 1983, Phys. Rev. Lett. **51**, 312.

Desclaux, J.P., and A.J. Freeman, 1984, in: Handbook on the Physics and Chemistry of the Actinides, Vol. 1, eds A.J. Freeman and G.H. Lander (North-Holland) p. 1.

Doniach, S., and S. Engelsberg, 1966, Phys. Rev. Lett. **17**, 750.

Edelstein, A.S., C.J. Tranchita, O.D. McMasters and K.A. Gschneidner Jr, 1974, Solid State Commun. **15**, 81.

Einzel, D., P.J. Hirschfeld, F. Gross, B.S. Chandrasekhar, K. Andres, H.R. Ott, J. Beuers, Z. Fisk and J.L. Smith, 1986, Phys. Rev. Lett. **56**, 2513.

Eklund, P.C., D.M. Hoffman, L.E. Delong, E.T. Arakawa, J.L. Smith and Z. Fisk, 1987, Phys. Rev. B **35**, 4250.

Eliashberg, G.M., 1962, Zh. Eksp. & Teor. Fiz. **43**, 1005 [1963, Sov. Phys.-JETP **16**, 780].

Elliott, R.O., and H.H. Hill, 1970, J. Less-Common Met. **22**, 123.

Fay, D., and J. Appel, 1985, Phys. Rev. B **32**, 6071.

Fedders, P.A., and P.C. Martin, 1966, Phys. Rev. **143**, 245.

Felten, R., F. Steglich, G. Weber, H. Rietschel, F. Gompf, B. Renker and J. Beuers, 1986, Europhys. Lett. **2**, 323.

Fenton, E.W., 1985, Solid State Commun. **54**, 709.

Fenton, E.W., and G.C. Aers, 1985, Solid State Commun. **56**, 337.

Fischer, H.E., E.T. Swartz and R.O. Pohl, 1986, Bull. Am. Phys. Soc. **31**, 444, and private communication.

Fisk, Z., 1984, unpublished.

Fisk, Z., 1985, unpublished.

Fisk, Z., and G.W. Webb, 1976, Phys. Rev. Lett. **36**, 1084.

Fisk, Z., G.R. Stewart, J.O. Willis, H.R. Ott and F. Hulliger, 1984, Phys. Rev. B **30**, 6360.

Fisk, Z., J.L. Smith, H.R. Ott and B. Batlogg, 1985, J. Magn. & Magn. Mater. **52**, 79.

Flouquet, J., 1986, private communication.

Fournier, J.-M., and J. Beille, 1979, J. Phys. Colloq. **40**(C4), 145.

Fournier, J.-M., and R. Troć, 1985, in: Handbook on the Physics and Chemistry of the Actinides, Vol. 2, eds A.J. Freeman and G.H. Lander (North-Holland, Amsterdam) p. 29.

Franse, J.J.M., P.H. Frings, F.R. de Boer, A. Menovsky, C.J. Beers, A.P.J van Deursen, H.W. Myron and A.J. Arko, 1982, Phys. Rev. Lett. **48**, 1749.

Franse, J.J.M., P.H. Frings, A. de Visser, A. Menovsky, T.T.M. Palstra, P.H. Kes and J.A. Mydosh, 1984, Physica B **126**, 116.

Franse, J.J.M., A. Menovsky, A. de Visser, C.D. Bredl, U. Gottwick, W. Lieke, H.M. Mayer, U. Rauchschwalbe, G. Sparn and F. Steglich, 1985, Z. Phys. B **59**, 15.

Frick, B., J. Schoenes, O. Vogt and J.W. Allen, 1982, Solid State Commun. **42**, 331.

Frings, P.H., 1984, Thesis (University of Amsterdam).

Frings, P.H., J.J.M. Franse, F.R. de Boer and A. Menovsky, 1983, J. Magn. & Magn. Mater. **31–34**, 240.

Fukase, T., M. Tachiki, N. Toyota and Y. Muto, 1976, Solid State Commun. **18**, 505.

Fulde, P., J. Keller and G. Zwicknagel, 1987, to appear in Solid State Physics, eds F. Seitz, D. Turnbull and H. Ehrenreich.

Gandra, F., S. Schultz, S.B. Osheroff, Z. Fisk and J.L. Smith, 1985, Phys. Rev. Lett. **55**, 2719.

Geshkenbein, V.B., and A.I. Larkin, 1986, Pis'ma v Zh. Eksp. & Teor. Fiz. **43**, 306 [JETP Lett. **43**, 395].

Giorgi, A.L., Z. Fisk, J.O. Willis, G.R. Stewart and J.L. Smith, 1984, in: Proc. 17th Int. Conf. on Low Temperature Physics, eds U. Eckern, A. Schmid, W. Weber and H. Wühl (North-Holland, Amsterdam) p. 229.

Golding, B., B. Batlogg, D.J. Bishop, W.H. Haemmerle, Z. Fisk, J.L. Smith and H.R. Ott, 1985a, in: Proc. 2nd Int. Conf. on Phonon Physics, eds J. Kollar, N. Kroo, N. Ményhard and T. Siklos (World Scientific, Singapore) p. 406.

Golding, B., D.J. Bishop, B. Batlogg, W.H.

Haemmerle, Z. Fisk, J.L. Smith and H.R. Ott, 1985b, Phys. Rev. Lett. **55**, 2479.
Goldman, A.I., S.M. Shapiro, D.E. Cox, J.L. Smith and Z. Fisk, 1985, Phys. Rev. B **32**, 6042.
Goldman, A.I., S.M. Shapiro, G. Shirane, J.L. Smith and Z. Fisk, 1986a, Phys. Rev. B **33**, 1627.
Goldman, A.I., G. Shirane, G. Aeppli, B. Batlogg and E. Bucher, 1986b, Phys. Rev. B **34**, 6564.
Grohs, H., H. Höchst, P. Steiner, S. Hüfner and K.H.J. Buschow, 1980, Solid State Commun. **33**, 573.
Gross, F., B.S. Chandrasekhar, D. Einzel, K. Andres, P.J. Hirschfeld, H.R. Ott, Z. Fisk, J.L. Smith and J. Beuers, 1986, Z. Phys. B **64**, 175.
Gschneidner Jr, K.A., and R. Smoluchowski, 1963, J. Less-Common. Met. **5**, 374.
Gunnarson, O., and K. Schönhammer, 1983, Phys. Rev. B **28**, 4315.
Gutzwiller, M.C., 1964, Phys. Rev. **137**, A1726.
Hadzic-Leroux, M., A. Hamzic, A. Fert, P. Haen, F. Lapierre and O. Laborde, 1986, Europhys. Lett. **1**, 579.
Han, S., K.W. Ng, E.L. Wolf, A. Millis, J.L. Smith and Z. Fisk, 1986, Phys. Rev. Lett. **57**, 238.
Heffner, R.H., D.W. Cooke, Z. Fisk, R.L. Hutson, M.E. Schillaci, J.L. Smith, J.O. Willis, D.E. MacLaughlin, C. Boekema, R.L. Lichti, A.B. Denison, J. Oostens, 1986, Phys. Rev. Lett. **57**, 1255.
Hewson, A.C., D.M. Newns, J.W. Rasul, N. Read, H.U. Desgranges and P. Strange, 1985, in: Theory of Heavy Fermions and Valence Fluctuations, eds T. Kasuya and T. Saso (Springer, Berlin) p. 134.
Hill, H.H., 1970, in: Plutonium 1970 and other Actinides, ed. W.N. Miner (The Metallurgical Society of the AIME, New York) p. 2.
Hirschfeld, P., D. Vollhardt and P. Wölfle, 1986, Solid State Commun. **59**, 111.
Jaccard, D., J. Flouquet, Z. Fisk, J.L. Smith and H.R. Ott, 1985a, J. Phys. Lett. **46**, L811.
Jaccard, D., J. Flouquet, P. Lejay and J.L. Tholence, 1985b, J. Appl. Phys. **57**, 3082.
Jayaraman, A., P.D. Dernier and L.D. Longinotti, 1975, High Temp.-High Pressures **7**, 1.
Johansson, B., 1974, Philos. Mag. **30**, 469.
Johansson, W.R., G.W. Crabtree, A.S. Edelstein and O.D. McMasters, 1983, J. Magn. & Magn. Mater. **31–34**, 377.
Johnson, S.M., A. Benoit, P.J. Brown, H. Capellmann, J. Flouquet, J.J.M. Franse, P.H. Frings and K.R.A. Ziebeck, 1986, Z. Phys. B **63**, 335.
Jones, B.A., and C.M. Varma, 1987, Phys. Rev. Lett. **58**, 843.
Joynt, R., and T.M. Rice, 1985, Phys. Rev. B **32**, 6074.
Joynt, R., T.M. Rice and K. Ueda, 1986, Phys. Rev. Lett. **56**, 1412.
Kasuya, T., 1956, Prog. Theor. Phys. (Kyoto) **16**, 45, 58.
Kasuya, T., and T. Saso, eds, 1985, Theory of Heavy Fermions and Valence Fluctuations (Springer, Berlin).
Kasuya, T., K. Kojima and M. Kasaya, 1977, in: Valence Instabilities and Related Narrow-Band Phenomena, ed. R.D. Parks (Plenum, New York) p. 137.
Kitaoka, Y., K. Ueda, T. Kohara and K. Asayama, 1984, Solid State Commun. **51**, 461.
Kittel, C., 1963, Quantum Theory of Solids (Wiley, New York) p. 353.
Klemm, R.A., and K. Scharnberg, 1986, J. Magn. & Magn. Mater. **54–57**, 381.
Koelling, D.D., 1982, Solid State Commun. **43**, 247.
Koelling, D.D., B.D. Dunlap and G.W. Crabtree, 1985, Phys. Rev. B **31**, 4966.
Kondo, J., 1964, Prog. Theor. Phys. **32**, 37.
Kotliar, G., and A.E. Ruckenstein, 1986, Phys. Rev. Lett. **57**, 1362.
Kuramoto, Y., 1985, in: Theory of Heavy Fermions and Valence Fluctuations, eds T. Kasuya and T. Saso (Springer, Berlin) p. 152.
Lambert, S.E., Y. Dalichaouch, M.B. Maple, J.L. Smith and Z. Fisk, 1986, Phys. Rev. Lett. **57**, 1619.
Lan, M.D., W.G. Clark, D.E. McLaughlin, C. Tien, J.L. Smith, Z. Fisk and H.R. Ott, 1985, Bull. Am. Phys. Soc. **30**, 356.
Landau, L.D., 1956, Zh. Eksp. & Teor. Fiz. **30**, 1058 [1957, Sov. Phys.-JETP **3**, 920].
Lander, G.H., J.F. Reddy, A. Delapalme and P.J. Brown, 1980, Phys. Rev. Lett. **44**, 603.
Landgren, G., Y. Jugnet, J.F. Morar, A.J. Arko, Z. Fisk, J.L. Smith, H.R. Ott and B. Reihl, 1984, Phys. Rev. B **29**, 493.
Lavagna, M., A.J. Millis and P.A. Lee, 1987, J. Magn. & Magn. Mater. **63/64**, 432.
Lawson, A.W., and Ting-Yuan Tang, 1949, Phys. Rev. **76**, 301.
Lee, P.A., T.M. Rice, J.W. Serene, L.J. Sham and J.W. Wilkins, 1986, Comments on Condensed Matter Physics **12**, 99.
Leggett, A.J., 1975, Rev. Mod. Phys. **47**, 331.
Lin, C.L., L.W. Zhou, J.E. Crow, R.P. Guertin and G.R. Stewart, 1986, J. Magn. & Magn. Mater. **54–57**, 391.
Lonzarich, G.G., and L. Taillefer, 1987, private communication.
Loong, C.K., M. Löwenhaupt and M.L. Urtis, 1986, Physica B **136**, 413.
Löwenhaupt, M., S. Horn, F. Steglich, E. Holland-Moritz and G.H. Lander, 1979, J. Phys. Colloq. **40**(C4), 142.
Lüthi, B., 1980, in: Dynamical Properties of Solids, Vol. 3, eds G.K. Horton and A.A. Maradudin (North-Holland, Amsterdam) p. 247, and refs. therein.
Lüthi, B., 1985, J. Magn. & Magn. Mater. **52**, 70.
Machida, K., T. Ohmi and M. Ozaki, 1985, J. Phys. Soc. Jpn. **54**, 1552.
Maple, M.B., J.G. Huber, B.R. Coles and A.C. Lawson, 1970, J. Low Temp. Phys. **3**, 137.
Maple, M.B., J.W. Chen, S.E. Lambert, Z. Fisk,

J.L. Smith, H.R. Ott, J.S. Brooks and M.J. Naughton, 1985, Phys. Rev. Lett. **54**, 477.
Maple, M.B., J.W. Chen, Y. Dalichaouch, T. Kohara, C. Rossel, M.S. Torikachvili, M.W. McElfresh and J.D. Thompson, 1986, Phys. Rev. Lett. **56**, 185.
Marshall, W., and S.W. Lovesey, 1971, Theory of Thermal Neutron Scattering (Clarendon, Oxford).
Martin, R.M., 1982, Phys. Rev. Lett. **48**, 362.
Mattenberger, K., O. Vogt, J.C. Spirlet and J. Rebizant, 1986, J. Magn. & Magn. Mater. **54–57**, 539.
Mayer, H.M., U. Rauchschwalbe, C.D. Bredl, F. Steglich, H. Rietschel, H. Schmidt, H. Wühl and J. Beuers, 1986, Phys. Rev. B **33**, 3168.
McLaughlin, D.E., C. Tien, W.G. Clark, M.D. Lan, Z. Fisk, J.L. Smith and H.R. Ott, 1984, Phys. Rev. Lett. **53**, 1833.
Meisner, G.P., A.L. Giorgi, A.C. Lawson, G.R. Stewart, J.O. Willis, M.S. Wire and J.L. Smith, 1984, Phys. Rev. Lett. **53**, 1829.
Misch, L., 1935, Metallwirtschaft **14**, 897.
Misiuk, A., J. Mulak and A. Czopnik, 1973, Bull. Acad. Pol. Sci. Ser. Sci. Chim. **21**, 487.
Miyake, K., and C.M. Varma, 1986, Phys. Rev. Lett. **57**, 1627.
Miyake, K., T. Matsuura, H. Jichu and Y. Nagaoka, 1984, Prog. Theor. Phys. **72**, 1063.
Miyake, K., S. Schmitt-Rink and C.M. Varma, 1986, Phys. Rev. B **34**, 6554.
Mock, R., B. Hillebrands, H. Schmidt, G. Güntherodt, Z. Fisk and A. Meyer, 1985, J. Magn. & Magn. Mater. **47/48**, 312.
Moser, M., P. Wachter, J.J.M. Franse, G.P. Meisner and E. Walker, 1986, J. Magn. & Magn. Mater. **54–57**, 373.
Mulak, J., and A. Misiuk, 1971, Bull. Acad. Pol. Sci. Ser. Sci. Chim. **19**, 207.
Müller, V., D. Maurer, E.W. Scheidt, Ch. Roth, K. Lüders, E. Bucher and H.E. Bömmel, 1986a, Solid State Commun. **57**, 319.
Müller, V., D. Maurer, K. de Groot, E. Bucher and H.R. Bömmel, 1986b, Phys. Rev. Lett. **56**, 248.
Murasik, A., S. Ligenza and A. Zygmunt, 1974, Phys. Status Solidi a **23**, K163.
Naegele, J.R., L. Manes, J.C. Spirlet and J.-M. Fournier, 1980, Appl. Surf. Sci. **4**, 510.
Nozières, P., 1974, J. Low Temp. Phys. **17**, 31.
Nozières, P., 1985, Ann. Phys. **10**, 19.
Oguchi, T., and A.J. Freeman, 1985a, J. Magn. & Magn. Mater. **52**, 174.
Oguchi, T., and A.J. Freeman, 1985b, Physica B **135**, 46.
Ohkawa, F.J., and H. Fukuyama, 1984, J. Phys. Soc. Jpn. **53**, 4344.
Onuki, Y., Y. Furukaya and T. Komatsubara, 1984, J. Phys. Soc. Jpn. **53**, 2734.
Ott, H.R., 1984, Physica B **126**, 100.
Ott, H.R., and Z. Fisk, 1985, unpublished.
Ott, H.R., H. Rudigier, Z. Fisk and J.L. Smith, 1983, Phys. Rev. Lett. **50**, 1595.
Ott, H.R., O. Marti and F. Hulliger, 1984a, Solid State Commun. **49**, 1129.
Ott, H.R., H. Rudigier, P. Delsing and Z. Fisk, 1984b, Phys. Rev. Lett. **52**, 1551.
Ott, H.R., H. Rudigier, T.M. Rice, K. Ueda, Z. Fisk and J.L. Smith, 1984c, Phys. Rev. Lett. **52**, 1915.
Ott, H.R., H. Rudigier, Z. Fisk and J.L. Smith, 1984d, in: Moment Formation in Solids, ed. W.J.L. Buyers (Plenum, New York) p. 305.
Ott, H.R., H. Rudigier, Z. Fisk and J.L. Smith, 1985a, Phys. Rev. B **31**, 1651.
Ott, H.R., H. Rudigier, Z. Fisk and J.L. Smith, 1985b, J. Appl. Phys. **57**, 3044.
Ott, H.R., H. Rudigier, E. Felder, Z. Fisk and B. Batlogg, 1985c, Phys. Rev. Lett. **55**, 1595.
Ott, H.R., H. Rudigier, Z. Fisk, J.O. Willis and G.R. Stewart, 1985d, Solid State Commun. **53**, 235.
Ott, H.R., E. Felder, H. Rudigier and Z. Fisk, 1985e, unpublished.
Ott, H.R., Z. Fisk and J.L. Smith, 1985f, unpublished.
Ott, H.R., H. Rudigier and Z. Fisk, 1985g, unpublished.
Ott, H.R., H. Rudigier, E. Felder, Z. Fisk and J.L. Smith, 1986a, Phys. Rev. B **33**, 126.
Ott, H.R., H. Rudigier, Z. Fisk and J.L. Smith, 1986b, unpublished.
Ott, H.R., E. Felder and Z. Fisk, 1986c, unpublished.
Ott, H.R., E. Felder, H. Rudigier, Z. Fisk and J.L. Smith, 1986d, unpublished.
Ott, H.R., H. Rudigier, E. Felder, Z. Fisk and J.D. Thompson, 1987a, Phys. Rev. B **35**, 1452.
Ott, H.R., E. Felder, C. Bruder and T.M. Rice, 1987b, Europhys. Lett. **3**, 1123.
Ott, H.R., E. Felder, A. Bernasconi, Z. Fisk, J.L. Smith, L. Taillefer and G.G. Lonzarich, 1987c, submitted to LT-18, Kyoto, Japan.
Palstra, T.T.M., A.A. Menovsky, J. van den Berg, A.J. Dirkmaat, P.H. Kes, G.J. Nieuwenhuys and J.A. Mydosh, 1985, Phys. Rev. Lett. **55**, 2727.
Parks, R.D., M.L. den Boer, S. Raaen, J.L. Smith and G.P. Williams, 1984, Phys. Rev. B **30**, 1580.
Pauling, L.C., 1950, quoted in: A.F. Schuch and J.H. Sturdivant, J. Chem. Phys. **18**, 145.
Penney, T., J. Stankiewicz, S. Von Molnar, Z. Fisk, J.L. Smith and H.R. Ott, 1986, J. Magn. & Magn. Mater. **54–57**, 370.
Percheron, A., J.C. Achard, O. Gorochov, B. Cornut, D. Jérome and B. Coqblin, 1973, Solid State Commun. **12**, 1289.
Pethick, C.J., and D. Pines, 1987, Proc. 4th Int. Conf. on Recent Progress in Many-Body Theories, eds P. Siemens and R.A. Smith (Springer, Berlin), in print.
Pethick, C.J., D. Pines, K.F. Quader, K.S. Bedell and G.E. Brown, 1986, Phys. Rev. Lett. **57**, 1955.
Rajan, V.T., 1983, Phys. Rev. Lett. **51**, 308.
Rao, V.U.S., and R. Vijayaraghavan, 1968, J. Phys. Chem. Solids **29**, 123.

Rauchschwalbe, U., U. Ahlheim, F. Steglich, D. Rainer and J.J.M. Franse, 1985, Z. Phys. B **60**, 379.

Rauchschwalbe, U., F. Steglich, G.R. Stewart, A.L. Giorgi, P. Fulde and K. Maki, 1987a, Europhys. Lett. **3**, 751.

Rauchschwalbe, U., C.D. Bredl, F. Steglich, K. Maki and P. Fulde, 1987b, Europhys. Lett. **3**, 757.

Ravex, A., D. Jaccard, J. Flouquet, J.L. Tholence and A. Meyer, 1987, J. Low Temp. Phys., to be published.

Razafimandimby, H., P. Fulde and J. Keller, 1984, Z. Phys. B **54**, 111.

Read, N., 1985, J. Phys. C **18**, 2651.

Read, N., and D.M. Newns, 1984, Solid State Commun. **52**, 993.

Read, N., D.M. Newns and S. Doniach, 1984, Phys. Rev. B **30**, 384.

Reihl, B., 1984, private communication.

Reinders, P.H.P., M. Springford, P.T. Coleridge, R. Boulet and D. Ravot, 1986, Phys. Rev. Lett. **57**, 1631.

Remenyi, G., D. Jaccard, J. Flouquet, A. Briggs, Z. Fisk, J.L. Smith and H.R. Ott, 1986, J. Phys. (France) **47**, 367.

Renker, B., F. Gompf, W. Reichardt, H. Rietschel, J.B. Suck and J. Beuers, 1985, Phys. Rev. B **32**, 1859.

Ribault, M., A. Benoit, J. Flouquet and J. Palleau, 1979, J. Phys. Lett. **40**, 413.

Rice, T.M., and K. Ueda, 1985, Phys. Rev. Lett. **55**, 95.

Rice, T.M., and K. Ueda, 1986, Phys. Rev. B **34**, 6420.

Rice, T.M., K. Ueda, H.R. Ott and H. Rudigier, 1985, Phys. Rev. B **31**, 594.

Rice, T.M., K. Ueda and H.R. Ott, 1986, J. Magn. & Magn. Mater. **54–57**, 317.

Robinson, R.A., J.D. Axe, A.I. Goldman, Z. Fisk, J.L. Smith and H.R. Ott, 1986, Phys. Rev. B **33**, 6488.

Rodriguez, J.P., 1985, Phys. Rev. Lett. **55**, 250.

Samorukov, O.P., V.N. Kostrynkov, F.A. Kostylev and V.A. Tumbakov, 1974, At. Energ. **37**, 28 [1975, Sov. At. Energy **37**, 705].

Sauls, J., Z. Zou and P.W. Anderson, 1986, preprint.

Scharnberg, K., and R.A. Klemm, 1985, Phys. Rev. Lett. **54**, 2445.

Schlabitz, W., F. Steglich, C.D. Bredl and W. Franz, 1980, Physica B **102**, 321.

Schlabitz, W., J. Baumann, J. Diesing, W. Krause, G. Neumann, C.D. Bredl and U. Rauchschwalbe, 1984, in: Abstracts Int. Conf. on Valence Fluctuations, Cologne, Fed. Rep. Germany, unpublished.

Schlottmann, P., 1983, Z. Phys. B **51**, 223.

Schmitt-Rink, S., K. Miyake and C.M. Varma, 1986, Phys. Rev. Lett. **57**, 2575.

Schneider, W.D., and C. Laubschat, 1981a, Phys. Rev. Lett. **46**, 1023.

Schneider, W.D., and C. Laubschat, 1981b, Phys. Rev. B **23**, 997.

Schneider, W.D., B. Reihl, N. Mårtensson and A.J. Arko, 1982, Phys. Rev. B **26**, 423.

Schoenes, J., and J.J.M. Franse, 1985, Physica B **130**, 69.

Schoenes, J., and J.J.M. Franse, 1986, Phys. Rev. B **33**, 5138.

Shamir, N., M. Melamud, H. Shaked and M. Weger, 1978, Physica B **94**, 225.

Shapiro, S.M., 1986, Physica B **136**, 365.

Shivaram, B.S., Y.H. Jeong, T.F. Rosenbaum and D.J. Hinks, 1986, Phys. Rev. Lett. **56**, 1078.

Sigrist, M., R. Joynt and T.M. Rice, 1987, Europhys. Lett. **3**, 629.

Smith, J.L., Z. Fisk, J.O. Willis and R.G. Haire, 1983, in: Plutonium Chemistry, ACS Symposium Series, Vol. 216, eds W.T. Carnall and G.R. Choppin (American Chemical Society, Washington, DC) p. 65.

Smith, J.L., Z. Fisk, J.O. Willis, B. Batlogg and H.R. Ott, 1984a, J. Appl. Phys. **55**, 1996.

Smith, J.L., Z. Fisk and G.R. Stewart, 1984b, unpublished.

Smith, J.L., Z. Fisk, J.O. Willis, A.L. Giorgi, R.B. Roof, H.R. Ott, H. Rudigier and E. Felder, 1985, Physica B **135**, 3.

Sommerfeld, A., and H.A. Bethe, 1967, Elektronentheorie der Metalle, Heidelberger Taschenbücher (Springer, Berlin).

Stassis, C., J. Arthur, C.F. Majkrzak, J.D. Axe, B. Batlogg, J. Remeika, Z. Fisk, J.L. Smith and A.S Edelstein, 1986, Phys. Rev. B **34**, 4382.

Steglich, F., J. Aarts, C.D. Bredl, W. Lieke, D. Meschede, W. Franz and H. Schäfer, 1979, Phys. Rev. Lett. **43**, 1892.

Steglich, F., U. Rauchschwalbe, U. Gottwick, H.M. Mayer, G. Sparn, N. Grewe, U. Poppe and J.J.M. Franse, 1985, J. Appl. Phys. **57**, 3054.

Stewart, G.R., 1984, Rev. Mod. Phys. **56**, 755.

Stewart, G.R., and R.O. Elliott, 1981, in: Actinides 1981, ed. N. Edelstein (Lawrence Berkeley Laboratory, University of California) p. 206.

Stewart, G.R., and A.L. Giorgi, 1984, J. Low Temp. Phys. **56**, 379.

Stewart, G.R., A.L. Giorgi, B.L. Brandt, S. Foner and A.J. Arko, 1983, Phys. Rev. B **28**, 1524.

Stewart, G.R., Z. Fisk, J.L. Smith, J.O. Willis and M.S. Wire, 1984a, Phys. Rev. B **30**, 1249.

Stewart, G.R., Z. Fisk, J.O. Willis and J.L. Smith, 1984b, Physica B+C **127**, 448.

Stewart, G.R., Z. Fisk, J.O. Willis and J.L. Smith, 1984c, Phys. Rev. Lett. **52**, 679.

Stewart, G.R., Z. Fisk, J.L. Smith, H.R. Ott and F.M. Müller, 1984d, in: Proc. Int. Conf. on Low Temperature Physics, eds U. Eckern, A. Schmid, W. Weber and H. Wühl (North-Holland, Amsterdam) p. 321.

Stewart, G.R., Z. Fisk and M.S. Wire, 1984e, Phys. Rev. B **30**, 482.

Sticht, J., and J. Kübler, 1985, Solid State Commun. **54**, 389.

Strange, P., and B.L. Gyorffy, 1985, Physica B **130**, 41.

Suhl, H., 1985, Phys. Rev. Lett. **55**, 2999.

Sulpice, A., P. Gandit, J. Chaussy, J. Flouquet, D. Jaccard, P. Lejay and J.L. Tholence, 1986, J. Low Temp. Phys. **62**, 39.

Tachiki, M., and S. Maekawa, 1984, Phys. Rev. B **29**, 2497.

Taillefer, L., R. Newbury, G.G. Lonzarich, Z. Fisk and J.L. Smith, 1987, J. Magn. & Magn. Mater. **63/64**, 372.

Takegahara, H., H. Harima and T. Kasuya, 1985, J. Magn. & Magn. Mater. **47/48**, 263.

Thompson, J.D., Z. Fisk, J.M. Lawrence, J.L. Smith and R.W. Martin, 1983, Phys. Rev. Lett. **50**, 1081.

Thompson, J.D., Z. Fisk and H.R. Ott, 1986, J. Magn. & Magn. Mater. **54–57**, 393.

Trainor, R.J., M.B. Brodsky and G.S. Knapp, 1975a, in: Plutonium 1975 and other Actinides, ed. H. Blank and R. Linder (North-Holland, Amsterdam) p. 475.

Trainor, R.J., M.B. Brodsky and H.V. Culbert, 1975b, Phys. Rev. Lett. **34**, 1019.

Trainor, R.J., M.B. Brodsky, D.B. Dunlap and G.K. Shenoy, 1976, Phys. Rev. Lett. **37**, 1511.

Troć, R., W. Trzebiatowski and K. Piprek, 1971, Bull. Acad. Pol. Sci. Ser. Sci. Chim. **19**, 427.

Ueda, K., and T.M. Rice, 1985, Phys. Rev. B **31**, 7114.

Valls, O.T., and Z. Tesanovic, 1984, Phys. Rev. Lett. **53**, 1497.

van Daal, H.J., K.H.J. Buschow, P.B. van Aken and M.H. van Maaren, 1975, Phys. Rev. Lett. **34**, 1457.

van Vucht, J.H.N., and K.H.J. Buschow, 1965, J. Less-Common Met. **10**, 98.

Varma, C.M., 1984a, Bull. Am. Phys. Soc. **29**, 404.

Varma, C.M., 1984b, in: Moment Formation in Solids, ed. W.J.L. Buyers (Plenum, New York) p. 83.

Varma, C.M., 1985a, Phys. Rev. Lett. **55**, 2723.

Varma, C.M., 1985b, Comments on Solid State Physics **11**, 221.

Varma, C.M., K. Miyake and S. Schmitt-Rink, 1986, Phys. Rev. Lett. **57**, 626.

Volovik, G.E., and L.P. Gor'kov, 1984, Pis'ma v Zh. Eksp. & Teor. Fiz. **39**, 550 [JETP Lett. **39**, 674].

Volovik, G.E., and L.P. Gor'kov, 1985, Zh. Eksp. & Teor. Fiz. **88**, 1412 [Sov. Phys.-JETP **61**, 843].

Volovik, G.E., and D.E. Khmel'nitskii, 1984, Pis'ma v Zh. Eksp. & Teor. Fiz. **40**, 469 [JETP Lett. **40**, 1299].

Wilkins, J.W., 1985, private communication.

Willis, J.O., 1985, private communication.

Willis, J.O., J.D. Thompson, Z. Fisk, A. de Visser, J.J.M. Franse and A. Menovsky, 1985a, Phys. Rev. B **31**, 1654.

Willis, J.O., J.L. Smith, Z. Fisk and H.R. Ott, 1985b, J. Appl. Phys. **57**, 3079.

Willis, J.O., Z. Fisk and H.R. Ott, 1985c, unpublished.

Willis, J.O., Z. Fisk, G.R. Stewart and H.R. Ott, 1986, J. Magn. & Magn. Mater. **54–57**, 395.

Wire, M.S., J.D. Thompson and Z. Fisk, 1984, Phys. Rev. B **30**, 5591.

Wuilloud, E., Y. Baer, H.R. Ott, Z. Fisk and J.L. Smith, 1984, Phys. Rev. B **29**, 5228.

Yanase, A., 1983, J. Magn. & Magn. Mater. **31–34**, 453.

Yoshimori, A., and H. Kasai, 1983, J. Magn. & Magn. Mater. **31**, 475.

Yoshizawa, M., B. Lüthi, T. Goto, T. Suzuki, B. Renker, A. de Visser, P. Frings and J.J.M. Franse, 1985, J. Magn. & Magn. Mater. **52**, 413.

Yu, C.C., and P.W. Anderson, 1984, Phys. Rev. B **29**, 6165.

Zachariasen, W.H., 1949, quoted by Lawson and Ting-Yuan Tang (1949).

Zhang, F.C., and T.K. Lee, 1987, Phys. Rev. Lett. **58**, 2728.

Zou, Z., and P.W. Anderson, 1986, Phys. Rev. Lett. **57**, 2073.

Handbook on the Physics and Chemistry of the Actinides, edited by A.J. Freeman and G.H. Lander

Chapter 3

THE EFFECT OF HIGH PRESSURES ON ACTINIDE METALS

U. BENEDICT
Commission of the European Communities, Joint Research Centre,
European Institute for Transuranium Elements,
Postfach 2340, D-7500 Karlsruhe, Fed. Rep. Germany

Contents

1. Introduction

Actinides, by the singular behaviour of their 5f electrons, allow the solid state scientist to study phenomena which are not found, or not so clearly found, in other groups of the Periodic Table of the elements. More insight into the physical properties of actinides will contribute to broadening the basis of knowledge of the solid state.

Often a good way of investigating a situation is to change it and, from the effects observed, to draw conclusions about the initial state. This is exactly the case of high-pressure studies: the changes occurring under the action of pressure in a solid can give rise to spectacular effects which in turn can give more information about electronic structure and physical properties under ambient pressure. In particular, high pressure is the most efficient, and at the same time the 'cleanest' way of influencing the interatomic distances in a lattice. A number of physical properties are sensitive to variations of these distances. Applying high pressure to a solid can for instance lead to transfer of electrons from one energy band to another. In the well known example of the lanthanide cerium, the 4f electron is thought to become a bonding electron under pressure, increasing the metallic valence, decreasing the atomic volume, and giving rise to crystal structures of low symmetry. The appearance of low-symmetry phases and the so-called 'volume collapse' in the metals americium, curium, berkelium and californium at high pressures are also interpreted as being connected to increased participation of f electrons – here the 5f electrons – in the metallic bond.

Interatomic distances can be measured by X-ray diffraction (XRD). The close relationship between these distances and certain physical properties makes high-pressure X-ray diffraction (HPXRD) a preferred tool for studying pressure effects in actinides. Another advantage of this method is that the compressibility, which is a very useful quantity in theoretical and experimental considerations, can be determined from the observed variation of the volume with pressure.

Practically all other physical properties change under pressure as well. Since phenomena such as the unusual phase transformations observed under pressure are not yet fully understood, it is desirable to characterise them more completely by measuring other physical properties under pressure. Whereas X-ray diffraction still constitutes a major part of the high-pressure work carried out on actinides with a high specific radioactivity, a number of other physical measurements have been applied to uranium, thorium and their compounds under pressure.

Those actinide elements of which sufficient amounts for solid state studies are available can be divided into two parts according to the behaviour of the 5f electrons. Actinide (An) metals from Am to Cf have localised 5f electrons, their properties resemble those of the lanthanides which have localised 4f electrons. In the lighter part of the actinide series, the metals Pa to Pu are characterised by the presence of itinerant 5f electrons, which explains some of their unusual properties and makes them similar to the transition metals with partly filled shells with itinerant 3d, 4d and 5d electrons, respectively.

Experimental activity in the field of high-pressure actinide research can be guided by two mechanisms. In a more pragmatical way, one can look at the known experimental results and try to fill gaps which exist, e.g., by investigating an element or compound in a group of materials where the neighbouring materials have already been studied, or by applying a new method to a material which has already been explored by other experimental methods. The other way consists in guidance through the results of theoretical calculations. Starting from the results obtained on lanthanides and other transition metals (and their compounds), and using the existing, but still less complete experimental data on actinides as fixed points, solid state theory has developed tools to calculate electronic structure and bulk properties of the actinides and actinide compounds. These tools are sophisticated computer programs, based for instance on band structure theory or atomic orbital theory. Predictions based on these calculations are not always confirmed by subsequent experimental work. But it can be useful to take them as a guide for choosing an experimental line of investigation.

The experimental approach often consists of applying known physical methods of measurement to actinide microsamples, which are contained in the central hole of the gasket of a diamond anvil cell. Extended high-pressure activity was made possible by the recent availability of efficient diamond anvil cells which are capable of routinely reaching pressures of the order of 50 GPa (500 kbar). By small modifications in cell design, study of samples at pressures beyond 100 GPa (1 Mbar) is within the reach of present possibilities. In the latter pressure range, the volumes of some solids can be reduced to less than half their original values. This corresponds to a decrease in interatomic distances of more than 20% Similar variation in interatomic distances can also be achieved in the 'chemical' way by varying the non-actinide partner in compounds. For example, going from UN to UBi in the uranium monopnictide series increases the lattice distances by the same amount. But changing the chemical partner introduces an additional parameter which complicates interpretation of the results. This is why the physical way of changing the volume, i.e., application of pressure, was called the 'cleanest' one.

Together with the gaseous elements, actinides are among those materials whose intense study under pressure has started only recently. Interest of solid state physicists in the high-pressure properties of actinides arose mainly on the basis of high-pressure studies of the so-called 4f elements, the lanthanides (Ln). The 5f shell being successively filled in the actinide (An) series, comparison of both series promised to give information on similarities and differences between the behaviour of the 4f and the 5f electrons.

The strong radioactivity of most of the actinides has in the past restricted their study to a few specialised laboratories. This is not the case for thorium and uranium. Rather extensive work was thus reported for uranium and its compounds, while the proportion of results for the highly radioactive actinides is relatively small.

For actinides heavier than einsteinium (fermium, mendelevium, nobelium and lawrencium) the available amounts are too small to allow preparation of pure solids suitable for solid state studies. Nuclides of these elements are formed in appreciable quantities in an isotope-producing reactor and have half-lives roughly of the order of seconds for No and Lr, of minutes for Md, and of hours for Fm. Einsteinium-253, having a half-life of 20 days, can in principle be studied by solid state microtechniques immediately after its isolation. But its high specific activity and the resulting self-heat and possible reactions with the surrounding materials have up to now prevented high-pressure study. In addition, any measurement that takes more than a few hours will deal with a composition varying in time due to the fast buildup of the daughter, berkelium-249. Measurements on the latter nuclide are also subject to the phenomenon of varying composition, though on a more extended time scale, the half-life of Bk-249 being 320 days. High-pressure X-ray diffraction (HPXRD) measurements on Bk-249, e.g., for which a series with a sufficient number of pressure steps takes between 1 and 3 weeks, will always end on a product richer in californium-249 than they started on.

Actinium, which is in general discussed together with the actinides, is extremely difficult to handle due to the high alpha-, beta- and gamma-activity of Ac-227 and its daughter nuclides. Many data of actinium are missing for this reason, and no high-pressure studies have been reported on this element and its compounds. Experimental data on high-pressure behaviour are thus limited up to now to the remaining nine actinides: thorium, protactinium, uranium, neptunium, plutonium, americium, curium, berkelium and californium. Even theoretical calculations are scarce for elements heavier than californium.

2. Subgroups of the actinide metals

The solid state properties of the actinide metals are controlled by the dualism of the localised and the itinerant configuration of the 5f electrons (Johansson 1976, Freeman 1980, Johansson et al. 1980, Johansson and Skriver 1982). At ambient pressure, this dualism allows to define the two main subgroups of actinide metals.

The first subgroup, elements with atomic number 91 to 94 (protactinium, uranium, neptunium, plutonium), is characterized by 5f electrons in an itinerant (delocalised) state. This means they are of band type, hybridise with the conduction electrons and thus contribute to the metallic bonding. Magnetic order, which in the lanthanide metals is limited to the presence of localised 4f electrons, is consequently not observed in this subgroup. The strengthening of the metallic bond by the 5f contribution leads to small atomic volumes, a high cohesive energy, and a low compressibility. Low-symmetry (orthorhombic and monoclinic) crystal structures are found whose formation is probably related to the particular directional properties of the hybridised orbitals including a contribution from 5f electrons.

The second subgroup, elements with atomic number 95 to 98 (americium, curium, berkelium, californium), is characterised by localised 5f electrons. In terms of energy, this means that these electrons have well defined energy levels and do not contribute to the metallic bond. In a spatial sense, it means that a particular 5f electron is fixed ('localised') to a particular actinide atom. In contrast to the first subgroup, the localised 5f electrons contribute to the appearance of magnetic order in curium, berkelium and californium. The atomic volumes are larger, thus closer to those of the trivalent lanthanide metals. The cohesive energies are in general lower than those of the 'itinerant' 5f metals. The compressibilities are also of the same order as those of the trivalent lanthanides. The crystal structure of the four metals of the second subgroup at ambient pressure and temperature is double hexagonal close-packed (dhcp), thus of relatively high symmetry. Their metallic valency is in general assumed to be 3.

Thorium, the first member of the actinide series, is in fact not a real actinide metal because its 5f levels are practically unoccupied in the ground state configuration. Filling of the 5f shell, which according to atomic number should begin at thorium, is delayed and starts only at the following element, protactinium. Thorium should thus be considered as a subgroup of its own, differing from the following subgroup (Pa to Pu) by e.g., a high structural symmetry (cubic close-packed, ccp*), high atomic volume and higher compressibility. Together with actinium, it has some ressemblance with the d-transition metals, actinium being related to the trivalent ones (Sc, Y, La), and thorium having some similarity to the tetravalent ones (Ti, Zr, Hf).

The fourth subgroup is made up by the five heaviest actinide elements, atomic numbers 99 to 103 (einsteinium, fermium, mendelevium, nobelium, lawrencium). With the exception of the crystal structure and the melting point of einsteinium, no solid state properties are known for these elements. There is some indication that their solid state properties are different from those of the other actinide subgroups. Einsteinium, the first member of the subgroup, differs from the preceding four elements (Am to Cf) by the crystal structure (ccp, instead of dhcp), by a much higher atomic volume, and by a metallic valence of 2 instead of 3. Also for the following four actinide elements there is a slight tendency to more stable lower valencies in solution, with respect to the subgroup Am to Cf.

The most remarkable effect of pressure on the actinide metals is that due to closer contact between the lattice atoms, localised 5f electrons can become itinerant, hybridise with the conduction electrons, and participate in the metallic bond. The subgroup with localised 5f electrons can thus under pressure acquire properties which, at ambient pressure, are characteristic for the subgroup Pa to Pu.

3. Experimental methods

A brief introduction into the techniques of generating and measuring pressure and of determining various physical properties under pressure seems necessary for understanding some of the results of such measurements on actinides. A few of the high-pressure devices and methods which have been applied to actinide metals will be briefly described. Details on high-pressure techniques can be found in previous reviews such as those of Klement and Jayaraman (1966) (devices for static pressures), Jayaraman (1983) (diamond anvil cells), Peggs (1983) (determination of pressure), and Van Thiel (1977) and Walsh et al. (1957) (dynamic pressure techniques).

*This designation is preferred throughout this chapter to 'fcc' (face-centred cubic, which, for elements, is equivalent to 'ccp').

3.1. *Measurements under static pressure*

The term 'static pressure' refers to devices which allow subjecting a sample to pressure for virtually unlimited time, while 'dynamic pressure' (see section 3.2.) means that a sample is under pressure for fractions of a second only under the effect of shock waves.

3.1.1. *Generation of static pressure*

A device for subjecting a solid to static pressure contains two or more parts that can be shifted with respect to each other to reduce the volume enclosing the solid. Tightness of the device is obtained either by the moving parts themselves, or by special gaskets inserted between them. The parts are in general shifted by a relatively simple mechanical system.

3.1.1.1. *Piston–cylinder devices.* As the name indicates, the sample, located in a cylinder closed at one end, is compressed by a piston that slides inside this cylinder. Such devices have mainly been used for compression studies, determining the volume from the known cross-section and the displacement of the piston, and the pressure from the applied force. Bridgman (1948) called this method 'piezometry' (from the Greek piezo, 'I press'). By suitable insertion of current leads, resistance measurements can be carried out in such apparatus (McWhan et al. 1963), and addition of heating devices allows differential thermal analysis (DTA) (see e.g., Stephens 1966). Relatively large samples are required for piston-cylinder methods to allow accurate determination of volumes and detection of heats of transformation.

3.1.1.2. *Anvil devices.* These consist of two or more so-called 'anvils' (mechanical parts of high compressive strength) which compress the sample between them. The extremities of the anvils are in general tapered. This increases the pressure they can exert with a given applied force, and creates the so-called 'massive support' to the anvil's pressurising face which extends the range of application to pressures higher than the compressive strength of the anvil material.

Bridgman (*or 'opposed'*) *anvil devices* have two anvils with parallel pressurising faces and can be used with or without a gasket. With use of gaskets, the sample is contained in the central hole of the gasket. The gasket is compressed between the two pressurising faces and, in addition to ensuring tightness, acts at the same time as container for the sample. The best known and most used opposed anvil device is the *diamond anvil cell* (*DAC*). Since its introduction around 1959, it has progressively developed into the most versatile and powerful instrument for generating high static pressures. The culet tips of two diamonds of gem quality are flattened, and the so-formed small surfaces are pressed against each other by a relatively simple mechanical device. These surfaces are eight- or sixteen-sided polygons with areas in the range of 0.01 to 0.5 mm^2 (0.1 to 0.8 mm width between opposing sides), while the back surface of the anvil, the so-called 'table', has a surface 25 to 1000 times larger.

This, in turn, is mounted in a metallic part of still much larger cross-section. Thus with a moderate force acting on the back of this metallic part, very high pressures can be achieved in the sample, which is still much smaller than the culet surface of the diamonds. The small sample size makes the DAC technique a genuine micro-technique, the sample loading procedure requires use of a microscope, and a typical sample weight is of the order of 10 microgram.

There are quite a number of different DAC designs suitable for different types of measurements and for different pressure ranges. Certain cell types can routinely reach maximum pressures of the order of 30, 50, or 80 GPa, respectively. The highest pressure reported for a sample in a DAC is 147 GPa (Mao et al. 1985), and a DAC with specially selected diamonds, having a culet surface of only 0.1 mm width, was reported by Xu et al. (1986) to have achieved 550 GPa (5.5 Mbar) on a ruby splinter which had been placed between the anvils for pressure determination by an extrapolation of the ruby fluorescence method (see section 3.1.8).

In addition to their high compressive strength, diamond anvils have the important advantage of being transparent to visible light and to hard X-rays. Transparency to light enables the user to study a number of optical properties 'in situ', i.e., under sustained pressure, and to have visual control of the sample at all pressures. Transparency to X-rays has made X-ray diffraction in situ at very high pressures possible; introduction of the DAC has thus given rise to a wealth of information on high-pressure crystal structures and volume compression data. In addition to optical and X-ray diffraction study, the DAC has been successfully used for electrical resistance measurements (Reichlin 1983). With special heating and cooling techniques, temperature can be introduced as an additional parameter into DAC work.

3.1.2. *Determination of volumes and interatomic distances*

For large samples, and for pressures up to about 10 GPa, volumes can be determined in piston-cylinder apparatus from the cross-section and the displacement of the piston (see e.g., Bridgman 1948). For microsamples, volumes are obtained from the interatomic distances determined by high-pressure X-ray diffraction (HPXRD), if the crystal structure of the sample is known (see e.g., the HPXRD work described in sections 6.1 to 6.4, as well as figs. 7 and 10). This method of volume determination can, with suitable DAC's, be extended well over 100 GPa (see e.g., Mao et al. 1985). In the lower pressure range to about 3 GPa, neutron diffraction can be used in the same way, but much larger samples than for HPXRD are needed, and a different type of pressure cell is used (Fournier et al. 1980).

3.1.3. *Determination of crystal structures*

High-pressure crystal structures are derived from the spectra obtained by X-ray or neutron diffraction under pressure (see e.g., the high-pressure structures described in sections 6.1 to 6.4). The methods for evaluating the spectra are in principle the same as for diffraction at ambient pressure, but the spectra are in general of lower precision and resolution. This makes unambiguous structure determination more difficult. An example where this situation leads to some ambiguity in structure determination is discussed in section 6.7.

3.1.4. *Mechanical properties*

The compressibility K of a material, defined by

$$K = -(\partial V/\partial p)_T/V$$

or its inverse, the bulk modulus

$$B = -V(\partial p/\partial V)_T$$

as well as the pressure derivative $B' = \partial B/\partial p$ can be determined from the pressure–volume relationship obtained through a piezometric method or from HPXRD (examples for actinide metals are given in sections 5.5 and 6.9). This is in general done by fitting the pressure and volume data to a semi-empirical equation of state which contains the inverse of the initial compressibility, the bulk modulus B_0, and its pressure derivative B'_0. Examples of these equations are those proposed by Murnaghan (1937), Birch (1947) and Keane (1954). B_0 and B'_0 are taken to be those values for which least squares fit of the data is best. A more detailed discussion of this method is found in section 4. The bulk modulus can also be calculated from the elastic constants determined by measuring the ultrasound velocity in the material.

3.1.5. *Thermal properties*

Differential thermal analysis (DTA) can be carried out in piston–cylinder apparatus (see e.g., Stephens 1966). Chu and Knapp (1973) report on a.c. calorimetry on a uranium single crystal in a pressure cell with a liquid pressure medium.

3.1.6. *Electrical and magnetic properties*

Electrical resistance measurements have been carried out in nearly all types of pressure-generating apparatus. For the DAC, resistance measurements on a gasketed sample in a pressure medium were reported up to about 20 GPa, and up to 40 GPa for a dry sample (Reichlin 1983). Techniques for measuring resistance under pressure at cryogenic temperatures were developed, and these allowed determination of the superconducting transition temperature under pressure (see e.g., the references of section 5.2.3. for applications to uranium metal).

Some magnetic properties can be measured in gas-pressure apparatus (to about 1 GPa; see e.g., Louis et al. 1986) and in piston–cylinder devices (to 1–2 GPa) (see e.g., Maple and Wohlleben 1972 for an application to uranium metal). No reports are available on susceptibility measurements in DAC's.

3.1.7. *Spectroscopic methods*

The optical transparency of diamonds has led to the use of DAC's for optical absorption, reflection and luminescence measurements under pressure. Applications to actinide compounds were described by Haire et al. (1985b) and Peterson et al. (1985) for optical absorption, and by Syassen et al. (1986) for reflection. Gregorian and Holzapfel (1984) have reported on luminescence of Pr^{3+} under pressure in diamond anvil cells.

X-ray absorption can be measured with the X-rays passing either through the diamond anvils (Staun Olsen et al. 1986), or for absorption phenomena at low X-ray energies through a beryllium gasket (Röhler 1986). Tungsten carbide anvils are used for Mössbauer spectroscopy under pressure (to about 12 GPa) (Kalvius et al. 1985).

3.1.8. *Pressure determination*

A variety of methods has been applied for the determination of the pressure to which a sample was subjected in a pressure device. In piston–cylinder equipment, a rough pressure determination can be made from the applied force and the cross-section of the sample. A number of substances such as bismuth, having phase transformations at known pressures, which can be detected by some method in the pressure cell, have been used to pressure-calibrate equipment of different types. When X-ray diffraction in situ can be used in a specific cell, the pressure can be determined from the observed lattice parameters of a 'marker', or 'crystallographic standard' which is added to the sample. Substances such as sodium chloride, cesium chloride, silver, and gold have been used for that purpose, which have the common feature that the pressure dependence of their lattice parameters is relatively well known in certain ranges of pressure.

The most versatile method for DAC's is the *ruby fluorescence* method. The Cr^{3+} ions in a ruby crystal close to the sample are excited to emit fluorescent radiation at 694.24 nm (R_1 line at 298 K, ambient pressure) by a strong light source such as a laser. The fluorescent line shifts considerably in wavelength with pressure, and from the known wavelength–pressure relationship, the pressure can be determined. The R_1 line and its satellite line R_2 correspond to transitions between the ^2E and $^4A_2$ crystal field split states of the Cr^{3+} ion (Forman et al. 1972, Mao et al. 1978).

3.2. *Measurements under dynamic pressure*

Very high pressures can be generated in a sample for a few microseconds by shock waves originating, e.g., from the impact of a projectile, or from a nuclear explosion. Typical equipment for such 'dynamic' pressure tests is a 'shock wave gun', i.e., a tube of at least 10 m length in which a projectile driven by a chemical explosive hits the target sample. Large samples are required for this type of experiments. Elaborate techniques have been developed for determining the pressure and volume changes which occur in the few microseconds during which the shock wave passes through the sample. The raw pressure–volume data are on the so-called Hugoniot curve and have to be corrected for the heating of the sample caused by the shock wave in order to be comparable to results from static pressure tests. Examples of shock wave investigation into actinide materials are given in sections 4 and 5.2.1.

4. Actinium and thorium

As mentioned above, no experimental high-pressure activity has been reported for *actinium.* There are some theoretical calculations of bulk properties of actinium and actinide metals, such as the bulk modulus and the cohesive energy, which can be relevant for the behaviour of actinium under pressure (Brooks 1984, Brooks et al.

1984, Johansson and Rosengren 1975a, Johansson and Skriver 1982, Skriver et al. 1978). The band structure, Fermi surface, superconductivity, and resistivity of actinium under high pressure were calculated by Dakshinamoorthy and Iyakutti (1984).

Compression data for *thorium* were reported from shock wave tests up to 140 GPa and from X-ray diffraction up to 68 GPa. About 30% volume compression was reached at 43 GPa in shock wave tests by Walsh et al. (1957). The Hugoniot pressure–volume relationship up to the latter pressure is given in their paper. Van Thiel's compendium of shock wave data reports compressions to about 54% of the initial volume at 140 GPa (Van Thiel 1977). Macroscopic pressure–volume data to about 10 GPa had earlier been determined with a piezometric technique by Bridgman (1948). A theoretical equation of state for thorium up to 40 GPa was calculated by Skriver and Jan (1980). Godwal and Sikka (1982) computed a theoretical Hugoniot equation of state using the pseudopotential method.

No phase transition was observed in thorium metal at room temperature up to 68 GPa by HPXRD studies with either energy-dispersive (Bellussi et al. 1981) or angle-dispersive (Benjamin et al. 1981) techniques.

The bulk modulus of thorium and its pressure derivative were determined from some of these compression data. It is interesting to compare the compressibilities determined from the different sets of data, and to compare them to the bulk modulus determined by the ultrasonic method and to several values of B_0 and B_0' calculated from theoretical models (table 1). It is probably correct to assume the ultrasonic values of B_0 are the most accurate ones. The spread of the other

TABLE 1

Bulk modulus B_0 and pressure derivative B_0' of thorium from experimental determination and theoretical calculation. The figures include decimals given in the original publications, but it has to be kept in mind that in general the error is a few GPa on B_0, and can be up to 30% on B_0'.

Method	B_0(GPa)	B_0'	Refs.[a]
Ultrasound velocity	57.8	–	[1]
	61	–	[2]
Piezometry	54.3	6.6	[3] (data from [4])
HPXRD	57.3	4.2	[5] (Birch eq.)
	72	2.0	[6] (Murnaghan eq.)
	67	3.0	[6] (Birch eq.)
LMTO calculation	50.3	3.4	[3]
	63.2	2.5	[7]
	66[b]	–	[8]

[a]References:
[1] Armstrong et al. (1959).
[2] Reynolds (1951).
[3] Glötzel (1978).
[4] Bridgman (1948).
[5] Bellussi et al. (1981).
[6] Benjamin et al. (1981).
[7] Skriver and Jan (1980).
[8] Brooks et al. (1984).

[b]Read from graph.

experimental values around those is quite characteristic of a situation as it is often found with Birch and Murnaghan fits for data from different authors. Differences up to 10% are frequent. To minimise these differences, it is recommended not to rely upon one way of fitting the data, but to use both equations, and, if they give different results, either to choose the fit of better quality, or if they are of comparable quality, to use mean values for B_0 and B'_0. It can also be helpful to fit data for different pressure ranges (e.g., 0–20 GPa versus 0–50 GPa) and in case of disagreement on B_0 and B'_0, to adopt the results for the smaller pressure range, provided the latter comprises a statistically significant number of data points; an empirical minimum number of data points for this case is 7 to 8. On the other hand, the example of the compressibility data of UO_2 shows that with particularly good quality data, agreement between independent determinations can be very good, i.e., of the order of 2%. This compound was independently and nearly simultaneously studied by HPXRD in two different groups (Benjamin et al. 1981, Benedict and Dufour 1982). Benjamin et al. used an angle-dispersive technique and determined $B_0 = 210(10)$ GPa, $B'_0 = 7(2)$. Benedict and Dufour, using an energy-dispersive technique, found $B_0 = 207$ GPa and $B'_0 = 7.2$. Theoretically calculated values can, according to the model chosen and to the experimental fixed points selected, also deviate in a relatively large range. Recommended values for all lighter actinide metals are plotted in fig. 4.

The electrical resistivity at room temperature does not show an anomaly in the pressure range investigated ($\leqslant 16$ GPa). The superconducting transition temperature was measured in the same pressure range. The authors observed an unusual pressure dependence of T_c which they believed to be due either to a pressure-induced change in the Fermi surface topology or to a crystallographic phase change near 7 GPa at very low temperatures (Fertig et al. 1972). Palmy et al. (1971) measured T_c and the critical magnetic field of thorium at pressures up to 2.4 GPa and deduced its electronic Grüneisen constant. Vijayakumar (1985) measured the variation of the room temperature thermopower with pressures up to 10 GPa. Sahu, Yousuf and Govinda Rajan observed, by measuring the electrical resistivity of cold-worked thorium, that major recovery of defects occurs sharply at 2 GPa in this metal (Sahu et al. 1984, Govinda Rajan 1984, Yousuf 1986).

5. Protactinium, uranium, neptunium and plutonium

These four metals are discussed together because all of them are characterised by itinerant 5f electrons at ambient conditions.

By far the most investigated among them is uranium. This is due to its technical and military importance in the nuclear field, to its large-scale availability and to its relatively low radioactivity that facilitates handling. The nuclear use has also generated a large amount of work on plutonium. Protactinium and neptunium, in contrast, have been less studied due to lack of technological interest and, in the case of protactinium, to restricted availability.

5.1. *Protactinium*

Protactinium, being only a minor constituent of uranium ores such as pitchblende, has been extracted in limited amounts – only a bit more than 100 gram – from the residues of ore treatment, and is made available at relatively high prices. For these reasons, protactinium is a somewhat 'neglected' metal in the group of the lighter actinides. This is the case for actinide studies in general, but is particularly true for high-pressure investigation. The only high-pressure study known up to now is a HPXRD study by Benedict et al. (1982). Up to 53 GPa, the highest pressure reached, protactinium kept its body-centred tetragonal crystal structure. If one thinks of Pa as having itinerant 5f electrons, this result is not surprising. While metals with localised 5f electrons (see section 6) tend to phase transformations under pressure (to so-called 'collapsed' phases) due to the delocalisation of these electrons, the lighter actinide metals Pa to Pu are already in the 'collapsed' state at ambient pressure.

In the same study, the bulk modulus of Pa was determined as $B_0 = 157(5)$ GPa, with a pressure derivative $B_0' = 1.5(5)$. No other experimental determination of the compressibility has been made. The bulk modulus determined contrasted with previous LMTO calculations (Skriver et al. 1978, Johansson and Skriver 1982) which yielded B_0 approximately equal to 100 GPa. Considerations on the variation of compressibility and B_0' in the actinide series, and on the relationship between bulk modulus and cohesive energy (see section 5) seem to support the experimental result. A possible explanation for the low calculated value is that the calculation was made for the cubic allotrope of protactinium and for zero temperature.

A theoretical equation of state to about 6 GPa was calculated by Skriver (1980).

5.2. *Uranium*

5.2.1. *Pressure–volume relationship and compressibility*

Macroscopic volume decrease of uranium with pressure was determined up to about 10 GPa by Bridgman (1948). Van Thiel's compendium (Van Thiel 1977) gives pressure–volume data up to about 660 GPa obtained by shock wave techniques. Later experiments using a planar shock generated in the vicinity of an underground nuclear explosion (Ragan 1980) extended the pressure range to 6.7 TPa; Hugoniot data were obtained for this pressure, and the density of uranium at this pressure was determined as 59.0 Mg/m^3 (about 3.1 times the density at ambient pressure). Examples of theoretical calculations of shock Hugoniot curves are the results from the SESAME library (Bennet et al. 1978) and the work of Pénicaud (1984). The latter work also includes calculation of the structure of the conduction band for densities 1.2 ρ_0, 1.5ρ_0 and 1.7 ρ_0 (ρ_0 being the density at ambient pressure), and of the variation of the edges of the 6s, $6p_{1/2}$, $6p_{3/2}$ and cônduction bands up to twofold compression. A theoretical equation of state for uranium to 15 GPa was given by Skriver (1980).

Another application of shock wave techniques is the work of Michel (1975) and Hornung and Michel (1979) in which information on the electronic density of states is derived from the entropy of uranium in the shocked state on the basis of the shock vaporisation technique.

HPXRD techniques were applied to uranium by three different groups of workers, the highest pressures reached being 20 GPa (Merx and Moussin 1980), 42 GPa (Benedict and Dufour 1985), and 50 GPa (Akella et al. 1985), respectively. The uranium used by Merx and Moussin was slightly alloyed with 0.2 at% of vanadium. No phase transition was observed in this range of pressures. This is in line with the reasoning on the presence of itinerant 5f electrons developed above for protactinium. Preliminary data obtained at about 71 GPa by Akella et al. (1985) show additional diffraction lines which may represent a high-pressure phase. This observation seems to indicate that at very high pressures, actinide metals with itinerant 5f electrons may undergo another phase transition, which is of a different nature than the 'delocalisation transition' described below for Am, Cm, Bk and Cf. No data having been obtained for uranium between 50 and 71 GPa, the transition would occur at some pressure intermediate to these two pressures. Reports on a volume discontinuity around 50 GPa observed during shock wave tests (Viard 1962), which could be another indication for a phase transition, are interesting in this connection.

The compressibility of uranium was determined from each of these HPXRD data sets. Table 2 compares the bulk moduli and pressure derivatives, and also gives a comparison with bulk moduli determined by the ultrasonic method (Simmons and Wang 1971) and calculated from theory by the LMTO method (Skriver et al. 1978, Johansson and Skriver 1982, Pénicaud 1984).

TABLE 2

Experimental and calculated bulk moduli B_0 and pressure derivatives B'_0 of uranium. The figures include decimals given in the original publications, but it has to be kept in mind that in general the error is a few GPa on B_0, and can be up to 30% on B'_0.

Method	B_0(GPa)	B'_0	Refs.[a]
Ultrasound velocity	113(2)	–	[1]
HPXRD (U–0.2 at% V)	143.4	3.8	[2]
(pure U)	147(3)	2.8(3)	[3]
(pure U)	125(2.5)	6.2	[4]
Piezometry (Bridgman data)	100.7	–	[5]
LMTO calculation	115[b]	–	[6]
	145[b]	–	[7]
	117	–	[8]
	152[b]	–	[9]

[a]References:
[1] Simmons and Wang (1971).
[2] Merx and Moussin (1980).
[3] Benedict and Dufour (1985).
[4] Akella et al. (1985).
[5] Gschneidner (1964).
[6] Skriver et al. (1978).
[7] Johansson and Skriver (1982).
[8] Pénicaud (1984).
[9] Brooks et al. (1984).

[b]Read from graph.

There is very good agreement between two of the data sets from HPXRD; the third HPXRD data set deviates, and all three data sets give a bulk modulus considerably higher than that determined from ultrasonic data. The latter, though, has considerable weight as it has been found by two independent determinations, and because the ultrasonic method has the reputation to be rather accurate. On the other hand, the differences seem too large to be accounted for by differences in impurity content of the samples. Additional work on the compressibility of uranium is desirable.

5.2.2. *Pressure–temperature phase diagram*

Asami et al. (1961) studied the phase diagram of uranium at pressures up to 8 GPa and temperatures up to 1473 K by measuring the change in resistance at the phase boundaries. Differential thermal analysis (DTA) was applied for investigating the phases of uranium by Klement et al. (1963), Lloyd et al. (1966), and Ganguly and Kennedy (1973). The pressure–temperature phase diagram of uranium in fig. 1 is based on this DTA work. Pressure favours the densest allotrope U(I) (α-uranium) at the expense of U(II) (β-uranium), which eventually disappears at about 3.5 GPa. Klement et al. (1963) report that the density increases in the sequence U(III)–U(II)–U(I). The phase transitions are shifted to higher temperatures with increasing pressure. Some controversy exists about the melting line. The full lines in the

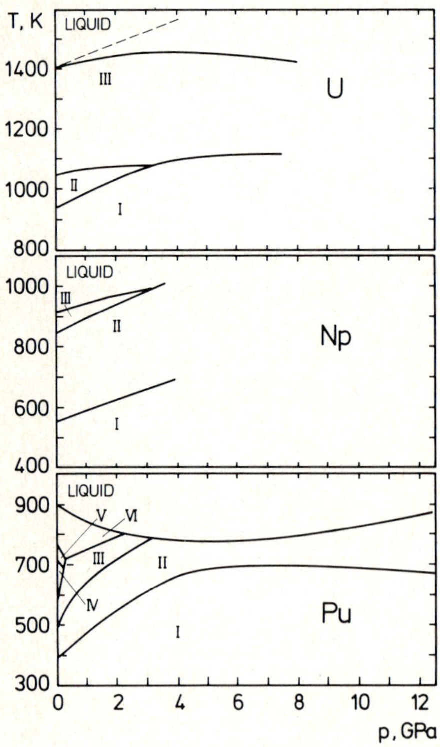

Fig. 1. Pressure–temperature phase diagrams of U, Np and Pu; dashed line: linear melting curve of uranium after Ganguly and Kennedy (1973). Equivalence between nomenclatures in Roman numerals and Greek letters: I = α, II = β, III = γ, IV = δ, V = δ', VI = ε.

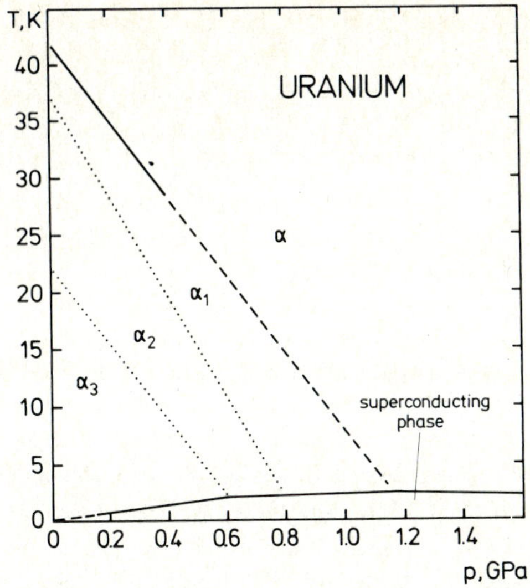

Fig. 2. Phase diagram of uranium at low temperatures. (Adapted from Smith and Fisher 1973.)

uranium phase diagram of fig. 1 are based on the work of Asami et al. (1961) and include a melting line which first increases and then decreases with pressure. The dashed straight line is the linearly increasing melting line given by Ganguly and Kennedy (1973). Lloyd et al. (1966) also studied the changes in microstructure induced by different pressure–temperature treatments.

5.2.3. *Other physical measurements*

Most of these measurements are in relation with the low-temperature phase transitions of uranium. First indications for such a transition were reported in 1961 by Fisher and McSkimin (1961). Later, evidence was obtained for the existence of three low-temperature phases of uranium, α_1, α_2 and α_3. At first, these did not seem to exhibit crystallographic structures different from that of α-uranium. But in 1980 two independent research groups (Smith et al. 1980, Marmeggi and Delapalme 1980) reported on weak additional reflections observed in single-crystal neutron diffraction, and Marmeggi et al. (1982) gave an interpretation in terms of charge-density waves. A survey of the experimental situation has been given by Smith and Lander (1984).

The $T_c(p)$ curves for polycrystalline uranium obtained by Gardner and Smith (1967), Maple and Wohlleben (1972), and Smith and Fisher (1973) are in relatively good agreement in the pressure range 0 to 2.4 GPa. Maple and Wohlleben also measured the magnetic susceptibility at room temperature and pressures up to 1.6 GPa and gave the $T_c(p)$ curve up to about 8.5 GPa. Smith and Fisher (1973) found a significantly different $T_c(p)$ curve for single-crystal uranium; they ascribed the particular features of this curve to the phase transformations observed in uranium at ambient pressure and temperatures below 43 K and on this basis suggested a low-temperature phase diagram to 1.4 GPa (fig. 2). Palmy and Fisher

(1970) measured T_c, the critical field, and the magnetisation of a uranium single crystal at pressures up to 0.8 GPa.

All these authors agree that superconductivity is enhanced by pressures up to 1.2 GPa: T_c increases to 2 K from its ambient pressure value 0.1 K. When the pressure further increases, T_c decreases again and reaches about 0.5 K at 8.5 GPa.

Chu and Knapp (1973) used an a.c. calorimetric technique to investigate an alpha-U single crystal at pressures to 1.2 GPa. They observed that increasing pressure reduces the temperature of the phase transition which occurs around 37 K at ambient pressure, and finally suppresses the transition. The pressure dependence of the single-crystal elastic constants of uranium under hydrostatic helium pressure up to 0.4 GPa was studied by Fisher and Dever (1970). The temperature (about 43 K) at which marked anomalies in the elastic constants, indicative of some kind of phase transition, occur at ambient pressure, decreased with increasing pressure. This also indicates that pressure suppresses the low-temperature phases; it has been determined that their volume is larger than that of the alpha phase (Barrett et al. 1963). Neutron scattering experiments (Smith and Lander 1984) have further confirmed the suppression of this phase transition by pressure: the superlattice reflections characteristic of the structure of uranium below 43 K disappeared at 0.2–0.3 GPa (nearly commensurate reflections) and 0.6 GPa (incommensurate reflections), respectively.

Since the charge-density wave transition in alpha-U is driven by the strong electron–phonon coupling and the subsequent collapse of an optic phonon, neutron experiments were also performed to measure the phonon frequency as a function of pressure (Smith and Lander 1984). At room temperature there is a marked stiffening of the phonon up to about 0.5 GPa and this effect is even more dramatic at lower temperature. Unfortunately, the large size of the crystals (0.1 cm^3) required by this technique precludes major advances in mapping out the p, T phase diagram.

Bader et al. (1975) measured the heat capacity of uranium between 0.1 and 2 K at ambient pressure and discussed the relation of their results to the pressure dependence of T_c. Tindall and Steinitz (1984) studied the low-temperature phase transitions in uranium by capacitance dilatometry and also discussed the relationship of these phases to the variation of T_c with pressure. Schirber and Arko (1980) were able to generate pressures up to about 0.95 GPa in uranium metal by careful freezing of helium around the coil-sample assembly used and determined the Fermi surface of uranium by carrying out de Haas–van Alphen measurements on the sample under these conditions.

Vijayakumar (1985) measured the variation of room temperature thermopower with pressure up to 10 GPa. Leman (1970) investigated the pressure dependence of the electrical resistance of uranium metal and did not observe an indication for a phase transition.

5.3. *Neptunium*

Compression and DTA measurements on neptunium up to 3.5 GPa were reported by Stephens (1966). Based on the DTA results, Stephens could construct the phase diagram of neptunium shown in fig. 1. Similar as for uranium, pressure increases the

TABLE 3

Bulk modulus B_0 and pressure derivative B_0' of neptunium. The figures include decimals given in the original publications, but it has to be kept in mind that in general the error is a few GPa on B_0, and can be up to 30% on B_0'.

Method	B_0(GPa)	B_0'	Refs.[a]
Piezometry	73.5	–	[1]
HPXRD	118(2)	6.6(6)	[2]
LMTO calculation	92[b]	–	[3]
	110[b]	–	[4]
	95[b]	–	[5]

[a]References:
[1] Stephens (1966).
[2] Dabos et al. (1987).
[3] Skriver et al. (1978).
[4] Johansson and Skriver (1982).
[5] Brooks et al. (1984).
[b]Read from graph.

temperatures of the transitions to the less dense high-temperature allotropes II (beta) and III (gamma). The temperature range of Np(III) narrows with increasing pressure, and eventually vanishes at about 3.2 GPa. McWhan et al. (1963) had studied the transition from Np(I) to Np(II) at 5 pressures between 0.4 and 1.1 GPa by measuring the electrical resistivity. Their transition temperatures agree satisfactorily with those determined by Stephens; an apparent difference in slope of the transition line is not necessarily significant.

The pressure–volume relationship at room temperature up to 52 GPa was determined by HPXRD (Dabos et al. 1987). As expected from the itinerant character of the 5f electrons in neptunium metal, no phase transition was observed up to this relatively high pressure (also compare section 5.1). However, the observation of phase transitions in the two neighbouring metals U (around 60 GPa) and Pu (at 40–45 GPa, see section 5.4) indicates that there may be a phase transition in neptunium not too far above the highest pressure investigated, i.e., around 60 GPa. Table 3 compares the compressibilities determined in the latter work, by Stephens (1966), and by theoretical calculations (Skriver et al. 1978, Johansson and Skriver 1982).

5.4. *Plutonium*

5.4.1. *Pressure–volume relationship and compressibility*

The first study of plutonium under solid state pressure was undertaken as early as 1945 by Bridgman. This metal, isolated for the first time in 1942, then was of interest in classified research. The results of Bridgman's work were published in 1959 (Bridgman 1959). They include compression data from piezometric work up to about 10 GPa and the pressures and volume changes at the I to II (alpha to beta) transition at 163 and 205°C.

Roof (1981a) described HPXRD work up to 21 GPa at room temperature on the pure metal and on a Pu–6.5 at % Ga alloy. No phase transition was found in this

TABLE 4

Bulk modulus B_0 and pressure derivative B_0' of plutonium. The figures include decimals given in the original publications, but it has to be kept in mind that in general the error is a few GPa on B_0, and can be up to 30% on B_0'.

Method	B_0(GPa)	B_0'	Refs.[a]
Ultrasound velocity	42(3)	–	[1]
	49(1)	–	[2]
	52	–	[3]
	47(1)	–	[4]
Piezometry	52(2)	10(2)	calc. from [5]
HPXRD	42.2(7)	10.5(2)	[6]
	40	16	[7]
LMTO calculation	55[b]	–	[8]
	40[b]	–	[9]
	52[b]	–	[10]

[a] References:
[1] Laquer (1961).
[2] de Cadenet (1961).
[3] Robinson (1961).
[4] Linford and Kay (1967).
[5] Bridgman (1959).
[6] Roof (1981a).
[7] Dabos and Benedict (1986).
[8] Skriver et al. (1978).
[9] Johansson and Skriver (1982).
[10] Brooks et al. (1984).

[b] Read from graph.

investigation. Dabos and Benedict (1986) extended the pressure range to 59 GPa and observed a phase transition at 40–45 GPa to an up to now undetermined structure. A theoretical equation of state, giving the Wigner–Seitz radius versus pressure, was calculated by Skriver (1980). The partial contributions of the s, p, d, and f orbitals to the equation of state were given by Skriver et al. (1978). The experimental compressibilities are compared to calculated compressibilities in table 4. For the Pu–6.5 at% Ga alloy, Roof determined $B_0 = 26(2)$ and $B_0' = 10.1(5)$. There may be a relatively large uncertainty on the bulk modulus of the alloy because there are fewer data for low pressures than for the pure metal, and because compression starts on the delta phase, which converts to alpha only after some compression.

5.4.2. *Pressure–temperature phase diagram*

The phase diagram of plutonium shown in fig. 1 is based on the following investigations:

Stephens (1963) studied the phase relationships up to 3.5 GPa by DTA and by observation of volume discontinuities on transition. His phase diagram was extended to 7.5 GPa by Liptai and Friddle (1966a,b) who also used the DTA technique. All but the I and II (alpha and beta) phases disappear at pressures above 3 GPa at all temperatures. The low-pressure region of the diagram up to 1 GPa was determined in more detail by Morgan (1970).

Other phase diagram studies are those by Loasby (1961) up to 1.2 GPa by adiabatic calorimetry, and by Prunier et al. (1970) up to 1 GPa by thermal analysis. Liptai and Friddle (1966a) measured the variation in the volume change accompanying the beta-to-alpha transformation in plutonium with pressures up to 4.8 GPa. McWhan et al. (1963) used changes in the electrical resistance to study the alpha-to-beta transition in plutonium to 1.8 GPa.

Liptai and Friddle (1967) also published a DTA study of the phase diagrams of a Pu–3 at% Ga alloy and of Pu alloyed with 1, 3 and 6 at% Ti, up to about 3 GPa. These phase diagrams are similar in nature to that of pure plutonium, with relatively small shifts of the phase boundaries. Although at least the gallium-containing alloy is reported to have been in the delta phase initially, the transformation pressures from delta to the other phases were not determined. It appears from the diagram that they are probably lower than 0.2 GPa. Roof (1981a) also reported the delta-to-alpha transformation at room temperature to take place below 1 GPa. Roux et al. (1971), in contrast, report stability of the delta phase in Pu–6 at% Ga alloy up to about 1.7 GPa at room temperature. The latter authors used the results of resistance measurements to determine pressure–temperature phase diagrams up to 4 GPa and the melting points of Pu–Ga alloys containing 2, 4, 6, and 8 at% Ga. According to their results, increasing the gallium content extends the phase field of the delta phase to higher pressures.

5.5. *Compressibility and cohesive energy of the light actinide metals*

Figure 3 shows the compression curves of the ambient-pressure phases of the light actinide metals. Thorium is seen to be the most compressible of these, but plutonium

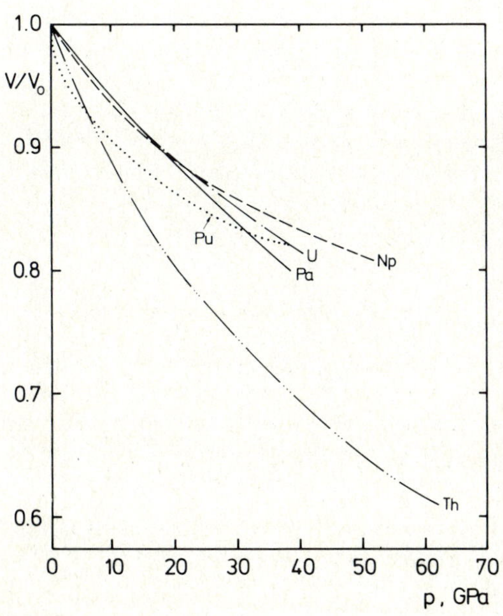

Fig. 3. Compression curves (relative volumes V/V_0 versus pressure) for the ambient-pressure phases of the light actinide metals from HPXRD measurements. V_0: volume at ambient pressure. Th: Benjamin et al. (1981), Pa: Benedict et al. (1982), U: Benedict and Dufour (1985), Np: Dabos et al. (1987), Pu: Dabos and Benedict (1986).

has the highest *initial* compressibility. The remaining three metals are 'stiffer', i.e., less compressible at beginning compression, but independent of the initial compressibility, the four actinide metals with itinerant 5f electrons (Pa, U, Np, Pu) end at about the same compression (18 to 20%) in the pressure range 38–52 GPa.

A plot of bulk moduli of actinides compared to those of lanthanides and d-transition metals (Benedict et al. 1982) shows a parallelism between the light actinide metals and the d-transition metals. In the 3d, 4d, and 5d transition metal series, as well as the sub-series of the lighter actinide metals, a marked maximum in the bulk modulus is observed in the middle of the series. This variation of compressibility with Z is linked to the band character of the electrons in the respective incomplete shell.

Figure 4 compares the bulk moduli and their pressure derivatives for the light actinide metals from thorium to plutonium. Plotted are the B_0 and B_0' values which correspond to the initial slope of the curves of fig. 3. It appears from this figure that high bulk moduli, i.e., low compressibilities, are accompanied by low-pressure derivatives, and vice versa. This means that an actinide metal which is initially very compressible undergoes a relatively fast change in compressibility with compression, whereas a material which is already stiff at ambient pressure has a smaller potential

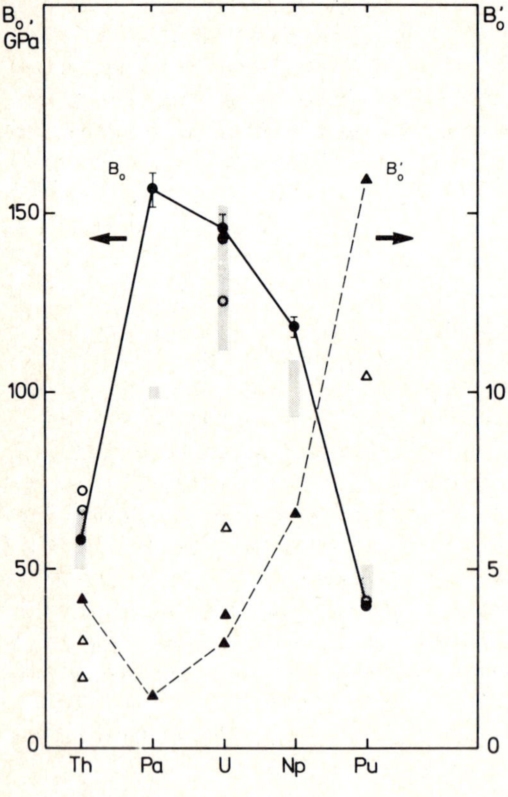

Fig. 4. Bulk moduli B_0 (——, circular symbols) and their pressure derivatives B_0' (– – –, triangular symbols) for the light actinide metals as determined by HPXRD (from tables 1 to 4; value for Pa from Benedict et al. 1982); full symbols: energy-dispersive HPXRD; open symbols: angle-dispersive HPXRD; shaded areas: range of theoretically calculated bulk moduli. For consistency, the curves are based on the values obtained by the energy-dispersive method because this is the only one applied to all of the light actinide metals.

for further stiffening. The compression curve (pressure–volume curve) has a marked curvature in the first case, and can approach a straight line in an extreme case of the latter kind (protactinium) (cf. fig. 3). These considerations indicate that even unusually low (1.5 for Pa) or high (10.5 or 16 for Pu) values of the pressure derivative, which are far from the range $4 \leqslant B'_0 \leqslant 6$ that is sometimes considered as 'regular', are not necessarily due to errors in experiment or evaluation, but can follow a systematic trend in relation with the value of the initial compressibility.

Plutonium is an exception in the group with itinerant 5f electrons. Its bulk modulus is of the same order as those of the heavier actinides and of the lanthanides. Brooks et al. (1984) relate the low bulk modulus of Pu to its relatively high atomic volume which is increased above the value expected from f-bonding considerations by spin–orbit coupling.

There is a general correlation between compressibility and cohesive energy. Plots of cohesive energy versus Z in the 3d, 4d and 5d transition metal series (Gschneidner 1964) have a maximum in the middle of the series in a similar way as was mentioned above for the bulk moduli. Expressing this correlation in a simple way, we can say that a high cohesive energy means strong interatomic attraction, thus dense structures. It is, of course, more difficult to further compress dense structures than less dense ones. Calculated cohesive energies of the light actinide metals (Johansson and Skriver 1982, and unpublished work of M.S.S. Brooks quoted therein) have a maximum at protactinium as the bulk moduli shown in fig. 4. For the experimental values of cohesive energy, there is unfortunately a rather large difference between two determinations of different authors. One author (Bradbury 1981) gives a cohesive energy of about 165 kcal/mol, which is by far the highest value determined for an actinide metal, and thus the plot of cohesive energy versus Z would have its maximum at protactinium, as has the plot of the bulk moduli. The other determination (Ward 1985) gives about 135 kcal/mol, a value slightly lower than that of the neighbouring element thorium, and would thus result in a cohesive energy maximum at uranium. The general context – maximum of cohesive energy found at protactinium by two different ways of theoretical calculation, maximum of experimental bulk moduli at protactinium, general tendency in transition metal series to have correspondence between high bulk moduli and high cohesive energies – gives support to the higher value for the cohesive energy of Pa.

The high cohesive energy of some of the light actinide metals reflects a strong participation of the 5f electrons in the metallic bond, as do the low atomic volumes and interatomic distances observed in the same metals. In this way, the particularly high value of the bulk modulus of protactinium can be interpreted as a sign of a particularly strong 5f contribution to bonding in this metal.

There seems to be a systematic difference between the experimental bulk moduli of the light actinides and those previously calculated by the LMTO method. The former ones are considerably higher than the theoretical ones, for protactinium by about 40%, for uranium by about 25%, and for neptunium by about 28%. The situation is less clear for plutonium, but there seems to be a tendency for LMTO calculations to over-estimate compressibility in the actinide metals with the strongest 5f participation in bonding.

6. Americium, curium, berkelium and californium

While the 5f electrons are itinerant in the four actinide metals preceding americium in the actinide series, at americium the contribution of the spin polarisation energy to the total energy becomes large enough to stabilise the localised state of the 5f electrons (Johansson and Skriver 1982). The calculated f-bonding energies for the actinide metals are lower than the localisation energies for Pa to Pu, but higher for the following elements (fig. 5). The localised state should thus be the stable one at ambient pressure for the four elements discussed in this section. Figure 5 also shows that the localisation energy has a minimum for curium. This is in relation with the $5f^7$ configuration which represents half-filling of the 5f shell and thus, according to Hund's rule, the maximum possible number of unpaired 5f electrons.

6.1. *Americium*

Much of the high-pressure work on americium was stimulated by the discovery of high-pressure phase transitions in lanthanum, cerium and other lanthanides and the observation of similarities in electronic structure and physical properties between americium and the lanthanides which suggested similar phase transitions might exist in americium. Though similar considerations are valid for the three following elements, Cm, Bk and Cf, interest focused on americium, probably because this element was the most easily available of the four.

The first published solid state study of americium under pressure seems to be the resistance measurements of McWhan et al. (1963). Up to 3.6 GPa and 450°C they did not detect any reproducible resistance anomaly. Stephens et al. (1968) extended the range of resistance measurements up to 12 GPa. They found a linear resistance increase to 8 GPa and a much steeper nonlinear increase above 9 GPa (fig. 6). They suggested the possibility of a first-order phase transition around 9 GPa, but did not exclude the possibility that the anomaly around 9 GPa was connected with residual strain since it did not appear in a run at 50°C. Transformation from dhcp to ccp under pressure having been observed in La, Pr and Nd (Piermarini and Weir 1964),

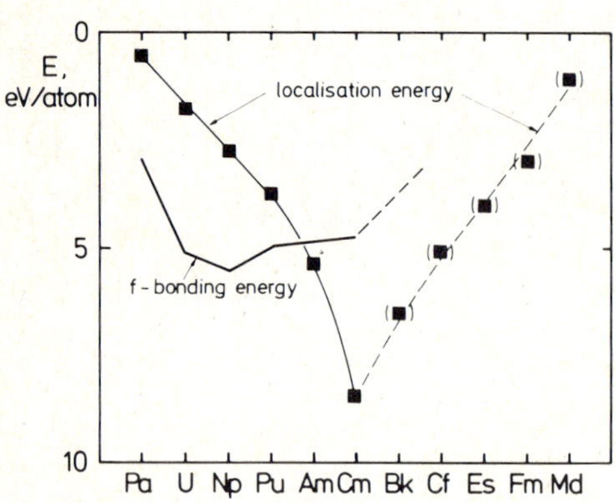

Fig. 5. Energy of 5f electrons in the localised and itinerant states. (Adapted from Johansson et al. 1980.)

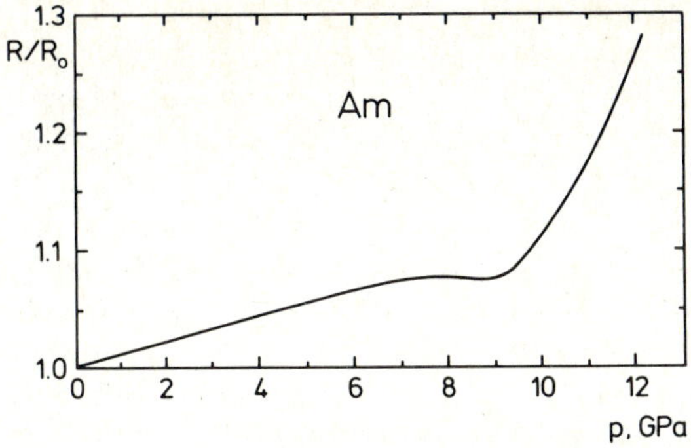

Fig. 6. Relative electrical resistance of americium versus pressure; R_0: resistance at ambient pressure. (Adapted from Stephens et al. 1968.)

Stephens et al. estimated that in the case of a pressure-induced phase transition in Am this would also be from dhcp to ccp in view of the rare-earth-like character of this metal.

Referring to the work of Stephens el al. (1968), Johansson (1978) suggested that still higher pressures than those reached by Stephens et al. might be necessary to delocalise the 5f electrons in americium. A correspondence between the high-volume, 5f-localised gamma-Ce and Am on one hand, and between plutonium and the low-volume alpha-Ce which was assumed to have itinerant 5f electrons, on the other hand, was suggested by Skriver et al. (1978).

The next experimental study is that of Akella et al. (1979a) who studied americium by HPXRD to 14 GPa. They indeed observed that Am changed to the ccp-structure above 6.5 GPa. Pressures in this case were determined with reference to the X-ray diffraction lines of a nickel standard. Later the same group of authors (Akella et al. 1979b) reported on two additional phase transitions in Am at 11 and 16 GPa, but could not yet determine the structures of the high-pressure phases. The results were summarised in another paper (Akella et al. 1980) where the transition pressures are given as 5, 10 and 15 GPa, with an error of ± 1 GPa. Samples of 99.5% purity (5000 ppm Si as the main impurity) and of 99.9% purity (200 ppm Ca as the main impurity) were used for the experiments. The highest pressure reached was 18 GPa.

In connection with these results, Skriver et al. (1980), using the LMTO method, calculated a pressure–volume relationship for americium which indicated a volume collapse of the order of 32% around 10 GPa. This collapse was calculated to occur at a Wigner–Seitz radius of 2.98 a.u. The authors suggested that above this transition americium could exist in crystal structures which are otherwise typical for light actinides, and that superconductivity could be suppressed. Brooks (1983) calculated relativistic corrections to the volume of 'collapsed' americium. These relativistic volumes (for which spin polarisation was not taken into account) are higher than those calculated by Skriver et al. (1980) and thus reduce the calculated volume collapse on delocalisation to 22%. All volume collapses are given in this section as a percentage of the volume immediately before collapse.

Roof et al. (1980) reported on another HPXRD study to 18 GPa of americium containing 2000 ppm of ytterbium and less than 100 ppm of other impurities. The addition of ytterbium had stabilised the ccp phase. These authors confirmed the existence of 3 high-pressure phases and identified the phase existing at the highest pressures as being of the alpha-uranium type. The appearance of this light-actinide structure was interpreted as an indication of the participation of americium's 5f electrons in the metallic bond under pressure. This structure was observed by Roof et al. in americium at about 15, 16 and 18 GPa. Another high-pressure structure observed at 11 and 12.5 GPa could not yet be determined. The alpha-uranium type structure was obtained at compressions above 18%, and the volume of the unknown high-pressure phase, lying between the volumes of the fcc and the alpha-uranium type phase, can be estimated to correspond to compressions between 14 and 18%. The authors compared this compression to the density increase of 18% which they calculated for the case when the 5f charge density midway between americium atoms reaches that of plutonium at ambient density. This comparison was taken to confirm 5f participation in bonding.

The allotropes of americium are best designated by Roman numerals. Am(I) is the dhcp room-temperature ambient-pressure form, and Am(II) the fcc (ccp) form which exists at elevated temperature and/or pressure. The new phases observed under high pressure at ambient temperature are designated Am(III) and Am(IV), in the order of increasing pressure. Two attempts were made in 1980–1981 to determine the structure of Am(III). Both investigations concluded with a monoclinic structure. Roof (1981b) supplemented the earlier work (Roof et al. 1980) by a study of Am without the Inconel gasket and the aluminium pressure standard previously used, and thus found an additional diffraction line which had been masked by the (111) line of Al. The combined diffraction data from 4 different pressures allowed him to determine a monoclinic structure in space group $P2_{1/m}$. In contrast, Smith et al. (1981) proposed a C-centred monoclinic structure with different lattice parameters. Roof (1982b) showed that these two structures are similar. He also discussed the structural relationships between the phases I to IV of americium (Roof 1981c, 1982a).

Neither of the two experimental groups observed a large volume decrease at any of the phase transitions. This discrepancy with respect to the volume 'collapse' calculated from theory for the onset of itinerancy was explained by Johansson et al. (1981) along the following lines: (1) The model used for the calculations contained some imperfection in the description of the itinerant state; the volume calculated for plutonium was also 15% lower than the experimental one. (2) The calculations are made for 0 K whereas experiments are at room temperature; f-band metals, like the neighbouring element plutonium, have a particularly large thermal expansion between 0 K and room temperature, and it is possible that this also applies to americium. Both factors would contribute to reducing the difference between experimental and calculated volume change on transition. HPXRD measurements at cryogenic temperatures could give an answer to this problem. Theory, on the other hand, should look for possibilities to calculate the pressure dependence of the americium volume at room temperature.

The structural phase transitions of americium under pressure were re-investigated

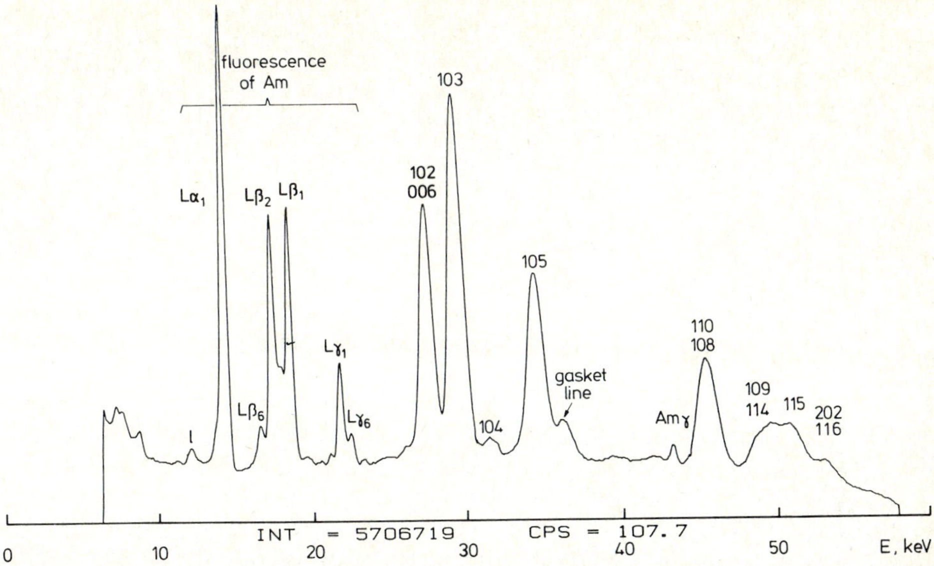

Fig. 7. Energy-dispersive diffraction spectrum of Am(III) at 20.3 GPa, hexagonal indexing with $a=312.9$, $c=1533.7$ pm (hkl values marked on top of the diffraction lines). L X-ray fluorescence lines excited in Am are marked in the left third of the spectrum. In addition, the spectrum contains the (111) diffraction line of the Inconel gasket and a peak from the gamma radiation of ^{241}Am.

by Benedict et al. (1985a, 1986a). While 18 GPa was the highest pressure attained in previous work, the pressure range was extended up to 52 GPa in the latter study. Important differences with previous studies were observed in three respects:

(1) Phase Am(III) could not be indexed in the monoclinic structure assigned to it by Roof (1981b). The resulting volumes are too large and would correspond to a widening of the lattice by an increasing-pressure phase transition. This contradiction could be lifted when Am(III) was indexed as a trigonal distortion of the cubic close-packed Am(II) (see fig. 7 and table 5), a structure which had been proposed for La, Pr and Y (Grosshans et al. 1982a, Vohra et al. 1984). As Table 5 shows, there is a good fit in d values and a less good fit in relative intensities, especially for the (103) and (104) lines. Preferred orientation effects which are frequent in the microsamples in diamond anvil cells, may contribute to the low quality of the intensity fit.

The latter structure and Roof's monoclinic structure are further discussed in section 6.7. If one can assume that the two structures proposed for Am(III) are of similar thermodynamic stability, it is conceivable that small changes in conditions, such as differing impurity content, alternatively favour one or the other of these structures. Indications for this type of behaviour also exist for cerium metal (Staun Olsen et al. 1985a).

(2) 5f delocalisation was suggested to occur only at 23 GPa, instead of the 10–12 GPa assumed on the basis of previous theoretical and experimental study This view was supported by three facts: the absence of a volume discontinuity at the

TABLE 5
Diffraction data of Am(III) at 20.3 GPa, indexed according to the trigonal structure of the work by Vohra et al. (1984). $a=312.9$ pm, $c=1533.7$ pm. Calculated intensities obtained with the program ENDIX (Hovestreydt et al. 1987) for the energy–dispersive diffraction procedure used.

hkl	d_{calc} (pm)	I_{calc} (a.u.)	d_{obs} (pm)	I_{obs}[a]
100	271.0	5	–	–
101	266.9	2	267.8	w
006	255.6	18	256.6	s
102	255.5	103		
103	239.5	2	239.5	s
104	221.3	66	224.6	w
105	203.1	29	204.7	s
106	185.9	2	186.2	w
107	170.4	25	168.6	w
009	170.4	12		
108	156.5	5	155.9	m
110	156.5	20		
112	153.3	4	–	–
113	149.6	5	–	–
114	144.8	2	143.3	m
109	144.3	1		
115	139.4	1	139.6	m
116	133.5	6	133.8	w
202	133.4	7		

[a] w: weak, m: medium, s: strong.

transition from Am(II) (ccp) to Am(III) (trigonal) at 13.5 GPa, a volume decrease of approximately 6% at the transition from Am(III) to Am(IV) (orthorhombic) at 23 GPa; the crystallographic similarity between Am(II) and Am(III) and the fact that the structure assigned to the latter phase also occurs in yttrium and lanthanum under pressure which both have no f electrons.

(3) The three phase transitions were observed at higher pressures than previously reported. The largest difference is found for the Am(III) to Am(IV) transition which was observed at 23 GPa by Benedict et al. (1985, 1986a) while earlier work placed it at 15 GPa.

The pressure–volume relationship for Am according to Benedict et al. (1985a) is given in fig. 8. Figure 9 shows a possible pressure–temperature phase diagram of Am. The only investigated parts of this diagram are (1) room temperature up to 52 GPa, (2) ambient pressure up to 1500 K, (3) the area between 1300 and 1530 K up to 3.5 GPa studied by Stephens et al. (1968). Figure 9 gives a possible distribution of phase fields on the basis of these results and by analogy with better known pressure–temperature phase diagrams of some lanthanides (Johansson and Rosengren 1975b). The phase transition temperature from Am(I) to Am(II) at ambient pressure is given as 1042 K, according to the high-temperature study of Shushakov et al. (1983) who did in situ study by X-ray diffraction, thermal analysis and electrical resistivity. The speculative lines of this diagram are dashed. Two versions of the phase fields are shown which represent the two different sets of

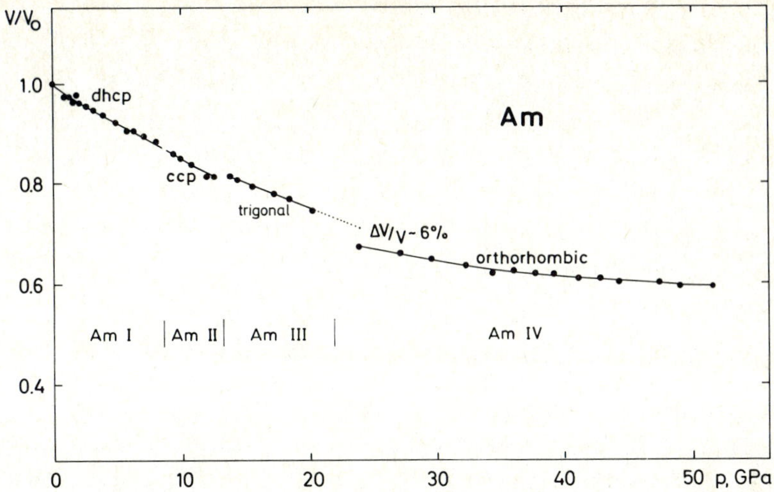

Fig. 8. Pressure–volume relationship of americium; $\Delta V/V$: volume decrease upon delocalisation of the 5f electrons. (Adapted from Benedict et al. 1985a.)

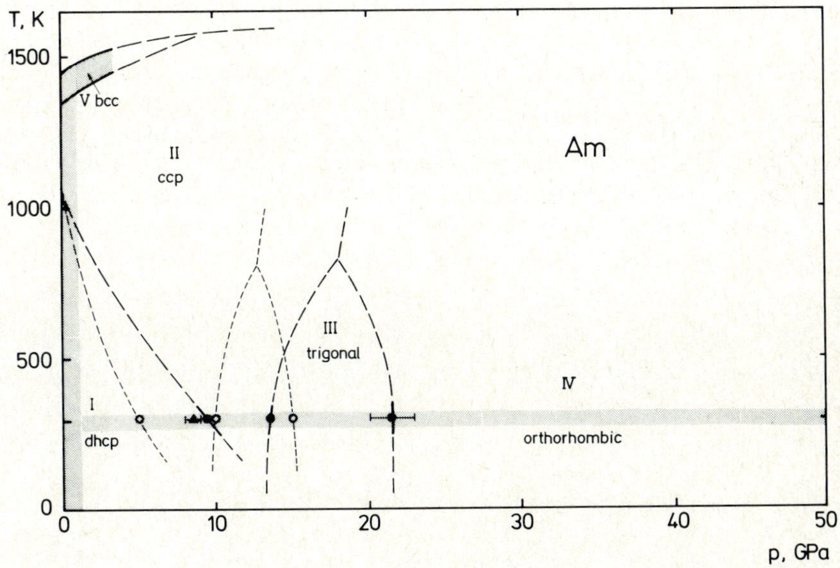

Fig. 9. Possible distribution of phase fields in a pressure–temperature phase diagram for americium. Only the shaded zones were experimentally investigated. The phase field of bcc Am(V) is based on the DTA study of Stephens et al. (1968). For the other phase fields, two versions, based on the work of Benedict et al. (1985a) and Stephens et al. (1968) on one hand (thicker lines), and on the work of Akella et al. (1980), on the other hand (thinner lines) are indicated. The phase transitions observed with increasing pressure at room temperature are marked by the following symbols: ●: Benedict et al. (1985a); ▲: Stephens et al. (1968); ○: Akella et al. (1980).

transition pressures of Akella et al. (1980), on one hand, and of Benedict et al. (1985a) and Stephens et al. (1968), on the other hand. Further studies would be useful to definitely decide between these two sets of transition pressures.

6.2. *Curium*

Much less high-pressure work was reported for curium than for americium. The most abundant isotope Cm-244 has only a half-life of 18 years, so the less abundant Cm-248 (half-life: 35 000 years) is preferred for the high-pressure studies. The first published work describes a HPXRD study up to 20 GPa (Reichlin et al. 1981). Data for pressures around 18 GPa suggested a possible structural change. The same group of workers somewhat later found that around 12 GPa the superlattice lines that determine the description of curium as double hexagonal close-packed (dhcp) disappear and the structure should be described as the simple hcp lattice (Akella 1983). Another HPXRD study (Benedict et al. 1985b, Haire et al. 1985a) covered the pressure range up to 52 GPa. On compressing the dhcp Cm(I), a cubic close-packed (ccp) structure, Cm(II), was formed at 23 GPa, and this transformed to another high-pressure phase, Cm(III) at 43 GPa. The latter phase could be indexed as an orthorhombic cell of the alpha-uranium structure type. At the transition from Cm(II) to Cm(III), the volume decreased by 21%. This decrease and the formation of a low-symmetry crystal structure suggest that the 5f electrons of curium become itinerant at 43 GPa. Previous band calculations (Johansson et al. 1981) had indicated a delocalisation pressure of about 45 GPa in curium, but another estimate (Johansson and Skriver 1982) indicated that pressures in the Megabar region would be required before the itinerant state is reached (1 Megabar = 100 GPa). The pressure–volume relationship of curium at room temperature is shown in fig. 10a. Skriver (1980) calculated the variation of the Wigner–Seitz radius of curium with pressures up to 60 GPa.

6.3. *Berkelium*

The only available isotope of Bk is the high-specific-activity beta-emitter Bk-249 which decays to Cf-249 with a half-life of 320 days. Cf-249 thus accumulates in Bk-249 at a rate of the order of one percent per week, and if a given study cannot be carried out within a few hours from separation of the pure Bk, it will always deal with berkelium having an appreciable californium content.

Berkelium was studied by HPXRD up to 57 GPa (Benedict et al. 1984, Haire et al. 1984). Similar to curium, this metal, with increasing pressure, first transforms to a ccp phase, Bk(II), then to an orthorhombic structure, Bk(III), which can be satisfactorily indexed in the alpha-uranium structure type. The transition pressures are lower than those observed for curium: 9–10 GPa for the transition from Bk(I) (dhcp) to Bk(II), and 32 GPa for the transition from Bk(II) to Bk(III). An isostructural phase transition due to a valence change from +3 to +4 had been suggested to occur by band calculations (Johansson et al. 1981) below 45 GPa. This type of

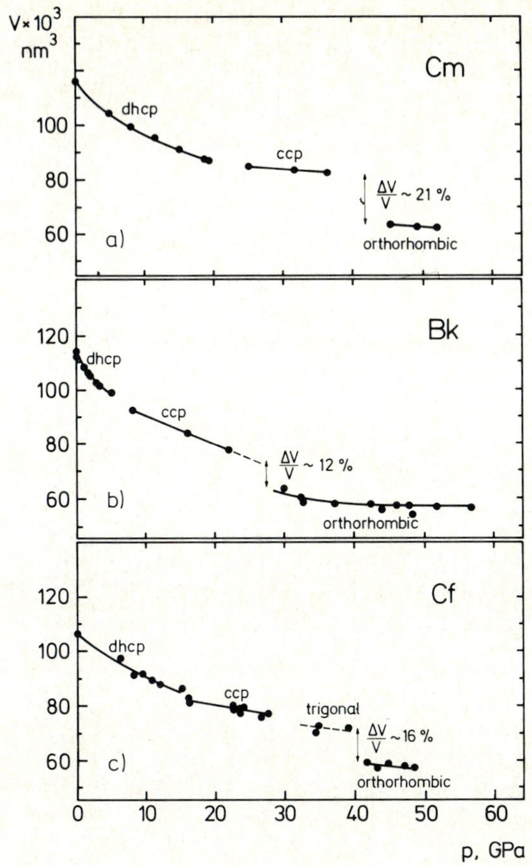

Fig. 10. Pressure–volume relationships (normalised unit cell volumes) of curium, berkelium and californium at room temperature; $\Delta V/V$: volume decrease upon delocalisation of the 5f electrons. (Adapted from Benedict et al., 1984, 1985b.)

transition was not confirmed by the experiment. The same band calculations had suggested that the 5f electrons delocalise only at very high pressures which are beyond the reach of routine DAC's. The described experimental study has shown that the critical pressure for delocalisation (delocalisation pressure) is considerably lower. The experimental pressure–volume relationship of berkelium metal at room temperature is shown in fig. 10b.

The theoretical variation of the Wigner–Seitz radius of berkelium between −20 and +5 GPa was calculated by Skriver (1980).

6.4. *Californium*

For this metal, the isotope Cf–249, an alpha-emitter with a half-life of 350 years, is available. Its decay is sufficiently slow to avoid the problem of ingrowth of daughter products as was mentioned above for berkelium.

Burns and Peterson (1978) studied Cf by HPXRD up to 14 GPa; the dhcp phase was conserved at all pressures applied. In later work on californium up to 16 GPa (Peterson et al. 1983), indications for a phase transition to a ccp phase were obtained. This transition was confirmed by Benedict et al (1984); the pressure at which this transition to the ccp phase Cf(II) occurs was determined as 17 GPa. This in turn transformed to phase Cf(III) around 30 GPa. Cf(III) could be described as having the trigonal structure ('distorted ccp') mentioned above for Am(III). Cf(III) transformed at 41 GPa to Cf(IV) which could be satisfactorily indexed in the orthorhombic alpha-uranium structure type, as was the case for the highest-pressure phases of the two elements preceding in the series, Cm(III) and Bk(III). Cf(IV) was stable up to the highest pressure applied, i.e., 48 GPa. The experimental pressure–volume relationship of californium metal at room temperature is shown in fig. 10c.

6.5. *Binary interactinide alloys*

If we look at certain properties of the actinide metals as a function of their atomic number, abrupt changes can be found between neighbouring elements. Examples are the strong volume increase when going from Pu to Am, and the maximum in transition pressures and volume change on delocalisation observed for curium (see section 6.2). Studies of binary alloys between neighbouring actinide elements can possibly yield more detailed information on these changes, which can then be considered as a function of electron concentration on a more resolved scale.

6.5.1. *Americium–curium alloys*

Haire et al. (1987) studied Am–Cm alloys containing 30, 50 and 80% Cm by HPXRD up to about 52 GPa. The three alloy compositions each exhibit three phase transitions with increasing pressure as the pure americium metal. The first transition is to a ccp phase, but the structures of the other two high-pressure phases have not yet been determined. The corresponding area of fig. 13 is correct as to the pressures of the observed phase transitions, but has to be considered as tentative with respect to the structures of the high-pressure phases. For the same reason, no volume decreases can be given for these alloys in the upper part of fig. 11. The bulk moduli of the dhcp phase of the alloys and their pressure derivatives are listed in table 7.

6.5.2. *Berkelium–californium alloys*

It was mentioned in section 6.4 that Cf-249 accumulates rapidly in Bk-249 by radioactive decay. Since californium is formed on an atomic scale at randomly distributed lattice sites, a perfectly homogeneous Bk–Cf solid solution is obtained. The homogeneity of this alloy should be better than that of most alloys obtained by the usual alloying procedures such as melting and sintering.

Itié et al. (1985) report on a HPXRD study, up to about 47 GPa, of Bk–Cf alloys at two compositions obtained by radio-active decay. With increasing pressure, the berkelium-rich alloy (about 35% Cf) underwent the same phase changes as (more or less pure – see section 6.3) berkelium, and the californium-rich alloy (about 60% Cf) exhibited the same phases as pure californium. An approximately linear increase of transition pressures with californium content was observed. The volume changes on delocalisation are of the same order as in the constituent metals.

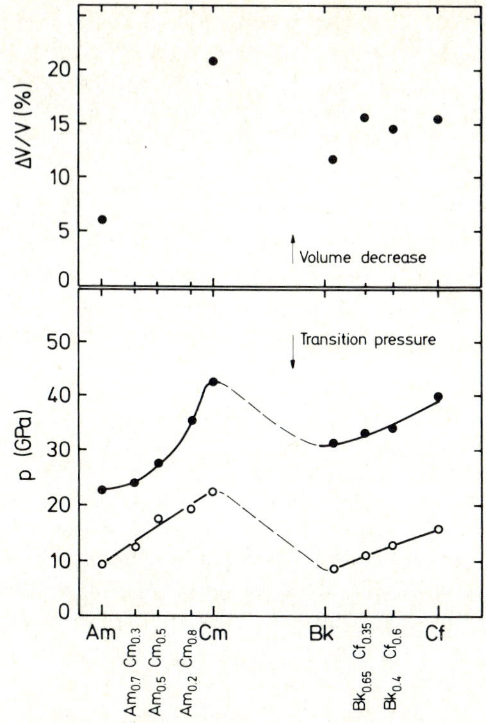

Fig. 11. Variation of transition pressures and volume changes on delocalisation in the heavy actinides.

6.6. *Transition pressures and volume decrease on delocalisation for the heavy actinides*

As can be seen in fig. 11, the pressure of the dhcp-to-ccp transition in curium is not in line with those for the same transitions in Am, Bk and Cf. Similar anomalies for curium are found in the plots of the delocalisation pressures and of the decreases in volume accompanying delocalisation. An anomaly is also found for curium at ambient pressure: its volume and atomic radius (as also those of CmN) are larger than those of americium, although in principle the 'actinide contraction' (in analogy to the lanthanide contraction is thought to start at americium.

In curium the 5f shell is half filled (configuration $5f^7$) and, according to Hund's rule, curium is therefore the actinide metal with the largest number of unpaired 5f electrons. Correspondingly, the gain in energy when the 5f's assume a localised character (see fig. 5) is greater in curium than in the other actinide metals: the spin polarisation (or 'localisation') energy has a maximum (in absolute value) at curium (Johansson et al. 1980). Curium thus exhibits a minimum of 5f electron participation in its metallic bonding under ambient conditions, and this explains why its atomic volume and lattice parameters are larger than those of the neighbouring actinide elements. It can also be inferred that higher pressure must be applied to curium in order to force the 5f electrons out of their localised state, i.e., a plot of the delocalisation pressures has a maximum at curium.

The dhcp-to-ccp transition in curium also occurs at a higher pressure, and also at higher compression, than in the neighbouring actinide elements. In the lanthanide metals, this transition is thought to be controlled by increasing 5d occupation (Duthie and Pettifor 1977). It can be assumed that as in the lanthanide, alkaline-earth, and alkali metals, the actinide metals also have their d-occupation number (6d shell) increased by application of pressure. However, there is no apparent reason why this increase should be delayed in curium with respect to its actinide neighbours. The fact that both transitions [(Cm(I)–Cm(II) and Cm(II)–Cm(III)] are shifted to higher pressures suggests some common mechanism may be contributing to this shift. In the case of the delocalisation transition, the effects of such a mechanism and of the energy minimum for the localised state of curium could be cumulative.

6.7. *The question of the so-called 'distorted fcc' phase*

High-pressure X-ray diffraction patterns obtained in diamond anvil cells can in general not reach the high standards of accuracy and resolution that can be met by X-ray diffraction at ambient pressure and temperature. This is inherently due to factors such as insufficient orientation statistics in the microsample studied – which affects relative intensities of the diffraction lines – and pressure gradients in the sample which broaden the diffraction lines. In addition, due to the mechanical requirements in the construction of diamond anvil cells (see section 3) there are in general less diffraction lines in a high-pressure diffraction spectrum than can be obtained in a good spectrum at ambient pressure, which makes the definition of the unit cell more difficult. The small sample size leads to low diffracted intensity, with respect to the incident intensity on one hand, and with respect to background intensity on the other; diffraction lines with a low relative intensity will often not be visible under these conditions. Even in the case of X-ray diffraction at ambient pressure, the exact description of complicated structures is sometimes difficult and can lead to differences in structural interpretation from different authors. It is therefore not surprising that such cases of conflicting interpretation also occur in high-pressure X-ray diffraction (see fig. 8 for an example of a diffraction pattern taken in a DAC).

In the field of actinide metals, this is especially true for the structures of Am(III) (see section 6.1) and of Cf(III) (see section 6.4). Three different structures have been proposed for Am(III) (two monoclinic ones and one trigonal), and there are two different proposals for the structure of Cf(III) (one trigonal and one triclinic). Two of the suggested structures have first been described for rare earth metals, and in view of the similarities between lanthanides and the heavy actinides Am to Cf, it may be useful to give a short account of work on these structures which mainly pertains to rare earth metals, but in a later stage also involves americium and californium.

Grosshans et al. (1982a,b) and Vohra et al. (1981), when further compressing the ccp high-pressure phases of Y, La and Pr, observed a second-order transition to a new phase which they described in a trigonal structure derived from a frozen soft phonon mode of the ccp structure. This structure was presented in more detail in a later paper by Vohra et al. (1984). The diffraction data of Cf(III) (Benedict et al.

1984) and of Am(III) (Benedict et al. 1985a) could also be fitted to this structure, though some difference in relative intensities exists between the experimental patterns of these two phases. A diffraction pattern of Am(III) is given in fig. 8.

A structural sequence hcp→Sm type→dhcp→ccp is observed in the lanthanide metals when the atomic number Z decreases at constant pressure, or when a lanthanide metal is compressed. McMahan and Young (1984) analysed this sequence in terms of layer packing and found that an additional type of simple layer packing is theoretically possible which had never been described before for a lanthanide metal. They called it 'triple hexagonal close-packed' (thcp). Smith and Akella (1984) suggested this structure as another possibility for the high-pressure phase Pr(III) which had previously been described as having the above-mentioned trigonal structure (Grosshans et al. 1982a). Nd(III) was also interpreted in the trigonal structure by Grosshans and Holzapfel (1984), but described as thcp in the papers of Akella and Smith (1986) and Akella et al. (1986). Grosshans and Holzapfel (1984), in addition, reported Gd and Tb to assume the trigonal structure under pressure. Staun Olsen et al. (1985b) observed for Tb a phase that could be indexed in the thcp structure, in a pressure range intermediate to those of the dhcp and the ccp phases.

Sikka et al. (1985) report that the trigonal structure fits to a $La_{0.8}Th_{0.2}$ alloy at ambient pressure, and that the observed intensities of the 100 and 103 lines do not agree with those required by the thcp structure. Smith and Akella (1985) interpret these diffraction data as a superposition of lines from fcc La–Th solid solution and impurity ThO_2, in which case they would be without relevance to the question discussed here. This argument was rejected by Sikka et al. (1986).

Sikka and Vijayakumar (1986) discussed these and a few other structures as possibilities for the description of high-pressure phases formed by pressure increase from ccp phases of rare earth and actinide metals. Smith and Akella (1986) introduced a second version of the thcp structure (thcp II) and compared, for the example of Pr(III), the fit of relative intensities with the calculated patterns for the thcp I, thcp II, and trigonal indexing; none of these showed excellent agreement between calculated and observed intensities. For the case of Cf(III), Roof (1986) suggested a triclinic structure as an alternative to the trigonal indexing of Benedict et al. (1984). The average deviation between calculated and observed d-spacings is approximately the same for both alternatives.

6.8. *'Pressure-quenched' high-pressure phases*

Little or no hysteresis to transformation is observed when pressure is reduced on the orthorhombic high-pressure phases of Am, Cm, Bk and Cf. They readily retransform to the ccp or trigonal high-pressure phases. In contrast, if pressure is further decreased, there is a strong hysteresis in Cm, Bk and Cf to ccp transforming back to the dhcp allotrope, at least at room temperature. The latter allotrope, which is thermodynamically stable at ambient pressure, and room temperature, does not appear, or appears only as a small fraction of the sample, even when pressure is completely released (Haire et al. 1986). By analogy with the quenching of samples

TABLE 6
Atomic volumes of the ccp and dhcp allotropes of Am, Cm, Bk and Cf. (After Haire et al. 1986).

Element	Atomic volumes × 10^3, nm^3, at room temperature		
	dhcp	ccp	
		pressure-quenched	temperature-quenched
Am	29.27	–	29.30
Cm	29.98	30.0	31.99
Bk	27.96	28.0	31.19
Cf	27.37	27.3	30.13

from high temperatures, this phenomenon has been called 'pressure-quenching'. Retention of the ccp phase in Cm, Bk and Cf metals after release of pressure provides an opportunity to determine the ambient-pressure lattice parameters (and possibly other ambient-pressure properties that can be determined on micro-samples) of the high-pressure ccp phase. As can be seen in table 6, the atomic volumes of the pressure-quenched ccp phases are virtually identical to those of the dhcp phases (Haire et al. 1986). Similar volumes for the two phases can be expected because there is no discontinuous volume change when dhcp is transformed to ccp under pressure. Americium cannot be compared in this respect because it re-transforms back to dhcp on complete pressure release.

The ccp phase is also formed – as shown from the example of the p, T diagram of americium (fig. 9) – when the first four transplutonium metals are quenched from high temperature. The atomic volume of temperature-quenched ccp americium is also equal to that of dhcp americium. In contrast, when ccp curium (resp. berkelium, californium) is quenched from high temperature, its room-temperature atomic volume is about 7% (resp.12%, 10%) higher than that of the dhcp allotrope (see table 6). A possible explanation is that lattice strain is created in Cm, Bk and Cf during quenching from high temperatures.

6.9. *Compressibility of the heavier actinides*

Table 7 and fig. 12 give the bulk moduli and their pressure derivatives for the elements from Am to Cf and some of their alloys. The compressibilities of these metals are much lower than those of Pa, U and Np, but of the same order as those of the lanthanide metals which are shown for comparison in the figure (see below).

The compressibilities of these metals have not been determined to date by the ultrasonic method because of their high radio-activity and/or the small amounts available. Calculating the bulk modulus for these soft materials (steep initial slope, but relatively strong curvature of the p–V curve) by fitting pressure–volume data to the Birch and Murnaghan equations can only be an accurate method if a sufficient number of data at pressures in the range up to about 2 GPa is available. This is usually not the case with diamond anvil cells for use at very high pressures.

TABLE 7

Bulk moduli and pressure derivatives for the heavier actinide elements Am to Cf and for some of their alloys. For reasons mentioned in the text, relatively large errors have to be taken into account (of the order of ±5 GPa for B_0 and of the order of ±2 for B_0').

Element	B_0(GPa)	B_0'	Refs.[a]
Am, HPXRD (Birch fit)	44.8	6.1	[1]
Am, HPXRD (Murnaghan fit)	45.1	5.6	[1]
Am, LMTO calculation	45[b]	–	[2]
Am, LMTO calculation	38[b]	–	[3]
Am, piezometry	36	6	[4]
$Am_{0.7}Cm_{0.3}$, HPXRD	21	8	[5]
$Am_{0.5}Cm_{0.5}$, HPXRD	28	7	[5]
$Am_{0.2}Cm_{0.8}$, HPXRD	34	4	[5]
Cm, HPXRD (Birch fit)	33(5)	6.5	[6]
Cm, LMTO calculation	40[b]	–	[3]
Bk, HPXRD (Birch fit)	25(5)	6(2)	[7]
Bk, LMTO calculation	35[b]	–	[3]
$Bk_{0.65}Cf_{0.35}$, HPXRD	33(5)	6	[9]
$Bk_{0.4}Cf_{0.6}$, HPXRD	46(5)	2.3	[9]
Cf, HPXRD (Birch fit)	50(5)	2(1)	[8]
Cf, HPXRD (Birch fit)	48(5)	3.2	[9]
Cf, HPXRD (Murnaghan fit)	49(5)	2.8	[9]

[a] References:
[1] Benedict et al. (1985a).
[2] Johansson et al. (1981).
[3] Brooks et al. (1984).
[4] Stephens et al. (1968).
[5] Haire et al. (1987).
[6] Haire et al. (1985a).
[7] Haire et al. (1984).
[8] Peterson et al. (1983).
[9] Dabos et al. (1986).

[b] Read from graph.

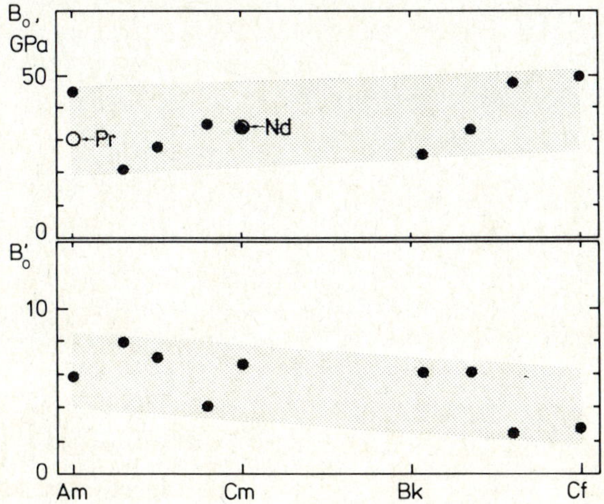

Fig. 12. Bulk moduli and their pressure derivatives for the heavier actinide metals; ●: actinides; ○: Pr, Nd for comparison, plotted according to the 'shifted homologous relationships' of Ward (1985).

This type of cell having been used for the HPXRD work from which the experimental data of table 7 were derived, there is a relatively large uncertainty on these bulk moduli. No curve has been drawn through the data points of fig. 12 for this reason, but the range of values is indicated by a shaded band. The bulk moduli of praseodymium and neodymium (taken from Gschneidner 1964) are shown for comparison as two examples of lanthanide metals which have the same structural sequence under pressure as the actinides Am to Cf.

7. Einsteinium

While the preceding four metals Am, Cm, Bk and Cf exhibit the dhcp structure at ambient conditions, this structure has not been observed for einsteinium. Two fcc phases with widely different lattice parameters, 541 and 575 pm, and a hexagonal close-packed (hcp) phase were reported for this metal. The atomic volumes of these allotropes are much higher than those of the preceding, trivalent, metals, indicating divalency for Es. As for the divalent lanthanide Eu, a particularly low bulk modulus can be expected for divalent Es. A tendency to divalency is also expected from the excitation energy for the transition from configuration $f^{10}s^2$ to configuration f^9ds^2, which is slightly above the critical value separating the divalent and trivalent metallic states (Johansson et al. 1981). Einsteinium can thus be expected to change to a trivalent, lower-volume phase under pressure and would be an interesting element to study by high-pressure techniques, e.g., HPXRD. As mentioned in the Introduction, the problems associated with the fast radioactive decay of Es have up to now discouraged its study under high pressure.

8. General diagram of the phase relations of actinide metals under pressure

The experimental data on the high-pressure structural behaviour of the actinide metals can be visualised together in a graphic presentation which consists of interconnected binary phase diagrams of neighbouring actinide metals (fig. 13). The figure refers to data taken at room temperature only. The solid dots on this diagram represent the phase transitions which have been experimentally observed. Two-phase regions were omitted to avoid complication of the diagram.

The binary diagrams are known only for the systems Am–Cm and Bk–Cf (see section 6.5). This leads to some uncertainty in the phase boundaries for the other binary systems, especially for the lighter elements Ac to Pu where wide pressure ranges without phase transitions are found. Phase limits in the left part of the diagram are consequently drawn as dashed straight lines which give only very approximate positions for the phase fields. It can be expected that these phase fields will be better defined in the not too distant future by the study of the corresponding binary alloys. Extension of the whole diagram to still higher pressures, with possible discovery of additional high-pressure phases, is now also in the range of technical possibilities due to the recent development of so-called 'Megabar' DAC's.

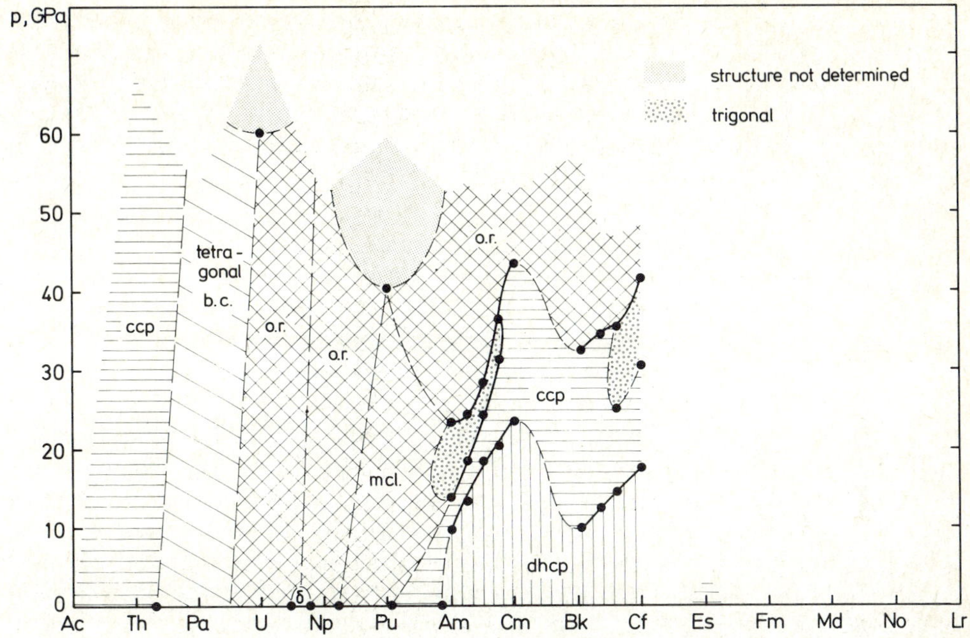

Fig. 13. General diagram of the phase relations of actinide metals under pressure: ●: experimentally observed phase transitions; o.r.: orthorhombic; mcl.: monoclinic; the other symbols have their usual meaning. See text (section 8) for more detailed explanation.

The orthorhombic and monoclinic phase fields have been marked with the same type of shading, thus indicating that despite having distinct crystal structures they belong to the same group of low-symmetry phases that are correlated with itinerant 5f electrons. The bottom line indicates the phase limits at ambient pressure, room temperature and represents the intersection with the interconnected binary phase diagram representation as a function of temperature by Smith and Kmetko (1983). Smith and Kmetko's and the present diagram, if intersecting each other perpendicularly along the bottom lines, constitute the limiting planes of a spatial Z,p,T diagram for $T>298$ K, which has to be filled by future high-temperature high-pressure investigations. Extension to temperatures below 298 K is of course also highly desirable. Low-temperature, high-pressure studies could, for example, help solving the discrepancy between theoretical and experimental volume decrease in americium (see section 6.1).

The diagram quite clearly shows the unique behaviour of curium. As described in section 6.2, the transition pressures of this metal are considerably higher than the corresponding ones of both neighbour elements. The diagram seems to suggest that the stability region of the dhcp phase is like a peninsula in a ccp lake, the lake extending from Pu to Es, and the peninsula being limited to the lower pressure region around Am, Cm, Bk and Cf. This hypothesis has certainly to be considered with some caution because ccp Am, Cm, Bk and Cf phases are trivalent, but the ccp Es is probably divalent, and there is some doubt whether they can really belong to a common phase field in this diagram. It is interesting to note in this context that the transition from dhcp to ccp, which occurs under the action of pressure in heavy actinides and light lanthanides, also occurs with decreasing atomic number at ambient pressure, and probably also at moderately elevated pressure. This fact has somewhat escaped attention up to now because at ambient pressure, the ccp structure does not appear in an elemental actinide metal, but only in the Pu–Am alloys. This situation strongly reminds us of the lanthanides, where the same structural sequence hcp→Sm type→dhcp is found with increasing pressure for constant Z and with decreasing Z at ambient pressure. At elevated pressure, this lanthanide structural sequence even continues to the ccp structure, the 'distorted ccp' structure, and to low-symmetry structures (Grosshans and Holzapfel 1984), in a close analogy to what is observed in the actinide metals. Otherwise said, one discovers the same structural sequence in the lanthanide and the actinide metals, but its first two members, hcp and Sm type, are missing in the actinide metals, and the phase transitions are shifted to lower pressures and higher atomic numbers in the actinides. This analogy is seen when one compares the righthand part of fig. 13 (between Pu and Es) to the lefthand part of the room-temperature section of the 'intra rare earth generalized binary phase diagram' for the lanthanide metals established by Gschneidner (1985). One main difference between lanthanide and actinide metals is the S-shape of the limit between the dhcp and ccp fields in the actinide case which is linked to the particular behaviour of curium, and which does not find an analogy in the lanthanide case. Gschneidner's diagram, which covers the range only up to 24 GPa, does not show the 'distorted ccp' phases which have also been observed in some lanthanide metals (Grosshans and Holzapfel 1984) and thus make the analogy between the two series of elements even more complete. This comparison between the two diagrams is also an illustration of the principle of 'shifted homologous relationships' between the transplutonium and the lighter rare-earth metals as suggested by Johansson (1975) and Ward (1985), though in this case it is not possible to establish a one-to-one relationship between the metals of the two series.

Looking at the two areas of the trigonal phase around Am and Cf, one can easily imagine a trigonal phase field extending as a band in the high-pressure region of the ccp phase field, but which for some unknown reason is interrupted around Cm and Bk. This interruption could be another phenomenon linked with the half-filling of the 5f shell, together with the high transition pressures and high volume 'collapse' in curium.

The systematics of the high-pressure phases of the actinides, as reflected by these interconnected binary phase diagrams, was recently compared to the corresponding systematics of the lanthanides (Benedict et al. 1986b). One main part of this comparison was a plot of the Wigner–Seitz radii R_{ws}, scaled with the orbital radii of the last filled shell of the trivalent ions (R_{5p} for the lanthanides, and R_{6p} for the actinides) (fig. 14). It is found that in the lanthanides (fig. 14a), a given type of phase transition occurs at the same radius ratio R_{ws}/R_{5p} for all elements. This is not the

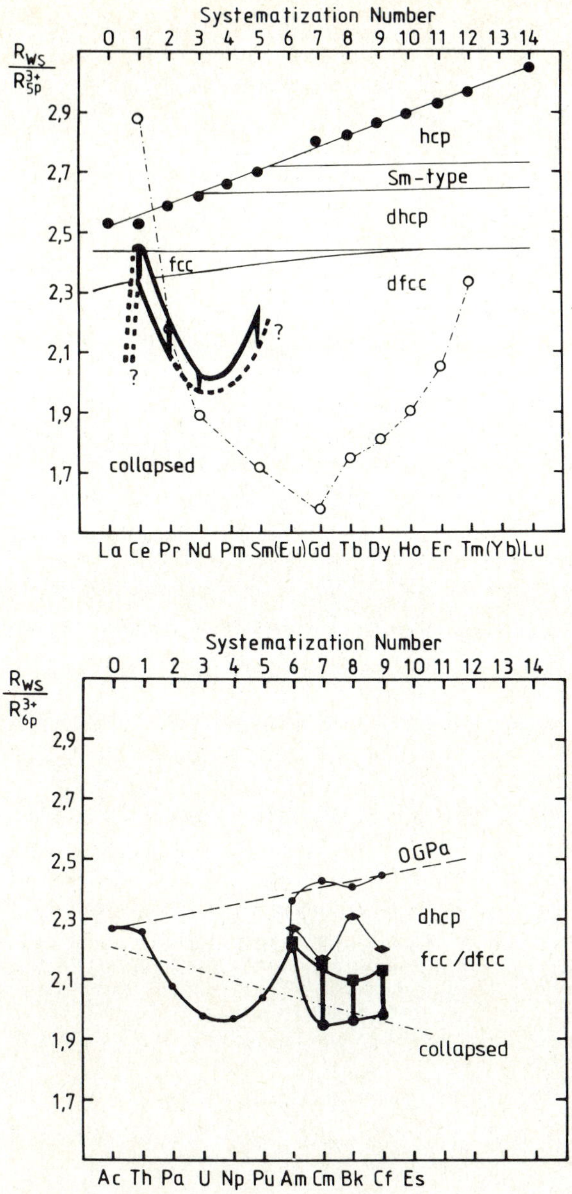

Fig. 14. High-pressure phases in the lanthanide (a) and the actinide (b) metals in relation to the scaled Wigner–Seitz radii. (From Benedict et al. 1986b.) (d)hcp: (double) hexagonal close-packed, fcc: face-centred cubic, dfcc: distorted face-centred cubic.

case in the actinides (fig. 14b). But there is another interesting agreement between the two series of f-elements. The radius ratios R_{ws}/R_{6p} for the actinides Th to Pu are quite close to the radius ratios R_{ws}/R_{5p} at which the corresponding lanthanide elements Ce to Sm undergo the transition to the 'collapsed' phase with itinerant 4f electrons. This seems to show that these radius ratios can be used as a criterion to distinguish between localised and itinerant states of the f-transition elements.

Acknowledgements

Thanks are due to S. Dabos for help with the critical evaluation of some of the compressibility data. The valuable comments of R.G. Haire, W.B. Holzapfel and G.H. Lander were much appreciated.

References

Akella, J., 1983, Rep. Workshop Actinides under Pressure, Karlsruhe, April 21/22, 1983 (Eur. Inst. Transuranium Elements, Karlsruhe) p. 5.

Akella, J., and G.S. Smith, 1986, J. Less-Common Met. **116**, 313–316.

Akella, J., Q. Johnson, W. Thayer and R.N. Schock, 1979a, J. Less-Common Met. **68**, 95–97.

Akella, J., Q. Johnson and R.N. Schock, 1979b, High-Pressure Science and Technology, 7th AIRAPT Conf., Le Creusot, France, July 1979.

Akella, J., Q. Johnson and R.N. Schock, 1980, J. Geophys. Res. B **85**, 7056–7058.

Akella, J., G.S. Smith and H. Weed, 1985, J. Phys. Chem. Solids **46**, 399, 400.

Akella, J., J. Xu and G.S. Smith, 1986, Physica B **139&140**, 285–288.

Armstrong, P.E.,.O.N. Carlson and J.E. Smith, 1959, J. Appl. Phys. **30**, 36.

Asami, N., M. Yamada and J. Takahashi, 1961, Trans. Jpn. Inst. Metals **31**, 389–394.

Bader, S.D., N.E. Phillips and E.S. Fisher, 1975, Phys. Rev. B **12**, 4929–4940.

Barrett, C.S., M.H. Mueller and R.L. Hitterman, 1963, Phys. Rev. **129**, 625.

Bellussi, G., U. Benedict and W.B. Holzapfel, 1981, J. Less-Common Met. **78**, 147–153.

Benedict, U., and C. Dufour, 1982, unpublished.

Benedict, U., and C. Dufour, 1985, in: TUSR-39, Programme progress report Nuclear Fuels and Actinide Research, Jan.–June 1985, CEC document 4201 (CEC, Karlsruhe) pp. 79, 80.

Benedict, U., J.C. Spirlet, C. Dufour, I. Birkel, W.B. Holzapfel and J.R. Peterson, 1982, J. Magn. & Magn. Mater. **29**, 287–290.

Benedict, U., J.R. Peterson, R.G. Haire and C. Dufour, 1984, J. Phys. F **14**, L43–L47.

Benedict, U., J.P. Itié, C. Dufour, S. Dabos and J.C. Spirlet, 1985a, in: Americium and Curium Chemistry and Technology, eds N.M. Edelstein, J.D. Navratil and W.W. Schulz (Reidel, Dordrecht/Boston) pp. 213–224.

Benedict, U., R.G. Haire, J.R. Peterson and J.P. Itié, 1985b, J. Phys. F **15**, L29–L35.

Benedict, U., J.P. Itié, C. Dufour, S. Dabos and J.C. Spirlet, 1986a, Physica B **139 & 140**, 284.

Benedict, U., W.A. Grosshans and W.B. Holzapfel, 1986b, Physica B **144**, 14.

Benjamin, T.M., G. Zou, H.K. Mao and P.M. Bell, 1981, Carnegie Institution of Washington, Year Book **80**, 280–283.

Bennet, B.I., J.D. Johnson, G.I. Kerley and G.T. Rood, 1978, Rep. LA-7130 (Los Alamos Scientific Laboratory, Los Alamos, NM).

Birch, F., 1947, Phys. Rev. **71**, 809–824.

Bradbury, M.H., 1981, J. Less-Common Met. **78**, 207–218.

Bridgman, P.W., 1948, Proc. Am. Acad. Arts Sci. **76**, 55–70.

Bridgman, P.W., 1959, J. Appl. Phys. **30**, 214–217.

Brooks, M.S.S., 1983, J. Phys. F **13**, 103–108.

Brooks, M.S.S., 1984, J. Phys. F **14**, 1157–1171.

Brooks, M.S.S., B. Johansson and H.L. Skriver, 1984, in: Handbook on the Physics and Chemistry of the Actinides, Vol. 1, eds A.J. Freeman and G.H. Lander (North-Holland, Amsterdam) pp. 153–269.

Burns, J.H., and J.R. Peterson, 1978, Inst. Phys. Conf. Series **37**, 52–54.

Chu, C.W., and G.S. Knapp, 1973, Phys. Lett. A **46**, 33–35.

Dabos, S., and U. Benedict, 1986, to be published.

Dabos, S., J.R. Peterson, R.G. Haire and U. Benedict, 1986, unpublished.

Dabos, S., C. Dufour, U. Benedict and M. Pagès, 1987, J. Magn. & Magn. Mater. **63/64**, 661–663.

Dakshinamoorthy, M., and K. Iyakutti, 1984, Phys. Rev. B **30**, 6943–6950.

de Cadenet, J.J., 1961, in: Plutonium 1960, eds E. Grison, W.B.H. Lord and R.D. Fowler (Cleaver-Hume Press, London) pp. 106, 107.

Duthie, J.C., and D.G. Pettifor, 1977, Phys. Rev. Lett. **38**, 773.

Fertig, W.A., A.R. Moodenbaugh and M.B. Maple, 1972, Phys. Lett. A **38**, 517,518.

Fisher, E.S., and D. Dever, 1970, Solid State Commun. **8**, 649–653.

Fisher, E.S., and H.J. McSkimin, 1961, Phys. Rev. **124**, 67.

Forman, R.A., G.J. Piermarini, J.D. Barnett and S. Block, 1972, Science **176**, 284,285.

Fournier, J.-M., J. Beille, A. Boeuf, C. Vettier and A. Wedgwood, 1980, Physica B **102**, 282–284.

Freeman, A.J., 1980, Physica B **102**, 3.

Ganguly, J., and G.C. Kennedy, 1973, J. Phys. Chem. Solids **34**, 2272–2274.

Gardner, W.E., and T.F. Smith, 1967, Phys. Rev. **154**, 309–315.

Glötzel, D., 1978, J. Phys. F **8**, L163, L205.

Godwal, B.K., and S.K. Sikka, 1982, J. Phys. F **12**, 655–660.
Govinda Rajan, K., 1984, Curr. Sci. **53**, 1115–1127.
Gregorian, T., and W.B. Holzapfel, 1984, J. Phys. Colloq. **45**(C8), 41, 42.
Grosshans, W.A., and W.B. Holzapfel, 1984, J. Phys. Colloq. **45**(C8), 141–143.
Grosshans, W.A., Y.K. Vohra and W.B. Holzapfel, 1982a, Phys. Rev. Lett. **49**, 1572–1575.
Grosshans, W.A., Y.K. Vohra and W.B. Holzapfel, 1982b, J. Magn. & Magn. Mater. **29**, 282–286.
Gschneidner Jr, K.A., 1964, in: Solid State Physics, Vol. 16, eds H. Ehrenreich, F. Seitz and D. Turnbull (Academic Press, New York) p. 275.
Gschneidner Jr, K.A., 1985, J. Less-Common Met. **110**, 1–10.
Haire, R.G., J.R. Peterson, U. Benedict and C. Dufour, 1984, J. Less-Common Met. **102**, 119–126.
Haire, R.G., U. Benedict, J.R. Peterson, C. Dufour and J.P. Itié, 1985a, J. Less-Common Met. **109**, 71–78.
Haire, R.G., U. Benedict, J.P. Young, J.R. Peterson and G.M. Begun, 1985b, J. Phys. C **18**, 4595–4601.
Haire, R.G., U. Benedict, J.R. Peterson, C. Dufour and S. Dabos, 1986, Physica B **144**, 19–22.
Haire, R.G., S. Dabos and U. Benedict, 1987, to be published.
Hornung, K., and K.W. Michel, 1979, J. Phys. Chem. Solids **40**, 899–910.
Hovestreydt, E., E. Parthé and U. Benedict, 1987, Rep. 10874 EN (Comm. Eur. Communities, Brussels).
Itié, J.P., J.R. Peterson, R.G. Haire, C. Dufour and U. Benedict, 1985, J. Phys. F **15**, L213–L219.
Jayaraman, A., 1983, Rev. Mod. Phys. **55**, 65–108.
Johansson, B., 1975, Phys. Rev. B **11**, 2836.
Johansson, B., 1977, in: Proc. 2nd Int. Conf. on the Electronic Structure of the Actinides, Wrocław, 1976, eds J. Mulak, W. Suski and R. Troć (Polish Academy of Sciences, Ossolineum, Wrocław) p. 49.
Johansson, B., 1978, J. Phys. Chem. Solids **39**, 467.
Johansson, B., and A. Rosengren, 1975a, Phys. Rev. B **11**, 1367–1373.
Johansson, B., and A. Rosengren, 1975b, Phys. Rev. B **11**, 2836–2857.
Johansson, B., and H.L. Skriver, 1982, J. Magn. & Magn. Mater. **29**, 217–229.
Johansson, B., H.L. Skriver, N. Mårtensson, O.K. Andersen and D. Glötzel, 1980, Physica B **102**, 12.
Johansson, B., H.L. Skriver and O.K. Andersen, 1981, in: Physics of Solids under High Pressure, eds J.S. Schilling and R.N. Shelton (North-Holland, Amsterdam) p. 245–262.
Kalvius, G.M., W. Potzel, J. Moser, F.J. Litterst, L. Asch, J. Zänkert, U. Potzel, A. Kratzer, M. Wunsch, J. Gal, S. Fredo, D. Dayan, M.P. Dariel, M. Bogé, J. Chappert, J.C. Spirlet, U. Benedict and B.D. Dunlap, 1985, Physica B **130**, 393–411.
Keane, A., 1954, Aust. J. Phys. **7**, 322–333.
Klement Jr, W., and A. Jayaraman, 1966, Prog. Solid State Chem. **3**, 289–376.
Klement Jr, W., A. Jayaraman and G.C. Kennedy, 1963, Phys. Rev. **129**, 1971–1975.
Laquer, H.L., 1961, in: The Metal Plutonium, eds A.S. Coffinberry and W.N. Miner (American Society for Metals, Chicago) p. 157.
Leman, G., 1970, FOA 4 Rep. C4434-23 (Försvarets Forskningsanstalt, Stockholm, Sweden).
Linford, P.F.T., and A.E. Kay, 1967, in: Plutonium Handbook, Vol. 1, ed. O.J. Wick (Gordon & Breach, New York) p. 66.
Liptai, R.G., and R.J. Friddle, 1966a, J. Nucl. Mater. **20**, 48–56.
Liptai, R.G., and R.J. Friddle, 1966b, J. Less-Common Met. **10**, 292–294.
Liptai, R.G., and R.J. Friddle, 1967, J. Nucl. Mater. **21**, 114–116.
Lloyd, L.T., R.G. Liptai and R.J. Friddle, 1966, J. Nucl. Mater. **19**, 173–180.
Loasby, R.G., 1961, in: Plutonium 1960, eds E. Grison, W.B.H. Lord and R.D. Fowler (Cleaver-Hume Press, London) p. 97.
Louis, E., A. de Visser, A. Menovsky and J.J.M. Franse, 1986, Physica B **144**, 48–53.
Mao, H.K., P.M. Bell, J.W. Shaner and D.J. Steinberg, 1978, J. Appl. Phys. **49**, 3276–3238.
Mao, H.K., P.M. Bell and R.J. Hemley, 1985, Phys. Rev. Lett. **55**, 99–102.
Maple, M.B., and D. Wohlleben, 1972, Phys. Lett. A **38**, 351,352.
Marmeggi, J.C., and A. Delapalme, 1980, Physica B **102**, 309–312.
Marmeggi, J.C., A. Delapalme, G.H. Lander, C. Vettier and N. Lehner, 1982, Solid State Commun. **43**, 577–581.
McMahan, A.K., and D.A. Young, 1984, Phys. Lett. A **105**, 129.
McWhan, D.B., B.B. Cunningham and J.C. Wallmann, 1962, J. Inorg. Nucl. Chem. **24**, 1025.
McWhan, D.B., P.W. Montgomery, H.D. Stromberg and G. Tura, 1963, J. Phys. Chem. **67**, 2308–2311.
Merx, H., and C. Moussin, 1980, in: High-Pressure Science and Technology, eds B. Vodar and Ph. Marteau (Pergamon, Oxford) pp. 204–206.
Michel, K.W., 1975, Ber. Bunsenges. Phys. Chem. **79**, 1063.
Morgan, J.R., 1970, in: Plutonium 1970 and Other Actinides, ed. W.N. Miner (Met. Soc. AIME, New York) p. 669.
Murnaghan, F.D., 1937, Am. J. Math. **59**, 235–260.
Palmy, C., and E.S. Fisher, 1970, Solid State Commun. **8**, 655–657.

Palmy, C., R. Flach and P. de Trey, 1971, Physica **55**, 663–668.

Peggs, G.N., 1983, High Pressure Measurement Techniques (Applied Science Publishers, London/New York).

Pénicaud, M., 1984, in: Shock Waves in Condensed Matter – 1983, eds J.R. Asay, R.A. Graham and G.K. Straub (North-Holland, Amsterdam) pp. 61–64.

Peterson, J.R., U. Benedict, C. Dufour, I. Birkel and R.G. Haire, 1983, J. Less-Common Met. **93**, 353–356.

Peterson, J.R., J.P. Young, R.G. Haire, G.M. Begun and U. Benedict, 1985, Inorg. Chem. **24**, 2466–2468.

Piermarini, G.J., and C.E. Weir, 1964, Science **144**, 69.

Prunier, C., S. Hay and B. Hocheid, 1970, in: Plutonium 1970 and Other Actinides, ed. W.N. Miner (Met. Soc. AIME, New York) pp. 1001–1012.

Ragan, C.E., 1980, in: High-Pressure Science and Technology, eds B. Vodar and Ph. Marteau (Pergamon Press, New York) pp. 993–999.

Reichlin, R.L., 1983, Rev. Sci. Instrum. **54**, 1674–1677.

Reichlin, R.L., J. Akella, G.S. Smith and M. Schwab, 1981, paper presented at Conf. on Actinides 1981, Asilomar, CA, Sept. 10–15, 1981.

Reynolds, M.B., 1951, Rep. AECD-3242 (U.S. Atomic Energy Commission, Washington, DC).

Robinson, A.C., 1961, quoted in: Linford and Kay (1967).

Röhler, J., 1986, Physica B **144**, 27–31.

Roof, R.B., 1981a, Adv. X-ray Anal. **24**, 221–230.

Roof, R.B., 1981b, J. Appl. Crystallogr. **14**, 447–450.

Roof, R.B., 1981c, paper presented at Conf. on Actinides 1981, Asilomar, CA, Sept. 10–15, 1981.

Roof, R.B., 1982a, Z. Kristallogr. B **15**, 307–312.

Roof, R.B., 1982b, personal communication.

Roof, R.B., 1986, J. Less-Common Met. **120**, 345–349.

Roof, R.B., R.G. Haire, D. Schiferl, L.A. Schwalbe, E.A. Kmetko and J.L. Smith, 1980, Science **207**, 1353–1355.

Roux, C., P. Leroux and M. Rapin, 1971, J. Nucl. Mater. **40**, 305–310.

Sahu, P.C., M. Yousuf and K. Govinda Rajan, 1983, Rep. Workshop Actinides under Pressure, April 21/22, 1983 (Eur. Inst. Transuranium Elements, Karlsruhe) p. 8.

Schirber, J.E., and A.J. Arko, 1980, Phys. Rev. B **21**, 2175–2179.

Shushakov, V.D., A.G. Seleznev, N.S. Kosulin and T.V. Shushakova, 1983, Phys. Met. Metallogr. **55**(2), 188–190.

Sikka, S.K., and V. Vijayakumar, 1986, Physica B **144**, 23–26.

Sikka, S.K., B.K. Godwal and V. Vijayakumar, 1985, Phys. Lett. A **108**, 83, 84.

Sikka, S.K., B.K. Godwal and V. Vijayakumar, 1986, submitted to Phys. Lett. A.

Simmons, E., and H. Wang, 1971, Single-Crystal Elastic Constants and Calculated Aggregate Properties: a Handbook, 2nd Ed. (MIT Press, Cambridge, MA).

Skriver, H.L., 1980, personal communication.

Skriver, H.L., and J.P. Jan, 1980, Phys. Rev. B **21**, 1489–1496.

Skriver, H.L., O.K. Andersen and B. Johansson, 1978, Phys. Rev. Lett. **41**, 42–45.

Skriver, H.L., O.K. Andersen and B. Johansson, 1980, Phys. Rev. Lett. **44**, 1230.

Smith, G.S., and J. Akella, 1984, Phys. Lett. A **105**, 132,133.

Smith, G.S., and J. Akella, 1985, Phys. Lett. A **112**, 47.

Smith, G.S., and J. Akella, 1986, Phys. Lett. A **118**, 136–138.

Smith, G.S., J. Akella, R. Reichlin, Q. Johnson, R.N. Schock and M. Schwab, 1981, paper presented at Conf. on Actinides 81, Asilomar, CA, Sept. 1981.

Smith, H.G., and G.H. Lander, 1984, Phys. Rev. B **30**, 5407.

Smith, H.G., N. Wakabayashi, W.P. Crummett, R.M. Nicklow, G.H. Lander and E.S. Fisher, 1980, Phys. Rev. Lett. **44**, 1612.

Smith, J.L., and E.A. Kmetko, 1983, J. Less-Common Met. **90**, 83–88.

Smith, T.F., and E.S. Fisher, 1973, J. Low Temp. Phys. **12**, 631–641.

Staun Olsen, J., L. Gerward, U. Benedict and J.P. Itié, 1985a, Physica B **133**, 129–137.

Staun Olsen, J., S. Steenstrup and L. Gerward, 1985b, Phys. Lett. A. **109**, 235–237.

Staun Olsen, J., U. Benedict, S. Dabos, L. Gerward and J.P. Itié, 1986, Physica B **144**, 56–60.

Stephens, D.R., 1963, J. Phys. Chem. Solids **24**, 1197–1202.

Stephens, D.R., 1966, J. Phys. Chem. Solids **27**, 1201–1204.

Stephens, D.R., H.D. Stromberg and E.M. Lilley, 1968, J. Phys. Chem. Solids **29**, 815–821.

Syassen, K., H. Winzen and U. Benedict, 1986, Physica B **144**, 91.

Tindall, D.A., and M.O. Steinitz, 1984, J. Phys. F **14**, 1131–1136.

Van Thiel, M., 1977, Compendium of Shock Wave Data, Rep. UCRL-50108, Vol. 1, 1st revision (University of California Radiation Laboratory, Lawrence Radiation Laboratory, Livermore, CA).

Viard, J., 1962, in: Les Ondes de Détonation (Centre National de la Recherche Scientifique, Paris) pp. 383–389.

Vijayakumar, V., 1985, J. Phys. Chem. Solids **46**, 17–20.

Vohra, Y.K., H. Olijnyk, W. Grosshans, W.B. Holzapfel, 1981, Phys. Rev. Lett. **47**, 1065–1067.

Vohra, Y.K., V. Vijayakumar, B.K. Godwal and S.K. Sikka, 1984, Phys. Rev. B **30**, 6205–6207.

Walsh, J.M., M.H. Rice, R.G. McQueen and F.L. Yarger, 1957, Phys. Rev. **108**, 196–216.

Ward, J.W., 1985, in: Americium and Curium Chemistry and Technology, eds N.M. Edelstein, J.D. Navrati and W.W. Schulz (Reidel, Dordrecht) pp. 135–145.

Xu, J.A., H.K. Mao and P.M. Bell, 1986, Science **232**, 1404–1406.

Yousuf, M., 1986, Thesis (Indian Institute of Science, Bangalore, India).

Handbook on the Physics and Chemistry of the Actinides, edited by A.J. Freeman and G.H. Lander

Chapter 4

ACTINIDES IN SILICATE GLASSES*

B.W. VEAL, J.N. MUNDY and D.J. LAM
Materials Science Division, Argonne National Laboratory, Argonne IL 60439, USA

Contents

1. Introduction

A large and rapidly growing body of knowledge has been accumulated to explain the physics and chemistry of silicate-based glasses. The simplest such glass is SiO_2, a material with substantial commercial and technological importance. When alkali metal oxides are added to SiO_2, the glass commonly shows a remarkably high capability for the dissolution of other metal oxides. This synergism underscores the complexity and great variety of silicate glasses. The list of known silicate glasses, many with commercial importance, is greatly expanded when silicates containing glass formers such as B, P, or Ge and modifiers (alkalis, alkaline earths, and numerous other metals) are included.

Because of the high solubilities for an extensive range of metal oxides (and other favorable properties), silicate or borosilicate glasses have recently received consideration as possible media for long-term immobilization of radioactive wastes. This technology depends on the dissolution and immobilization of actinide and other metal oxides into an aqueous-leach-resistant matrix. The successful utilization of silicate glass matrix materials for this technology must exploit the knowledge base already acquired for the silicate-based glasses, but additional detailed studies of

*Work supported by the US Department of Energy, Basic Energy Sciences – Materials Science, under contract W-31-109-Eng-38.

silicates containing actinide solutes will be required. Until the stimulus provided by this technology appeared, however, very little was known about the physics and chemistry of silicate-based glasses containing dissolved actinides. Reported studies are still quite limited in number and scope. In this chapter, we consider the physics and chemistry of silicates containing dissolved actinides, drawing, as needed, upon the knowledge base established for actinide-free silicates and crystalline minerals. In general, the review will be confined to studies of relatively simple glasses where specific measurements or systematic studies provide a reasonably clear picture of mechanisms governing properties behavior.

The first actinide element incorporated into glasses for practical application was uranium. It has found use as a component in glass-to-metal graded seals. In the nineteen thirties, uranium oxide was an important coloring agent for silicate glasses. Because the glasses were normally melted in air, uranium usually occurred in the hexavalent form which, in dilute quantity, imparted a yellow-green color to the transmitted light (associated with a strong absorption band at 420 nm). When the glass was melted in a reducing atmosphere, a proportion of the uranium was left in the tetravalent state. The color of the glass could be changed from yellow-green to green depending on the ratio of hexavalent and tetravalent ions in the glass (Weyl 1951, Smith and Cohen 1963). Another attractive feature of the uranium colored glass was its visible fluorescence. This has been studied by Kroger et al. (1948) who suggested that the fluorescence is associated with the uranyl group in the glass. Because of the safety hazards associated with its radioactivity, uranium is no longer being used as a glass coloring agent and the study of uranium glasses was practically stopped until the advent of the development of long-term nuclear waste disposal technology in the sixties.

There are two types of radioactive wastes, if the nuclear fuel is reprocessed, that require long-term isolation from the biosphere for safe disposal, i.e., the high level waste (HLW) and the transuranic waste (TRU). The HLW is generated from the reprocessing of spent fuel for the separation and recovery of the remaining useful uranium and plutonium. Ninety-eight percent or more of the contained uranium and plutonium will be removed in the reprocessing cycle. The residual (HLW) will contain alpha-emitting isotopes and virtually all of the fission products. A considerable inventory of HLW in the form of liquid solution (HLLW) exists in the world. The TRU wastes result from the fabrication of nuclear explosives and from fabrication of mixed oxide fuel for nuclear reactors. TRU wastes are generated as solids and emit low levels of penetrating radiation. These wastes can be stored or transported in sealed metal containers without heavy shielding. The method for long-term isolation of these radioactive wastes that has received the most effort to date is disposal in an underground cavity in a well-chosen geological repository. In this concept, five barriers would be placed between the nuclear waste and the biosphere. First, the waste would be immobilized in a solid borosilicate glass form. Second, the glass solid would be encased in corrosion resistant metal containers. Third, the canister would be sealed in overpack, in a second metal canister. Fourth, a barrier such as clay would be packed around the canister to limit the flow of any groundwater that may leak into the repository. Finally, the repository would be sited thousands of meters below the earth's surface in a well-chosen geological formation.

Vitrification processes for the solidification of HLLW have been under investigation in several Western European countries since the early sixties (Heimerl 1979) and in the United States since the mid-seventies. The incorporation of waste constituents into a borosilicate glass was selected in France and the United Kingdom from the beginning of their vitrification programs, whereas a phosphate glass was initially chosen in Germany. The Germans later abandoned the phosphate glass in favor of a borosilicate glass in order to conform with practices in the other European countries. The borosilicate glass waste form has also been adopted in the United States to immobilize HLW generated by the military nuclear energy programs. The compositions of borosilicate glasses used by different glass waste from producers vary because of variations in the composition of the HLLW and the different geological environments of the nuclear waste repositories.

As noted above, the HLW contains virtually all of the fission products and alpha-emitting isotopes. Since most of the fission products are beta–gamma emitters with half-lives of a few decades or less, and since the decay of a beta–gamma emitter generally leads immediately to a stable isotope, HLW radioactivity will be reduced to a very low level within about 1000 years of its production. The transuranic isotopes are the major contributors to the long-term (greater than 1000 years) radiation hazard of nuclear waste. This occurs since the transuranic isotopes are alpha emitters. While they have half-lives comparable to those of beta–gamma emitters, the decay of an alpha emitter generally leads to another radioactive alpha-emitting isotope. The long term buildup of alpha-emitting daughters is the main contributor to the potential risk for periods longer than 1000 years.

The viability of this waste disposal concept, while reinforced by the multiple layers of shielding from the biosphere and storage in a deep repository, will be enhanced by preparing the primary storage medium, the glass, in the most effective manner. Understanding of the bonding characteristics of transuranium elements and daughters in silicate- and borosilicate-based glasses thus constitutes an important task in the nuclear waste management program. In this chapter, we focus on the transuranic solutes in silicate-based glasses.

2. Structure of covalent glasses

Amorphous solids differ from crystalline solids in their lack of long range order or structural periodicity. Amorphous solids do, however, show short range order. The earliest observations of the short range order in glasses came from X-ray measurements of the radial distribution function. The radial distribution function $\rho(r)$ is a measure of the probability of finding a neighboring atom at a distance between r and $r+\mathrm{d}r$ from a central atom. In a crystalline solid $\rho(r)$ has well-defined peaks at values of r equal to each shell of neighboring atoms. In glasses, as in liquids, $\rho(r)$ shows a few broad peaks at small r (short range order), but the absence of structure in $\rho(r)$ at larger r reflects the absence of long range order.

Zachariasen (1932), noting the similarity of both structure and internal energy between crystalline and glassy oxides, chose to model glass structure as a modification of the crystalline structure. He proposed the random network model to explain oxide glasses such as silicates. The model has also been extensively applied to other covalently bonded glasses such as germanates and borates and to amorphous silicon and germanium. The continuous random network has become the most commonly accepted model for the structure of oxide glasses, and in the following we will discuss other models largely in terms of their challenge to that of a continuous random network. While the present discussion will center around silicate glasses, the same concepts apply to all oxide glasses. Short range order in oxide glasses occurs within the glass-forming oxygen polyhedra that form around cations, such as silicon, that are known as network formers. In the amorphous and crystalline forms of both silica and silicates, the silicon atom normally resides at the center of a tetrahedron of oxygen atoms and the tetrahedra are linked to one another at their corners. The network, while possessing short range order, lacks the periodicity of the crystalline state and has no long range order. Figure 1 shows a two-dimensional representation of the random network model for SiO_2. The radial distribution function for such a model is the same as that found from X-ray or neutron diffraction studies of oxide glasses.

It should be noted at this point that the continuous random network model does not deal with the important consideration of how a glass is formed. While amorphous solids can be formed in a number of ways (rapid quenching, vapor condensation, and sol-gel processing) the most common method of forming oxide glasses is by solidification from the melt. Materials in a liquid state generally show a continuous decrease in volume upon cooling. At a particular temperature (freezing

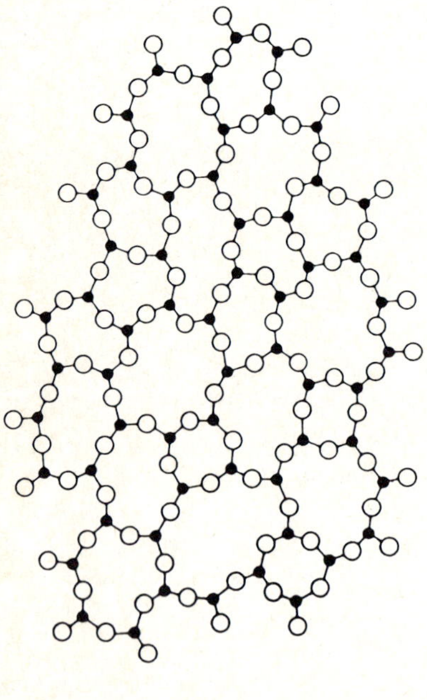

Fig. 1. A two-dimensional representation of the random network model of SiO_2 as proposed by Zachariasen (1932). Solid circles are silicon atoms, open circles are oxygen atoms.

point) some materials show a sharp decrease in volume and form crystalline solids. Upon further cooling the material shows a lower thermal expansion coefficient than that found in the liquid. Other materials, upon cooling from the liquid state, enter a temperature region where the thermal expansion coefficient changes continuously from that representative of a liquid to that of a solid. The location of this temperature region, which is referred to as the glass transformation range, will depend upon the cooling rate. The lower the cooling rate, the lower the temperature range where the transformation occurs and the greater the time for the atomic configurations to relax and result in a glass of lower entropy and higher density. It is this (unrelaxed) atomic configuration that is frozen and that is described by the random network model. It is a metastable state that is separated from the crystalline state by an energy barrier which cannot be overcome at temperatures below the glass transformation range because of the high viscosities in the glassy state. Thus, although the random network model is used to describe all these glassy states, the structure of a particular glass is dependent not only upon its composition but also on the details of its formation. Comparison of measurements of physical properties of glasses is often complicated by the fact that different samples may experience different degrees of relaxation resulting from variations in cooling rates. Cooling rate considerations also complicate the comparison of experimentally measured properties with molecular dynamic simulations since computer time limitations have required the use of extremely high quench rates in the modeling studies.

The present picture of the structure of oxide glasses has been largely developed by X-ray and, more recently, by neutron diffraction studies. In amorphous silica, the near interatomic distances (mean values) are well established at Si–O: 1.62 Å, O–O: 2.65 Å, Si–Si: 3.12 Å (Mozzi and Warren 1969, DaSilva et al. 1975, Uhlmann and Wicks 1979), and the Si–O–Si bond angles are distributed over a range of angles within $\pm 15\%$ of $150°$ (see also Wright et al. 1982, Kirkpatrick et al. 1985, Greaves et al. 1981, and Greaves 1983). Three-dimensional ball and stick random network models (Bell and Dean 1972) have qualitatively reproduced the density and the X-ray and neutron diffraction data of amorphous silica. Molecular dynamic simulations of amorphous silica (Woodcock et al. 1976, Soules 1982) have also shown good agreement with the diffraction data in spite of the high quench rates used in the simulations. The simulations assume interatomic pair potentials of the Born–Meyer type. The resulting tetrahedral configuration of oxygen around the silicon would suggest that this coordination is more a function of the oxygen/silicon radius ratio (Kingery et al. 1976) than the presence of covalent bonding.

X-ray and neutron diffraction studies of the alkali silicate glasses (Warren 1937, Wicks 1974, Misawa et al. 1980) generally show that Si–O, O–O, and Si–Si interatomic distances are comparable to corresponding distances found in amorphous silica. However, the addition of alkali oxides to silica glass provides extra oxygen atoms to the continuous random network of silica tetrahedra with the result that some oxygen atoms are bonded to only one silicon atom. These oxygens are referred to as nonbridging oxygens (NBO) in contrast to the bridging oxygens which link one tetrahedron to the next. The number of nonbridging oxygens is in direct proportion to the number of alkali ions added. The bridging and nonbridging

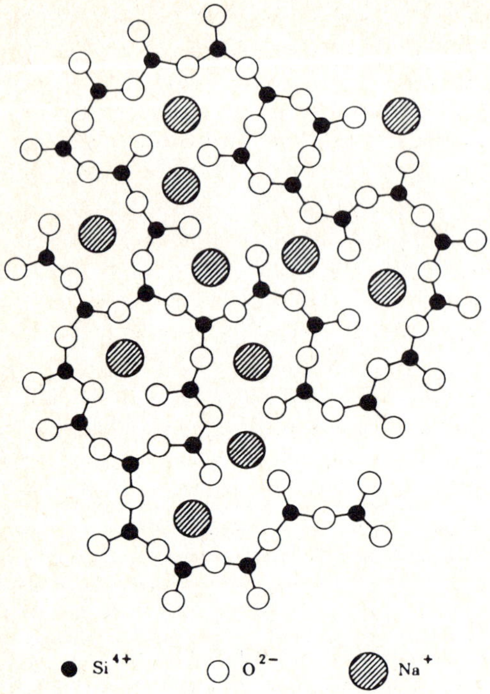

Fig. 2. A two-dimensional representation of the random network model of sodium silicate glass. (After Warren 1941.)

oxygen atoms can be directly observed using X-ray photoelectron spectroscopy (XPS) (Bruckner et al. 1980, Jen and Kalinowski 1979, Veal and Lam 1978), nuclear magnetic resonance (NMR) (Kirkpatrick et al. 1985) and Raman spectroscopy (Brawer and White 1977) techniques. For reasons of charge compensation, the alkali ions which modify the silica network are located in the vicinity of the nonbridging oxygens. Molecular dynamic simulations have also been made for alkali silicate glasses (Soules 1979, Soules and Varshneya 1981). Again, agreement with the diffraction data is generally good for coordination numbers, interatomic spacings and the angular distribution of the Si–O–Si bond angles. Figure 2 shows a two-dimensional representation of sodium disilicate glass in the random network model (from Warren 1941).

The agreement between the diffraction data and the molecular dynamic simulations which generate a random silica network would suggest a strong confirmation of Zachariasen's random network model for the structure of oxide glasses. However, prior to Zachariasen's work, Lebedev (1921) had proposed that the structure of glass was made up of small crystallites and the controversy between these models has remained in some form to the present time (Zallen 1983, Kreidl 1983). The crystallite model (Valenkov and Poroi-Koshits 1936) is based on the fact that the broad peaks found in the X-ray diffraction patterns of glass could be the result of the known broadening of peaks obtained from polycrystalline samples when the crystallites are smaller than $\approx 10^{-5}$cm. The classic X-ray diffraction studies of Warren et al. (Warren 1937, Warren et al. 1938) on sodium silicate glasses were often used to support the random network model, although Warren himself pointed out that his data were also consistent with SiO_2 island groups in a Na_2O-rich silicate structure. The controversy between these and other models was particularly strong in the fifties and sixties and certainly promoted the implementation of many careful experiments including small angle X-ray scattering (SAXS) and electron microscopy measurements. The introduction of these new techniques did not resolve the controversy (Kreidl 1983, Porai-Koshits 1985, Phillips 1982) but provided a better definition of the possible magnitudes of the inhomogeneities and provided the tools for examining phase separated glasses.

Phase separation can occur in a glass when the liquid from which it forms separates into two or more liquids upon cooling. The separation occurs because of liquid immiscibilities and results in glasses with phase separated regions 30 to 1000 Å in extent. The heterogeneity produced by phase separation has been examined in many oxide glasses (Kingery et al. 1976, Wong and Angell 1976) largely by use of SAXS and electron microscopy techniques. An example of phase separated glass would be $x\mathrm{Na_2O}:(1-x)\mathrm{SiO_2}$ in the molar fraction range $0.01 \lesssim x \lesssim 0.2$. In this range of x, two-phase structures are found with islands of sodium silicate in a silica-rich matrix at low x changing with increasing x to silica-rich islands in a matrix of sodium silicate (Kreidl 1983). Only for $x \geqslant 0.2$ could one expect to find a random distribution of sodium throughout a silica network.

Recent molecular dynamic simulations of $x\mathrm{Na_2O}:(1-x)\mathrm{SiO_2}$ glass with $x=0.05$, 0.10, 0.15, 0.20 and 0.33 (Mitra and Hockney 1982) have also shown evidence for the clustering of sodium ions at low x. The high quench speeds used in such simulations may create glasses which are less relaxed than their laboratory counterparts but the advent of faster computers should allow the continuing development of valuable simulation studies of microheterogeneity. Much of the literature for and against microheterogeneities in glass has been reviewed recently by Kreidl (1983). The review suggests that in alkali silicate glasses there are a number of experiments where a nonrandom distribution of alkali is indicated.

Much of the work involving actinides dissolved in silicate glasses pertains to alkali-rich silicate solvent glasses, including sodium disilicate. Sodium disilicate devitrifies to a crystalline form with well-characterized crystal structure (Pant and Cruickshank 1968). The sodium disilicate glass contains, on average, one nonbridging oxygen (NBO) atom per SiO_4^{4-} tetrahedral unit. In the extreme case, only one nonbridging site appears on any given tetrahedron, and each tetrahedron has three corner bonds to bridge neighboring tetrahedra. Numerous mineral silicates consisting of tetrahedral units with one nonbridging oxygen atom are known (Deer et al. 1962). Ubiquitously, the tetrahedra join to form silica sheets separated by a layer of alkali or other metal atoms. The tetrahedra join to form rings (usually six-membered but sometimes four- or eight-membered) but the three corner bonds apparently impose a sheet-like continuity as well.

Inference of glass structure from a knowledge of crystalline analogs would suggest that a remnant of this sheet-like layering might also be expected in alkali-rich silicate glasses. Support for this extrapolation procedure from crystalline to glassy sodium disilicate is provided by measurement of the distribution of nonbridging

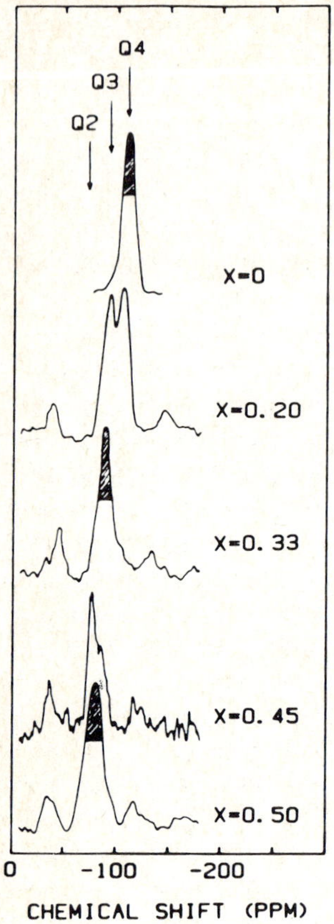

Fig. 3. Rotating sample Si NMR spectra for sodium silicate glasses. Q's designate the number of bridging oxygen atoms bonded to Si. These results show that the distribution of nonbridging oxygen sites is nonrandom. To good approximation, SiO_2 ($x=0$) has only Q4 sites, sodium disilicate ($x=0.33$) has only Q3 sites and sodium metasilicate ($x=0.5$) has only Q2 sites. (After Dupree et al. 1984.)

oxygen atoms among SiO_4 tetrahedra. Recent 'magic-angle sample-spinning' (MASS) NMR studies utilizing the silicon resonance at the .tetrahedral center confirm that the distribution of nonbridging oxygens among the tetrahedra, for lithium and sodium silicate glasses, is, indeed, nonrandom (Dupree et al. 1984, Grimmer et al. 1984, Schramm et al. 1984, Selvaray et al. 1985). This work shows that, to high precision, each tetrahedron of sodium disilicate glass has just one nonbridging oxygen atom. NMR data (Dupree et al. 1984) are shown in fig. 3 for sodium silicate glasses $(Na_2O)_x(SiO_2)_{1-x}$ with compositions $x=0$, 0.20, 0.33, 0.45, and 0.50. Nuclear magnetic resonances from silicon atoms within tetrahedra containing 4, 3, and 2 bridging oxygens (0, 1, and 2 NBO's) are indicated as Q4, Q3, and Q2, respectively. For the sodium disilicate glass ($x=0.33$), the NMR spectrum is almost totally dominated by the Q3 signal, that is, from silicon atoms located within SiO_4 tetrahedra that contain a single nonbridging oxygen atom.

Since all of the SiO_4 tetrahedra in the amorphous disilicates contain a single NBO and since the mineral silicates that contain these structural units ubiquitously form silica sheet structures, it is natural to infer the existence of a layering tendency in the glasses as well. Silica sheet-like regions may exist as intermediate-range structural entities of the sodium disilicate glass.

In a glass the sheets might double back onto themselves to form a complex polymeric-type material. However, limitations in the range of permitted bond angles at the tetrahedral corners may require that the region of space over which this 'doubling-back' occurs is large. On the other hand, significant distortions from the ordered crystalline sheets observed in the minerals must occur in the glasses. [Such distortions produce a broadening of the silicon resonance line (MASS) of the sodium disilicate glass relative to the crystal. The broadening represents a distribution of Si environments associated with the loss of long range order.] Arguably, the sheets meander about in the glasses, but in the process preserve identifiable layered regions of silica separated by layers of alkali. For more alkali-dilute glasses (e.g., sodium trisilicates), some of the SiO_4 tetrahedra must contain a complete shell of bridging oxygens. However, essentially no tetrahedra would be expected to contain more than one NBO. Thus, in the extreme alkali-dilute limit, isolated NBO sites occur. As the alkali concentration increases, alkali-rich layers begin to form. At the sodium disilicate composition, the alkali layers are separated by silica sheets comprised of SiO_4 units each containing one NBO. (For more alkali-rich glasses, two or more NBO sites appear within a tetrahedron and chains of bridged tetrahedra now appear.)

Incorporation of this intermediate range order represents a modification of the random network model as applied to alkali silicates. Variations of the modified model, for alkali silicate and alkali phosphate glasses, have been proposed. Greaves and co-workers (Greaves et al. 1981 and Greaves 1985) proposed a model comprised of two interlacing sublattices: network regions constructed from network formers and internetwork regions made up of glass modifiers such as the alkali. The overall structure (illustrated in fig. 4 in a two-dimensional representation) has islands of network surrounded by channels or layers of modifier. The extent of these channels in which the alkali ion transport would occur depends on the total alkali content of the glass. Using this model together with percolation theory, Greaves (1985) was able to explain qualitatively the ionic transport properties of alkali glasses. Greaves' model, however, does not incorporate the constraint, for sodium disilicate, that each tetrahedral unit must contain a single nonbridging oxygen atom.

The combination of XPS and extended X-ray absorption fine structure (EXAFS) measurements (Lam et al. 1983, Knapp et al. 1984a) was used to suggest a model with extended network and modifier regions for silicate glasses containing uranium. In glasses containing U^{6+} ions, it was argued that uranium is present in extended uranyl sheets surrounded by layers of alkali and SiO_4 network with sheet structures similar to those found in some alkali silicate crystalline compounds (Liebau 1961a, Imaoka et al. 1983). These workers concluded that the sheet structures were characteristic of the alkali silicate solvent glass and that its layered character was compatible for easy dissolution of uranyl, also a layered material.

To explain NMR measurements of sodium metaphosphate glasses, Bunker et al. (1987) also proposed a similar modified random network model in which elements of the sheet structure, found in the devitrified metaphosphate, was preserved in the vitreous form. The glass viscosity increased as trivalent nitrogen was added to the glass, apparently cross-linking the phosphate sheets. This cross-linking behavior and its affect on viscosity was also noted by Greaves (1985) and by Lam et al. (1983).

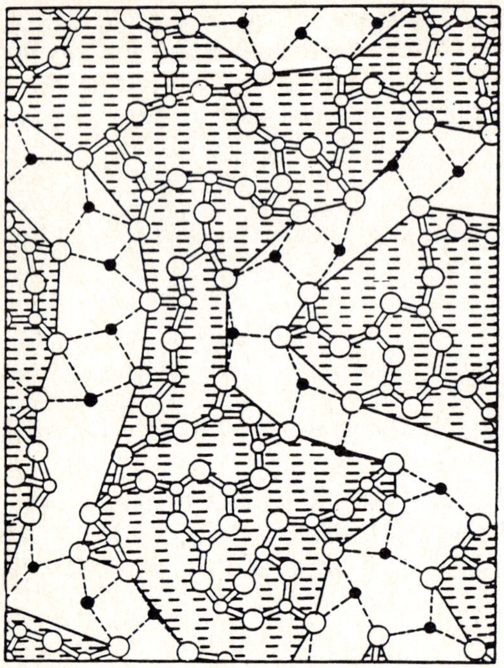

Fig. 4. A modified random network model of alkali silicate glass showing intermediate range ordering. Small open circles are silicon, large open circles are oxygen, and black circles are alkali atoms. (After Greaves 1985.)

3. Actinides in silicate-based glasses

3.1. *Actinides – chemical considerations*

Actinide ions can occur in a multiplicity of valence states with highly variable properties associated with the different valences. Chemical properties are, of course, also related to electronegativity and ionic size. For elements thorium through plutonium, outer electron atomic configurations and Pauling electronegativities are summarized in table 1. The configurations show somewhat erratic behavior in the

TABLE 1
Atomic configurations of neutral actinide atoms and Pauling electronegativities.

Atom	Atomic configuration	Pauling electronegativity
Th	$6d^2 7s^2$	1.3
Pa	$5f^2 6d^1 7s^2$	1.5
U	$5f^3 6d^1 7s^2$	1.7
Np	$5f^4 6d^1 7s^2$	1.3
Pu	$5f^6 6d^0 7s^2$	1.3

TABLE 2
Ionic radii of actinide ions in different valence states and coordinations.

Valence	Coordination	Ionic radius (Å)				
		Th	Pa	U	Np	Pu
3	VI	–	1.04	1.025	1.01	1.00
4	VI	0.94	0.90	0.89	0.87	0.84
5	VI	–	0.78	0.76	0.75	0.74
6	VI	–	–	0.73	0.72	0.71
4	VIII	1.05	1.01	1.00	0.98	0.96

systematics of the filling of the 5f shell. In the solid state, the 5f electrons of light actinides are often sufficiently delocalized (i.e., of sufficient radial extent) so that hybridizing or mixing of electron wave functions of neighboring atoms commonly occurs. Progressing to higher Z (atomic number), the 5f orbital radii become increasingly more spatially contracted and, consequently, more atomic-like in their behavior (Brooks et al. 1984). This tendency is illustrated in table 2 where ionic radii (Shannon 1976) are tabulated for elements Th through Pu in their different possible valence states when cation-to-ligand coordination numbers equal six. Radii are also shown for the tetravalent eight-coordinated cations. For any given valence state and coordination, the ion size is progressively reduced with increasing Z, a manifestation of the 'actinide contraction'. The light actinides show rich chemistries since the 5f electrons have considerable versatility in their ability to participate in chemical bonding. The large number of possible valence states for elements Pa through Pu provides an indication of this complex chemical behavior. This chemical versatility is considerably reduced, however, for the heavier actinides where the more localized atomic-like 5f electrons resist active participation in bond formation. For $Z > 95$ (americium), only valence states 2 through 4 are observed with a strong preference for formation of the trivalent state. For U through Pu, valence states 3 through 6 (and some valence 7) can occur.

For highly ionic compounds, 5f electrons show varying degrees of participation in bond formation, depending on cation valence state and electronegativity (or ionic size) of the neighboring anions. One approach for monitoring this participation is afforded by Mulliken charge density analyses of the populated orbitals obtained from local-density molecular cluster calculations configured to simulate the solid state. To illustrate:

We compare calculations for UO_2 (Gubanov et al. 1977) and for the uranyl structure UO_2^{2+} (Walch and Ellis 1976). The former contains tetravalent uranium in which four of the six uranium valence electrons are nominally involved in formation of a relatively ionic bond. Thus, one would expect that the 5f electron configuration would be $5f^2$, a result which is confirmed to high accuracy by measurements that are sensitive to the unpaired spins or residual magnetic moment of the system (Faber and Lander 1976). On the other hand, the cluster calculations for UO_2 show a Mulliken 5f population of 2.7 and a total uranium residual charge of about 3.8

electrons (Gubanov et al. 1977). The computation provides a description of the system that represents the charge density in terms of a linear combination of atomic orbitals. Thus, the charge density of UO_2 is represented in this approach by two molecular orbitals that are essentially pure 5f in character and by a number of other molecular orbitals (of predominately ligand p character) that contain an admixture of hybridized 5f atomic orbitals. This hybridized 5f character may be viewed as 5f electrons which participate in bond formation. Applying this perspective to the U^{6+} ion in the uranyl, we note that a residual charge of ~3 electrons (Walch and Ellis 1976) occurs at the U site, with about 2.3 electrons being attributed to U 5f electrons. Thus, the Mulliken analysis suggests that the U^{4+} and U^{6+} ions might be viewed as being $U^{2.2+}$ and U^{3+} with 0.7 and 2.3 of the U 5f electrons, respectively, participating in bond formation.

We note that the molecular cluster calculations represent a model approach where a very small fragment (containing only a few atoms) of the solid is investigated. No knowledge of band dispersion in momentum space is obtained. Results may be influenced by cluster size and stoichiometry, as well as basis set and self-consistency considerations. At the same time, band calculations (see Brooks et al. 1984) suffer from limited ability to treat the highly localized 5f levels (their inclusion as hybridized band states in UO_2 will show metallic behavior). Thus the role played by 5f electrons in chemical bond formation in UO_2 has not been clearly resolved. The problem of 5f participation in bond formation in UO_2, including a review of relevant experiments, has been discussed by Baer (1984).

A characteristic of light actinide oxygen compounds, apparently associated with the peculiar nature of the actinide 5f electrons, is their tendency to form actinyl structures (Eisenstein and Pryce 1955). This is a highly unusual 3-atom unit containing an actinide ion surrounded by two oxygen atoms arranged in a nearly linear chain. The actinyl structure is prevalent in hexavalent uranium compounds (uranyls) and also occurs in Np^{5+} and Np^{6+} compounds and in Pu compounds. The uranyl structure readily forms when uranium is dissolved in alkali silicate glasses and profoundly influences the properties of the glass. The neptunyl structure also occurs in alkali silicates.

3.2. *Solubilities*

The solubility of uranium (and actinide ions, in general) in silicate-based glasses prepared by melting depends on the valence state of the actinide ion. The valence state of the dissolved actinide may be controlled during glass preparation by controlling the atmosphere above the melt. A reducing (oxidizing) atmosphere is conducive to formation of a lower (higher) valence state. Valence states could be expressible as a phase field over oxygen partial pressure (for a given melt temperature and quench condition). However, data of this kind are very limited for the glass silicates. Solubility data are available only for glasses containing an actinide species in a single, isolated valence state obtained by preparing the glass in an extreme oxidizing or reducing atmosphere. Most studies have been done on silicate glasses with base compositions $0.33\,Na_2O \cdot 0.67\,SiO_2$ (sodium disilicate) or

$0.25\,Na_2O \cdot 0.75\,SiO_2$ (sodium trisilicate) into which actinide oxides in varying quantity have been dissolved.

Dissolution of UO_2 or UO_3 in sodium silicate glasses in an atmosphere of oxygen or air produces a glass in which uranium appears only in the hexavalent state (Lam et al. 1979). Trisilicate glasses were melted at 1250°C in a flowing atmosphere and were cooled from melt temperature in the argon atmosphere of the containment glove box. To obtain dissolved tetravalent uranium ions, it was necessary to melt the glass in very reducing conditions; e.g., in a CO-rich atmosphere (Lam et al. 1983). Borosilicates predominantly containing tetravalent U ions were melted in an argon atmosphere in graphite crucibles (Antonini et al. 1984). The melting of NpO_2 (at 1250°C) into silicate glass under flowing oxygen produces a glass with pentavalent Np ions; the reducing atmosphere leaves the Np ions in the tetravalent state (Veal et al. 1986). In contrast, dissolved Pu ions appear in the tetravalent state, apparently quite independent of the atmosphere under which the glass was melted (Lam et al. 1983).

The available solubility data for actinide ions dissolved in sodium trisilicate glasses are summarized in table 3. Results are given as values of x in the formula $(0.25\,Na_2O \cdot 0.75\,SiO_2)_{1-x} \cdot (AnO_y)_x$ where An represents the actinide. The O/An ratio y is determined by the valence state assumed by the actinide ion. These solubilities were established by monitoring dissolution of dioxides in base glasses. Changes of the valence state of the actinide ion during preparation are charge compensated by adjustment of the oxygen stoichiometry. Solubility limits of about 2 wt% were reported by Eller et al. (1985), for dioxides of Th, Np, Pu, and Am and more than 10 wt% for UO_2 in more complex borosilicate glasses melted in air at 1000°C. In general, actinide ion solubilities are substantially greater for the higher valence states indicating profound differences in the chemical behavior of the actinide ions in the glasses. It should be noted that solubility limits depend on melt temperatures and base glass composition. However, reported results often depend on the preparation techniques employed.

Solubilities of the 4+ species indicate relative chemical similarity for this group of ions. The low solubilities suggest that the 4+ ions do not enter the silicate lattice by substituting for the Si^{4+} ions in the lattice. Considering a large ionic size mismatch between Si^{4+} and the An^{4+} ions, ionic substitution is not expected in any case. The Si ions are coordinated to 4 oxygens, while the actinides will normally be coordinated to six or more oxygen neighbors.

TABLE 3
Solubilities (x) of actinide (An) oxides: $(0.25Na_2O \cdot 0.75SiO_2)_{1-x}(AnO_y)_x$.

Valence	Oxide			
	Th	U	Np	Pu
4+	0.05 ± 0.01	0.05 ± 0.005	0.03 ± 0.005	0.02 ± 0.005
5+	–	–	0.09 ± 0.005	–
6+	–	0.2	–	–

In general, the dissolution of metal oxides into alkali silicate glasses is closely associated with the amount of alkali present; the dissolution capability often increases with increased alkali content. While pure silica glass is not a generally effective solvent for metal oxides, alkali silicates are. However, the relatively low solubilities of the 4+ actinide ions may suggest, for these ions, a rather weak dependence on alkali content. Some information is available on this point from studies of solubility limits established for the sodium disilicates (Veal et al. 1986, Aldred 1980) that complement the sodium trisilicate data cited above. Expressible as $(0.33\,Na_2O \cdot 0.67\,SiO_2)_{1-x}(AnO_2)_x$, values of x comparable to those in table 3, for the trisilicates, were obtained for 4+ ion solubilities. [Note that some compositions in the literature have been specified as $(Na_2O \cdot 2SiO_2)_{1-y}(AnO_2)_y$. In the work by Aldred (1980, for example, for dissolved PuO_2, $y=0.06$ which corresponds to $x=0.021$ when specified as $(0.33\,Na_2O \cdot 0.67\,SiO_2)_{1-x}(AnO_2)_x$.] Since melting points are lower for the high alkali glasses, melting temperatures for the disilicates were generally lower than for the trisilicates. For the high valence An ions (e.g., U^{6+}, Np^{5+}), solubility limits have not been clearly established for the disilicates. Thus, it might be expected that solubilities of actinide oxides into silicates might be quite variable, depending on the chemical state of the dissolved ions and the chemical makeup of the glass. A very limited amount of data are available establishing solubility limits for different solvent glasses.

3.3. *Ionic size and solubility*

The random network model of glass structure has been most extensively used to explain those properties that relate to structure. One application of this model to deal with the issue of metal-oxide solubilities is to argue that the solute ions are randomly situated in large interstitial spaces or holes of the amorphous alkali–silicate network. Variations of bond angles between SiO_4 tetrahedra give rise to the holes of irregular shape and size (see Lam et al. 1983). An implication of this simple model view is that solubilities of incorporated elements should be determined, in large part, by the ionic sizes of the added elements with solubilities increasing as ionic sizes become smaller. The solubility data and ionic sizes (Shannon 1976) for eightfold coordinated An^{4+} ions are shown in fig. 5. We note that the solubility of the 4+ ions decreases with Z as does the ionic size (actinide contraction) in contradiction to the simple model prediction. On the other hand, the relatively large 4+ ions have the expected low solubilities, while the much smaller 5+ and 6+ ions have dramatically higher solubilities. For hexavalent uranium, where the molar (UO_3) solubility in sodium trisilicate is 20%, the ion size is considerably smaller (for all coordinations; see Shannon 1976) than the U^{4+} ion and the solubility increase is dramatic. Furthermore, the U^{6+} ion, smaller than Np^{5+}, also shows a relative solubility increase.

It may be argued that, if solubilities are determined by ion size, then, for the actinyl glasses, one should more properly consider the size of the actinyl structure, rather than the actinide ion. If both the U^{4+} and UO_2^{2+} ions are situated in the interstitial holes of the glass network, then the solubility of UO_2^{2+} should be smaller,

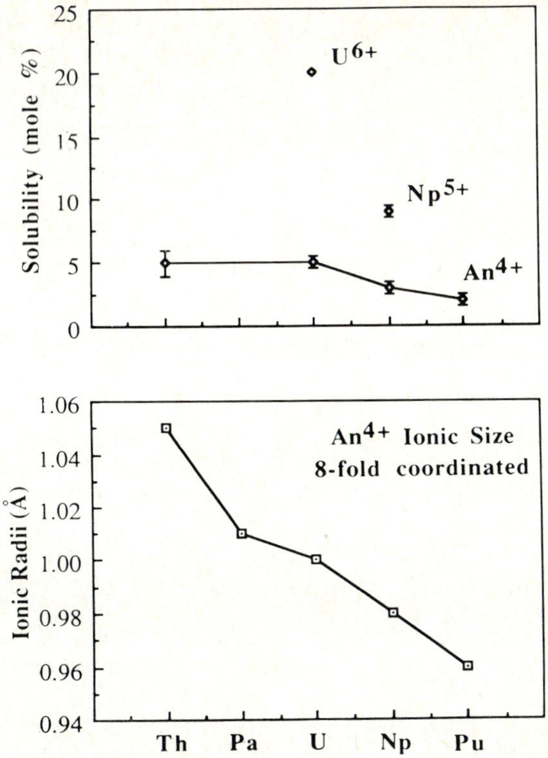

Fig. 5. (a) Solubility data for actinide oxides in sodium trisilicate glass. Np^{5+} and U^{6+} solubilities are substantially higher than the actinide 4+ solubilities. (b) Ionic radii of actinide 4+ ions in eightfold coordination.

instead of much larger than the U^{4+} ion. Thus, it appears that packing or bonding considerations other than simple ion size are needed to explain actinide ion solubilities.

It may be that, for the 6+ glass, uranium solubility limits are largely determined by the alkali content with the more alkali-rich glasses able to dissolve more metal oxide. [Such behavior is also observed in other silicates; e.g., iron solubility in sodium silicates (Ballard and Pye 1976).] While U^{6+} solubility studies have not extensively been reported for solvent glasses of differing alkali content, it is found that the solubility limits (at 1250°C) do increase from about 20 mol% in trisilicate glasses $[0.25\,Na_2O \cdot 0.75\,SiO_2]_{0.8}[UO_3]_{0.2}$ to slightly more than 25% in disilicate glasses $[0.33\,Na_2O \cdot 0.67\,SiO_2]_{0.75}[UO_3]_{0.25}$ (Lam 1980). In each of these cases, the ratio of sodium to uranium concentration Na/U = 2.0. In any case, if the presence of large holes in the glass structure is the key to the high solubilities, then we must infer that alkali ions are needed in the structure to permit formation of these large holes.

3.4. *Thorium (and An^{4+} ions) in silicate glass*

Since thorium ions are found in a single, 4+ valence state, Th glass measurements

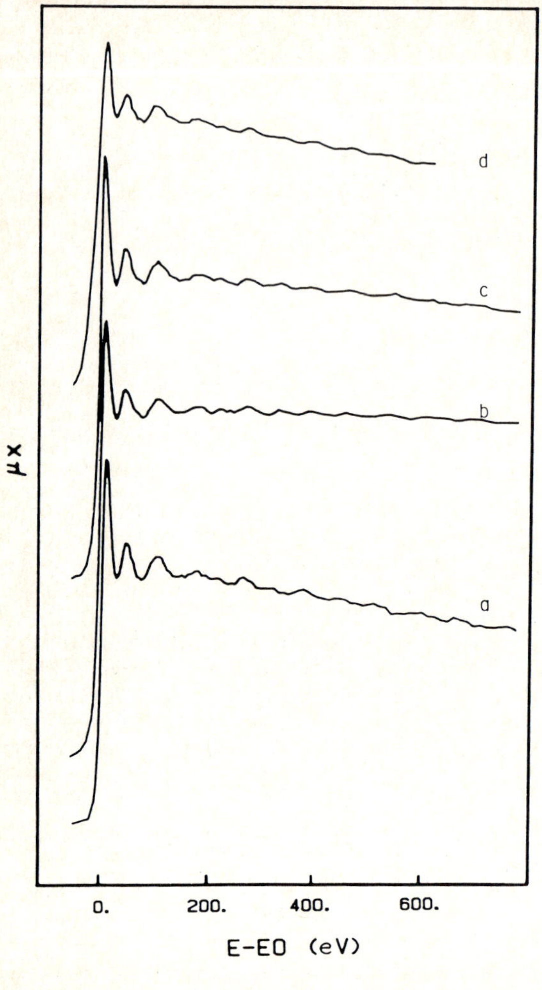

Fig. 6. L_{III} EXAFS spectra for (a) ThO_2, (b) UO_2, (c) NpO_2, and (d) PuO_2 with absorption edges (i.e. EO) aligned. μ is the absorption coefficient and x is the sample thickness.

provide a valuable baseline for a more general study of actinide ions dissolved in silicate glasses where a variety of valence states may be encountered. In this section, emphasis will be placed on EXAFS measurements. EXAFS is a particularly valuable technique for investigating actinide ion near-neighbor environments in glassy silicates. The technique provides the means for measuring interatomic distances between an absorber atom and its near neighbors. Information can also be obtained about the number of atoms within a given neighboring shell. Because some of these (and other results in sections 3.5 through 3.7) have not been previously published, the EXAFS results will be discussed in some detail.

Figure 6 shows actinide L_{III} EXAFS spectra, presented with absorption edges aligned, for dioxides of Th, U, Np, and Pu. Since all of these materials occur in the cubic fluorite structure, the spectra are remarkably similar. (Unit cell parameters systematically contract with increasing actinide atomic number, however.) Figure 7

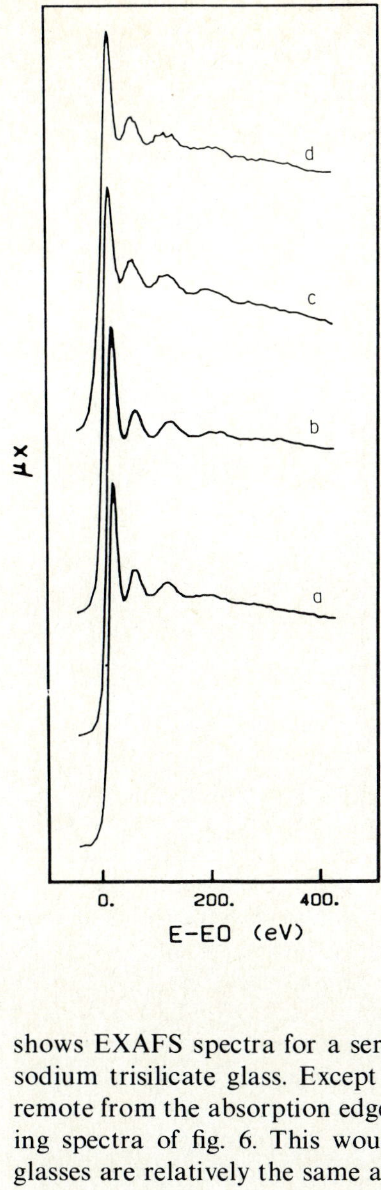

Fig. 7. L_{III} EXAFS spectra for tetravalent (a) Th, (b) U, (c) Np, and (d) Pu ions dissolved in sodium trisilicate glasses. Absorption edges (EO) are aligned. Note the similarity to spectra of fig. 6.

shows EXAFS spectra for a series of actinide absorber atoms that are dissolved in sodium trisilicate glass. Except for attenuation of the spectral features at energies remote from the absorption edge, the spectra in fig. 7 are very similar to corresponding spectra of fig. 6. This would suggest that near-neighbor environments in the glasses are relatively the same as they are in the crystalline dioxides. This point will become particularly apparent when spectra for U^{6+} and Np^{5+}, with distinctly different appearances, are compared (see sections 3.5 and 3.6).

Analyses of the spectra in figs. 6 and 7 yield the actinide-to-near-neighbor oxygen distances ($An–O_I$) shown in fig. 8. In the upper curve, EXAFS results are compared to the distances obtained from X-ray diffraction studies. This comparison is provided as a test of the EXAFS analyses, considered appropriate since EXAFS is a relatively new technique and few studies have been performed using heavy absorber atoms. For these analyses, the oxygen backscattering amplitude and phase functions

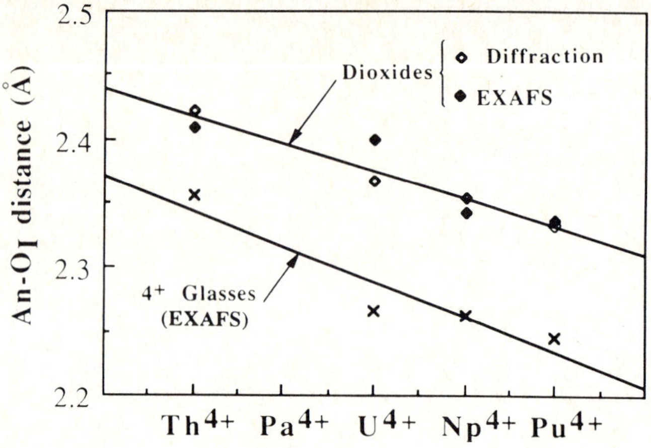

Fig. 8. EXAFS measurements of the separation between actinide ions and near-neighbor oxygens (lower curve) for An^{4+} ions dissolved in sodium silicate glass. Results are compared to An–O separations in the actinide dioxides as determined by diffraction measurements. EXAFS measurements for the dioxides are also shown.

were calculated using a curved wave formalism (McKale et al. 1986). An empirical central atom phase shift function with the form

$$\phi(k) = a + bk + ck^2$$

was used. Here, k is the scattered electron wave vector. The curvature (parameter c) was taken from a fit to the calculation by Teo and Lee (1979) for lead. Parameters a and b were adjusted to obtain a single central atom phase shift function that was used for all the EXAFS analyses (Th–Pu) shown in fig. 8. Using the single set of EXAFS functions, the measured magnitudes of An–O_I and their variation with Z are found to be in satisfactory agreement with the diffraction results. While the measurements of figs. 6 and 7 are qualitatively very similar, analyses of the spectra show that An–O_I distances in the glass are reduced by about 0.07–0.10 Å relative to corresponding distances in the crystalline dioxides. This observation is in agreement with results reported by Eller et al. (1985) for thorium-containing borosilicates. Eller et al. noted that the reduced Th–O interatomic separation might result in a decrease (from 8 to about 6.5) in the coordination number of oxygens around the thorium atom. Results in fig. 8 show that the tetravalent U, Np, and Pu glasses behave similarly.

Generally, for glassy structures, relatively little detailed structural information is available in an EXAFS experiment, since disorder in interatomic distances and bond angles normally restricts information content to the first neighbor shell surrounding the absorber. Note, for example the Th L_{III} spectra in figs. 6 and 7 (curves a). In ThO_2, strong oscillations persist to energies more than a kilovolt from the absorption edge but in the Th glass, structure is confined to energies relatively near the edge. Similarly, in fig. 9, where measurements are presented as $k\chi(k)$, the oscillatory

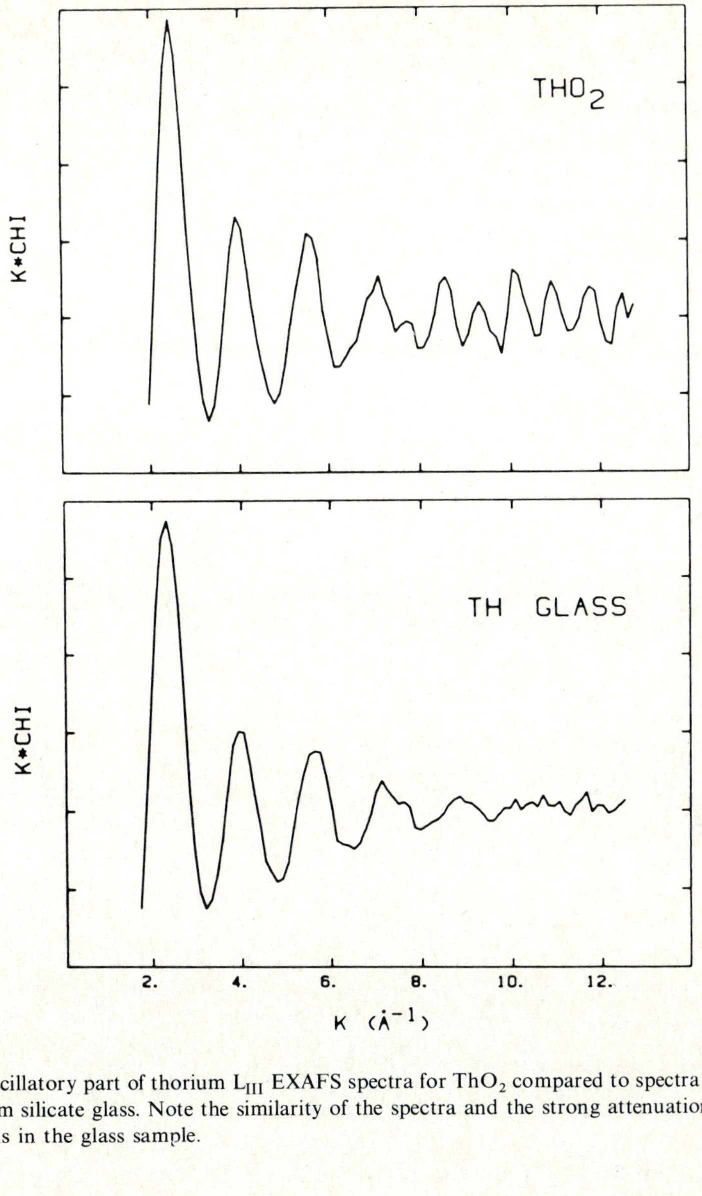

Fig. 9. The oscillatory part of thorium L_{III} EXAFS spectra for ThO_2 compared to spectra for a thorium-bearing sodium silicate glass. Note the similarity of the spectra and the strong attenuation, at high k, of the oscillations in the glass sample.

behavior is dramatically damped at high wave vector (k), for the glass sample (see also Eller et al. 1985). [$\chi(k)$ is the oscillatory part of the μx]. Peaks in the Fourier transform (the 'EXAFS radial distribution function'), shown in fig. 10, give approximate distances between the absorber atom and neighboring atomic shells. The peak at approximately 2 Å corresponds to the first thorium–oxygen shell and the second peak near 4 Å corresponds to the first Th–Th shell. Note that, for both samples, the Th–O peak is strong, indicating a substantial degree of short range

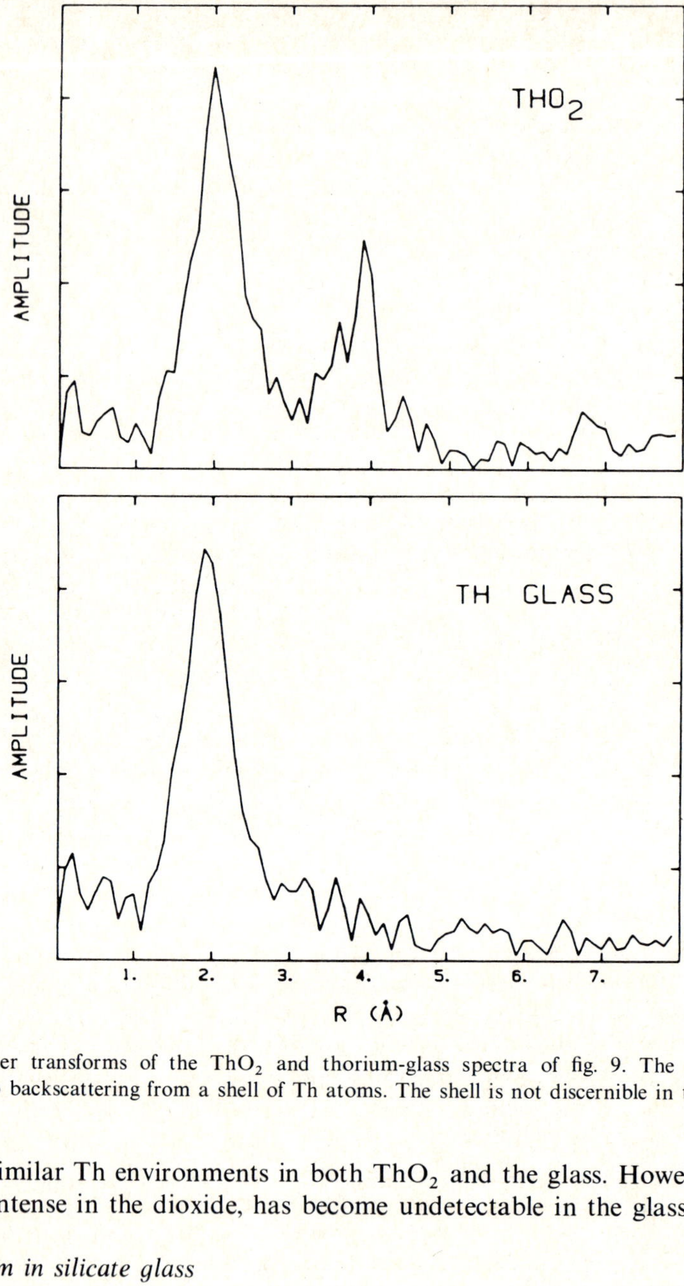

Fig. 10. Fourier transforms of the ThO_2 and thorium-glass spectra of fig. 9. The peak near 4 Å corresponds to backscattering from a shell of Th atoms. The shell is not discernible in the glass.

order and similar Th environments in both ThO_2 and the glass. However, the Th–Th signal, intense in the dioxide, has become undetectable in the glass.

3.5. *Uranium in silicate glass*

The unusually high solubility of hexavalent uranium in alkali silicate glasses makes it a candidate for special study. Optical (Kroger et al. 1948, Karraker 1982), EXAFS (Knapp et al. 1984a), and XPS (Lam et al. 1979) studies have concluded that the uranium ion is contained within a uranyl structure. This conclusion was

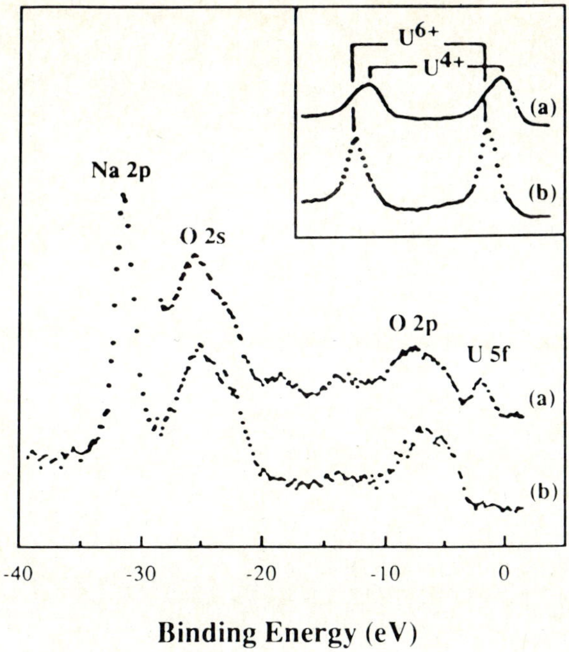

Fig. 11. XPS data near the Fermi level for uranium sodium disilicate glasses nominally containing 20 mol% UO_2. Sample (a) contains U^{4+} ions and sample (b) contains U^{6+} ions. Inset shows the chemically shifted U 4f core levels. A U 5f peak appears in the reduced glass. (After Lam et al. 1979.)

initially reached from optical studies of glasses with very dilute uranium concentrations. Glasses with high uranium concentrations were subsequently studied using the XPS technique. Shallow U $6p_{3/2}$ electron core levels were examined. Prior studies on a large number of uranyl compounds with systematically varying near-neighbor U–O separations showed that the U $6p_{3/2}$ splitting systematically increased as the U–O separation was reduced. As the axial electrical field at the U site increased, the uranyl structure became more clearly defined (Veal et al. 1975). The initial U-glass studies were performed on $Na_2O \cdot 2SiO_2$ containing dissolved 6+ uranium. The valence state could be monitored with XPS at the U 4f level where a relative binding energy shift between the 4+ and 6+ states was observed (see inset fig. 11). Also, a nonbonding U 5f peak, absent for U 6+, could be observed at the 4+ ion (fig. 11). Splitting of the 6p level closely approximated that observed in γ-UO_3 suggesting the formation of linear U–O chains with near-neighbor U–O separation of about 1.8 Å. However, the XPS measurements characterizing the uranyl were relatively imprecise since the photoemitted Si 3s–3p valence electrons interfered with the U 6p signal and difference techniques were employed to isolate the U 6p signal.

The most direct evidence for the detection of the uranyl structure in glasses comes from EXAFS measurements at the uranium L_{III} edge (Knapp et al. 1984a,c). Figure 12 shows EXAFS spectra for a series of sodium trisilicate glasses containing hexavalent uranium. For these glasses, the spectra within about 100 eV of the absorption edge are very similar but are very unlike the spectra of the 4+ ions (figs. 6 and 7). The glass spectra near the absorption edge also closely resemble the

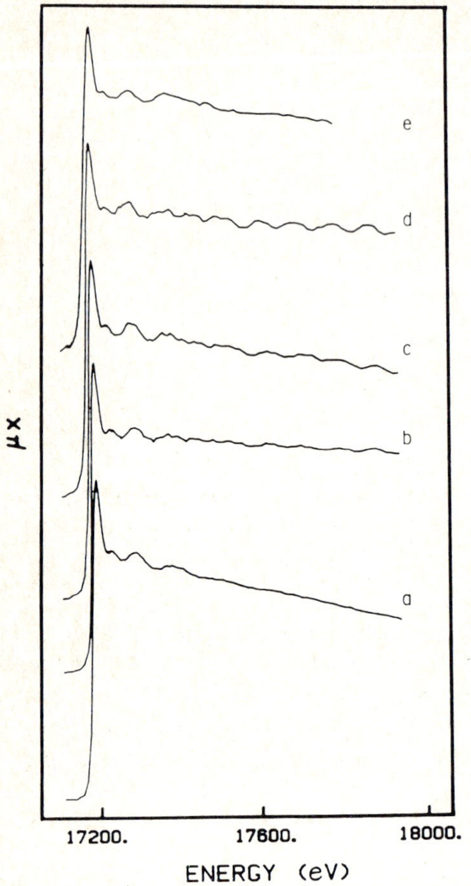

Fig. 12. Uranium L_{III} EXAFS spectra for hexavalent uranium glasses containing: (a) 2 mol%, (b) 5 mol%, (c) 7 mol%, and (d) 10 mol% UO_3 in a sodium trisilicate base. Spectra near the absorption edge are similar in appearance and resemble spectra of the uranyl compound (e) $Li_2U_2O_7$. Note that EXAFS oscillations in the glasses become progressively more intense as the UO_3 content increases.

spectra of $Li_2U_2O_7$ (fig. 12), a crystalline material which contains hexavalent uranium in the uranyl configuration with a U–O_I distance believed to be about 1.83 Å (Veal et al. 1975). Also, the spectra of uranyl nitrate have a similar appearance (Eller et al. 1985). These 'fingerprint' results suggest that a strong similarity exists between the uranium environments in the sodium silicate glass and in the uranyl compounds. The suggestion is confirmed by analysis of the glass spectra to quantitatively determine the structural parameters. Analysis yields a U–O_I distance of about 1.85 Å, and a second-neighbor (U–O_{II}) distance of about 2.25 Å (Knapp et al. 1984a). Also, in contrast to the An^{4+} glasses, the U^{6+} glasses (particularly the more U rich samples) show strong EXAFS oscillations at high k resulting from uranium backscattering. The near uranium environments show no discernable dependence on uranium concentration in the U^{4+} glasses.

For EXAFS measurements, the fact that actinides are high-Z elements is helpful for identifying the atoms which comprise the atomic shells neighboring the actinide absorber atom. The k-dependence of the EXAFS amplitude by a given backscattering

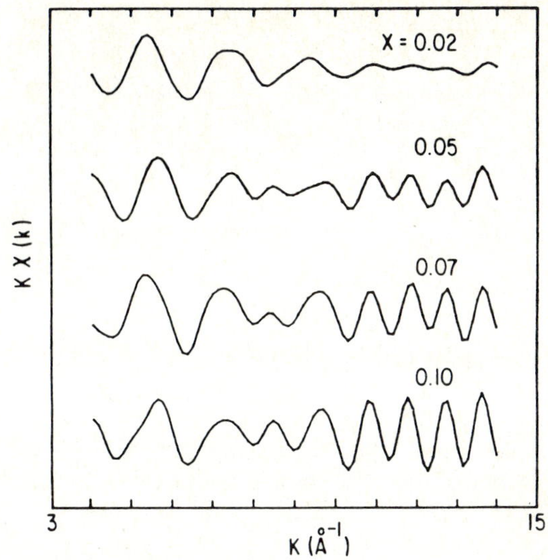

Fig. 13. The oscillatory part of the EXAFS spectra for the hexavalent uranium glasses in fig. 12 filtered to restrict information to radial distances less than 4 Å from the uranium absorber. Note the dramatic increase in intensity of the oscillations for high values of k as the uranium content increases, resulting from electron backscattering from a neighboring uranium shell. The strong uranium backscattering signal, even in the relatively uranium-dilute glasses, suggests that uranium-rich clusters are formed. (After Knapp et al. 1984a.)

shell is largely governed by the backscattering amplitude function appropriate for the atoms in that shell (see Teo and Lee 1979). The backscattering functions are dependent on Z with contrast between two such functions generally becoming increasingly larger as the separation in Z increases. [Backscattering and phase shift functions for the actinides have been calculated by Chan and Knapp (1984).] Because of their large separation in Z, EXAFS signals from An and O neighbor shells can be unambiguously identified by their characteristic k-dependence. For example, the strong EXAFS signal in ThO_2 at high k (fig. 9) is a consequence of Th backscattering. The rapidly damped signal from the Th glass (fig. 9) results almost exclusively from a single oxygen backscattering shell.

Figure 13 shows the U^{6+} L_{III} EXAFS spectra of fig. 12, here presented as $k\chi(k)$, the oscillatory part of the absorption, filtered to restrict information to the near-oxygen ($U–O_I$ and $U–O_{II}$) and the first uranium backscattering shells (Knapp et al. 1984a). Like the crystalline dioxides (e.g., ThO_2 in fig. 9), a remarkably strong EXAFS signal exists for $k > 10\ Å^{-1}$, an unmistakable signature of scattering from a highly ordered shell of uranium atoms surrounding the uranium absorber. The highly ordered shell of uranium atoms is located about 3.3 Å away from the central U absorber. The number of uranium atoms observed in this shell depends on the concentration of uranium in the glass. However, a high U neighbor concentration was observed, even for the low $x = 0.05$ doping in the sodium trisilicate glasses $(0.25\,Na_2O \cdot 0.75\,SiO_2)_{1-x}(UO_3)_x$. The high concentration of U neighbors suggests a strong tendency for the U atoms to form clusters in the glass. (Extreme dilution would suggest that essentially no U neighbors would be in near proximity to the absorber.) For the most U-concentrated glass ($x = 0.1$), the number of uranium neighbors was found to be approximately four. Additional evidence for a high degree of local order in the glass was found from examination of the EXAFS

Debye–Waller-like factor which showed that the thermal and static disorder of atoms in the first few shells about the U absorber observed in the U^{6+} glasses is comparable to the disorder found in the neighboring shells of stoichiometric UO_2.

The results of Knapp et al. (1984a) indicate that U^{6+} glasses can display a remarkable amount of crystallographic order. Detailed structural information was extracted for three shells of neighboring atoms surrounding the U absorber, a highly unusual circumstance for solutes in glassy materials. It is unusual to observe strong backscattering from distant shells in amorphous materials because disorder imposes a strong damping term on the EXAFS signal. Typically, only the nearest-neighbor shell is observed.

Thornley et al. (1986) and Greaves (1986) have recently presented glancing angle U L_{III} EXAFS data for virgin and aqueous-leached borosilicate glasses with weight compositions 53.7% SiO_2, 16.7% B_2O_3, 21.0% Na_2O, 5.6% Al_2O_3 and 3.0 wt% UO_2. Emphasis was directed to concentration and chemical changes in the corroding surface region. Features in the EXAFS radial distribution function of the virgin glass were attributed to a uranyl structure with a U–O_I distance of about 1.8 Å orthogonal to four planar UO_{II} oxygens located about 2.2 Å from the central uranium atom (Greaves 1986). Aqueous leaching led to an enhanced uranium concentration at the glass surface with increased clustering of the uranyl radicals in the near surface region.

Petiau et al. (1984) have also reported U L_{III} EXAFS measurements on borosilicate glasses containing hexavalent uranium (glass compositions are not reported). They also observe that uranium is bound in a uranyl configuration with U–O_I = 1.77 Å. The second oxygen shell was observed to contain 4.7 atoms located 2.18 Å distant from the uranium absorber. The uranium valence state was determined by monitoring the energy of the uranium absorption edge. The edge corresponded closely to the U edge in uranyl nitrate (U^{6+}) but was shifted by several electronvolt from the U edge in UO_2 (U^{4+}).

X-ray data show that, for uranyls, primary (UO_I) and secondary (UO_{II}) bond lengths are correlated with coordination; coordination numbers and U–O_{II} distances tend to decrease as U–O_I distances increase. Results for a series of uranyls, taken from the tabulated data of Veal et al. (1975) are shown in fig. 14. Also, EXAFS results for uranium doped sodium silicate glass (Knapp et al. 1984a) are displayed in the figure. We see that the relationship between first and second neighbor shell distances, as well as coordination numbers, for the U^{6+} glasses are typical of uranyl structures. EXAFS measurements show that hexavalent uranium, when dissolved in a variety of alkali silicate and complex borosilicate glasses, has a tendency to form well-defined (and often very similar) uranyl structures.

While it appears that the occurrence of a uranyl structure has been reliably established in several hexavalent uranium glasses, precise characterization of the structure is difficult because of the large number of parameters needed to fit the EXAFS data. Reported distances vary between 1.75–1.85 Å for U–O_I and 2.15–2.25 Å for U–O_{II}. The measurements were taken on glass samples with compositions that vary significantly. In contrast, EXAFS measurements reported by Eller et al. (1985) on uranium-bearing borosilicates with base compositions (wt%) 61.2 SiO_2,

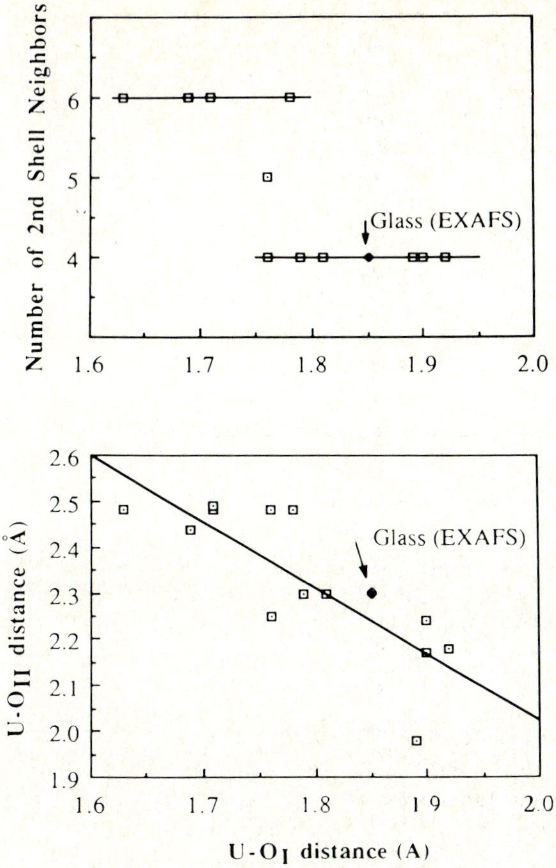

Fig. 14. Properties of the uranyl structure (number of second-shell oxygen atoms neighboring uranium and the uranium-to-second-shell distance) plotted vs the near U–O separation for a series of uranyl compounds. EXAFS measurements of these properties are also included for hexavalent uranium in sodium trisilicate glass.

13.4 Na_2O, 9 Li_2O, 9.5 B_2O_3, 5.3 CaO, 0.4 TiO_2, 0.3 $Al_2O_3 \cdot 0.1 K_2O$ were reported to contain hexavalent uranium that did not appear in a uranyl bonding configuration. Figure 15 shows L_{III} X-ray absorption near-edge structure for uranyl nitrate, for UO_2 and for a 4 wt% uranium glass. The edge shift (and optical absorption data) establish that uranium is hexavalent. However, the similarity (discussed above), frequently observed between spectra of U^{6+} glasses and spectra of uranyl compounds, is not observed in fig. 15. EXAFS measurements indicate that a single oxygen shell occurs 2.09 Å distant from the uranium absorber. To more precisely characterize the uranium environment in the U^{6+} glasses and to establish dependence on glass composition, additional studies, including more extensive testing of analytical procedures with well characterized uranyl standards, are needed.

As suggested by the solubility study (see section 3.2), the chemistry of U^{4+} ions in alkali silicates is very different from the U^{6+} chemistry. The near U environment in the 4+ glass appears to be similar to that in UO_2 (Antonini et al. 1984, Knapp et al. 1984a,b). Like UO_2, in a CaF_2-type crystal structure, the first neighbor shell of

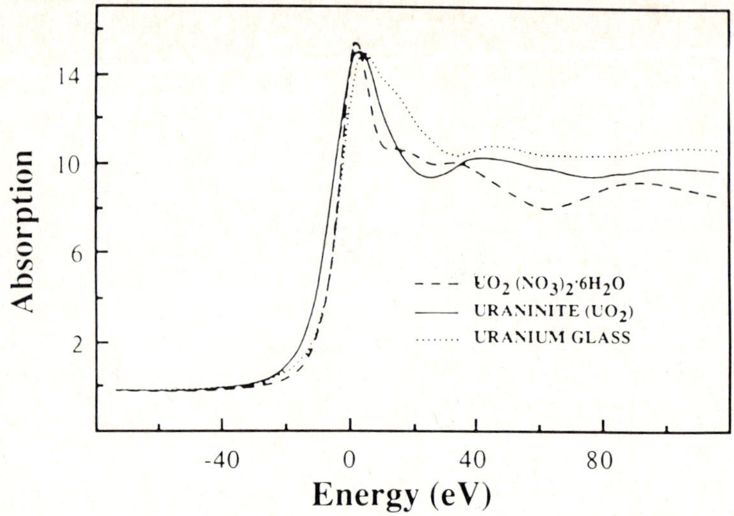

Fig. 15. Uranium L_{III} X-ray absorption near-edge measurements for a complex uranium borosilicate glass compared to measurements of UO_2 (U^{4+}) and uranyl nitrate (U^{6+}). The uranium in the glass is hexavalent but the spectra are not similar in appearance to the uranyl spectra. Note the edge shift that occurs with change of uranium valence. (After Eller et al. 1985.)

oxygen atoms is apparently eightfold coordinated to the uranium absorber (Knapp et al. 1984b) but the U–O separation is somewhat reduced from the 2.37 Å separation observed in UO_2. Like the thorium glass in figs. 7 and 9, Knapp et al. could reliably observe only one neighboring shell in the An 4+ glasses with EXAFS. Since no uranium backscattering signal was observed, EXAFS data gave no indication that the 4+ uranium atoms cluster in the glass. However, Antonini et al. (1984) show EXAFS data for aluminum-rich sodium silicates containing uranium believed to be predominantly tetravalent. A strong uranium backscattering signal was observed. Since the U was relatively dilute (2.5 and 5 wt%), either substantial clustering (with strong local order) must occur in these glasses or the UO_2 starting material was inadequately dissolved in the glass.

3.6. *Neptunium in silicate glass*

Early studies of neptunium doped silicate glasses, largely based on XPS measurements, reported that the Np ions were tetravalent, whether the glasses were prepared in reducing or oxidizing environments (Lam et al. 1979, Karraker 1982, Lam et al. 1983, Karim et al. 1982). Subsequent work, utilizing XPS, Mössbauer, and optical-absorption measurements (Veal et al. 1986) demonstrated that, like uranium glasses, the valence state and solubility of Np ions in alkali silicates is very dependent on preparation conditions of the glass. Two valence states, Np^{4+} and Np^{5+}, associated with glasses prepared under reducing and oxidizing conditions, respectively, were identified. Both optical absorption and Mössbauer spectroscopy showed strong

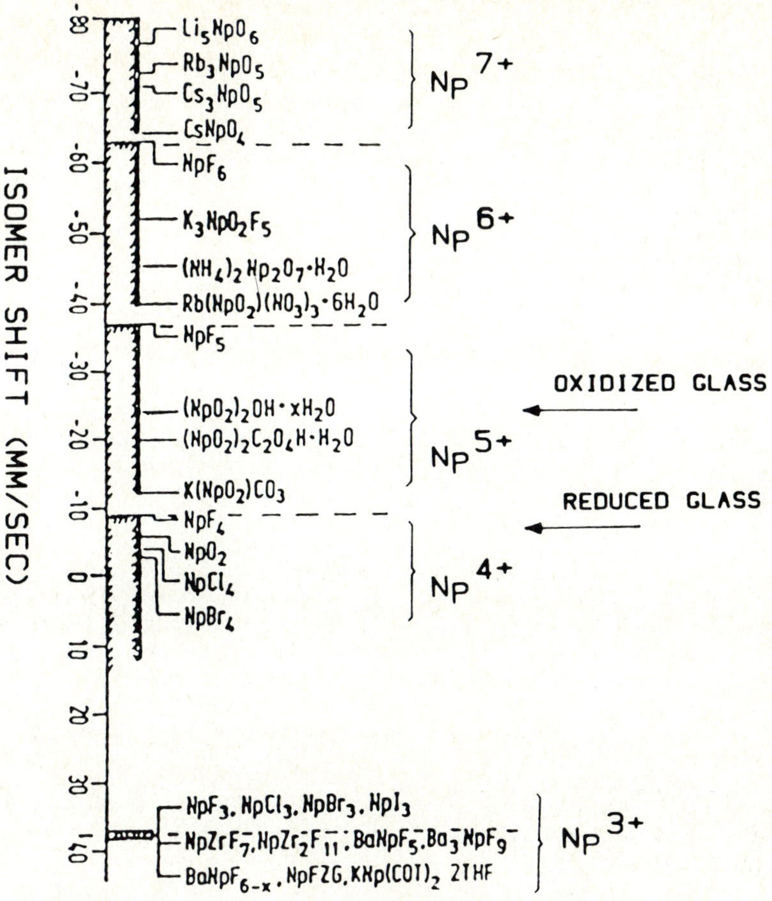

Fig. 16. ^{237}Np Mössbauer isomer shift measurements of reduced and oxidized Np-bearing sodium trisilicate glasses compared to isomer shifts in representative Np compounds. Measurements are shown relative to $NpAl_2$ (0 mm/s). These measurements indicate that the reduced glass contains tetravalent Np and the oxidized glass contains pentavalent Np. (After Veal et al. 1986.)

sensitivity to variation in the valence state. XPS consistently showed only the Np^{4+} state, apparently a consequence of chemical reduction in the near-surface region of the sample that took place during the course of measurement.

Figure 16 shows Mössbauer isomer shift data (Dunlap and Kalvius 1985, Veal et al. 1986) for a series of Np compounds containing 4+, 5+, 6+, and 7+ Np ions. The compounds with a given Np valence state are nicely grouped within a specific range of isomer shift values. Isomer shift data for Np glasses prepared under oxidizing and reducing conditions are also shown in fig. 16. These data provide an unambiguous identification of the Np valence states in the glasses.

Optical absorption spectra (Veal et al. 1986) for the oxidized glasses also indicate the presence of Np^{5+} ions. Optical absorption spectra, showing the contrast

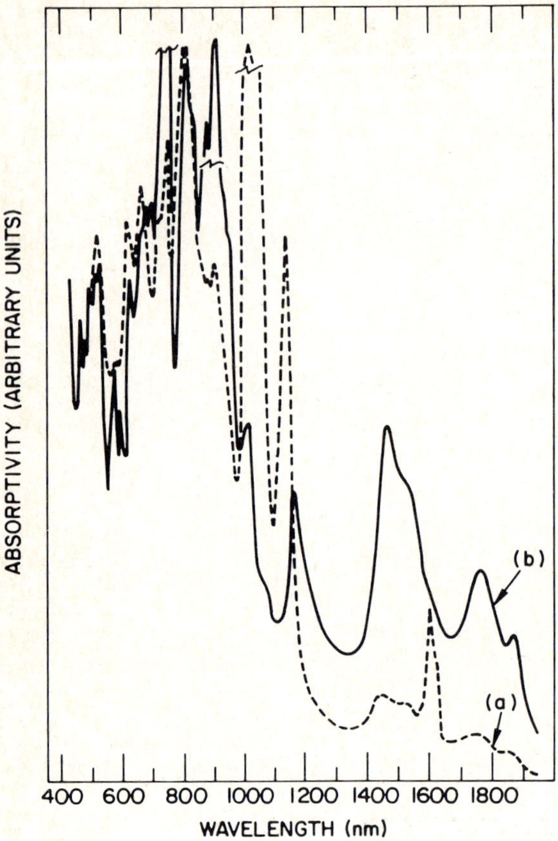

Fig. 17. Optical absorption spectra for the oxidized (a) and reduced (b) glass of fig. 16. Consistent with the Mössbauer results, features in spectrum (a) near 1000, 1100, and 1600 nm are characteristic of pentavalent Np. In spectrum (b) the feature near 1200 nm and the broad absorption band between 1400–2000 nm are indicative of tetravalent Np. (After Veal et al. 1986.)

between Np^{4+} and Np^{5+} glasses, are shown in fig. 17. Contrast between these spectra is most prominent for wavelengths greater than 100 nm. The intense features in the oxidized glass [spectrum (a)] at 1000, 1100, and 1600 nm are characteristic of pentavalent Np in the NpO_2^+ structure and the broad absorption bands between 1400–2000 nm in the reduced glass [spectrum (b)] strongly resemble spectra for the tetravalent Np compound $Np(OH)_2SO_4$. Correlation with spectrum (b) also exists for Np^{4+} ions in aqueous solution.

We note that valence state information for Np ions can probably be reliably obtained from X-ray absorption near-edge measurements as was done for uranium ions by Petiau et al. (1984) and by Eller et al. (1985). Such measurements will, of course, require careful examination and intercomparison of X-ray edges of well characterized neptunium compounds that include the different valence states of interest.

In neptunium compounds, Np^{5+} and Np^{6+} ions frequently occur in the actinyl structure (Dunlap and Kalvius 1985). A number of Np^{7+} compounds are also known. Some evidence suggests that the neptunyl structure is also found in the 7+ compounds (see Dunlap and Kalvius 1985, and references therein). Thus, like the

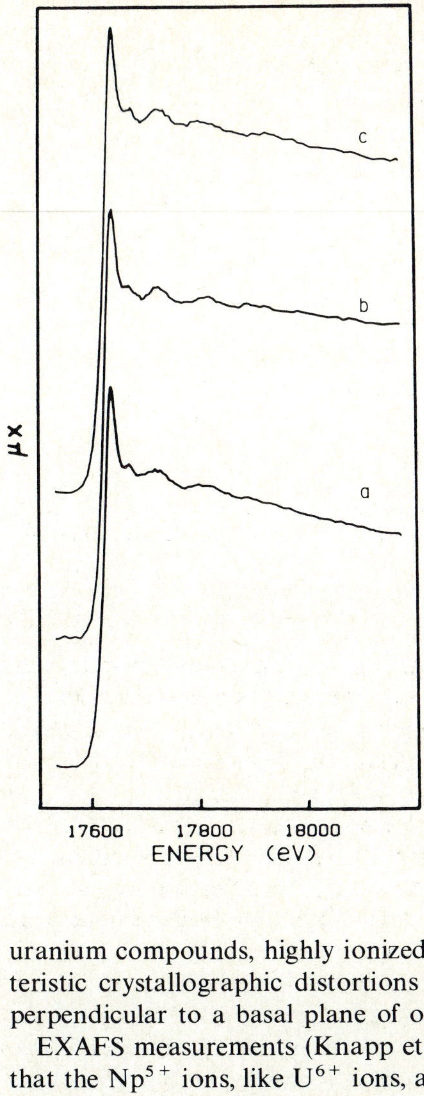

Fig. 18. Neptunium L_{III} EXAFS spectra for sodium trisilicate glasses containing: (a) 5 mol%, (b) 7 mol%, and (c) 10 mol% NpO_2. Near-edge spectra are very similar to the uranyl spectra of fig. 12 suggesting that neptunium appears in an actinyl configuration. Mössbauer data indicate that the glasses contain pentavalent Np.

uranium compounds, highly ionized neptunium compounds often show the characteristic crystallographic distortions which produce a linear NpO_2 chain oriented perpendicular to a basal plane of oxygen (or other ligand) atoms.

EXAFS measurements (Knapp et al. 1984b,c) on oxidized (5+) Np glass confirm that the Np^{5+} ions, like U^{6+} ions, are dissolved in the glass in the actinyl structure. Spectra for Np^{5+} glasses are shown in fig. 18. These spectra (near the absorption edge) show a remarkably strong similarity to those of the uranyl spectra from the hexavalent uranium glasses (fig. 12) and a dissimilarity to the An^{4+} spectra of figs. 6 and 7. [This simple 'fingerprint' test using EXAFS data may provide a simple and convenient method of monitoring valence state (and presence of the actinyl configuration) in hexavalent uranium and pentavalent neptunium glasses.] Analysis of the Np EXAFS data yields Np–O distances of 1.86 and 2.33 Å (using the procedure and parameters of section 3.4), quite comparable to the U–O distances observed in the uranyl glasses. This is not unexpected since the ionic sizes of Np^{5+} and U^{6+} ions, in the same coordination environments, are very comparable; e.g., 0.75 and 0.73 Å, respectively, in VI coordination (Shannon 1976).

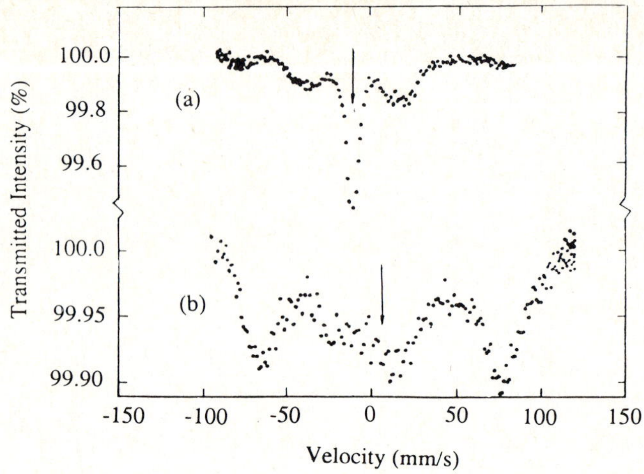

Fig. 19. Mössbauer spectra for the oxidized (Np^{5+}) (a) and reduced (Np^{4+}) (b) glasses of figs. 16 and 17. Np spin–spin interactions in the 5+ (actinyl) glass may be indicative of Np clustering, a phenomenon also observed in the uranyl glasses. (After Veal et al. 1986.)

If Np^{6+} glass could be fabricated, one might expect to find the neptunyl structure with a somewhat reduced An–O separation. One might also speculate that the Np solubility would then be increased, perhaps comparable to that of the U^{6+} glasses, or even greater. However, preparation of alkali silicate glasses containing Np^{6+} will require greater effort since the 6+ specie is apparently quite unstable.

The Mössbauer isomer shift for the Np^{4+} ion (dissolved in alkali silicate glass) is nearly the same as that of NpO_2 (fig. 16). Consistently, EXAFS measurements show that Np^{4+} ions are coordinated to a shell of 8 oxygen neighbors (Knapp et al. 1984b), although some reduction of An–O separation is found in the glass relative to the dioxide (see fig. 8). Figure 19 compares Mössbauer spectra for the Np^{4+} and Np^{5+} glasses. Both samples show line broadenings due to a distribution of Np ion environments in the glasses. The 5+ sample shows predominately electric quadrupole hyperfine splitting while the 4+ sample shows magnetic hyperfine splitting. It is suggested (Veal et al. 1986) that spin–spin interactions between Np ions in the 5+ glass induce sufficiently large spin relaxation rates to average out the magnetic hyperfine field and leave only a quadrupole splitting. The strong Np–Np interaction, even for the dilute glasses, would suggest that the Np ions are spatially correlated, or clustered, in the 5+ glass, a result similar to the U^{6+} glasses. Furthermore, EXAFS indicates that both the Np^{5+} and U^{6+} glasses exhibit the actinyl structure. We note that crystalline 5+ neptunyls show magnetic hyperfine structure that is absent in the glasses (Dunlap and Kalvius 1985). This may be a consequence of reduced Np–Np separation in the glasses relative to the compounds. Consistently, for the uranyl glass analogs, a reduced An–An separation (3.3 Å) was reported (Knapp et al. 1984a) relative to the U–U separation in typical uranyl compounds (see e.g., Wyckoff 1963). Thus the EXAFS and Mössbauer spectra suggest that the

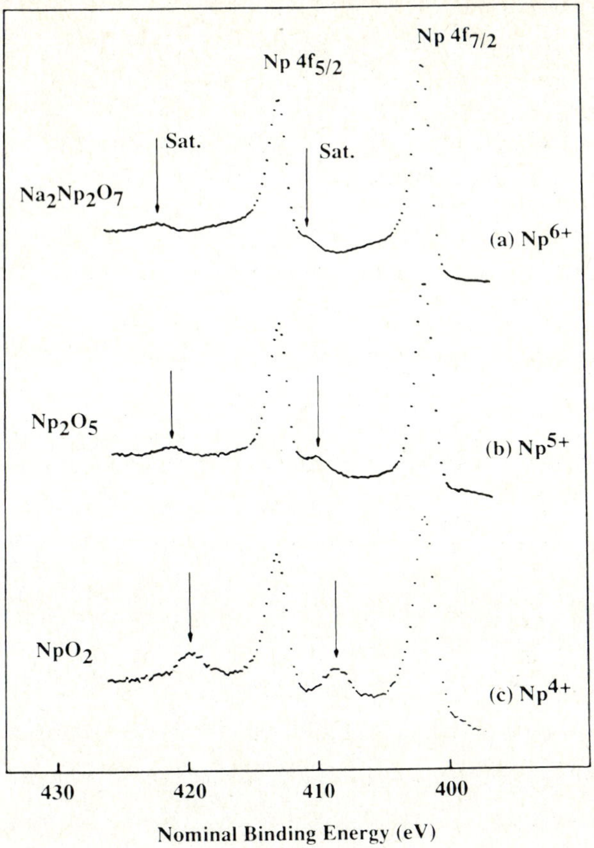

Fig. 20. XPS spectra from Np 4f core levels in: (a) $Na_2Np_2O_7$ (Np^{6+}), (b) Np_2O_5 (Np^{5+}), and (c) NpO_2 (Np^{4+}). Satellite separations from the main lines increase with increasing Np valence state.

6+ uranyls and 5+ neptunyls in the sodium silicate glasses show rather similar chemical behavior and coordination environments. In both cases, actinide clustering tendencies are suggested. In contrast, the Mössbauer data, like the EXAFS, do not suggest a strong Np–Np correlation in the 4+ glasses.

Karraker (1982) obtained a ^{237}Np Mössbauer spectrum for a Np-doped air melted multicomponent borosilicate glass which shows a single, though relatively broad, Mössbauer resonance with an isomer shift which he attributes to a Np valence state of +4. However, the work of Veal et al. (1986), based on studies of simple silicates, would suggest that the Np ions in air melted Np-doped glasses might appear in the 5+ state. These apparently differing results may, however, underscore the importance of chemical composition and preparation conditions as governing factors in the determination of the bonding of Np ions in silicate-based glasses. Eller et al. (1985) also reported the observation of Np^{4+} ions in a complex air-melted borosilicate glass. Their valence assignment was based on optical absorption measurements which, however, could not be unequivocally interpreted by comparison with spectra for Np ions in aqueous solution. We note that dominant spectral features (especially the intense unidentified absorption at 1021 nm) closely correspond to the oxidized (Np^{5+}) spectrum of fig. 17. Clearly, more systematic studies of Np-bearing silicate and borosilicate glasses with varied chemical compositions and preparation conditions are needed.

Chemical reduction of the high-valence Np ions in the near-surface region of the glass samples during the course of XPS measurement is indicative of the relative instability of these ionic species. Such instability is also observed, in varying degrees, in 6+ and 5+ Np oxides (Banks et al. 1980). The Np valence states in the oxides; and their in situ transformation, can be determined by monitoring satellite structure at the Np 4f XPS lines (see Veal and Paulikas 1985). Figure 20 shows typical results for Np 6+, 5+ and 4+ ions, here represented by the compounds $Na_2Np_2O_7$, Np_2O_5, and NpO_2. These satellite features are also useful for identifying valence states in the actinide glasses, particularly in those cases when binding energy shifts are insensitive to changes in valence state. However, lengthy data acquisition times, often needed for dilute samples, pose the risk that high valence ions will be chemically reduced during measurement.

3.7. *Plutonium in silicate glass*

XPS binding energy measurements of the Pu 4f core levels and their satellite structures provide a sensitive measure of Pu valence. Figure 21 shows XPS data from a complex borosilicate glass (Lam et al. 1983) containing one mole percent each of UO_3, NpO_2, and PuO_2. Two valence states for Pu (+4 and +3) and for U (+6 and +4) are discernable.

Like the Np glasses, however, early XPS measurements on Pu doped silicates

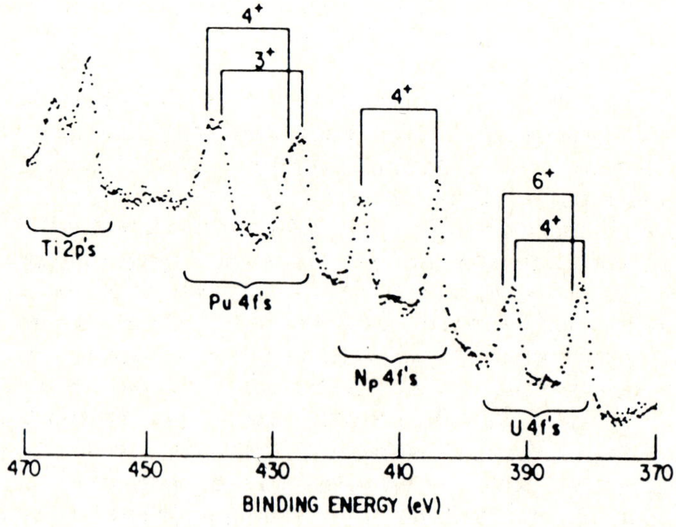

Fig. 21. XPS 4f core level spectra from a borosilicate glass containing dissolved UO_3, NpO_2, and PuO_2. Multiple valence states appear for U and Pu. Intensities of the low valence states are enhanced by in situ reduction during the XPS measurements. (After Lam et al. 1983.)

provided misleading results pertaining to the valence state mix in the bulk glass. Pu^{4+} and Pu^{3+} ions were reported (Karim et al. 1982) in air melted sodium silicate glasses. In later studies, the time evolution of Pu XPS core peaks was monitored to investigate sample reduction tendencies that occurred during the course of measurement (Lam et al. 1983). This work showed that Pu ions occur in the 4+ valence state apparently showing little dependence on the atmosphere in which the glass was melted. This work also showed that the reduction rate of the Pu ions could be affected by inclusion of small quantities of oxide additives such as ZnO, CaO, or B_2O_3. The curious reduction effect emphasizes the complex nature of the glasses including possible cooperative involvement of the multiple components. Similarly complex phenomena might influence leaching behavior in the complex, multicomponent glasses of interest for radioactive waste storage.

Studies of complex borosilicate glasses containing dissolved plutonium were also reported by Karraker (1982) and by Eller et al. (1985). In both studies (using optical absorption spectroscopy), only Pu^{4+} valence states were identified.

EXAFS measurements (Knapp et al. 1984b) for the 4+ plutonium doped silicates showed results that were consistent with the systematic behavior observed for the tetravalent U and Np doped glasses. A slight reduction in the An–O near-neighbor separation was observed in the glass relative to PuO_2 but the Pu-near environments in the glass and in the dioxide are expected to be similar.

Magnetic susceptibility measurements were reported for Pu doped sodium disilicate glasses (Aldred 1980). Unlike crystalline PuO_2, which has temperature independent susceptibility, the glasses showed Curie–Weiss behavior. Because of its crystal electric field, PuO_2, with a $5f^4$ ground state configuration (Raphael and Lallement 1968), has no ground state magnetic moment. For the 4+ Pu glasses, however, the susceptibility results suggest that, like the 4+ Np glasses (Veal et al. 1986), a distribution of local An environments must occur. While the coordination number of oxygen atoms about Pu appears to be comparable to that in PuO_2 (Knapp et al. 1984b) and the 5f electron configurations equivalent, local distortions from cubic symmetry result in a net magnetic moment observed at the Pu site in the glasses.

3.8. *Structural models*

Available structural modeling studies for alkali silicates containing dissolved actinides are largely based on a crystal chemistry approach. As discussed in section 2, local structures and short range ordering considerations are largely inferred (consistent with spectroscopic and bulk properties measurements) from crystal structures of silicates with chemical compositions approximating the glasses of interest. Structural features in the crystalline forms should be preserved, in varying degrees, in the glasses. For example, the SiO_4 tetrahedron is a structural element common to all of the silicate glasses. Of comparable importance is the possibility that an extended range of chains or sheets, readily observable in the crystalline silicates, may also appear in the glasses and play an important role in determining physical properties.

For modeling studies, it is instructive to consider the disilicate, particularly in its crystalline forms. This silicate consists of easily recognizable sheets that are assembled from SiO_4 tetrahedra. Three covalent bridging oxygen bonds join the tetrahedra with bond strengths probably comparable to those found in pure SiO_2. A single nonbridging oxygen on each tetrahedron provides a termination layer of alkali atoms on the sheet surface. Thus, the sheets, which contain strong internal bonds, are weakly joined through ionically bonded, monovalent alkali atoms. These sheet structures are found in the α- and β-phases of sodium disilicate and in the known phases of other alkali disilicates. At higher alkali concentrations, notably for the metasilicate composition ($Na_2O \cdot SiO_2$), a characteristic chain structure is observed resulting from the formation of two nonbridging oxygens on each SiO_4 tetrahedron. These sheet and chain structures are pervasive in silicate mineralogy.

It is common practice to divide silicate minerals into classes: sheet, chain and network structures (Zoltai 1960). For sheet structures, silica layers are preserved with additive metal ions typically appearing in the intersheet regions interspersed with alkali ions. It is noteworthy that, even for network silicates, the silica sheet structure may be quite evident. For example, to convert the compound $Li_2O \cdot 2\,SiO_2$ (Liebau 1961a), which displays weak (ionic) intersheet bonding, into the strongly bonded mineral petalite (Liebau 1961b), which has the composition $Li_2O \cdot Al_2O_3 \cdot 8\,SiO_2$, merely requires the replacement of three Li atoms with one Al, a process that is accomplished entirely within the intersheet region without significantly perturbing the sheet structure of the parent disilicate.

Another instructive example is afforded by the compound gillespite with chemical composition $BaFeSi_4O_{10}$. Again, an 'accordian-pleated' sheet structure of SiO_4 tetrahedra is separated by a layer of metal atoms. Fe^{2+} ions (0.76 Å radius) are pressed into planar tetrahedral sites, near-neighbors to the 'nonbridging' oxygen atoms while the much larger Ba^{2+} ions (1.35 Å radius) reside in twelvefold coordinated sites. Thus, even for these divalent metal ions with wide variability in ionic sizes, the silica sheet structure is preserved. Of course, properties of different silicate minerals vary dramatically depending upon the extent of cross-linkage bonding (between silica sheets or chains) involving polyvalent ions interspersed with alkalis or alkaline earths.

If residual silica sheet structures are preserved in alkali silicate glasses, one might similarly expect to find wide variability in the physical properties of the glass depending on the selection of additive metal oxides. Added metal atoms (of valence two or higher) that form strong cross links between these sheets will be expected to significantly affect the glass rigidity. Such hardening (analogous to the cross linking of chains in polymers) is commonly observed in glasses. One manifestation of the cross linking is increased viscosity, often apparent when polyvalent metals are added to simple silicate glasses (Morey 1954).

The concept of residual, highly distorted, silica sheets persisting in the sodium disilicate base glass provides a useful framework for modeling the behavior of An 4+, 5+, and 6+ dissolution in the glasses. During cooldown, SiO_4 tetrahedra link together to form a sheet-like structure with one nonbridging oxygen per tetrahedron. The sodium ions, being completely ionized (Ching et al. 1983, 1985), are situated between the sheets weakly bonding the sheets together. Arguably, tetravalent actinide ions are also located in the 'intersheet' regions but they bond with greater strength than alkalis to oxygens of the tetrahedra, thus crosslinking the distorted silica layers. This glassy structure will be frozen in as the sample is cooled. This 'strong crosslinking' model is consistent with the (qualitative) observation that, with addition of actinide 4+ ions, the glass viscosity increases (Lam et al. 1983).

It is noteworthy that, for all An^{4+} glasses, An–O separations were significantly reduced relative to the An–O separation in the corresponding dioxide (see fig. 8). If the actinide dioxides serve as prototype materials with 'normal' An–O interatomic separations for An 4+ ions, then the reduced interatomic separations in the glasses might suggest that substantial local strain is introduced in the vicinity of the An ion. This may affect the low solubility rates for the 4+ ions. Also the tendency for the large actinide 4+ ions to have a high coordination number (and the apparent satisfaction of this condition in the glasses) may contribute to the local strain.

A rather different picture is needed to explain the bonding of U^{6+} and Np^{5+} in alkali silicate glasses. As discussed above, these ions can be incorporated in the glasses with high solubility where they appear in the actinyl configuration. Thus, the bonding characteristics of the actinyl unit become the dominant consideration. Again, guided by crystal chemistry considerations, we note that alkali uranates containing hexavalent uranium in the uranyl structure form easily and in consider-

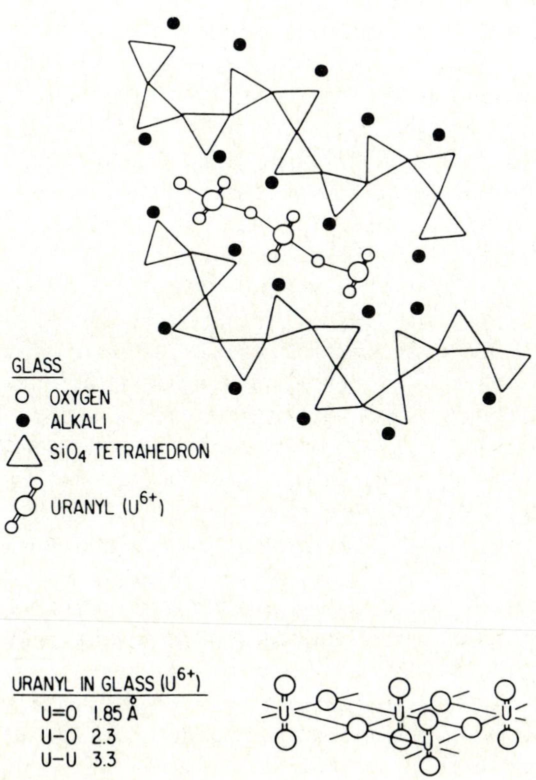

Fig. 22. Top: Proposed structural model of U^{6+} ions in alkali silicate glass. Layers of uranyl are interspersed between layers of silica. Bottom: A detailed view of the uranyl sheet. (After Knapp et al. 1984a.)

able variety (Veal et al. 1975). They often display sheet structures (sometimes distorted) with sheets separated by alkali or alkaline-earth atoms. These same features, sheets separated by alkalis, are characteristic of crystalline alkali disilicates. It thus appears likely that, during cooldown of U^{6+} glasses, uranyl structures form and bond into the interstice regions between distorted silica sheets. That is, as shown in fig. 22, a thin uranate layer, formed between silica sheets, is weakly bonded by intervening layers of alkali atoms. This dissolution mode permits a large U^{6+} solubility. Similar mechanisms probably control dissolution rates of Np and Pu actinyls also.

Since alkalis play the role of delineating silica and uranate sheets, structural rigidity should be weak. The relatively low viscosity of U^{6+} silicate glasses (a qualitative observation) supports this picture. Uranates are not stable at 1250°C where most of the glasses were prepared, so macroscopic phase separation of uranates is not expected. Consistently, X-ray diffraction data did not show the presence of crystalline second phase regions (Lam et al. 1983).

From solubility considerations, it appears that Np^{5+} ions display bonding characteristics intermediate between those of U^{4+} and U^{6+} ions. Probably this occurs because Np^{5+} ions also tend to form an actinyl structure. However, bonding properties are somewhat different from the UO_2^{2+} ion since the characteristic unit is now the singly-charged NpO_2^+ ion which is somewhat larger than UO_2^{2+} and has a different charge state.

4. Conclusions

Stimulated by its potential importance to the technology of long term immobilization of radioactive wastes, the study of the physics and chemistry of silicate-based glasses has recently commanded international attention. Necessarily, these studies build upon the extensive knowledge base accumulated for the simpler silicate glasses and minerals. Consequently, efforts draw heavily upon the pioneering work of Zachariasen and co-workers who established that SiO_4 tetrahedral structural units comprise fundamental building blocks that preserve their short range order. The random network model, and modifications of it, provide the essential base for modeling structural properties of silicates with dissolved actinide oxides.

Although short range order is preserved, it is harder to establish the extent to which an intermediate range of crystallographic ordering occurs in the glasses. This issue is the focus of much work in the present day study of covalent glasses. Interestingly, hexavalent uranium in alkali silicate glasses provides evidence for a remarkable degree of intermediate range order. It is suggested that highly distorted silica layers occur in the glass separated by a layer of alkali atoms that contain a well ordered uranyl layer. It has been noted (Greaves et al. 1981) that the lifting, in a glass, of the restrictions of long range order sometimes results in improved short range order. In the case of sodium trisilicate glasses containing U^{6+} (and possibly Np^{5+}) ions, discernible short range order extends to the third nearest neighbor of the actinide absorber.

Like many other metal oxides, actinide oxides can be dissolved into alkali silicate glasses, sometimes in considerable quantity. For sodium silicate glasses, large variability in solubilities has been observed as the actinide ion valence state or the sodium concentration is changed. Studies have been reported for sodium trisilicate glasses containing 4+ ions of Th, U, Np, and Pu. For these ions, solubilities are relatively low, varying systematically from about 5 to 2 mol%, and are relatively independent of the alkali content of the glass matrix.

In contrast, the solubilities of Np^{5+} and U^{6+} ions are much higher and probably show a greater dependence on alkali concentration. In sodium trisilicate, solubility limits are about 9 and 25 mol%, respectively. Both species occur in the actinyl structure.

A knowledge of leaching characteristics of the glasses under varied aqueous environments is needed for repository site selection and for assessment of projected nuclide migration at a given site. The solubility studies would suggest that Th glasses might show properties similar to the heavier An^{4+} glasses and thus could serve to simulate the active glasses containing 4+ ions. It is important to recognize, however, that Th-containing simulation glasses must not be considered to be representative of An^{5+} or An^{6+} glasses where chemistries are dramatically different.

Because of the great complexity and variety of silicate and borosilicate glasses, the efforts reviewed here must be considered as introductory to a problem of much greater potential scope. Even limiting the focus, as was largely done here, to alkali silicates, many additional questions arise.

How does actinide solubility, for various valence states, depend on alkali concentration? Does the alkali ion size play a role in actinide solubility? Solubility apparently depends on the size of the solute ion. Is some degree of ordering of the alkali ion involved in dissolution? The high solubilities of the actinyls might suggest the affirmative. (Also, as discussed in section 2, this may be more generally true for alkali silicates.)

Additional actinide ion valence states might be formed in the bulk glass. U^{4+} and U^{6+}, Np^{4+} and Np^{5+}, Pu^{4+} as well as Th^{4+} ions have been observed in the bulk glasses. Pu^{3+} has been observed in the surface region of the glass as a consequence of chemical reduction at the sample surface when placed in high vacuum in an X-ray flux. Np^{4+} has similarly been observed as a reduction product of a Np^{5+} glass. With suitable sample preparation procedures, the range of dissolved species contained in the bulk glasses might be increased. Possibilities include Np^{6+} (and perhaps 7+), Pu^{3+} and Pu^{6+}. The additional species should exhibit their own unique chemistries and bonding characteristics. They may, furthermore, be relevant to the problem of long-term aqueous leach resistance, existing in the surface regions of storage glasses as transient species under certain environmental conditions. If the higher valent Np and Pu glasses can be fabricated, it seems likely that the actinyl structure would again be observed and high solubilities in the alkali silicate glasses would be expected. Probably the character of the actinyl structure could be varied substantially by varying the glass composition and preparation conditions. (A large range of uranyls are known with widely varying U–O separations.)

A very limited amount of data are available establishing solubility limits for different solvent glasses. Solubilities may, for example, vary considerably if the alkali specie were changed. Large atoms typically have high oxygen coordination numbers, a circumstance that may make the packing of An^{4+} ions into silicates difficult. Is it possible that An^{4+} solubility would be higher in heavy alkali glasses (e.g. Cs) where the alkali ion size is large and probably also has a high coordination number?

The scope of the problem is dramatically increased if one considers the properties of silicates with added glass formers. Important to waste-form technology, of course, is the study of borosilicates with added actinide and other metal species, including a wide range of daughter products. This review represents a beginning in the study of silicate-based glasses designed for active waste isolation.

References

Aldred, A.T., 1980, J. Non-Cryst. Solids **40**, 347.

Antonini, M., A. Merlini and F.R. Thornley, 1984, in: EXAFS and Near Edge Structure III, Springer Proceedings in Physics, Vol. 2, eds K.O. Hodgson, B. Henner and J.E. Penner-Hahn (Springer, Berlin) p. 349.

Baer, Y., 1984, in: Handbook on the Physics and Chemistry of the Actinides, Vol. 1, eds A.J. Freeman and G.H. Lander (North-Holland, Amsterdam) p. 271.

Ballard, C.P., and L.D. Pye, 1976, J. Am. Ceram. Soc. **59**, 266.

Banks, R., W.T. Carnall, A.P. Paulikas, D.J. Lam and B.W. Veal, 1980, unpublished.

Bell, R.J., and P. Dean, 1972, Philos. Mag. **25**, 1381.

Brawer, S.A., and W.B. White, 1977, J. Non-Cryst. Solids **23**, 261.

Brooks, M.S.S., B. Johansson and H.L. Skriver, 1984, in: Handbook on the Physics and Chemistry of the Actinides, Vol. 1, eds A.J. Freeman and G.H. Lander (North-Holland, Amsterdam) p. 153.

Bruckner, R., H.U. Chun, H. Goretzki and M. Sammet, 1980, J. Non-Cryst. Solids **42**, 49.

Bunker, B.C., C.A. Dalfe, D.R. Tallant, R.J. Kirkpatrick, G.L. Turner and M.R. Reidmeyer, 1987, J. Am. Ceram. Soc., accepted for publication.

Chan, S.-K., and G.S. Knapp, 1984, in: EXAFS and Near Edge Structure III, Springer Proceedings in Physics, Vol. 2, eds K.O. Hodgson, B. Henner and J.E. Penner-Hahn (Springer, Berlin) p. 458.

Ching, W.Y., R.A. Murray, D.J. Lam and B.W. Veal, 1983, Phys. Rev. B **28**, 4724.

Ching, W.Y., Y.P. Li, B.W. Veal and D.J. Lam, 1985, Phys. Rev. B **32**, 1203.

DaSilva, J., D. Pinatti, C. Anderson and M. Rudee, 1975, Philos. Mag. **31**, 713.

Deer, W.A., R.A. Howie and J. Zussman, 1962, Rock Forming Minerals (Wiley, New York).

Dunlap, B.D., and G.M. Kalvius, 1985, in: Handbook on the Physics and Chemistry of the Actinides, Vol. 2, eds A.J. Freeman and G.H. Lander (North-Holland, Amsterdam) p. 329.

Dupree, R., D. Holland, P.W. McMillan and R.F. Pettifor, 1984, J. Non-Cryst. Solids **68**, 399.

Eisenstein, J.C., and M.H.L. Pryce, 1955, Proc. R. Soc. London Ser. A **229**, 20,31.

Eller, P.G., G.D. Jarvinen, J.D. Purson, R.A. Penneman, R.R. Ryan, F.W. Lytle and R.B. Greegor, 1985, Radiochimica Acta **39**, 17.

Faber, J.F., and G.H. Lander, 1976, Phys. Rev. B **14**, 1151.

Greaves, G.N., 1983, in: EXAFS and Near Edge Structure, Springer Series in Chemical Physics, Vol. 27, eds A. Bianconi, L. Incoccia and S. Stipcich (Springer, Berlin) p. 248.

Greaves, G.N., 1985, J. Non-Cryst. Solids **71**, 203.

Greaves, G.N., 1986, in: EXAFS and Near Edge Structure IV, to be published.

Greaves, G.N., A. Fontaine, P. Lagarde, D. Raoux and S.J. Gurman, 1981, Nature **293**, 611.

Grimmer, A.R., M. Mägi, M. Hähnert, H. Stade, A. Samoson, W. Wieker and E. Lippmaa, 1984, Phys. Chem. Glasses **25**, 105.

Gubanov, V.A., A. Rosen and D.E. Ellis, 1977, Solid State Commun. **22**, 219.

Heimerl, W., 1979, Techniques for High Level Waste Solidification in Europe, in: Scientific Basis for Nuclear Waste Management, Vol. 1, ed. G.J. McCarthy (Plenum Press, New York) pp. 21–29.

Iamaoka, M., H. Hasagawa and I. Yasui, 1983, Phys. Chem. Glasses **24**, 72.

Jen, J.S., and M.R. Kalinowski, 1979, J. Non-Cryst. Solids **38**, 21.

Karim, D.P., D.J. Lam, H. Diamond, A.M. Friedman, D.G. Coles, F. Bazan and G.L.

McVay, 1982, in: The Scientific Basis for Nuclear Waste Management, ed. S.V. Topp (Elsevier, New York) p. 67.
Karraker, D.G., 1982, J. Am. Ceram. Soc. **65**, 53.
Kingery, W.D., H.K. Bowen and D.R. Uhlmann, 1976, Introduction to Ceramics, 2nd Ed. (Wiley, New York) pp. 110–113.
Kirkpatrick, R.J., T. Dunn, S. Schramm, K.A. Smith, R. Oestrike and G. Turner, 1985, in: Structure and Bonding in Non-Crystalline Solids, eds A.G. Revesz and G.E. Walrafen (National Bureau of Standards, Washington) p. 303.
Knapp, G.S., B.W. Veal, D.J. Lam, A.P. Paulikas and H.K. Pan, 1984a, Mater. Lett. **2**, 253.
Knapp, G.S., B.W. Veal, A.P. Paulikas, A.W. Mitchell, D.J. Lam and T.E. Klippert, 1984b, in: EXAFS and Near Edge Structure III, Springer Proceedings in Physics, Vol. 2, eds K.O. Hodgson, B. Henner and J.E. Penner-Hahn (Springer, Berlin) p. 305.
Knapp, G.S., D.J. Lam, B.W. Veal, T.E. Klippert and A.P. Paulikas, 1984c, unpublished.
Kreidl, N.J., 1983, in: Glass Science and Technology, Vol. 1, eds N.J. Kreidl and D.E. Uhlmann (Academic Press, New York) pp. 122–126.
Kroger, F., J.M. Stevels and T.P.J. Borden, 1948, Philips Research Rep. **3**, 46.
Lam, D.J., 1980, unpublished.
Lam, D.J., B.W. Veal, H. Chen and G.S. Knapp, 1979, in: Scientific Basis for Nuclear Waste Management, Vol. 1, ed. G.J. McCarthy (Plenum Press, New York) pp. 97–107.
Lam, D.J., B.W. Veal and A.P. Paulikas, 1983, Plutonium Chemistry, ACS Symp. Series, No. 216, eds W.T. Carnall and G.R. Choppin (American Chemical Society, Washington, DC) p. 145.
Lebedev, N., 1921, Proc. State Opt. Inst. Leningr. **2**, 10.
Liebau, F., 1961a, Acta Crystallogr. **14**, 389.
Liebau, F., 1961b, Acta Crystallogr. **14**, 309.
McKale, A.G., G.S. Knapp and S.-K. Chan, 1986, Phys. Rev. B **33**, 841.
Misawa, M., D.C.L. Price and K.J. Susuki, 1980, J. Non-Cryst. Solids **37**, 85.
Mitra, S.K., and R.W. Hockney, 1982, in: Structure of Non-Crystalline Materials, eds P.H. Gaskell, J.M. Parker and E.A. Davis (Taylor and Francis, London) p. 316.
Morey, G.W., 1954, The Properties of Glass (Reinhold, New York).
Mozzi, R., and B. Warren, 1969, J. Appl. Crystallogr. **2**, 164.
Pant, A.K., and D.W.J. Cruickshank, 1968, Acta Crystallogr. B **24**, 13.
Petiau, J., G. Calas, T. Dumas and A.M. Heron, 1984, in: EXAFS and Near Edge Structure III, Springer Proceedings in Physics, Vol. 2, eds K.O. Hodgson, B. Henner and J.E. Penner-Hahn (Springer, Berlin) p. 291.
Phillips, J.C., 1982, in: Structure of Non-Crystalline Materials, eds P.H. Gaskell, J.M. Parker and E.A. Davis (Taylor and Francis, London) p. 123.
Porai-Koshits, E.A., 1985, J. Non-Cryst. Solids **73**, 79.
Raphael, G., and R. Lallement, 1968, Solid State Commun. **6**, 383.
Schramm, C.M., B.H.W.S. de Jong and V.E. Parziale, 1984, J. Am. Ceram. Soc. **106**, 4396.
Selvaray, U., K.J. Rao, C.N.R. Rao, J. Klinowski and J.M. Thomas, 1985, Chem. Phys. Lett. **114**, 24.
Shannon, R.D., 1976, Acta Crystallogr. A **32**, 751.
Smith, H.L., and A.J. Cohen, 1963, Phys. Chem. Glasses **4**, 173.
Soules, T.F., 1979, J. Chem. Phys. **71**, 4570.
Soules, T.F., 1982, J. Non-Cryst. Solids **49**, 29.
Soules, T.F., and A.K. Varshneya, 1981, J. Am. Ceram. Soc. **64**, 145.
Teo, B.K., and P.A. Lee, 1979, J. Am. Chem. Soc. **101**, 2815.
Thornley, F.R., N.T. Barrett, G.N. Greaves and G.M. Antonini, 1986, J. Phys. C **19**, L563.
Uhlmann, D.R., and G. Wicks, 1979, Wiss. Z. Friedrich-Schiller-Univ. Jena, Math. Naturwiss. Reihe **28**(2/3), 231.
Valenkov, A., and E.A. Porai-Koshits, 1936, Z. Kristallogr. **95**, 195.
Veal, B.W., and D.J. Lam, 1978, in: Proc. Int. Conf. on Phyics of SiO_2 and its Interfaces, ed. S. Pantelides (Pergamon, New York) p. 229.
Veal, B.W., and A.P. Paulikas, 1985, Phys. Rev. B **31**, 5399.
Veal, B.W., D.J. Lam, W.T. Carnall and H.R. Hoekstra, 1975, Phys. Rev. B **12**, 5651.
Veal, B.W., W.T. Carnall, B.D. Dunlap, A.W. Mitchell and D.J. Lam, 1987, in: High Tech Ceramics, ed. P. Vincenzini (Elsevier, Amsterdam) p. 2903.
Walch, P.F., and D.E. Ellis, 1976, J. Chem. Phys. **65**, 2387.
Warren, B.E., 1937, J. Appl. Phys. **8**, 645.
Warren, B.E., 1941, J. Am. Ceram. Soc. **24**, 256.
Warren, B.E., H. Krutter and O. Morningstar, 1938, J. Am. Ceram. Soc. **24**, 259.
Weyl, W.A., 1951, Colored Glasses (Soc. of Glass Tech., Sheffield, UK).
Wicks, G., 1974, Sc.D. Thesis (Massachusetts Institute of Technology, Cambridge, MA).
Wong, J., and C.A. Angell, 1976, Glass Structure by Spectroscopy (Marcel Dekker, New York).
Woodcock, L.V., C.A. Angell and P. Cheeseman, 1976, J. Chem. Phys. **65**, 1565.
Wright, A.C., G. Etherington, J.A.E. Desa, R. Sinclair, G.A.N. Connell and J.C.M. Mikkelsen Jr, 1982, J. Non-Cryst. Solids **49**, 63.
Wyckoff, R.W.G., 1963, Crystal Structures, Vol. I, 2nd Ed. (Interscience, New York) pp. 288, 321.
Zachariasen, W., 1932, J. Am. Ceram. Soc. **54**, 3481.
Zallen, R., 1983, Physics of Amorphous Solids (Wiley, New York) p. 43.
Zoltai, T., 1960, Am. Mineral. **45**, 960.

Handbook on the Physics and Chemistry of the Actinides, edited by A.J. Freeman and G.H. Lander

Chapter 5

MAGNETIC STUDIES OF TRANSPLUTONIUM ACTINIDES

PAUL G. HURAY and STANLEY E. NAVE
Department of Physics and Astronomy, University of Tennessee, Knoxville, TN 37996–1200, USA
and
Transuranium Research Laboratory, Oak Ridge National Laboratory, P.O. Box X, Oak Ridge, TN 37831, USA

Contents

1. Introduction

In most studies of the electronic properties of the actinides and their compounds, it is said (Smith and Kmetko 1983) that beyond plutonium the 5f electrons become localized while in the lighter end of the series they are delocalized into itinerant conduction electrons. This characterization is supported by crystallographic studies which show (Brooks et al. 1984, Haire and Baybarz 1973) two distinct tendencies:

(a) that for the lighter end of the series ($Z=90$–94) the volume per metal atom decreases parabolically with increasing atomic number like that of the 3d, 4d and 5d elements (whose electronic wave functions are usually described by bands), and

(b) that for the heavier actinides ($Z=95$–99) the volume per metal atom decreases monatonically with increasing atomic number like that of the 4f elements (whose electrons are often described by atomic states).

But we must ask if such crystallographic analogies are sufficient justification to define a sharp boundary between delocalization and localization? Furthermore is it proper to say that americium defines the boundary of localization for all actinide compounds? And even if the 5f electrons are localized they are certainly not isolated, so we must ask what cooperative phenomena do they exhibit and how do their properties change with temperature and magnetic field?

Electrons which move freely and independently in a lattice do not act cooperatively to align their spin moments in a magnetic field, and hence exhibit diminished or zero magnetic moments. Therefore, a measure of the degree of localization of the 5f electrons is determined by comparing the magnetic moment of the actinide metals and compounds to those expected for isolated ions.

When the 5f electrons are reasonably localized and exhibit magnetic moments, they nevertheless communicate with one another by slight overlap of wavefunctions with their actinide neighbors (direct exchange), by indirect interaction with their actinide neighbors via overlap with a common neighboring anion (superexchange), or by coupling through perturbations of the conduction electrons. The degree and nature of this communication can be examined through investigation of cooperative magnetic phenomena, i.e. a study of the ferromagnetic, ferrimagnetic, or antiferromagnetic transition temperatures and the low-temperature saturated magnetic moments. The strength of the 5f interaction with the localized electrons of neighboring atoms is also ascertained through the crystal field effects exhibited in low-temperature magnetic susceptibility data.

Over the past decade, instruments have been constructed with the sensitivity to make magnetic susceptibility measurements at low temperatures (4.2–400 K) and at high fields (up to 50 kG) on samples as small as 1 μg. These instruments have allowed the magnetic examination of the more stable heavy actinide metals and many of their compounds, samples of which have been produced in microgram or even milligram quantities. Most of the elements have been produced at the Oak Ridge National Laboratory in the high-flux isotope reactor (HFIR) and separated at the transuranium processing plant. The commonly used isotopes and the annual production amounts at the HFIR are shown in table 1. The actual size of samples obtained from the total production quantity is further limited by hazards from the radiation field, constraints due to chemical preparation techniques, containment requirements and measurement technique. In addition, the material is normally shared between national laboratories, universities and foreign collaborators of US investigators. The actual isotopic amounts available for a given experiment is thus usually several orders of magnitude smaller than those shown in table 1. For this

TABLE 1
HFIR production rates of the heavy actinide isotopes (elements 96–100).

Z	Isotope	Half-life (decay mode)	Production/year
96	Cm-248	3.397×10^5 y (alpha)	150 mg
97	Bk-249	320 d (beta)	50 mg
98	Cf-249	350.6 y (alpha)	From ^{249}Bk decay
99	Es-253	20.4 d (alpha)	2 mg
100	Fm-257	100.5 d (alpha)	1 pg

reason, no single crystals of the heavy actinides have been prepared for magnetic or neutron diffraction studies. The results presented in this work are therefore only for powder or polycrystalline studies.

In section 2 we present a brief review of the theory of paramagnetism for 5f electrons and the magnetic moments associated with localized electronic states. A discussion of the modifications in the magnetic susceptibility brought about by crystal field effects is briefly presented. In section 3 we discuss cooperative magnetic effects and the electronic interpretation of transition temperatures and saturated moments. To gain a perspective on how localized the 5f electrons are by comparison to their 4f homologues a brief review of the results for lanthanide metal and compounds is presented (section 4). We have limited the presentation to those lanthanide compounds for which the corresponding actinide counter-part has been studied. Since the recent measurements have been facilitated by a new generation of magnetic instruments we have included a section on these instruments, their advantages, or disadvantages (section 5).

The bulk of this work involves magnetic studies of metals, oxides, halides, and pnictides of Am, Cm, Bk, Cf, and Es, a survey of which is given in section 6.

2. Theory of paramagnetism

2.1. *L–S coupling and Hund's rule*

In a free atom, the largest terms in the Hamiltonian are those which describe:

(1) The kinetic energy of the electrons.

(2) The electrostatic interaction of the electrons with the nucleus.

(3a) The electrostatic interaction of the electrons with other core (filled-shell) electrons.

The effective potentials produced by these terms are central in nature and have no influence upon orbital, spin or angular momenta of valence electrons which are outside of a filled shell.

Another term in the Hamiltonian is the term which describes:

(3b) The residual electrostatic interaction of the electrons with other electrons in unfilled shells.

An additional term with a mathematical form similar to (3b) can be included which requires electrons with parallel spins to avoid (or repel) one another spatially, and preserves antisymmetry with respect to the exchange of all pairs of space and spin coordinates. This term is called:

(3c) The exchange correlation between electrons and other electrons in unfilled shells.

Because electrons in unfilled shells have magnetic moments produced by their spin and orbital angular momentum, additional terms can be included in a Hamiltonian which account for:

(4) The spin–orbit interactions.

(5) The spin–spin interactions.

(6) The orbit–orbit interactions.

Other terms which account for nuclear effects are smaller than these and will be neglected in this discussion.

For most atoms the exchange correlation term (3c) dominates the residual electrostatic interaction term (3b) and the third next important term in the nonspherically symmetric components of the Hamiltonian is the spin–orbit term (4).

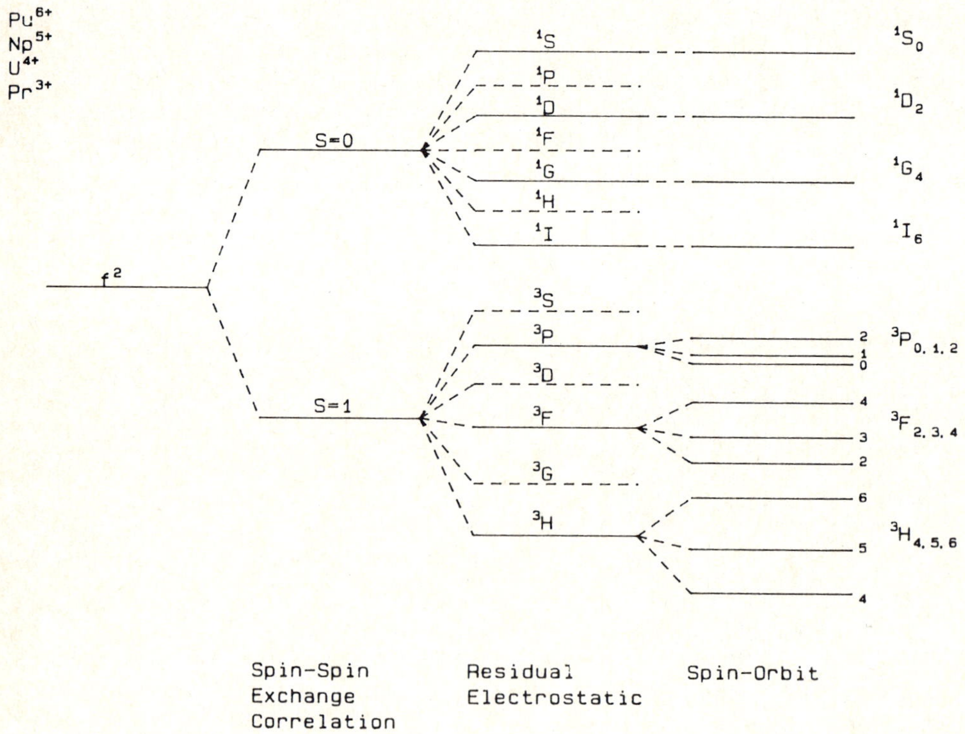

Fig. 1. Energy level diagram for the f^2 electronic configuration showing splitting due to Hamiltonian terms (3c), (3b) and (4) (see text).

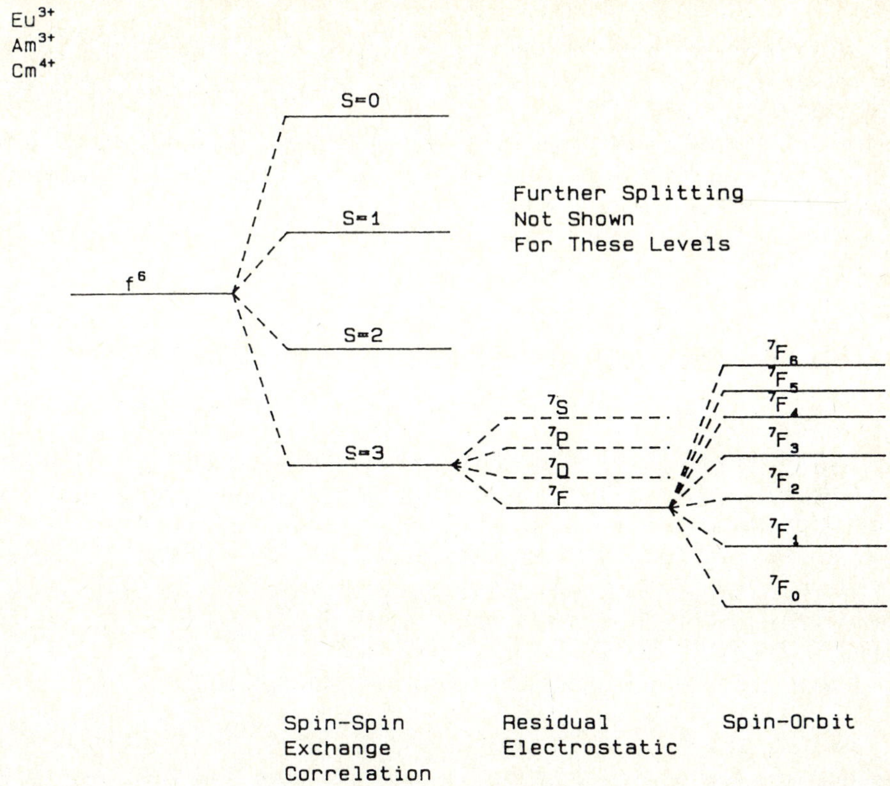

Fig. 2. Energy level diagram for the f^6 electronic configuration showing splitting due to Hamiltonian terms (3c), (3b) and (4) (see text).

Direct magnetic interactions between spin moments (5) and orbital moments (6) are the least significant.

Since terms (3b), (3c), (4), (5) and (6) do not involve central forces, the total orbital and spin angular momenta of the valence electrons (characterized by the quantum numbers L and S) influence the resultant energy of a particular electronic configuration. This order of important terms was first studied by Russell and Saunders (1925) and is called the L–S coupling model.

When the exchange correlation (3c) is the dominant term of the non-spherically symmetric components it is thus the first term considered in removing the degeneracy for particular electronic configurations of the valence electrons. The repulsive character of this term for electrons with parallel spins keeps individual orbitals as separated as possible and hence the minimum electrostatic energy of interaction occurs when the spins are all parallel (*largest possible* S, see figs. 1 and 2).

Since the orbitals of the individual electrons are not constants of the motion, the orbits (above) will not remain separated unless they revolve together in the same direction (have the same orbital angular momentum) about the nucleus. For electrons with the same spin this is forbidden by the Pauli exclusion principle. The residual electrostatic energy (3b) will nevertheless be a minimum when this condition is satisfied timewise to the maximum extent and hence the lowest lying states will occur when the individual orbital momenta are as close as possible to one another (*largest possible L*, see figs. 1 and 2).

Finally, if term (4) is considered as an additional perturbation upon the above levels, $\langle \boldsymbol{L}\cdot\boldsymbol{S} \rangle$ further influences the level structure. In terms of the total angular momentum $\boldsymbol{J}=\boldsymbol{L}+\boldsymbol{S}$, this term may be evaluated through

$$|\boldsymbol{J}|^2=|\boldsymbol{L}|^2+|\boldsymbol{S}|^2+2\boldsymbol{L}\cdot\boldsymbol{S}$$

so that

$$\langle \boldsymbol{L}\cdot\boldsymbol{S} \rangle=\tfrac{1}{2}\hbar^2\{J(J+1)-L(L+1)-S(S+1)\}.$$

Thus for given values of L and S (above), states of different J will have different energy. Depending upon the coefficient of $\langle \boldsymbol{L}\cdot\boldsymbol{S} \rangle$ in the Hamiltonian, the minimum energy will be produced for states of either maximum or minimum total angular momenta and ranging in integer steps between $J=L-S$ to $L+S$. *For less than a half-filled shell* the coefficient is positive and hence $J=L-S$ is lowest lying. *For more than a half-filled shell* the coefficient is negative and hence the multiplet of J values is inverted with $J=L+S$ lying lowest (see figs. 1 and 2).

These rules (underlined) constitute Hund's rule for determining the lowest energy state. A complete schematic diagram illustrating the splitting of a level corresponding to two equivalent f electrons in the valence shell (Pr^{3+}, U^{4+}, Np^{5+} or Pu^{6+}) is illustrated in fig. 1. Terms (energy levels) which are shown as dotted lines are forbidden by the exclusion principle. If the relative splittings between the singlet and triplet state are less than that shown in fig. 1 then the $^1I_6$ state might lie within the triplet multiplet. As will be seen later, this is the case for U^{4+}. Were the various final levels to be split by an additional field into different energies for each value of m_J, the multiplet would contain 91 independent states.

The rules also are applicable for multiple electrons such as that for six equivalent f electrons in the valence shell (Eu^{3+}, Am^{3+} or Cm^{4+}) as is illustrated in fig. 2. As can be seen, however, the multiplicity of states is considerably larger. Were all to be shown for each value of m_J, 3003 independent states would result.

Similar schemes have been constructed for the splitting of other combinations of $4f^n$ equivalent electrons and are found to be nominally in agreement with experimental data obtained for the 3+ lanthanide elements (Carnall and Crosswhite 1985) for energies below 1 eV (8065 cm^{-1}). These energy levels are shown in fig. 3, and are determined using the Hamiltonian described above with a set of four parameters (Racah integrals E^1, E^2, E^3 and spin–orbit coupling parameter ζ) chosen to most closely match peaks in optical absorption data.

2.2. *Intermediate coupling*

As one proceeds from light to heavy atoms the spin–orbit term (4) becomes larger until it is comparable in magnitude to the exchange correlation (3c) or the residual electrostatic interaction (3b). As this occurs, one often describes the resultant

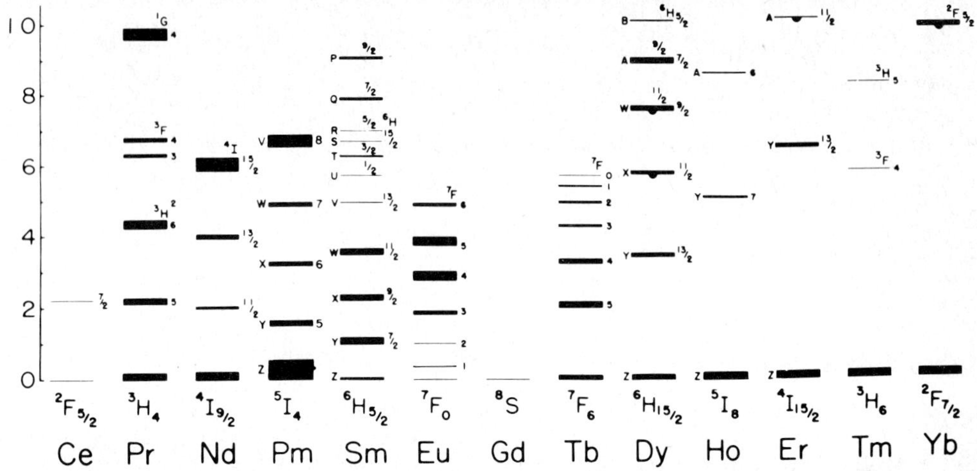

Fig. 3. Energy level diagram and spectroscopic labels of the excited states of the 3+ lanthanide ions. y-axis units are $1000\,cm^{-1}$. (From Carnall and Crosswhite 1985.)

perturbation scheme as becoming 'intermediate' between L–S coupling and a second extreme called j–j coupling. j–j coupling would occur in its ideal form if the spin–orbit term (4) were much larger than exchange correlation (3c) or the residual electrostatic interaction (3b). In that case the first term to be considered in removing the degeneracy for particular electronic configurations of the valence electrons would be the spin–orbit interactions. The energy levels corresponding to different combinations of j-values for the individual valence electrons (for f electrons $j=\frac{5}{2}$ or $\frac{7}{2}$) have been calculated by a relativistic Hartree–Fock wavefunction code (Nestor et al. 1966). The $5f_{7/2}$ and $5f_{5/2}$ energy eigenvalues typically differ by 1.3 eV ($10\,400\,cm^{-1}$). The lowest-lying of these levels will correspond to all of the electrons having their smallest j-value consistent with the exclusion principle. Finally, the electrostatic and correlation energies split each of the above levels into a number of levels characterized by different values of the total angular momentum J. Again the exclusion principle may limit the possible values of J.

The total number of levels in j–j coupling scheme is the same as for the L–S coupling scheme and the overall J-value designations are in correspondence. However, one cannot say which levels correspond in a one-to-one fashion between the two schemes. When calculations are performed ab initio and perturbation schemes are not employed, the wavefunctions and energy levels are never exactly describable in terms of ideal L–S or j–j coupling models.

Fortunately, most realistic cases are sufficiently close to L–S coupling that the notation is useful in describing eigenstates. Often a linear combination of the complete set of L–S eigenfunctions is employed to evaluate a given atomic wavefunction and to give some indication for the degree of departure from the ideal case. These computations are also sometimes called intermediate-coupling schemes

TABLE 2
Intermediate-coupling model wavefunctions for f^2 configurations of Ce^{2+}, Pr^{3+} and U^{4+}.

Nominal state	Largest and next-largest *L–S* components in the f^2 eigenstates		
	Ce^{2+}	Pr^{3+}	U^{4+}
$^3H_4$	$0.985\,^3H_4+0.167\,^1G_4$	$0.985\,^3H_4+0.167\,^1G_4$	$0.945\,^3H_4+0.314\,^1G_4$
$^3H_5$	$1.000\,^3H_5$	$1.000\,^3H_5$	$1.000\,^3H_5$
$^3H_6$	$0.999\,^3H_6-0.047\,^1I_6$	$0.999\,^3H_6-0.054\,^1I_6$	$0.973\,^3H_6-0.232\,^1I_6$
$^3F_2$	$0.989\,^3F_2+0.146\,^1D_2$	$0.988\,^3F_2+0.151\,^1D_2$	$-0.929\,^3F_2-0.316\,^1G_4$
$^3F_3$	$1.000\,^3F_3$	$1.000\,^3F_3$	$1.000\,^3F_3$
$^3F_4$	$0.778\,^3F_4-0.615\,^1G_4$	$0.806\,^3F_4-0.579\,^1G_4$	$0.706\,^3F_4-0.649\,^1G_4$
$^3P_0$	$0.994\,^3P_0+0.110\,^1S_0$	$0.996\,^3P_0+0.092\,^1S_0$	$0.966\,^3P_0+0.258\,^1S_0$
$^3P_1$	$1.000\,^3P_1$	$1.000\,^3P_1$	$1.000\,^3P_1$
$^3P_2$	$-0.965\,^3P_2-0.261\,^1D_2$	$-0.958\,^3P_2-0.287\,^1D_2$	$0.811\,^3P_2+0.567\,^1D_2$
$^1I_6$	$0.999\,^1I_6+0.047\,^3H_6$	$0.999\,^1I_6+0.054\,^3H_6$	$0.973\,^1I_6+0.232\,^3H_6$
$^1G_4$	$0.771\,^1G_4+0.627\,^3F_4$	$0.798\,^1G_4+0.591\,^3F_4$	$0.693\,^1G_4+0.702\,^3F_4$
$^1D_2$	$-0.954\,^1D_2+0.262\,^3P_2$	$-0.946\,^1D_2+0.288\,^3P_2$	$-0.740\,^1D_2+0.579\,^3P_2$
$^1S_0$	$0.994\,^1S_0-0.110\,^3P_0$	$0.996\,^1S_0-0.092\,^3P_0$	$0.966\,^1S_0-0.258\,^3P_0$

and have the additional value that they permit a simple estimation of the effective magnetic moments of atoms in ionic compounds from the *L–S* components.

A review of the eigenvalues and eigenfunctions for actinide ions in a great variety of compounds has recently been completed by Carnall and Crosswhite (1985). This treatment examines integrals of the radial wavefunctions using an ab initio single-configuration Hartree–Fock method which includes electrostatic, spin–orbit interactions and miscellaneous magnetic effects. A comparison is also made to a parametric model which substitutes variables for basic Hamiltonian terms with the value of the variables determined from a best fit to experimental optical absorption data in a least-squares fitting scheme. A comparison of the two sets of parameters is made to determine the magnitude of neglected corrections in the basic model. Generally speaking, the correction to the ab initio values is constant across the series. From this work, the spin–orbit interaction and radial integrals for various $5f^n$ configurations are evaluated for the actinide elements with Z between 91 and 100 and n between 2 and 11 (Pa, U, Np, Pu, Am, Cm, Br, Cf, Es and Fm) in a 3+ charge state. The calculated spin–orbit terms vary almost linearly from 1600 cm^{-1} to 4600 cm^{-1} across the series. By contrast these values are greater than their homologues from the lanthanide elements which have spin–orbit coupling constants between 700 cm^{-1} for $n=1$ and 2350 cm^{-1} for $n=11$ but are less than the electrostatic components (typically 20 000 cm^{-1}) in either series.

Thus, we would expect the *L–S* coupling scheme to give a good representation of the wavefunctions for light elements of the lanthanide series and to be less valid for heavy elements of the actinide series. This is seen to be true from table 2 for the f^2 example previously shown in fig. 1. In a similar way calculations of ground-state wavefunctions have been performed for combinations of $5f^n$ equivalent electrons. For energies below 1 eV (8065 cm^{-1}) the splittings are also nominally in agreement with terms predicted by the *L–S* coupling scheme (see fig. 4). These free-ion energy

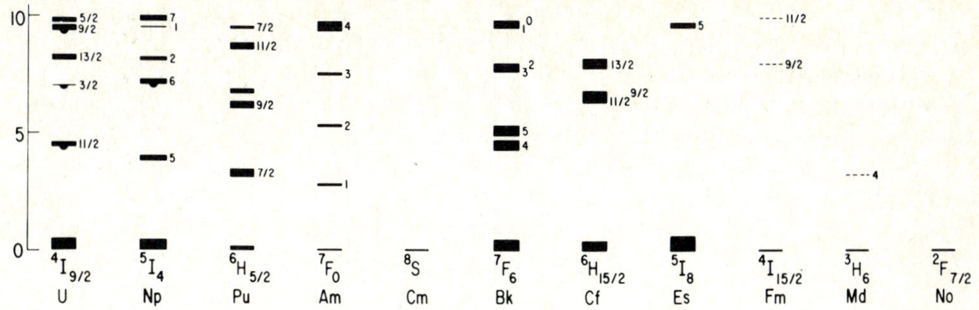

Fig. 4. Energy level diagram and spectroscopic labels of the excited states of the 3+ actinide ions. *y*-axis units are 1000 cm^{-1}. (From Carnall and Crosswhite 1985.)

levels for 3+ actinides are also obtained by Carnall and Crosswhite (1985) using model parameters that best match experimentally observed optical absorption peaks. A variety of other ab initio calculations have been done to determine the ground-state wavefunctions and have been expressed in terms of the *L–S* states. Table 3 gives the results of some of those calculations as reported by Chan and Lam (1974) for the light actinides and as calculated by the authors for curium and heavier actinides using Slater integrals and spin–orbit parameters as determined by Carnall and Crosswhite (1985) for these heavier actinides. As may be seen from this table, the heavier actinide atoms agree less well with the nominal ground-state configurations predicted by the simple *L–S* scheme than do lighter actinide atoms.

2.3. *Magnetic moments*

When an atom of well-characterized spin and orbital angular momentum is

TABLE 3

Major *L–S* components of intermediate-coupling model ground states of the actinide ions.

Atom	Nominal ground state	Major components of the Intermediate-coupling ground state
U^{4+}	$^3H_4$	89% ^{3}H; 10% ^{1}G
U^{3+}	$^4I_{9/2}$	83% ^{4}I; 15% ^{2}H
Np^{3+}	$^5I_4$	80% ^{5}I; 17% ^{3}H
Pu^{3+}	$^6H_{5/2}$	67% ^{6}H; 25% ^{4}G
Am^{3+}	$^7F_0$	47% ^{7}F; 38% ^{5}D; 13% ^{3}P
Cm^{3+}	$^8S_{7/2}$	79% ^{8}S; 17% ^{6}P
Bk^{3+}	$^7F_6$	72% ^{7}F; 22% ^{5}G; 4% ^{3}H
Cf^{4+}	$^7F_6$	67% ^{7}F; 25% ^{5}G; 5% ^{3}H
Bk^{2+}	$^6H_{15/2}$	71% ^{6}H; 23% ^{4}I; 4% ^{2}K
Cf^{3+}	$^6H_{15/2}$	69% ^{6}H; 25% ^{4}I; 4% ^{2}K
Cf^{2+}	$^5I_8$	75% ^{5}I; 22% ^{3}K
Es^{3+}	$^5I_8$	74% ^{5}I; 23% ^{3}K
Fm^{3+}	$^4I_{15/2}$	89% ^{4}I; 11% ^{2}K

placed in a modest applied magnetic field $\boldsymbol{H}$ (so that the interaction energy between the atomic moment and the applied field is significantly smaller than the spin–orbit, exchange correlation or residual electrostatic interaction) the ground state will be split by the applied field into $2J+1$ equally spaced energy levels. The magnetic moment created by the orbital motion, $\boldsymbol{\mu}_l = -(e/2m)\boldsymbol{L}$, combines vectorally with the magnetic moment created by the spin, $\boldsymbol{\mu}_S = -(e/m)\boldsymbol{S}$, to produce an effective atomic magnetic moment $\boldsymbol{\mu}_J$ and when a large set of N non-interacting, well-characterized atoms are placed in a modest external magnetic field $\boldsymbol{H}$ at a finite temperature T, a Maxwell–Boltzmann distribution of the atomic moments will occur according to

$$n(\boldsymbol{\mu}_J) = \exp\left[(\boldsymbol{\mu}_J \cdot \boldsymbol{H})/kT\right]. \tag{2.1}$$

This will lead to an effective magnetic moment per unit volume, M, which varies with temperature as

$$M = NgJ\mu_B B_J(x), \tag{2.2}$$

where

$$g = 1 + \frac{J(J+1) + S(S+1) - L(L+1)}{2J(J+1)}$$

is called the spectroscopic splitting factor, $\mu_B = e\hbar/2m$ is the Bohr magneton (9.27×10^{-21} erg/gauss) and

$$B_J(x) = B_J\left(\frac{gJ\mu_B H}{kT}\right) = \frac{2J+1}{2J}\coth\left[\frac{(2J+1)x}{2J}\right] - \frac{1}{2J}\coth\left[\frac{x}{2J}\right] \tag{2.3}$$

is called the Brillouin function. When its argument x is small compared to 1, the moment per unit volume is given by

$$M = NJ(J+1)\, g^2 \mu_B^2 H/kT. \tag{2.4}$$

For this case, the paramagnetic volume susceptibility χ (defined as M/H) can be expressed by the Curie law

$$\chi = C/T, \tag{2.5}$$

where

$$C = N\mu_{\text{eff}}^2/3k, \tag{2.6}$$

and

$$\mu_{\text{eff}} = [J(J+1)]^{1/2} g\mu_B. \tag{2.7}$$

Thus, for the ground-state configurations produced by equivalent $4f^n$ or $5f^n$ electrons, Hund's rule predicts the systematic behavior of the atomic magnetic moments in modest applied fields across the series of either lanthanides or actinides to be the same, and the results are given in table 4.

When the effective atomic magnetic moments are plotted against n, the familiar double-humped curve is produced, against which experimental results are often compared as a check for the applicability of this simple theory. This characteristic curve is shown by the solid line in fig. 5.

TABLE 4

Paramagnetic effective moments, based on L–S coupling model and Hund's rule, as a function of the number of equivalent f electrons n. (μ_{eff} in units of μ_B.)

n	μ_{eff} (μ_B)	n	μ_{eff} (μ_B)	n	μ_{eff} (μ_B)
0	0	5	0.84	10	10.60
1	2.54	6	0	11	9.59
2	3.58	7	7.94	12	7.57
3	3.62	8	9.72	13	4.54
4	2.68	9	10.63	14	0

In the event that excited states lie close in energy to the ground states, such as in the case for the Sm^{3+} or the Eu^{3+} states (see fig. 3), the Maxwell–Boltzmann distribution may include a significant population of those excited states. Especially at higher temperatures (such as room temperature where $kT = 0.025$ eV) the magnetic contribution from these states may alter the average effective magnetic moment per unit volume of a large set of ions. Van Vleck (1932) recognized that this population was significant for room-temperature measurements of lanthanide ions and modified the double-humped curve accordingly (dotted line in fig. 5). From fig. 4 it may likewise be seen that no excited states lie in close enough proximity to the ground states for actinide ions to be significantly populated even at room

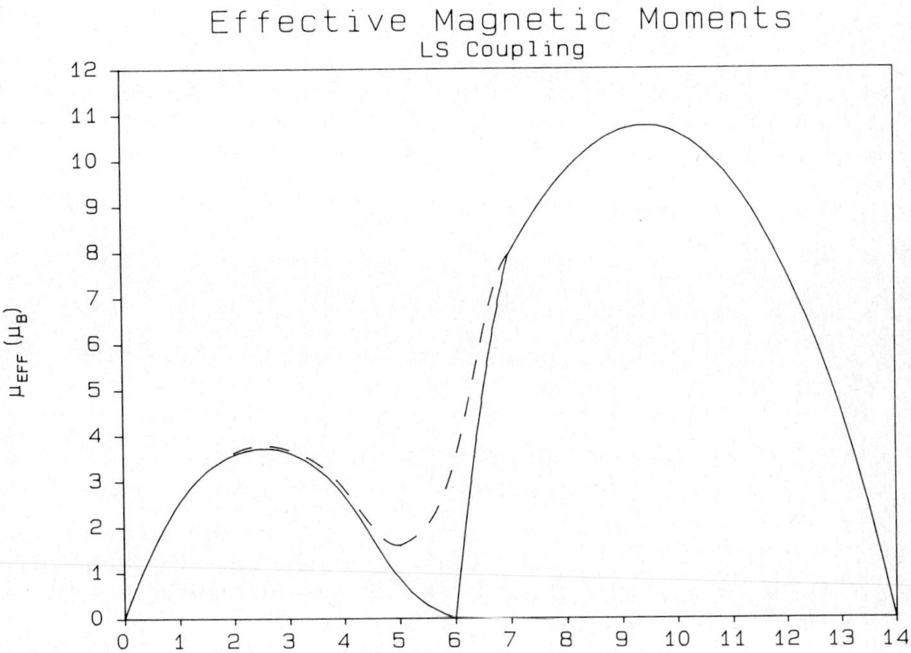

Fig. 5. Paramagnetic moments as a function of the number of f electrons assuming an L–S coupling model and the Hund's rule ground state. (The dashed line is the Van Vleck modification.)

temperature. Thus the solid curve of fig. 5 would be expected to adequately describe the effective magnetic moments of actinide ions if they are described by Hund's rule and L–S coupling.

In the cases where the ground-state moment is small or zero one must consider a second-order contribution to the susceptibility, which can be significant if an excited state is not too far above the ground-state multiplet. In this case a temperature-independent paramagnetism (TIP) results due to mixing of excited states into the ground state by the magnetic field, and is calculated using second-order perturbation theory. This leads to a temperature-independent term given by

$$\chi_{\mathrm{TIP}}=\sum_j \frac{|\langle\phi_i|\mu|\phi_j\rangle|^2}{E_i-E_j}, \tag{2.8}$$

where i is the ground state and j is an excited state.

Of course, as we have seen in the last section on intermediate coupling, such a description is not completely adequate due to the tendency toward j–j coupling. Expressing a ground-state wavefunction in terms of a linear combination of L–S wavefunctions (with the same value of J)

$$|\psi(\text{ground})\rangle=\sum_{L,S} a_{L,S}|\psi(L,S)\rangle, \tag{2.9}$$

then

$$\mu_{\mathrm{eff}}(\text{I.C.})=\sum_{L,S} a^2_{L,S}\,\mu_{\mathrm{eff}}(L,S). \tag{2.10}$$

As can be seen from table 3 this correction is more serious for the heavier actinides where the fraction of the Hund's rule ground state becomes significantly less than 1. The intermediate-coupling values for μ_{eff} are given in table 5 using the ground-state wavefunctions from table 3.

2.4. *Crystal-field effects*

Once a paramagnetic ion is chemically bound into a solid, its magnetic susceptibility will also depend upon the magnitude of the electric fields produced by

TABLE 5
Paramagnetic effective moments of actinide ions as a function of the number of equivalent f electrons, n, using the intermediate-coupling (I.C.) wavefunctions of table 3.

n	μ_{eff}(I.C.)	Ion	n	μ_{eff}(I.C.)	Ion
0	0	—	6	0	Am^{3+}
1	2.54	—	7	7.6	Cm^{3+}
2	3.63	U^{4+}	8	9.3	Bk^{3+}
3	3.79	U^{3+}	9	10.2	Cf^{3+}
4	2.89	Np^{3+}	10	10.3	Es^{3+}
5	1.24	Pu^{3+}	11	9.5	Fm^{3+}

TABLE 6
Ground state of level split by crystal field as function of the number of equivalent f electrons *n*.

n	Symmetry		*n*	Symmetry	
	Cubic	Octahedral		Cubic	Octahedral
0	Γ_1	Γ_1	8	Γ_2, Γ_3[a]	Γ_2, Γ_1[a]
1	Γ_8	Γ_7	9	Γ_7, Γ_8[a]	Γ_7, Γ_8[a]
2	Γ_1, Γ_5[a]	Γ_1	10	Γ_3, Γ_5[a]	Γ_3, Γ_1[a]
3	Γ_8	Γ_8, Γ_6[a]	11	Γ_7 Γ_6[a]	Γ_7, Γ_8[a]
4	Γ_1	Γ_1, Γ_5[a]	12	Γ_2, Γ_3[a]	Γ_2, Γ_1[a]
5	Γ_8	Γ_7	13	Γ_6, Γ_7[a]	Γ_6
6	Γ_1	Γ_1	14	Γ_1	Γ_1
7	Γ_6, Γ_8[a]	Γ_8, Γ_7[a]			

[a]Two states are given when either can be the ground state (depending upon the ratio B_4/B_6).

other atoms in the crystal relative to the spin–orbit, residual electrostatic or exchange correlation terms. For most of the 4f ions the crystal-field interactions are an order of magnitude smaller than the spin–orbit terms and are usually considered a perturbation upon the *L–S* ground state terms. For this case, the crystal field in cubic or octahedral crystals may be expressed in terms of the fourth- or sixth-order Stevens operators by

$$H_{\text{c.f.}} = B_4[O_4^0 + 5O_4^4] + B_6[O_6^0 - 21O_6^4],$$

and the sign and magnitude of the coefficients are determined by the crystal geometry. As in the case of an applied magnetic field, the ground state will be split by the crystalline electric field into $2J+1$ components but, depending upon the crystal symmetry, some may remain degenerate with others. Table 6 gives the Bethe notation of the resulting ground-state levels by a cubic or octahedral crystal field. A review of the magnetic susceptibility of light actinides in octahedral crystal fields (including intermediate coupling) has been presented by Chan and Lam (1974). A review of lanthanide elements in NaCl-type compounds (pnictides) has been presented by Hulliger (1978).

When the crystal field splitting is significant compared to kT, the ground-state levels of table 6 become the dominant terms populated in the Maxwell–Boltzmann distribution. In that case, part of the magnetic moment of the Hund's rule ground state, previously given in fig. 5, is quenched. This modification is largest at low temperature; e.g., Sm^{3+} has, in some compounds, a magnetic susceptibility at 4.2 K which is only 30% of that of the free ion. Also any singlet state (e.g., Γ_1 or Γ_2) is non-magnetic, so any non-Kramers ion in an orthorhombic field or sometimes in a cubic field may exhibit a χ which is independent of temperature. Since the symmetry of the crystal determines the exact nature of the splitting, there is no general rule for the magnetic moment of a given actinide element at low temperature. The results of crystal-field effects on actinide elements in various compounds are therefore discussed individually in each experimental section.

3. Cooperative phenomena in magnetism

When an atom is located within a lattice of other atoms it may experience not only crystal electric fields due to adjacent atoms but it may also experience magnetic fields in excess of those imposed externally. The additional fields may result from the direct wavefunction overlap of adjacent magnetic ions, indirect coupling via a second intermediary species (superexchange), or by spin modification of conduction electrons which may transmit information between actinide moments at distances substantially beyond nearest neighbors (RKKY oscillations). The effects may occur in pure materials or in compounds. Magnetic exchange fields were studied early by Pierre Weiss and often bear his name.

3.1. *Ferromagnetism*

The exchange fields H_e may be substantially higher than applied fields (up to 10^7 G) and may cause samples to exhibit spontaneous magnetic moments at low temperatures even in zero applied fields. Usually H_e is assumed to be proportional to sample magnetization M ($H_e = \lambda M$). At high temperature (the paramagnetic phase) the additional field leads to a modification of the Curie law (section 2.4) such that

$$\chi = \frac{C}{(T+\theta_p)} \quad \text{with } \theta_p = -\lambda C, \tag{3.1}$$

and is called the Curie–Weiss law. For positive values of λ, the magnetic susceptibility is thus expected to exhibit a singularity. At the singular temperature, T_c, and below there is said to exist spontaneous magnetization or ferromagnetism. The behavior of χ for $T \gg T_c$ is generally given by eq. (3.1) for most substances, but for temperatures in the region of T_c, χ actually behaves more like $\chi \propto (T-T_c)^{-4/3}$. Thus the observed T_c and the asymptotic constant, θ_p, are not usually found to be the same. At temperatures below T_c, H may be approximated (for ferromagnetic samples) by λM alone and, from section 2.4,

$$M = NgJ\mu_B B_J\left(\frac{gJ\mu_B\lambda M}{kT}\right). \tag{3.2}$$

Thus as the temperature is reduced to zero, the average moment per atom, $\sigma(0) = M(0)/N$ (called the saturation moment) should reduce to

$$\sigma(0) = gJ\mu_B. \tag{3.3}$$

For the ground-state configurations produced by equivalent $4f^n$ or $5f^n$ electrons, Hund's rule predicts the systematic behavior of the saturated magnetic moments at 0 K for either lanthanides or actinides to be the same. When intermediate coupling effects are included, the actinide ground-state wavefunctions contain additional admixtures of higher L–S states as was shown in table 3. In this case the saturated

TABLE 7
Saturated moments of actinide ions in both *L–S* and intermediate coupling models as function of the number of equivalent 5f electrons *n*.

n	$\sigma(L-S)$	σ(I.C.)	Ion	n	$\sigma(L-S)$	σ(I.C.)	Ion
0	0	0	—	8	9	8.6	Bk^{3+}
1	2.14	2.14	—	9	10.0	9.6	Cf^{3+}
2	3.20	3.24	U^{4+}	10	10.0	9.7	Es^{3+}
3	3.27	3.42	U^{3+}	11	9.0	8.9	Fm^{3+}
4	2.40	2.59	Np^{3+}	12	7.0	—	—
5	0.71	1.04	Pu^{3+}	13	4.0	—	—
6	0	0	Am^{3+}	14	0	0	—
7	7.0	6.7	Cm^{3+}				

moments are computed from

$$\sigma(\text{I.C.}) = \sum_{L,S} a_{L,S}^2 g(L,S) J \mu_B.$$

These values are given for the actinide elements in table 7.

There may be several reasons for the measured saturation moments to be different from the values given in table 7:

(1) Conduction electrons in metals may be polarized in the vicinity of the ions and effectively screen the local moments.

(2) The alignment of the atomic moments may not be parallel, but form some more complicated structure (e.g., a helical arrangement).

(3) The magnitude of the f electron moments may vary from site to site due to atomic inequivalency (ferrimagnetism).

The first of these explanations would be expected to give consistent values for both the saturation moment at low temperatures and the effective paramagnetic moment at high temperatures, even when they disagree with the theoretical values. An unambiguous analysis of the moment arrangement of atoms within a lattice requires neutron diffraction. As far as we are aware, only two neutron diffraction studies have been made on transplutonium elements (Fournier et al. 1977, Lander et al. 1977). Due to the unavailability of single crystals, both analyses were made on polycrystalline samples. Magnetic order determinations in other transplutonium studies have been based upon the characteristics of the bulk magnetic susceptibility as it changes with temperature. The experimental sections which follow thus present the interpretation of magnetic order in terms of ferromagnetism if the behavior is similar to that presented in section 3.1, in terms of ferrimagnetism if the behavior is like that in section 3.2 and in terms of antiferromagnetism if it is like that presented in section 3.3.

3.2. *Ferrimagnetism*

Ferrimagnetism was coined to describe the spin order which occurs in ferrites; the most familiar example of which is magnetite, Fe_3O_4 or $FeO \cdot Fe_2O_3$. In these compounds, one third of the iron ions are ferrous (Fe^{2+}) with a magnetic moment of $4\mu_B$ and two thirds are ferric (Fe^{3+}) with magnetic moments of $5\mu_B$. Were the moments to be parallel at low temperature, the average saturation moment should be $4.67\mu_B$. Neutron diffraction studies have shown that the ferric atoms form an antiferromagnetic (antiparallel) arrangement (no net saturation moment) while the ferrous atoms have parallel moments and thus the net average saturation moment is $1.33\mu_B$. By extension, the term ferrimagnetism has come to apply to any compound in which some ions have a moment antiparallel to others but in which a residual saturation moment exists at 0 K.

For the simplest of these compounds, large exchange fields may act upon atoms in either of two sublattices (A and B) such that $H_{E,A} = -\lambda M_B$ and $H_{E,B} = -\lambda M_A$. When the sign of λ is positive, the interaction energy will be a minimum when M_A is directed opposite to M_B. For complex compounds, a more complicated expression may be required, and under certain conditions, a spin array other than simple antiparallelism may produce even lower energy.

At high temperatures (the paramagnetic phase) the exchange fields lead to a modification of the Curie–Weiss law (section 3.1) such that

$$\chi = \frac{M_A + M_B}{H} = \frac{(C_A + C_B)T - 2\lambda C_A C_B}{T^2 - T_c^2}, \tag{3.4}$$

with $T_c = \lambda (C_A C_B)^{1/2}$, and

$$C_A = \frac{N_A \mu_{\text{eff},A}^2}{3k} \quad \text{and} \quad C_B = \frac{N_B \mu_{\text{eff},B}^2}{3k} \tag{3.5}$$

are Curie constants for the two sites.

At temperatures below T_c, each sublattice behaves according to a Brillouin function. Thus as the temperature is reduced to zero, the average moment per atom becomes

$$\sigma(0) = \frac{M_A(0)}{N} - \frac{M_B(0)}{N} = \left(\frac{N_A g_A J_A}{N} - \frac{N_B g_B J_B}{N}\right)\mu_B. \tag{3.6}$$

3.3. *Antiferromagnetism*

In an antiferromagnet the magnetic moments are arranged antiparallel, and the sublattices A and B have equal saturation magnetizations. Thus $C_A = C_B$ and the critical temperature (called the Néel temperature) is given by $T_N = \lambda C$ where C refers to a given sublattice.

At high temperatures (the paramagnetic phase) the magnetic susceptibility may be obtained from section 3.2 as

$$\chi = \frac{2CT - 2\lambda C^2}{T^2 - \lambda^2 C^2} = \frac{2C}{T + T_N} = \frac{N\mu_{\text{eff}}^2}{3k(T + T_N)}. \tag{3.7}$$

When general sublattice arrangements are considered (Anderson 1950) and an additional exchange field is introduced to also describe interactions within a sublattice, $H_{E,A} = \lambda_1 M_A - \lambda_2 M_B$ and $H_{E,B} = \lambda_1 M_B - \lambda_2 M_A$, then the magnetic susceptibility is modified slightly for high-temperature antiferromagnets to be

$$\chi = \frac{N\mu_{\text{eff}}^2}{3k(T+\theta_p)} \quad \text{and} \quad \frac{\theta_p}{T_N} = \frac{(\lambda_1+\lambda_2)}{(\lambda_2-\lambda_1)}. \tag{3.8}$$

Thus for antiferromagnetism, like ferromagnetism, the effective high-temperature Weiss temperature θ_p obtained from experimental data generally differs from the experimentally determined transition temperature T_N.

For temperatures below the Néel temperature the susceptibility is different when the applied field H is perpendicular to the sublattice moments than when parallel. For H perpendicular to M_A and M_B the interaction energy is a minimum when each moment is inclined by angle $\phi = H/2\lambda M$ from its original direction. The net sample moment ΔM from the application of H is thus $\Delta M = 2M \sin\phi \approx H/\lambda$ and hence $\chi_\perp(T) = \Delta M/H = 1/\lambda = C/T_N$. For H parallel to M_A and M_B the magnetic interaction energy does not change if the moments are displaced in angle from their original direction. Thus $\Delta M_{\|}(0) = 0$ and $\chi_{\|}(0) = 0$. As the temperature increases the parallel susceptibility increases smoothly from 0 to $\chi(T_N)$, while the perpendicular susceptibility is constant and equal to $\chi(T_N)$. For polycrystalline samples the individual crystalline domains will be randomly oriented with respect to H. The average susceptibility of these crystals will thus be a weighted average of $\chi_\perp$ and $\chi_{\|}$,

$$\chi_{\text{poly}}(T) = \tfrac{2}{3}\chi_\perp(T) + \tfrac{1}{3}\chi_{\|}(T). \tag{3.9}$$

This variation leads to the result: $\chi_{\text{poly}}(0) = \frac{2}{3}\chi_{\text{poly}}(T_N)$. For the polycrystals used in all of the studies reported here, such a signature is taken as indicative of simple antiferromagnetism.

3.4. *Helical order*

Helical or screwlike arrangements of spins have been observed in the rare earths. For example the lanthanide metals Tb, Dy and Ho have helical magnetic structures (Koehler 1972). Such arrays can be explained by nearest-neighbor interactions which are coupled by a ferromagnetic exchange and next-nearest-neighbors which are coupled by an antiferromagnetic exchange. Since the magnetic moments for the 4f and 5f series are expected to be similar, the lattice parameters are of the same magnitude, and the pure elements are metals, we might expect helical or more complex magnetic structures in the actinides. However, due to the limited quantities of these elements, it is unlikely that single crystals of sufficient mass will be produced for definitive neutron diffraction studies of the actinides.

4. Magnetism of the lanthanides

The magnetic properties of lanthanide ions have been classically used to demonstrate the applicability of L–S coupling (section 2.1), Hund's rule (section 2.2) and the Van Vleck modification (section 2.4) to the 4f element series. Since the spin–orbit term in the Hamiltonian is of smaller magnitude than the exchange correlation term or the residual electrostatic term, the complexity of intermediate coupling (section 2.3) may be neglected. Additionally, if one chooses to study lanthanides in compounds with well-isolated ions, complications brought about by crystal-field interactions (section 2.5) may also be ignored. Several summaries of the magnetic properties of lanthanide elements have been previously given and only a selected review of metals and compounds is presented here as a backdrop against which results of their actinide homologues will be examined later (section 6).

4.1. *Lanthanide metals*

The crystal and magnetic structure of the lanthanide metals exhibit a complex behavior with temperature, as is seen in table 8 and evidenced by the multitude of crystal forms for some elements and the appearance of two regions of differing magnetic character for some elements. Brooks et al. (1984) have used the atomic volume, the cohesive energy, the bulk modulus and the room-temperature entropy

TABLE 8
Magnetic data for lanthanide metals.

Metal	Crystal structure	Lattice constants (Å)		Melting or transition temperature (K)	$\mu_{eff}(\mu_B)$	$-\theta_p(K)^a$	$T_c(K)$	$T_N(K)$
		a	c					
La	dhcp	3.770	2.159	606	0.49			
	fcc	5.302		1160				
	bcc	4.26		1296				
Ce	fcc	4.85		95				
	dhcp	3.68	11.92	263	2.51	−38		12.5
	fcc	5.161		1021				
	bcc	4.11		1093				
Pr	dhcp	3.672	11.835	1088	3.56			25
	bcc	4.13		1231				
Nd	dhcp	3.658	11.799	1158	3.3	−15	7	20
	bcc	4.13		1320				
Pm				1376				
Sm	hcp	3.621	26.25	1213	1.74	−60		15
	bcc	4.07		1368				
Eu	bcc	4.582		1122	7.12	108		90
Gd	hcp	3.636	5.783	1560	7.98	317	293	
	bcc	4.06		1608				
Tb	hcp	3.601	5.694	1622	9.77	236	220	229
	bcc(?)			1661				
Dy	hcp	3.590	5.646	1703	10.67	151	85	179
Ho	hcp	3.577	5.616	1757	10.8	88	20	132
Er	hcp	3.559	5.587	1793	9.8	47	20	85
Tm	hcp	3.537	5.555	1841	7.6	20	22	55
Yb	fcc	5.486		1094	0.4			
	bcc	4.45		1120				
Lu	hcp	3.503	5.551	1948	0.2			

[a] $\chi = C/(T+\theta_p)$.

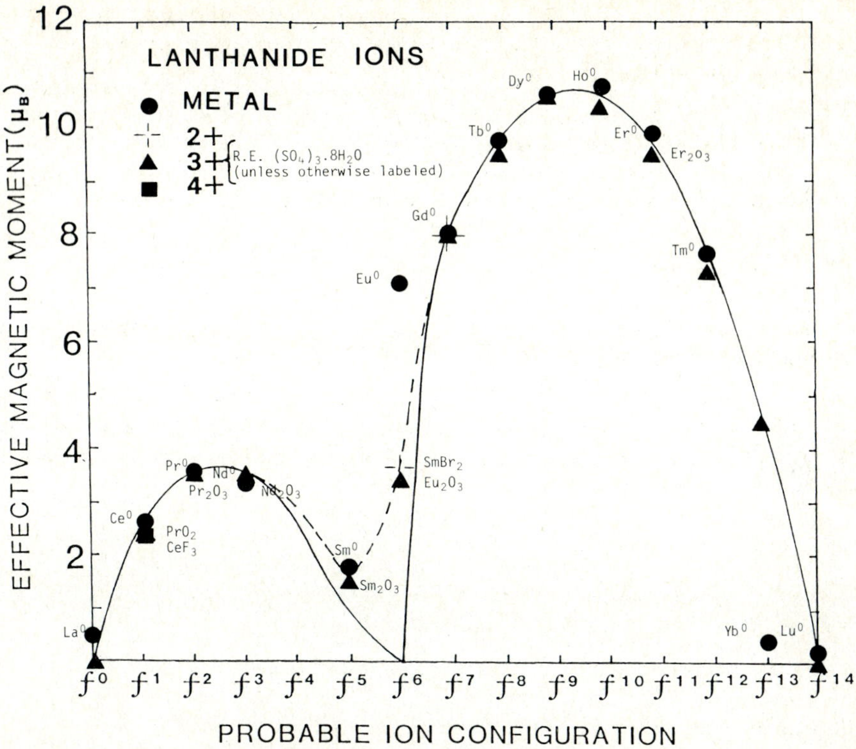

Fig. 6. The paramagnetic effective moments of lanthanide metals (solid circles) and trivalent compounds (solid triangles) as a function of the number of f electrons. The solid and dashed lines are from fig. 5.

to conclude empirically that most of the lanthanide elements may be considered as tri-positive $4f^n$ ions embedded in a sea of conduction electrons formed from the 5d and 6s valence electrons. They have reasoned that europium and ytterbium metals are an exception to this picture being considered divalent metals. Because the 4f electrons lie substantially within the outer closed shells they play only a small role in chemical bonding in the metals and (to a first approximation) behave magnetically as if they were free ions. The resultant effective magnetic moments of the metals at room temperature are shown in fig. 6 as solid circles. The probable ion configurations are chosen as 3+ for all except Eu and Yb which are chosen as 2+. As can be seen from fig. 6, these magnetic results are in agreement with the conclusions of Brooks et al. (1984). The room-temperature Van Vleck (1932) modification (the dashed line in fig. 5) is included in fig. 5 and is seen to be consistent with the experimental values plotted for europium and samarium.

Neutron diffraction results have shown a complex variety of magnetic ordering in lanthanide metals. These results are summarized by Sinha (1978) and Koehler (1972). For the purpose of comparing the systematic behavior of actinide metals, the transition temperatures T_N and T_c of the lanthanide metals, are shown in fig. 7 with

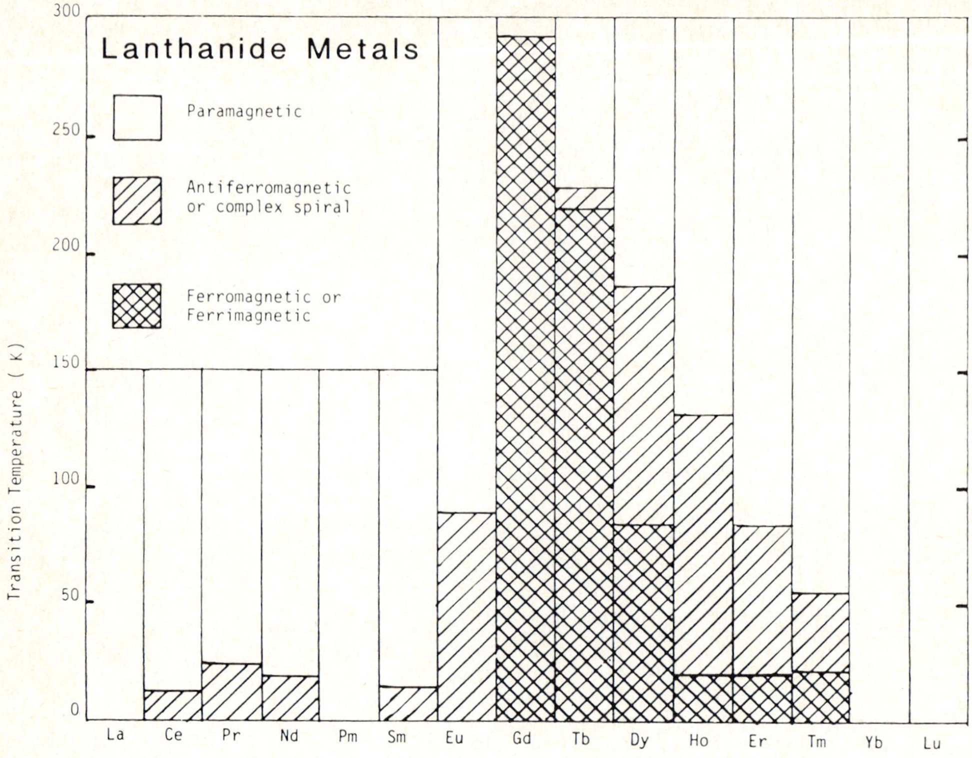

Fig. 7. Transition temperatures and regions of differing magnetic-ordering type for the lanthanide metals.

increasing atomic number. In some cases, the transition temperatures do not correspond to simple ferromagnetism or antiferromagnetism but rather to helical, longitudinal wave, conical, antiphase, or oblique helix spin structure.

4.2. *Lanthanide compounds*

The lanthanide elements have been systematically studied in a variety of compounds. Topp (1965) and Casey and Mitra (1976) have compiled the magnetic moments for the relatively separated tri-positive ions in sulfate and acetate compounds. These experimental moments are listed in table 9. The values for the sulfates are plotted in fig. 6 (solid triangles) to indicate moments of compounds in a trivalent configuration.

Schieber (1967) and Holtzberg et al. (1970) have summarized the magnetic properties for the oxides of the lanthanide elements. The only divalent compound for which the moment is reported in these works is for EuO ($\mu_{eff} = 7.94\mu_B$) with a ferromagnetic transition at $T_c = 77$ K. The trivalent sesquioxide compounds, for which moments are reported in table 10, all exhibit the body-centered-cubic (bcc) structure. Pr_2O_3 also exhibits a hexagonal (hcp) structure with $\mu_{eff} = 3.55\mu_B$ and

TABLE 9
Paramagnetic effective moments of lanthanide compounds.

Lanthanide	Compound		Lanthanide	Compound	
	$Ln_2(SO_4)_3 \cdot 8H_2O$	$Ln(C_5H_5)_3$		$Ln_2(SO_4)_3 \cdot 8H_2O$	$Ln(C_5H_5)_3$
	μ_{eff} (μ_B)	μ_{eff} (μ_B)		μ_{eff} (μ_B)	μ_{eff} (μ_B)
La	0	—	Tb	9.6	—
Ce	—	2.46	Dy	10.3	10.0
Pr	3.47	3.59	Ho	10.4	—
Nd	3.52	3.62	Er	9.4	9.44
Sm	1.58	1.54	Tm	7.0	—
Eu	3.54	—	Yb	4.3	4.0
Gd	7.9	7.95	Lu	0	—

TABLE 10
Magnetic properties of the Ln_2O_3 (bcc) compounds.

	μ_{eff} (μ_B)	θ_p[a] (K)	T_N (K)		μ_{eff} (μ_B)	θ_p(K)[a]	T_N (K)
La	—	—	—	Tb	9.67	−13	2.0
Ce	—	—	—	Dy	10.5	−21	—
Pr	3.59	−73	—	Ho	10.4	−14	1.8
Nd	3.66	−32	—	Er	9.44	−13	4.3
Sm	1.50	−150	—	Tm	7.28	−42	—
Eu	3.64	−135	—	Yb	—	−90	3.0
Gd	7.90	−18	—	Lu	—	—	—

[a] $\chi = C/(T+\theta_p)$

$\theta_p = +64$ K. As indicated in table 10, some of the sesquioxides exhibit an antiferromagnetic transition at low temperatures.

The magnetic properties of only a few lanthanide dioxide compounds (tetravalent) have been reported. Face-centered CeO_2 (MacChesney 1965) shows a temperature-independent paramagnetism. Face-centered-cubic PrO_2 (MacChesney 1964) shows an antiferromagnetic behavior with $\mu_{eff} = 2.47\,\mu_B$, $\theta_p = +105$ K and $T_N = 14$ K. Face-centered-cubic TbO_2 (Chapin et al. 1964) is antiferromagnetic with $\mu_{eff} = 7.90\mu_B$, $\theta_p = +15$ K and $T_N = 3$ K.

Grünberg and Maletta (1982) and Hulliger (1978) have also summarized the magnetic properties of the lanthanide pnictide compounds. These compounds are of particular interest because they all form in a simple NaCl structure; but with variable lanthanide–lanthanide separation, depending upon which pnictide (N, P, As, Sb or Bi) is used as the anion. For the lanthanide elements the separation between adjacent ions is sufficiently large that cooperative magnetic phenomena are observed only at low temperatures. However, the high conductivity of the compounds gives them a semimetal character which derives from a well-populated conduction band as was discussed by Brooks et al. (1984) for the light actinide elements. The lanthanide pnictide data are presented in table 11 as a backdrop against which the actinide pnictides may be compared.

TABLE 11
Magnetic data for mono-pnictides of the lanthanides.

		La	Ce	Pr	Nd	Sm	Eu	Gd	Tb	Dy	Ho	Er	Tm	Yb	Lu
N	a (Å)	3.75	3.552	3.58	3.63	3.570	3.540	3.524	3.490	3.461	3.449	3.42	3.401	3.376	
	μ_{eff} (μ_B)			3.57	3.7				9.7	10.5	10.6	9.4	7.32	4.93	
	$-\theta_p$ (K)[a]			−11	15			69	43	20	12	4	−21	−102	
	T_c (T_N)		(2.5)		29	(18)		69	42	26	18	5		<(1.7)	
P	a (Å)		4.17	4.167	4.120				4.013	3.997	3.970	3.956	3.93	3.919	
	μ_{eff} (μ_B)		2.66	3.77				7.7	9.56	9.9	10.2	9.32	7.6		
	$-\theta_p$ (K)[a]		−12	−4				0	1	6	4.2	0	0		
	T_c (T_N)		(8.5)		(11)	(1.7)		(15)	(9)	(8)	5.5	(3.1)		(<1.4)	
As	a (Å)		4.29	4.255	4.213	4.187		4.139	4.112	4.097	4.079	4.055	4.04	4.029	
	μ_{eff} (μ_B)		2.67	3.8				8.2	9.65	10.1	10.5	9.6	7.6		
	$-\theta_p$ (K)[a]		−8	−8				−12	−4	2	1	−1.5	0		
	T_c (T_N)		(7.5)		(13)	(1.8)		(2.5)	(12)	2	(4.8)	(3.5)		(<1.6)	
Sb	a (Å)	4.58	4.53	4.500	4.461	4.434		4.396	4.369	4.351	4.334	4.318	4.30	4.29	4.28
	μ_{eff} (μ_B)		2.56	3.63	3.75			8.1	9.7	10.7	10.8	9.8	7.6	4.24	
	$-\theta_p$ (K)[a]		5	−5	−3		−3	−42	−14	−4	−2.5	−3	−5	−70	
	T_c (T_N)		(18)		(16)	(2.1)		(28)	(14)	(9.5)	(5.5)	(3.7)			
Bi	a (Å)	4.64	4.59	4.559	4.53				4.43	4.41	4.395	4.38	4.37		4.35
	μ_{eff} (μ_B)		2.38	3.52	3.58			8.03	9.52	10.8		9.4			
	$-\theta_p$ (K)[a]		12	0	−1			−65	−33	−30		−6			
	T_c (T_N)		(25)		(25)	(8.6)		(28)	(18)	(12)	(6.7)	(3.9)			

[a] $\chi = C/(T+\theta_p)$

5. Magnetic measurements

5.1. *Micro-magnetic susceptometer*

A susceptometer which has the sensitivity necessary to measure the diamagnitism of metal samples as small as 1 μg ($V \sim 10^{-7}\,\mathrm{cm}^3$) has been described previously (Nave and Huray 1980). This instrument has been called a micro-magnetic susceptometer and is essential for magnetic work on transplutonium actinides due to limited sample quantities available. This susceptometer incorporates a commercially available device known as a superconducting quantum interference device (SQUID) to detect DC flux by means of a set of external superconducting pickup coils. In most respects the susceptometer is like commercially available systems using the SQUID technology but, in addition, it incorporates several design features for working with micro samples.

The basic principle for determining the magnetic moment of a sample is to record the DC flux change sensed by the SQUID due to the changing flux through a set of pickup coils as a sample is moved through the coils. The pickup coils consist of an asymmetric pair, i.e. identical coils wound in an opposite sense in order to make them less sensitive to changes in the uniform applied field used for magnetizing the sample. The output of the SQUID versus sample position is shown in fig. 8 and the peak-to-peak height is compared to a calibrating standard for determining the absolute susceptibility. The desired increase in sensitivity for small volume samples may be achieved by making the diameter of the pickup coils as small as possible. A typical coil diameter is 1.5 mm and a typical applied field is 50 kG. In order to maximize flux transfer between the pickup coils and the input coil of the SQUID (maximizing the instrument sensitivity) one matches the inductance of the two coils. In this case the measurement sensitivity varies as the inverse of the coil diameter squared. A practical limit arises because the number of turns on the pickup coils must increase in inverse proportion to the coil diameter in order to maintain an inductance match. Eventually, the wire size becomes too small for simple coil fabrication.

In the case of extremely small samples, the magnetic signal from the sampleholder is significantly larger than that of the sample. There are several techniques for dealing with this problem:

(1) Measure the empty holder at all fields and temperatures and subtract this result from measurement on the holder plus sample.

(2) Construct the sample holder from two or more materials that have cancelling susceptibility over some temperature range and then apply method (1) with less error.

(3) Make two identical holders and arrange to have one in one coil without a sample and one in the other with the sample and thus cancel the sample holder signal to first order (Roy et al. 1982).

(4) Choose a sampleholder effectively the size of the sample by compressing the sample between two uniform gold wires of such length that the ends extend well beyond the coils in both directions. By this method the sample holder is in essence

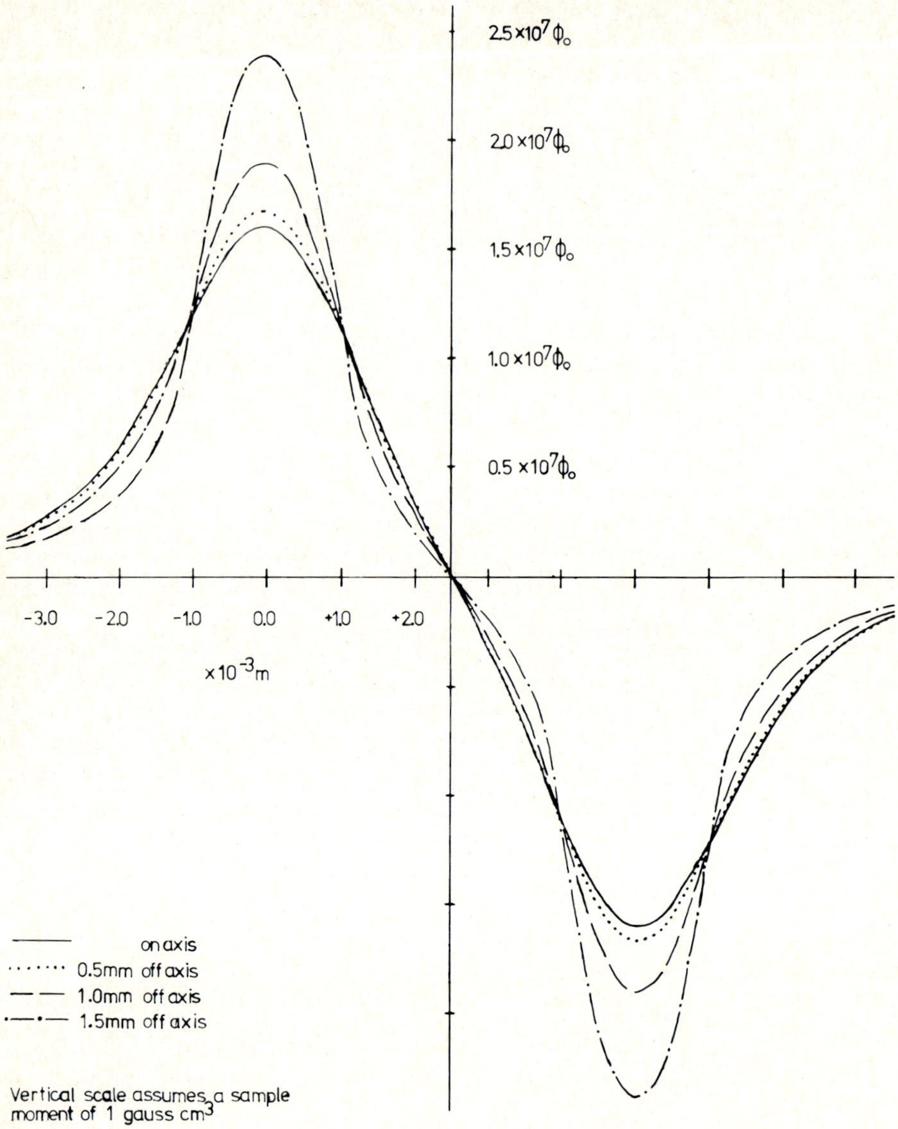

Fig. 8. Magnetic flux of a point dipole through the SQUID pickup coils as a function of position relative to the coil center.

the absence of gold over a region on the order of the volume of the sample. A scheme for holding samples in this fashion is shown in fig. 9.

Unfortunately, the toxic nature of the heavy actinides and their reactivity to atmosphere or contaminants often prevents the application of this simple scheme. In those cases, a sample is often contained inside a quartz capillary. A second empty capillary is butted against this one and the combination replaces the two gold wires.

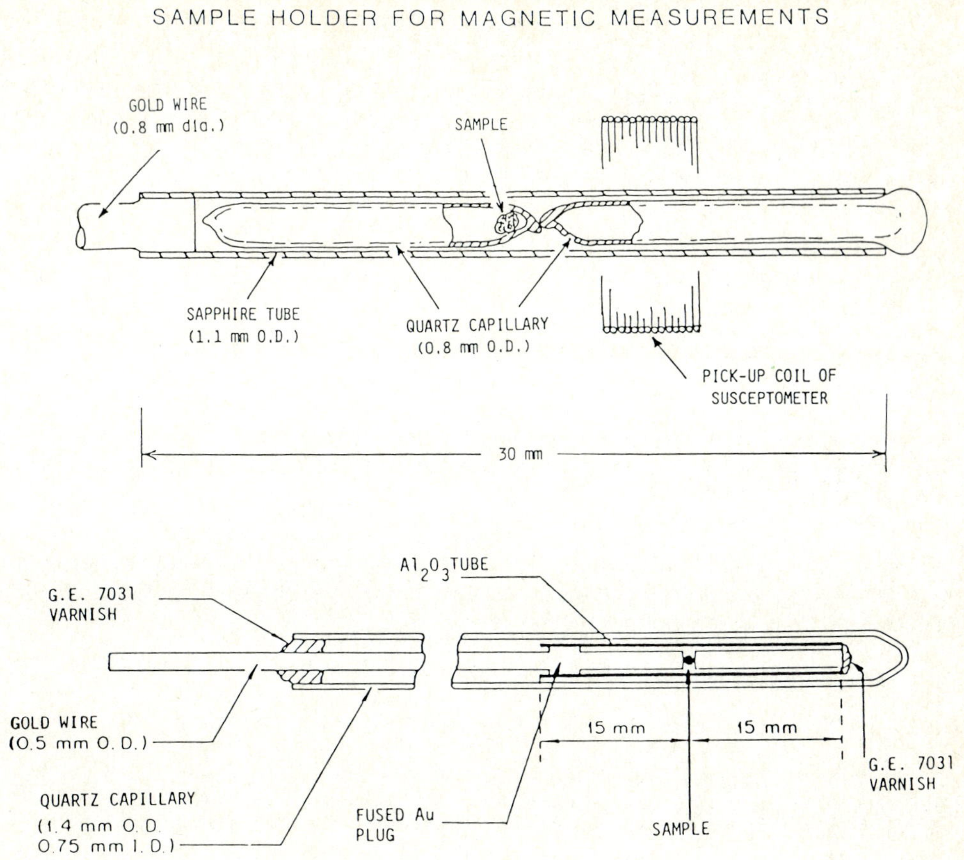

Fig. 9. Sample holders appropriate for small samples of radioactive materials.

This configuration results in a measurement of the magnetic moment of the ends of the quartz tubes and of the sample. Since the susceptibility of quartz is relatively temperature independent, it is possible to ascribe the temperature-dependent components of a measurement to be caused by the actinide sample. The technique has the disadvantage that the temperature-independent term in the sample susceptibility is masked by that of the holder. Thus reports of χ_0 for a particular measurement are often meaningless.

5.2. *Localized paramagnetic susceptibility*

As described in section 3, a system of slightly interacting local moments has a susceptibility given by the Curie–Weiss law [eq. (3.1)]. As was explained in section 5.1, the experimental data often must be fit by a modified Curie–Weiss law (MCW) which includes a temperature-independent term, χ_0, in addition to the Curie–Weiss term. This term may be due to contributions from saturated ferromagnetic impurities in the sample holder or impurities in the sample itself, but when such a temperature-independent paramagnetism (TIP) is determined to be characteristic of the pure material it will be designated χ_{TIP} and can originate from two sources. First, it can be due to an enhanced Pauli susceptibility of the conduction electrons (band model) in metals, semimetals or intermetallic compounds. Second, it can be due to a second-order Zeeman effect, an effect due to mixing of excited states into the ground state by the magnetic field and calculated using second-order perturbation theory. In this case χ_{TIP} is given by eq. (2.8), which is valid when the multiplet splitting is large compared to kT.

When applying the MCW law for analyzing data in a case where χ_{TIP} is large and μ_{eff} is small, a complication arises. It was shown in section 3.1 that the paramagnetic Curie temperature, θ_p, comes about when one considers the susceptibility of a collection of non-interacting local moments ($\chi = C/T$) within the framework of a mean-field approximation to account for the interactions. However, if χ_{TIP} is included in the expression for the susceptibility of the noninteracting moments ($\chi = \chi_{TIP} + C/T$), before applying a mean-field approximation, an expression identical to the MCW law is obtained but with renormalized values for χ_{TIP} and C (Amoretti and Fournier 1984). In this case the expression for the magnetic susceptibility can be written as

$$\chi^* = \chi_0^* + C^*/(T + \theta_p), \tag{5.1}$$

with

$$\chi_0^* = \chi_0/(1 - \chi_0\lambda), \qquad C^* = C/(1 - \chi_0\lambda), \qquad \theta_p = C\lambda/(1 - \chi_0\lambda),$$

where the asterisk denotes measured values, λ is the exchange constant and $\chi_0 = \chi_{TIP}$.

6. Magnetism of the heavy actinides

6.1. *Pure metals*

6.1.1. *Americium metal*

This metal has been observed to have a temperature-independent susceptibility with a value for χ_0 in the range from 680×10^{-6} emu/mole to 880×10^{-6} emu/mole (Brodsky 1971, McWhan 1961, Kanellakopulos et al. 1975). This result agrees with a localized ionic-core model for the metal with a trivalent, f^6 ground-state configuration which gives $J = 0$ in both the L–S and intermediate coupling models. The theoretical value for the local paramagnetic moment is zero for either case and the temperature-dependent component for the susceptibility is, consequently, also zero. The rather large value for the temperature independent paramagnetism, χ_{TIP}, is due to a second-order contribution to the susceptibility from the first excited state. It should be noted that a band model (Fournier and Troć 1985) can similarly account for the temperature-independent paramagnetism. In all of the reported measurements, a sharp increase in the susceptibility was observed at temperatures near 4 K. This may be attributed to paramagnetic impurities, to a real nonzero

TABLE 12
Curium metal (dhcp).

Isotope	θ_p (K)	μ_{eff} (μ_B)	Ref.[a]
^{244}Cm	135	8.15	[1, 2]
	224	7.85	[1, 2]
	286	8.01	[1, 2]
	348	7.96	[1, 2]
	354	8.02	[3]
	356	7.97	[3]
	384	8.10	[3]
	386	8.07	[4]
	560	8.8	[4]
	−72	6.0	[5]
^{248}Cm	259	7.6	[6]
	180	7.6	[6]

[a] References:
[1] Marei (1965).
[2] Marei and Cunningham (1972).
[3] Bansal (1966).
[4] Kanellakopulos et al. (1975).
[5] Fujita et al. (1976).
[6] Nave and Huray (1984).

moment value for the americium ions (intermediate-valence model) or to a small fraction of the americium ions with a non-f^6 configuration (mixed-valence model). In a neutron diffraction experiment a moment of 0.6 μ_B was determined for the americium sites in the intermetallic compound $AmFe_2$ (Lander et al. 1977) and a mixed-valence model was proposed to best explain the results.

6.1.2. *Curium metal*

Of the transplutonium actinide metals the most magnetic work has been done on curium. This is because it is the first magnetic metal of the actinide series and is available in multi-milligram quantities. It can exist in two crystal forms at room temperature and atmospheric pressure. The stable form is the dhcp crystal structure but a high-temperature fcc phase is metastable at room temperature when quenched in from the melt or, as has been observed recently, it can be formed by application of pressures greater than 15 GPa and remains in this form after relaxation of the pressure (Haire et al. 1985).

Magnetic measurements have been made on both crystal forms. All of the work prior to that of the authors has involved the isotope curium-244, which has a relatively short half-life of 18 years. ^{244}Cm is an intense alpha-emitter and magnetic measurements are complicated by radiation damage and/or self-heating (2.8 W/g). The isotope curium-248 has a half-life of 3.4×10^5 y and is free from these complications. Above about 150 K the susceptibility of the dhcp form follows a Curie–Weiss law. The effective-moment values are in reasonable agreement for all experimental measurements as shown in table 12 and also agree with the predicted value for an f^7 configuration. Assuming an L–S coupling model and a Hund's rule ground state one obtains $7.92\mu_B$ for a theoretical value (see table 4) while in an intermediate coupling model one obtains $7.55\mu_B$ (table 5). The values for θ_p are large and vary greatly from sample to sample. There is also an indication that θ_p even varies with time for ^{244}Cm samples. This probably results from the effect of radiation damage and/or increasing ^{240}Pu content. For microgram-size samples it is possible that a sample is composed of a few large crystallites and may not be a representative polycrystalline sample. If this is the case and if θ_p is different along the three crystal axes, then the majority axis could vary and would lead to a variation in θ_p from sample to sample. The f^7 configuration in the *L–S* model is a spatially isotropic *S* state, i.e. $L=0$, and hence should give an isotropic θ_p. Thus such an effect must be due to a breakdown of the free-ion *L–S* coupling model.

In the dhcp samples a maximum was observed in the susceptibility at 64 ± 1 K for ^{248}Cm and at 52 K for ^{244}Cm. This maximum is susceptibility indicates antiferromagnetic ordering, and simple antiferromagnetic ordering has been confirmed by neutron diffraction on a 120 mg multi-crystal sample of ^{244}Cm (Fournier et al. 1977). Figure 10 gives a representative plot of inverse susceptibility versus temperature for a 1 mg sample.

A single measurement has been reported for the fcc phase (Huray et al. 1980). There is a drastic deviation from the behavior observed for the dhcp metal and if

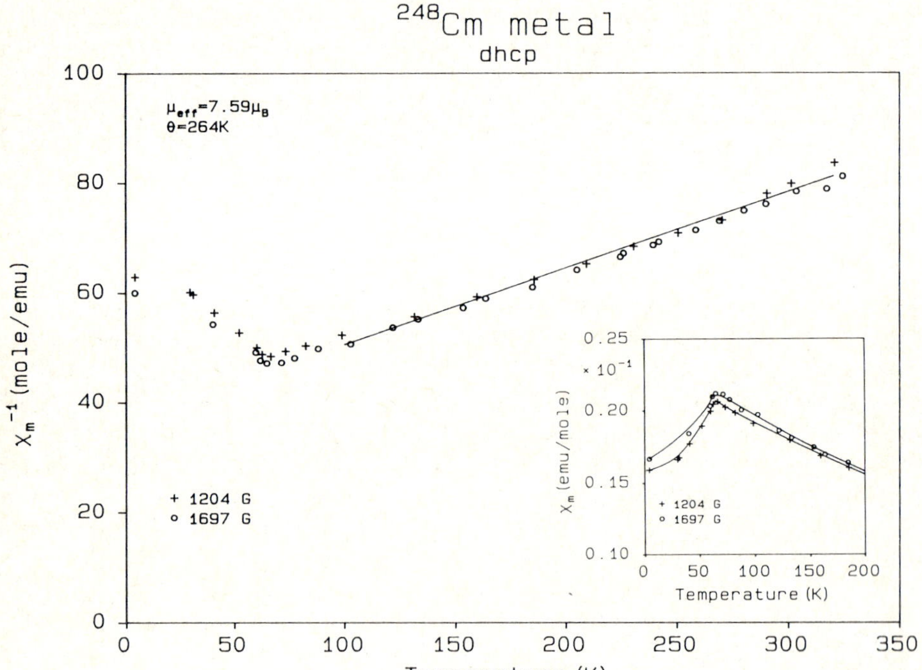

Fig. 10. Inverse susceptibility versus temperature for dhcp curium metal. The solid line is the result of a Curie–Weiss fit to the data with the values for μ_{eff} and θ_p as indicated. The inset indicates the transition in the susceptibility at low temperatures.

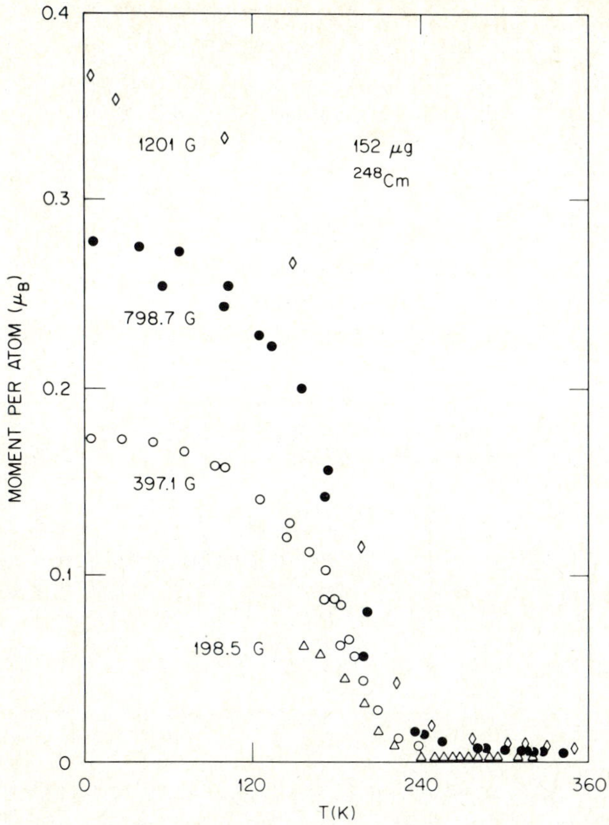

Fig. 11. The magnetic moment per atom versus temperature for fcc curium metal at the indicated magnetic fields.

confirmed by future work it may explain some of the variations in behavior observed in dhcp samples. The fcc phase apparently has a high-temperature transition, $T_N = 205$ K, as seen in fig. 11. The data for the inverse susceptibility as a function of temperature (fig. 12) is best fit by a ferrimagnetic two-site model. The solid line through the data is determined by this ferrimagnetic model using parameters obtained from a nonlinear least-squares fit to the data. Although all sites should be equivalent in this cubic crystal structure, this two-site model may approximate a spiral-type magnetic order as determined for some of the lanthanide metals by neutron diffraction. The effective moments determined from this model for each site are 5.93 μ_B and 6.13 μ_B. The average moment agrees with one experimental value reported for dhcp ^{244}Cm metal (Fujita et al. 1976). Clearly small amounts of an fcc phase present in a dhcp sample as a second phase could greatly affect the susceptibility due to the high temperature of the magnetic transition for the fcc metal. The fcc phase is metastable at room temperature and often two-phase samples are formed as indicated by extra lines in the X-ray pattern of a dhcp sample which are attributable to the fcc metal.

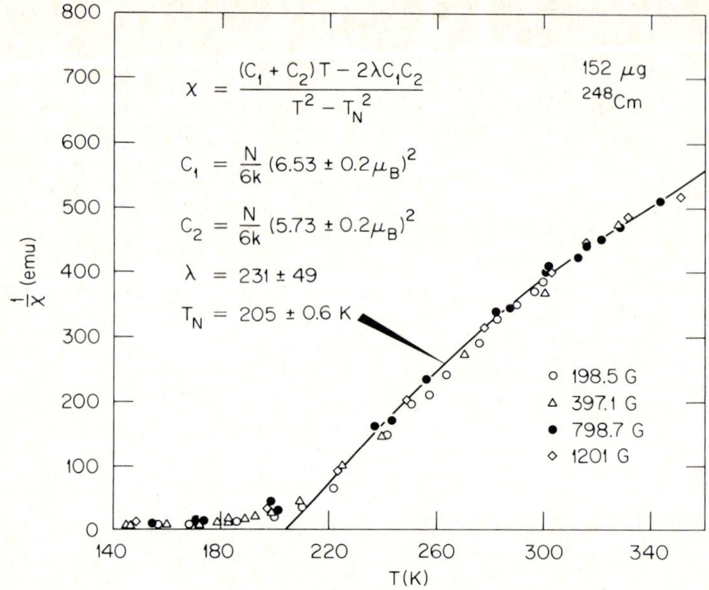

Fig. 12. Inverse susceptibility versus temperature for fcc curium metal. The solid line gives a fit to the data based on a ferrimagnetic two-site model.

6.1.3. *Berkelium metal*

Work on berkelium metal is difficult due to its relatively short half-life of 314 days. Its californium-249 daughter, which is created in the beta-decay, grows in as an impurity at the rate of 0.3 atomic percent per day and preparation of the metal takes at least a day. Nevertheless, magnetic measurements are reported for samples with as little as 2% Cf daughter.

Early measurements on samples of 1–6 μg were complicated by mixed phase samples and/or large ^{249}Cf content (Fujita 1969). More recent measurements (Nave et al. 1980) involved samples which were pure dhcp phase within the limits of detectability by X-ray diffraction analysis, involved masses as large as 340 μg and the samples contained as little as 2% Cf at the time of measurement. The results of these as well as the earlier measurements are given in table 13. It appears that the dhcp phase exhibits Curie–Weiss behavior down to about 60 K (fig. 13) with an effective moment of $9.71\mu_B$ for the two best characterized samples. This supports the hypothesis of a localized, nominally f^8 configuration for the metal, corresponding to a trivalent ion and a contribution of three electrons to the conduction band. The theoretical values for the f^8 configuration are $9.72\mu_B$ for the *L–S* model (table 4) and $9.4\mu_B$ for the intermediate-coupling model.

The dhcp and predominantly dhcp phase samples have a maximum in the susceptibility at 34 K, indicating antiferromagnetic ordering. The results of measurements for a typical sample are given in the inset of fig. 13 where the susceptibility is plotted versus temperature. It is interesting that the one

TABLE 13
Berkelium metal.

Structure	Mass (μg)	%Cf	μ_{eff} (μ_B)	θ_p (K)	Comments	Ref.[a]
dhcp	338.7	2	9.8	208	200 G: AF at 22 K; 800 G: Sharp transition to plateau at 33 K	[1]
dhcp	21.0	12	9.6	102	AF at 34 K	[1]
Predominantly dhcp	19.0	16	8.8	84	AF at 34 K	[1]
Predominantly fcc	1.669	20	8.23	−64.4	Ferromagnetic at 140 K	[2]
Predominantly dhcp	5.629	16	8.52	72.7	AF at 35 K	[2]
Equal amounts dhcp/fcc	1.725	1.7	8.83	33.0	Deviation from Curie–Weiss between 100 K and 10 K	[2]

[a] References:
[1] Nave et al. (1980).
[2] Fujita (1969).

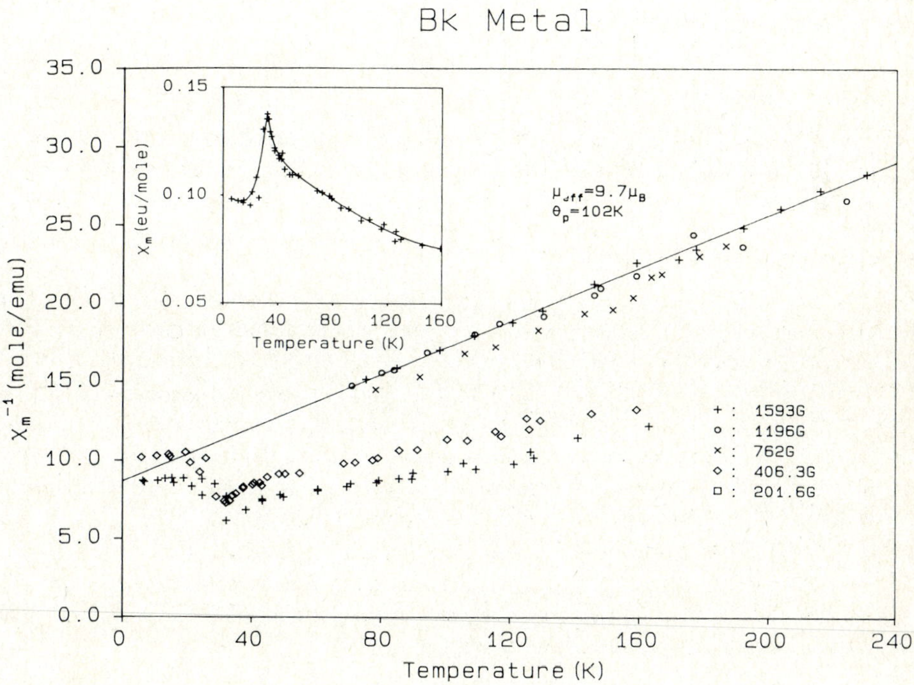

Fig. 13. Inverse susceptibility versus temperature for dhcp berkelium metal. The solid line is the result of a Curie–Weiss fit to the data with the values for μ_{eff} and θ_p as indicated. The inset indicates the transition in the susceptibility at low temperatures.

measurement by Fujita (1969) on a predominantly fcc phase sample indicated a ferro- or ferrimagnetic transition at a relatively high temperature (140 K), a behavior similar to fcc curium metal. The author in this case indicates that the 140 K transition may be due to an impurity, but further work is warranted on the fcc metals to confirm this dramatic deviation from the dhcp metal behavior.

6.1.4. *Californium metal*

Californium metal is the next in the series of actinides and is the heaviest element to be prepared as a substrate free metal. The isotope californium-249, on which all magnetic measurements have been made, has a long half-life of 351 y, but it emits a hard gamma-ray (388 keV) and is available in sample sizes of less than few hundred micrograms. The method of preparation involves collection of the metal on a substrate after effusion as a hot vapor from a tantalum crucible (Haire and Baybarz 1974). This leads to thin flakes with large surface area and in some cases the flakes have been combined in a tungsten press to form small disks.

Magnetic measurements have now been made on californium for temperatures in the range from 4.2 K to 340 K and in magnetic fields as high as 50 kG (Nave et al. 1985). It has Curie–Weiss behavior above 100 K with an effective moment of 9.7μ_B, as shown in fig. 14. The θ_p value of -40 K is smaller than that for the previous

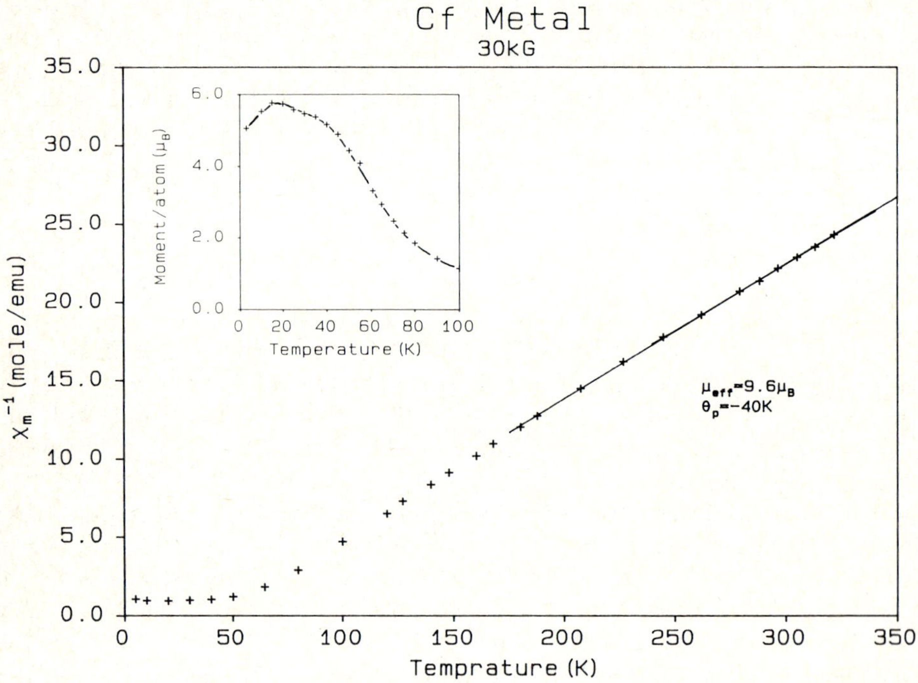

Fig.14. Inverse susceptibility versus temperature for dhcp californium metal. The solid line is the result of a Curie–Weiss fit to the data with the values for μ_{eff} and θ_p as indicated. The inset gives the moment per atom versus temperature in the transition region.

elements and indicates less interaction with near-neighbors, probably due to a more localized core wavefunction. This value is in accord with previous measurements by Fujita et al. (1976) on two small (9 μg) samples of fcc metal and is slightly lower than 10.2 μ_B calculated for an f^9 configuration (3+ state) based on intermediate coupling (table 5).

As shown in the inset of fig. 14 where the moment per atom is plotted versus temperature, there is a transition to an ordered state below 100 K. The size of the moment indicates that probably ferromagnetic ordering is attained even though the moment decreases slightly below the maximum at 15 K. This is also confirmed by the negative value for θ_p. From isomagnetization plots a value of 51 ± 2 K was determined for T_c. The saturation moment was determined to be 6.1μ_B by extrapolation of the graph of moment per atom at 15 K versus applied field to infinite field. Previous measurements (Fujita et al. 1976) on the fcc metal indicated no ordering down to 22 K, the lowest temperature reported. The Weiss temperature was small but negative for these samples and indicates the probability of a transition at lower temperatures.

6.1.5. *Einsteinium metal*

The first magnetic measurements were made on an einsteinium metal and involved a sample with a mass of only 0.25 μg. The sample was prepared by vapor deposition of the metal on the end of a gold wire during the actual metal preparation. This method, which is essentially the same as the method used to prepare californium metal (Haire and Baybarz 1974), unfortunately results in the possibility of formation of a Au–Es alloy, due to the high temperature of the gold wire. Over the limited temperature region from 4.2 K to 90 K a Curie–Weiss fit to the data gives a local moment value of 11.3μ_B which is somewhat higher than the 10.2μ_B theoretical value. However, the small sample size, large corrections for the sampleholder and an uncertainty in sample temperature due to self-heating, could lead to a large uncertainty in this experimental value.

6.2. *Oxides*

Since the 5f electronic wavefunctions are localized for transplutonium actinide compounds, the near free-ion behavior allows one to determine the applicability of an intermediate-coupling model through the determination of their local paramagnetic moments. Intermediate coupling becomes more important with increasing atomic number (section 2.3) so the near-ionic oxide compounds of heavy actinides provide a system through which to study this effect. In addition, the transamericium oxide compounds are ideal for investigating crystal-field interactions, due to the fact that both the dioxide and low-temperature sesquioxide provide sites with cubic symmetry. The compounds also allow the possibility of determining crystal-field parameters for the 3+ and 4+ actinide ions using the same sample since one can be readily converted to the other.

Americium lies on the boundary of two regions of behavior for actinide compounds. Below the boundary, ionic compounds with valence states higher than

3 are more stable, as for the light actinides uranium and neptunium, while above the boundary the trivalent state is more stable, as observed for all of the transamericium actinide compounds. Two crystal structures are known to exist for the sesquioxide of americium. The low-temperature form is the so-called C-type which has a body-centered-cubic structure. This form converts to the A-type with hexagonal symmetry above 650°C. For americium the dioxide is more stable than the sesquioxide at room temperature and crystallizes in the fluorite structure with face-centered-cubic symmetry.

For the transamericium oxides the trivalent sesquioxide is the more stable species. The sesquioxide exists in two crystal forms, the C-type and the B-type. The C-type with body-centered-cubic symmetry has two non-equivalent metal-ion locations, while the B-type with a monoclinic structure has three nonequivalent metal-ion sites. The bcc sesquioxide is converted irreversibly to the monoclinic form by heating to temperatures above 1100°C. The bcc sesquioxide can be converted reversibly to first an intermediate oxide (rhombohedral structure) with oxygen-to-metal ratio of 1.70 and then to the dioxide by oxidizing treatment of varying degrees. The intermediate oxide occurs as a rhombohedral distortion of the cubic structure.

6.2.1. *Americium oxides*

Trivalent americium as in the sesquioxide should nominally have a $^7F_0$ ground term, but as pointed out in section 2.3, an intermediate-coupling model should be more appropriate in view of the large spin–orbit coupling parameters observed for the actinides. Since the spin–orbit interaction only mixes states with the same J value as the ground state, a $J=0$ state is still obtained and μ_{eff} must be zero even in the intermediate-coupling model. It is possible for the crystal field to mix states with other J values into the ground state but this mixing is expected to be small (Hessler and Carnall 1980).

A similar situation occurs for ionic compounds of divalent samarium and trivalent europium which have a $4f^6$ configuration. This configuration also gives a $^7F_0$, nonmagnetic ground state. The experimental value for the moment is nonzero in these compounds but is also temperature dependent and is due to a contribution to the susceptibility arising from thermal population of an excited state as was suggested in section 2.4. The first excited state for the $4f^6$ state is 0.04 eV above the ground state and since it is significantly populated at room temperature ($kT=0.026$ eV), it contributes to the susceptibility at this temperature. The position of this level is determined from the optical absorption spectrum of the compounds. In the theoretical calculation of the energy levels, parameters are chosen to fit the positions of the experimental absorption maxima corresponding to the excited states which lie in the measurable region of the spectrum. The energy of the first excited state may then be calculated using these parameters. Application of the technique to AmF_3 results in determination that the first excited state lies 0.33 eV above the ground state. A typical spectrum is given in fig. 15 and the corresponding calculated energy levels based on a fit to experimental absorption spectra (Hessler and Carnell 1980) are indicated at the bottom by horizontal bars. This energy is too high to give significant thermal population at room temperature or below [$\exp(-E/kT)\approx 10^{-5}$] and does not contribute to the measured susceptibility.

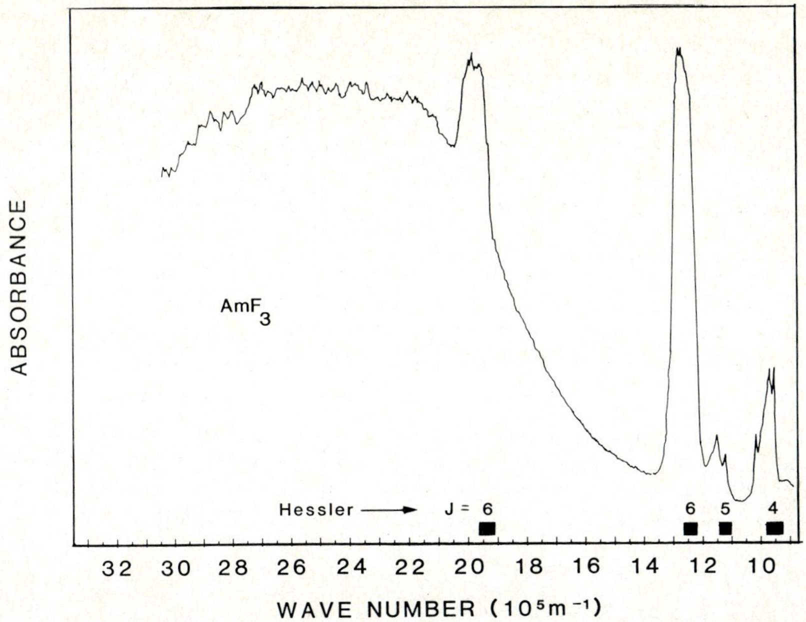

Fig. 15. Optical absorption of AmF_3 as a function of the reciprocal wavelength for photons with energy between 1 and 3 eV. The solid bars indicate the strong absorption bands for aqueous Am^{3+} ions and the J value indicates the corresponding total angular momentum quantum number of the excited state involved.

The only possible contribution to the paramagnetic susceptibility is then a second-order term due to mixing of excited states into the ground state by the magnetic field and calculated using second-order perturbation theory. This leads to a temperature-independent term given by eq. (2.8). A calculation of this quantity by Edelstein (1985) using parameters obtained from a fit to an optical absorption spectrum of Am^{3+} in $LaCl_3$ gives a value of 500×10^{-6} emu/mol while a similar calculation for Am^{3+} in $Am(C_5H_5)_3$ by Kanellakopulos et al. (1978) gives a value in the range $(520–717) \times 10^{-6}$ emu/mol depending on values chosen for the crystal-field parameters.

Recent measurements (Soderholm et al. 1986) have been reported on a large (13 mg) very pure sample of the hexagonal form of $^{243}Am_2O_3$. This sample was treated carefully to avoid oxidation, even though this crystal form should not oxidize easily. It exhibited TIP for the temperature range 5–300 K, with $\chi_{TIP} = (640 \pm 20) \times 10^{-6}$ emu (mol Am)$^{-1}$. No appreciable field dependence was observed. This value is slightly lower than that reported for the metal and slightly higher than the calculated values above.

Susceptibility measurements are reported (Karraker 1975a) for a 224 mg polycrystalline sample of $^{243}AmO_2$ but were limited to the temperature range from 2.2 to 97 K. Above 15 K the data was interpreted assuming two regions of Curie–Weiss behavior. These values are listed in table 14 and the behavior is attributed to

TABLE 14
Actinide oxides.

Compound (Structure)	Configuration	μ_{eff} (μ_B) (I.C.)	μ_{eff} (μ_B)	θ_p (K)	$\chi_0\left(10^{-6}\frac{emu}{mol}\right)$	T_N (K)	Comments	Ref.[a]
AmO_2 (fcc)	f^5	1.2	1.3	21	—	8.5	15–40 K (Γ_7)	[1]
			1.5	41	—	—	50–100 K (Γ_8)	
Am_2O_3 (hex)	f^6	0	0	—	640	—	TIP 5K–300 K	[2]
CmO_2 (fcc)	f^6	0	1.6–2.3	12–25	1900–4000	—	—	[3]
$BaCmO_3$ (bcc)	f^6	0	1.6–1.7	18–30	988	—	—	[3]
Cm_2O_3 (mono)	f^7	7.6	8.2	+149	—	13	20–80 K	[4]
			7.9	+130	—	—	100–300 K	
$^{244}Cm_2O_3$ (?)	f^7	7.6	6.1	22.5	—	—	2–50 K	[5]
Cm_2O_3 (mono)	f^7	7.6	7.5	72	—	—	100–300 K	[3]
Cm_2O_3 (mono)	f^7	7.6	7.7	130	—	18	50–300 K	[3]
Cm_2O_3 (bcc)	f^7	7.6	7.5	110	3700	—	50–300 K	[3]
BkO_2 (fcc)	f^7	7.6	7.8	250	—	40–50	50–200 K (3% Cf)	[3]
BkO_2 (fcc)	f^7	7.6	8.1	224	—	40	70–300 K (3% Cf)	[6]
$(BkO_2)ThO_2$ (fcc)	f^7	7.6	7.7	12	—	~3	10–95 K (Γ_6)	[7]
			5.8	−34	—		95–220 K (Γ_8) (<0.05% Cf)	
Bk_2O_3 (bcc)	f^8	9.3	9.0	141	—	32	60–300 K (2% Cf)	[6]
CfO_2 (fcc)	f^8	9.3	9.1	70	—	7	100–340 K	[8]
$BaCfO_3$ (bcc)	f^8	9.3	9.2	210	—	7	100–340 K	[8]
Cf_7O_{12} (rhom)	40% f^9 + 60% f^8	9.7	9.5	95	—	8	100–340 K	[8]
Cf_2O_3 (mono)	f^9	10.2	10.1	80	—	8	100–340 K	[8]
Cf_2O_3 (bcc)	f^9	10.2	9.8	115	—	19	100–340 K	[8]
Es_2O_3 (bcc)	f^{10}	10.3	10.5	53	—	—	4.2–180 K (13% Bk)	[9]

[a] References:
[1] Karraker (1975a).
[2] Soderholm et al. (1986).
[3] Nave et al. (1983).
[4] Morss et al. (1983).
[5] Kanellakopulos and Müller (1979).
[6] Nave et al. (1984).
[7] Karraker (1975b).
[8] Moore et al. (1986).
[9] Huray et al. (1983).

crystal-field splitting of the ground state. The lower temperature region gives a g value about 17% higher than that determined by EPR but the value is considered good enough to indicate the Γ_7 level as the ground state. The higher temperature region beginning at 50 K is interpreted as the start of population of the Γ_8 level about 35 cm^{-1} higher in energy. The data showed an antiferromagnetic transition at 8.5 ± 0.5 K; however, neither neutron diffraction (Boeuf et al. 1979) nor Mössbauer effect measurements (Kalvius et al. 1969) indicated any ordering.

6.2.2. *Curium oxides*

Magnetic measurements have been made on the sesquioxide of curium by several groups: Kanellakopulos and Müller (1979), Nave et al. (1983) and Morss et al. (1983). According to a free-ion model, assuming a trivalent metal ion, the electronic configuration should be $5f^7$. The theoretical value based on L–S coupling is $7.92\mu_B$ and when based on intermediate coupling is $7.55\mu_B$. From table 14, which summarizes the experimental results, it is seen that the values for μ_{eff} range from $7.5\mu_B$ to $8.2\mu_B$ in the high-temperature regions. This spread in experimental values prevents a determination of the validity of an intermediate-coupling model due to the small difference in the moment values for intermediate coupling and L–S coupling. The results for the curium-244 sesquioxide as reported by Kanellakopulos and Müller (1979) indicate significant curvature and the crystal structure is not reported. The inverse susceptibility versus temperature data (Nave et al. 1983) is given for the bcc form and the monoclinic form in fig. 16. There is a slight curvature in the data which is due to a small temperature-independent component (χ_0) in the susceptibility. The solid lines are a result of a MCW fit to the data with the parameters μ_{eff}, θ_p and χ_0 as given in table 14. It is seen from these plots that the monoclinic sample has an antiferromagnetic transition at 18 ± 5 K which is in accord with the results of Morss et al. (1983). The inverse susceptibility of the bcc form deviates sharply toward zero at 4.2 K.

The dioxide of curium is an interesting but difficult species to study. Since it involves a tetravalent state for curium, its electronic configuration should be close to that of trivalent americium as in Am_2O_3 and should be nominally a $5f^6$ configuration. As such, TIP is expected as discussed above for americium. Unlike americium, the more stable species of curium oxide compounds is the trivalent sesquioxide which has a very large paramagnetic moment. The presence of very small amounts of Cm^{3+} in the sample will be interpreted as a nonzero moment for the Cm^{4+} species. Thus questions of stiochiometry and purity are of utmost importance and will be included in the discussion below.

To date no one has observed TIP for CmO_2. The smallest measured value for μ_{eff} is reported (Nave et al. 1983) to be $1.63 \pm 0.04\,\mu_B$ for samples of curium-248. The temperature dependence of the inverse susceptibility is given in fig. 17, where it is seen that there is significant curvature in the data. This is explained by the presence of a large χ_0 or χ_{TIP} term in the susceptibility as well as a Curie–Weiss term. The smooth curve through the data points is determined by a non-linear least squares fit to the data with μ_{eff}, θ_p and χ_0 as parameters. The best-fit parameters are given in table 14 for each sample. Some contribution to χ_0 comes from the sample

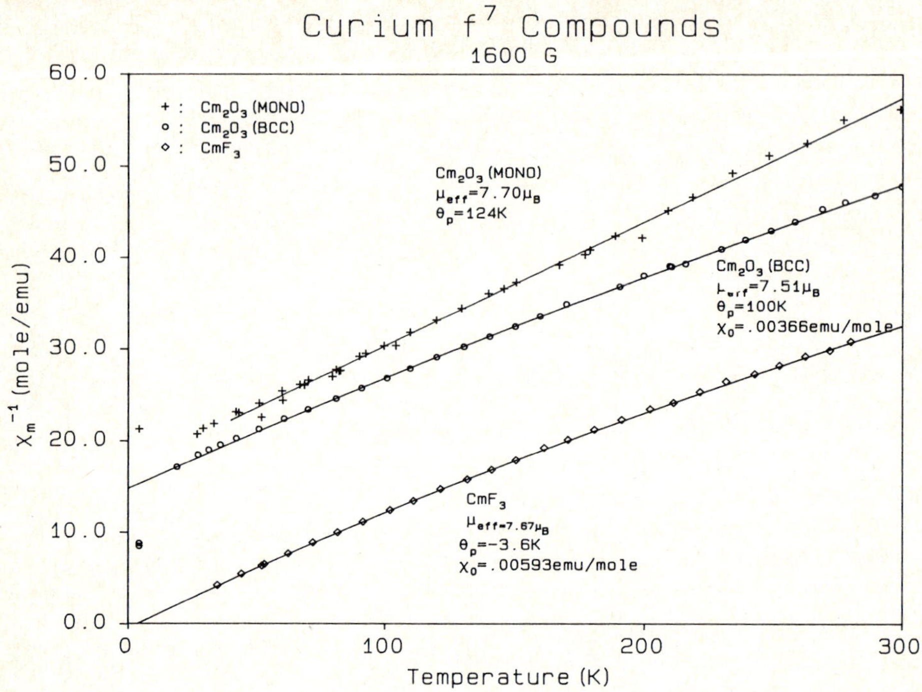

Fig. 16. Inverse susceptibility versus temperature for the nominal $5f^7$ curium compounds: Cm_2O_3(bcc), Cm_2O_3 (mono) and CmF_3. The solid lines are a result of a modified Curie–Weiss fit to the data with the parameters as indicated.

holder but most is due to the sample, and the value varies from sample to sample.

Efforts to find other compounds which better stabilize the tetravalent state have not led to smaller measured values for the moment. For example, the ternary compound $BaCmO_3$, which has a cubic perovskite structure, thermally decomposes at a higher temperature than the oxide and thus should provide better stabilization for the tetravalent state. Further evidence for better stabilization in this compound is given by the fact that TbO_2 can not be prepared in air or oxygen atmospheres but $BaTbO_3$ can be readily prepared in air. Nevertheless this compound has similar behavior to CmO_2 as seen in fig. 17 and has a value for the moment identical to that of the best dioxide sample. Another attempt along these lines involved ^{244}Cm diluted in $BaCeO_3$ at the 1% level (Bickel 1982). This work is complicated by the high alpha-radiation field of this isotope which appears to lead to time-dependent stoichiometry for the samples. The moment changed from $1.74\mu_B$ on day one to $2.11\mu_B$ on day six. This is most likely caused by reduction brought about by alpha-radiolysis. Extrapolating to $t=0$ gives a value of $1.39\mu_B$. Thus it appears that either stoichiometric samples of CmO_2, $BaCmO_3$ and $Ba(Ce_{0.99}Cm_{0.01})O_3$ have not been prepared, they do not exist at all, or the electronic configuration deviates significantly from pure $5f^6$.

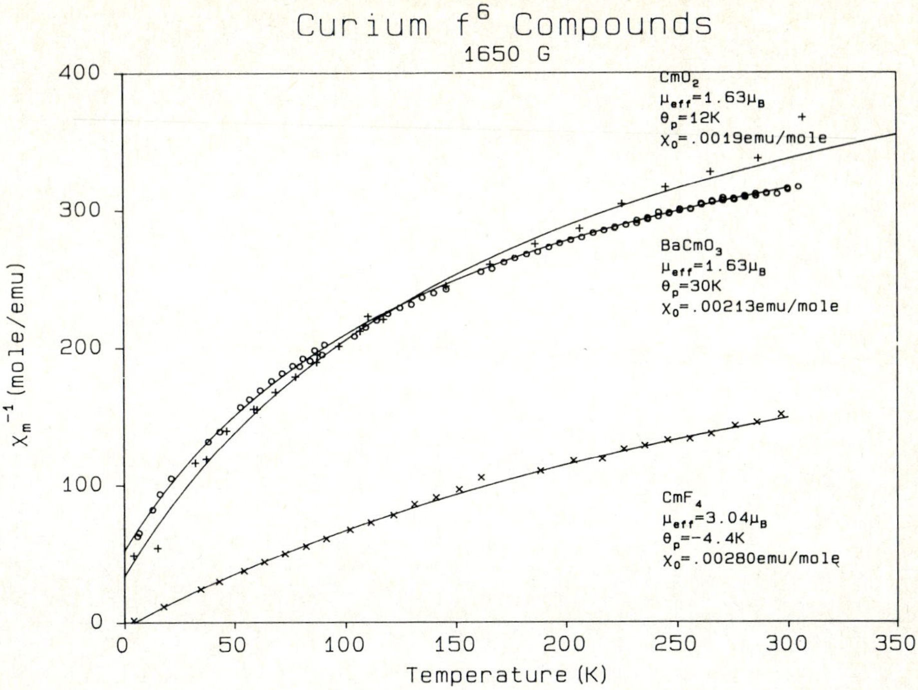

Fig. 17. Inverse susceptibility versus temperature for the nominal $5f^6$ curium compounds: CmO_2, $BaCmO_3$ and CmF_4. The solid lines are a result of a modified Curie–Weiss fit to the data with the parameters as indicated.

In the work by Nave et al. (1983) the samples were analyzed for impurity content by mass spectroscopy and radioassay. The total content of magnetic impurities was well below that necessary to account for the measured moment. X-ray analysis can be used to obtain information about the stoichiometry of the material in terms of the deviation of the lattice parameter from that of stoichiometric CmO_2; however, this depends on knowledge of the lattice parameter of the true stoichiometric compound and the assumption that only a single-phase solid solution of the form CmO_{2-x} is produced. The endpoint lattice parameter can be estimated by comparison to other actinide dioxides to be about 5.353 Å. The best value obtained in the preparations was 5.356 Å and indicates 2% Cm^{3+} for this sample, while the measured moment would require about 4%. The uncertainty of these values and the fact that perhaps as much as 7% of a second phase of Cm^{3+} may not be observable in an X-ray pattern make drawing any conclusions impossible at this point.

6.2.3. *Berkelium oxides*

Measurements have only been reported for the dioxide of berkelium (Nave et al. 1983, Karraker 1975b) although results of unpublished work of the authors on the bcc crystal form of the sesquioxide are given for completeness. Figures 18 and 19

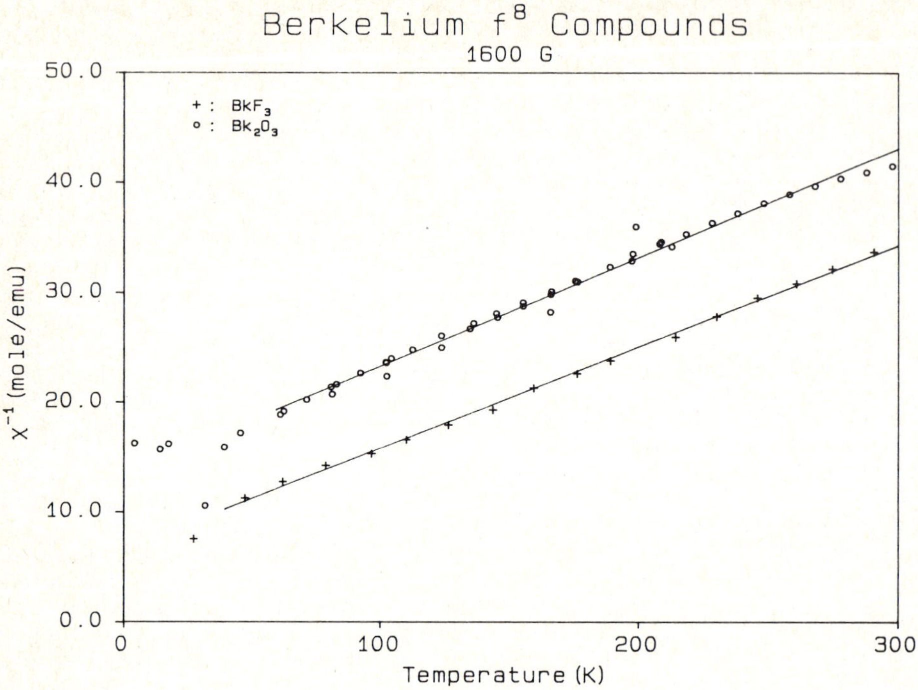

Fig. 18. Inverse susceptibility versus temperature for the nominal $5f^8$ berkelium compounds: Bk_2O_3 (bcc) and BKF_3. The solid lines are the result of a Curie–Weiss fit to the data with the values for μ_{eff} and θ_p as indicated.

give the temperature dependence of the inverse susceptibility for the sesquioxide and dioxide, respectively, at the highest fields of measurement ~1700 G. Although there is slight curvature in the region of 100 K to 300 K for the sesquioxide, a linear least-squares fit gives the values in table 14 for μ_{eff} and θ_p. This value of $9.1\mu_B$ for μ_{eff} is slightly lower than the theoretical value of $9.3\mu_B$ calculated for its nominally f^8 configuration based on an intermediate-coupling model. The data for BkO_2 give a better fit to a straight line in the Curie–Weiss region (70–300 K) than the sesquioxide and give the values for μ_{eff} and θ_p also listed in table 14. In this case the experimental values for μ_{eff} of $7.8\mu_B$ and $8.1\mu_B$ are closer to the theoretical value for an f^7 configuration based on an $L-S$ coupling model ($7.92\mu_B$, see table 4). Both the dioxide and sesquioxide order magnetically with T_N, the former at 40 K and the latter at 32 K.

The very large value for θ_p (~200 K) for BkO_2 indicates a significant exchange coupling between the local moments. Karraker considered BkO_2 diluted in the isostructural, nonmagnetic matrix of ThO_2 at the 1% doping level. This should remove any direct exchange coupling effects and reduce the values for θ_p. Indeed the value is an order of magnitude lower for the diluted case. The surprising difference between the pure and the diluted BkO_2 is the much larger crystal-field splitting in the latter. Karraker estimates the total splitting ($E_{\Gamma_7}-E_{\Gamma_6}$) of BkO_2–ThO_2 to be

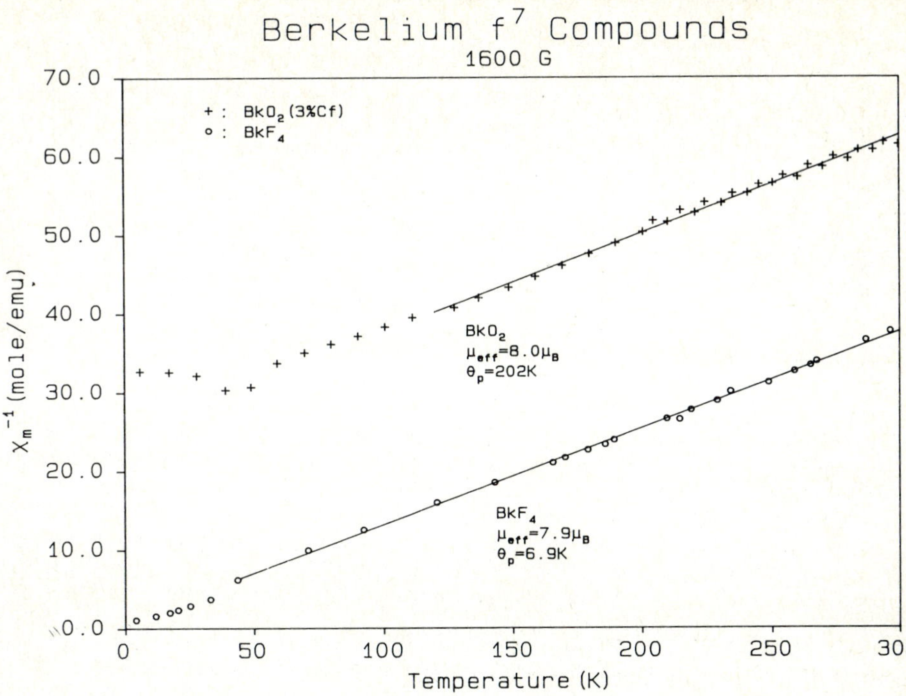

Fig. 19. Inverse susceptibility versus temperature for the nominal $5f^7$ berkelium compounds: BkO_2 and BkF_4. The solid line is the result of a Curie–Weiss fit to the data with the values for μ_{eff} and θ_p as indicated.

greater than $213\,cm^{-1}$ (or 307 K) based on their measurement. Clearly for pure BkO_2 the data indicates a splitting less than 70 K. It also should be noted that the antiferromagnetic transition attributed by Karraker to pockets of BkO_2 was at a temperature of ~3 K, a value much lower than that observed for pure BkO_2.

6.2.4. *Californium oxides*

For californium Moore et al. (1986) have reported results of measurements on the sesquioxide (both cubic and monoclinic forms), the dioxide, the intermediate oxide and the ternary compound $BaCfO_3$. All exhibit Curie–Weiss behavior at temperatures above 100 K and have a maximum in the susceptibility at temperatures in the range of 7–20 K. The temperature variation of the susceptibility is given in figs. 20–23, and table 14 lists the experimental values for μ_{eff} and θ_p determined from these graphs. Trivalent californium with a nominally $5f^9$ configuration has a theoretical value of $10.2\,\mu_B$ and tetravalent californium has a value of $9.3\,\mu_B$ (table 5). Assuming additive susceptibilities and assuming that Cf_7O_2 is a mixture of 40% Cf^{3+} and 60% Cf^{4+} gives a theoretical value of $9.7\mu_B$ for μ_{eff}. From table 14 it appeares that the experimental values are slightly lower than these theoretical values but follow the same trend in going from the 3+ to 4+ state.

All the compounds show evidence for antiferromagnetic ordering, the bcc Cf_2O_3

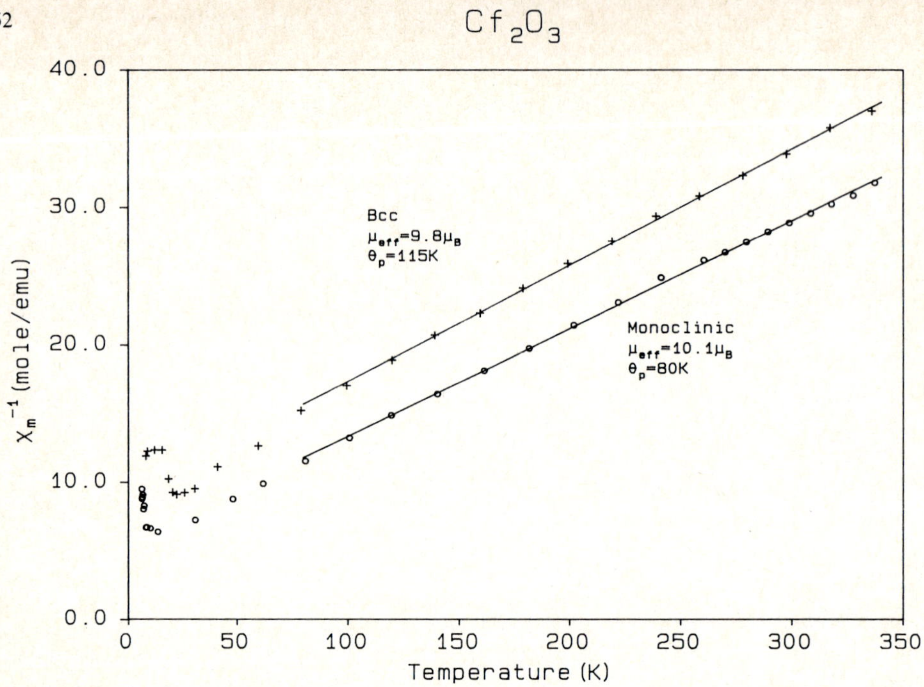

Fig. 20. Inverse susceptibility versus temperature for Cf_2O_3 (bcc) and Cf_2O_3 (monoclinic). The solid lines are the result of a Curie–Weiss fit to the data with the values for μ_{eff} and θ_p as indicated.

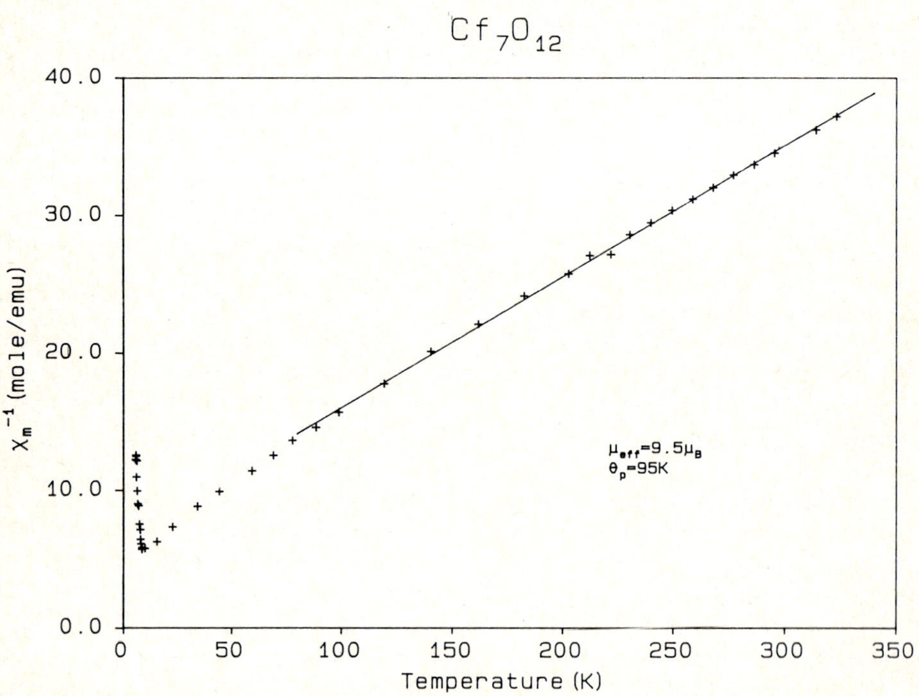

Fig. 21. Inverse susceptibility versus temperature for Cf_7O_{12}. The solid line is the result of a Curie–Weiss fit to the data with the values for μ_{eff} and θ_p as indicated.

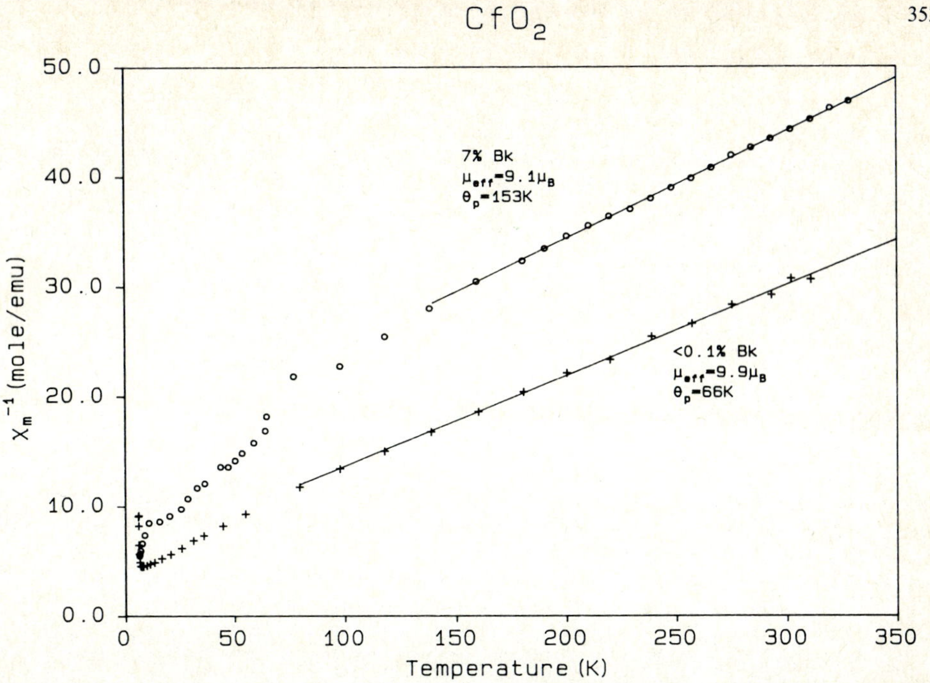

Fig. 22. Inverse susceptibility versus temperature for CfO_2 (7% Bk) and CfO_2 (<0.1%Bk). The solid lines are the result of a Curie–Weiss fit to the data with the values for μ_{eff} and θ_p as indicated.

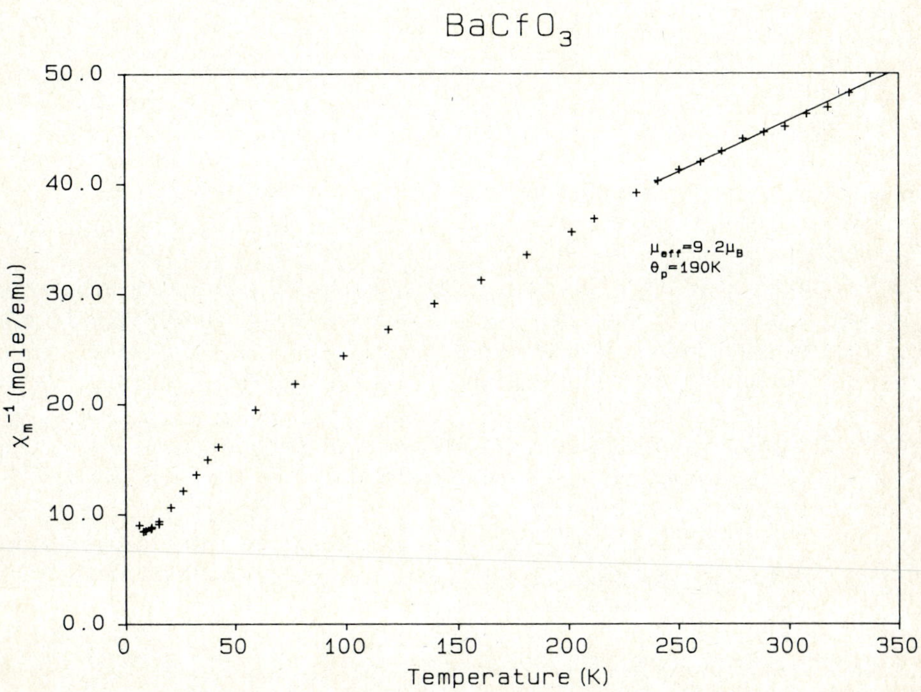

Fig. 23. Inverse susceptibility versus temperature for $BaCfO_3$. The solid line is the result of a Curie–Weiss fit to the data with the values for μ_{eff} and θ_p as indicated.

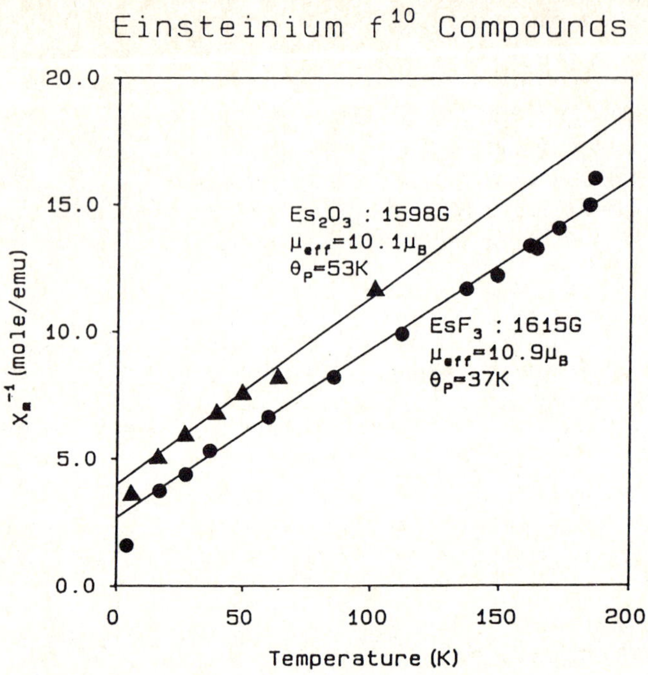

Fig. 24. Inverse susceptibility versus temperature for the nominal $5f^{10}$ einsteinium compounds: Es_2O_3 and EsF_3. The solid lines are the result of a Curie–Weiss fit to the data with the values for μ_{eff} and θ_p as indicated.

having the highest value of T_N of 19 K while the others have T_N falling in the range of 6–9 K. This is opposite to the behavior for curium or the rare earth terbium where the monoclinic sesquioxide has the higher antiferromagnetic transition temperature. Dysprosium, the homoloque of californium, has a T_N of 1.2 K for the sesquioxide.

Below 100 K deviations of the inverse susceptibilities from a linear variation with temperature, i.e. Curie–Weiss behavior, is evident for all the samples, as seen in figs. 20–23, and is indicative of crystal-field splitting of the ground state. This deviation is largest for the $BaCfO_3$ compound and one of the CfO_2 samples. CfO_2 was obtained from aged samples of BkO_2 as a result of the decay of the berkelium. One sample had 7% Bk at the time of measurement while the other of considerable age contained less than 0.1% Bk. Both gave the same value for μ_{eff} but the berkelium impurity appeared to greatly effect both θ_p and the behavior of the susceptibility below 100 K, as indicated in fig. 22 where a complicated structure is observed in this temperature range for the 7% sample.

6.2.5. *Einsteinium oxides*

A single measurement has been reported for Es_2O_3 (Huray et al. 1983) with data in the temperature range from 4.2 to 180 K and for fields less than 1600 G. X-ray analysis of the sample indicated it was amorphous (due to rapid damage to crystallinity from alpha-decay), but it was most likely of bcc form due to the low temperature of preparation. The results are shown in fig. 24, where the indicated straight line results in a value of $(10.1 \pm 0.3)\mu_B$ for μ_{eff} and 53 K for θ_p. At the time of measurement the 4.1 μg sample was 4 days old and thus contained 13% berkelium daughter. Assuming additive susceptibility and a moment of $9.7\mu_B$ for the berkelium leads to a corrected value of $10.5\mu_B$ for the sesquioxide. This value is consistent with that expected for a $5f^{10}$ configuration (Es^{3+} ion).

6.3. *Halides*

Reports of magnetic measurements on the halides have been mainly limited to the fluorides to date except for AmI_2 (Baybarz et al. 1972), $CfCl_3$ (Nave et al. 1987), and the complex compounds CmOCl (Marei and Cunningham 1972), $Cs_2NaAmCl_6$, $Cs_2NaCmCl_6$ and $Cs_2NaBkCl_6$ (Hendricks et al. 1974). The fluoride anion would be expected to produce a more ionic compound than the oxide based on electronegativity alone, although the crystal structure may have a significant effect. A summary of these results is given in table 15.

6.3.1. *Americium halides*

The results of magnetic measurements by the authors (Nave et al. 1983) on a 901.5 μg sample of AmF_3 are presented in fig. 25 as a plot of χT versus T. For this nominal f^6 configuration the measured paramagnetic moment is small, while χ_{TIP} is large. The slope of a linear least-squares fit to the data in this form (assuming a zero value for θ_p) gives a value of 714×10^{-6} emu/mol for χ_{TIP}. When this term is subtracted from the susceptibility, Curie–Weiss behavior is obtained with the values for μ_{eff} and θ_p listed in table 15. This value for χ_{TIP} is very close to that reported for another nominal f^6 compound, Am $(C_5H_5)_3$ (Kanellakopulos et al. 1978) where a χ_{TIP} of 715×10^{-6} emu/mol was obtained. These values are about 12% higher than the value reported for $Cs_2NaAmCl_6$ by Soderhom et al. (1986). A much higher value for this same compound was reported by Hendricks et al. (1974) (5400×10^{-6} emu/mol) but was complicated by the need to make a large correction for Cm^{3+} impurity. The calculated value for the free ion as discussed in section 6.2.1 is smaller than the lowest measured value and was determined to be 500×10^{-6} emu/mol. This calculated value is not appreciably sensitive to the crystal field and should not vary much in different compounds. It is interesting to note that all of these halide samples exhibited temperature-dependent paramagnetism at low temperatures. Solderholm et al. (1986) attributed the low-temperature behavior to a sampleholder effect while Hendricks et al. (1974) had to correct for 50 μg Cm^{3+} impurity, which would mask any such effect. Kannellakopulos et al. (1978) report a value of $0.27\,\mu_B$ for $^{243}Am(C_2H_5)_3$ but attribute it to Am^{2+} impurity – an uncommon oxidation state for americium. Due to the large moment for Am^{2+} ($7.6\,\mu_B$) the presence of only 0.12 atom percent of this charge state would give the appropriate value for μ_{eff} and there are no known analytical techniques to detect the presence of a second phase of americium at this level. Also at this impurity level it is more likely for another

TABLE 15
Actinide halides.

Compound (Structure)	Configuration	μ_{eff} (I.C.)	μ_{eff} (μ_B)	θ_p (K)	$\chi_0\left(10^{-6}\frac{\text{emu}}{\text{mol}}\right)$	Comments	Ref.[c]
AmI_2	f^7	7.6	6.7	25	—	—	[1]
AmF_3 (trig.)	f^6	0	0.63	1.8	714×10^{-6}	—	[2]
$Cs_2NaAmCl_6$	f^6	0	—	—	5400	—	[3]
CmF_4 (mono)	f^6	0	3.0–3.5	−1.2–4.4	328–2800	—	[2]
CmF_3 (trig)	f^7	7.6	7.7	−3.6	5928	—	[2]
$CmF_3 \cdot H_2O$	f^7	7.6	7.7	−5	—	—	[4]
$(CmF_3)LaF_3$	f^7	7.6	7.7	−22	—	—	[4]
CmOCl	f^7	7.6	7.6	−6	—	—	[4]
$Cs_2NaCmCl_6$[a]	f^7	7.6	{7.90	3.87	—	7.5–25 K	[4]
			7.48	1.15	—	25–48 K	[3]
BkF_4 (mono)	f^7	—	7.9	6.9	—	2% Cf	[2]
BkF_3 (trig.)	f^8	9.3	9.4	77.9	—	3% Cf	[5]
$Cs_2NaBkCl_6$[a]	f^8	9.3	—	—	192 000	10–40 K	[3]
$CfCl_3$ (hex)	f^9	10.2	10.1	13	—	60–350 K	[6]
$CfCl_3$ (ortho)	f^9	10.2	10.3	37	—	100–350 K	[6]
EsF_3	f^{10}	10.3	11.4	37	—	4.2–200 K (31% Bk)	[7]

[a] Doped into $Cs_2NaLuCl_6$ (1.6% by weight).
[b] Doped into $Cs_2NaLuCl_6$ (0.3% by weight).
[c] References:

[1] Baybarz et al. (1972).
[2] Nave et al. (1983).
[3] Hendricks et al. (1974).
[4] Marei and Cunningham (1972).
[5] Nave et al. (1981b).
[6] Nave et al. (1987).
[7] Huray et al. (1983).

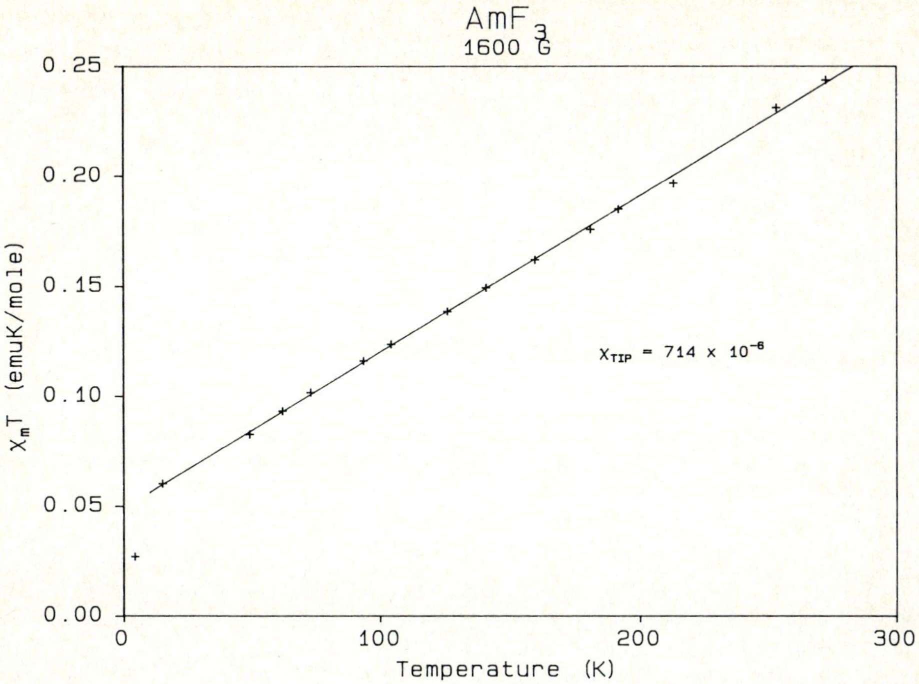

Fig. 25. A plot of susceptibility times temperature as a function of temperature for AmF_3. The slope of the fitted straight line gives the large temperature independent susceptibility term.

element with large paramagnetic moment (e.g. rare earth) to go undetected by mass spectroscopic analysis, especially when analyzing submilligram-size samples. The only measurements reported for divalent americium are for AmI_2 with values for μ_{eff} of $6.7\mu_B$ and θ_p of 25 K given (Baybarz et al. 1972). This value for μ_{eff} is considerably lower than that expected for the f^7 configuration ($7.55\mu_B$). No magnetic ordering was observed down to 4.2 K.

6.3.2. *Curium halides*

Measurements for both CmF_3 and CmF_4 have been reported (Marei and Cunningham 1972, Nave et al. 1983). The tetravalent species with its nominal f^6 configuration again fails to show pure TIP, however, in this case the deviation is more significant. A plot of inverse susceptibility versus temperature for CmF_4 is given in fig. 17 and illustrates the large curvature observed over the entire temperature range. Including both a large temperature-independent term and a Curie–Weiss term in a non-linear least-squares fitting procedure gives a very good fit to the data as indicated by the smooth curve. The TIP-part varied significantly from sample to sample [$\chi_0 = (328–2800) \times 10^{-6}$ emu/mol] but values for θ_p were small in all cases ($\theta_p = -1.2–4.4$ K). Values for μ_{eff} in the range $(3.0–3.5)\mu_B$ are obtained and would require the presence of $\sim 14\%$ of a second phase of trivalent curium or other large-moment impurity. Again all of the techniques as described for the oxides, i.e. X-ray analysis, mass spectroscopic analysis and optical absorption, were used to detect a second phase of trivalent curium and/or other large-moment impurity and each gave a negative result. At this level of impurity a second phase should show up in all of the analyses. Clearly there is some other effect besides stoichiometry or purity involved here and if the moment value represents the nominal f^6 electronic configuration the results indicate a substantial ligand modification of the f^6 configuration. No magnetic ordering has been observed down to 4.2 K.

Measurements were first reported for CmF_3 by Marei and Cunningham (1972), who also gave results for CmF_3 diluted in LaF_3 and the compound CmOCl. The measurements were repeated by the authors (Nave et al. 1983) for CmF_3 with the inverse susceptibility plot given in fig. 16. In all cases the values for μ_{eff} were in reasonable agreement with the theoretical intermediate-coupling-model value of $7.55\mu_B$ for the $5f^7$ configuration. It is surprising that values for θ_p are small in all cases, except for the diluted CmF_3–LaF_3 samples. No magnetic ordering was observed down to 4.2 K.

Hendricks et al. (1974) report measurements on $Cs_2NaCmCl_6$ as a dopant ($\sim$1.6 weight %) in $Cs_2NaLuCl_6$. Their measurements covered the limited range 3–50 K. In the range 7.5–25 K a Curie–Weiss behavior was observed with a value of $7.9\mu_B$ determined for μ_{eff}. Above 25 K the moment decreases slightly but the uncertainty in the measurements becomes large here. Below 7.5 K a deviation from Curie–Weiss behavior is attributed to crystal-field effects and suggests a total splitting of 5–10 cm^{-1} for the $J=\frac{7}{2}$ ground state, which splits into Γ_6, Γ_7 and Γ_8 level in the octahedral symmetry. The data does not permit a determination of the order of spacing of the individual levels.

6.3.3. *Berkelium halides*

Measurements are reported for both BkF_4 and BkF_3 (Nave et al. 1981b, 1983) with only 2% Cf daughter content for the former at the time of measurement and 3% Cf for the latter. The tetravalent BkF_4 with the nominal f^7 configuration has a value for μ_{eff} of $7.93\mu_B$ and is higher than that observed for CmF_3. On the other hand, the trivalent BkF_3 gave a value of $9.38\mu_B$ which is closer to the intermediate-coupling value of $9.3\mu_B$ for the f^8 configuration. No magnetic ordering was observed for either case, although for BkF_3 the measurements do not extend below 25 K. Deviations from Curie–Weiss behavior are seen below 50 K in the inverse susceptibility plots given in figs. 18 and 19. The trivalent compound has a much larger value for θ_p than the tetravalent compound (78 K versus 7 K).

Measurements on Bk^{3+} as a dopant in $Cs_2NaBkCl_6$ at the 0.3 wt% level are reported by Hendricks et al. (1974). A large temperature-independent paramagnetism was observed in the range 10–40 K, while above 40 K the susceptibility decreases slightly the $J=6$ ground state splits into Γ_1, Γ_2 and Γ_4 level in an octahedral crystal field. From a consideration of reasonable B_0^6/B_0^4 values, the authors conclude that the Γ_1 singlet must be the ground state, and to account for the large TIP term the next-higher level must be the Γ_4 with $E_{\Gamma_1}-E_{\Gamma_4}=85\,cm^{-1}$. A value of $\chi_{TIP}=192\,000\times10^{-6}$ emu/mol is reported.

6.3.4. *Californium halides*

Magnetic measurements are reported (Nave et al. 1987) for polycrystalline samples of californium trichloride in both a hexagonal [UCl_3 type; Cf(III) C.N. = 9] and orthorhombic [$PuBr_3$ type; Cf(III) C.N. = 8] crystal form. The compounds have Curie–Weiss behavior over different ranges of temperature (orthorhombic: 100–350 K, hexagonal: 60–350 K). The effective moments given in table 15 are in good agreement with an f^9 configuration, as expected. The extension of the Curie–Weiss region to lower temperatures for the hexagonal form indicates a smaller crystal-field splitting for this form than for the orthorhombic type. This behavior is observed in fig. 26 which gives the inverse susceptibility versus temperature for the two forms. The average values for several samples give a Weiss constant, θ_p, for the hexagonal form (13 ± 5 K) that is significantly lower than for the orthorhombic form (37 ± 10 K). Below 15 K a field-dependent transition is observed for both forms as shown in fig. 27. In a field of 5 kG an antiferromagnetic maximum in the susceptibility is observed at 7 K for the hexagonal form and at 13 K for the orthorhombic form. For 30 kG both types exhibit monotonically increasing susceptibilities with decreasing temperature and both have saturation type behavior as a function of field as shown in the inset of fig. 27. The hexagonal form has the closest approach to saturation in the maximum field with an extrapolated moment per atom of 6 μ_B.

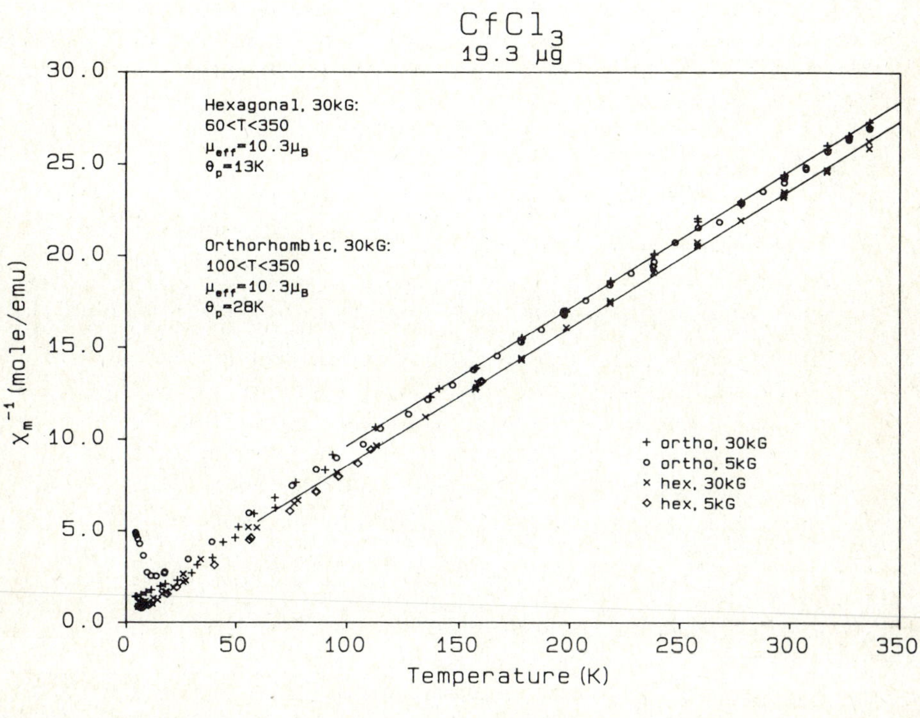

Fig. 26. Inverse susceptibility versus temperature for the hexagonal and orthorhombic forms of $CfCl_3$.

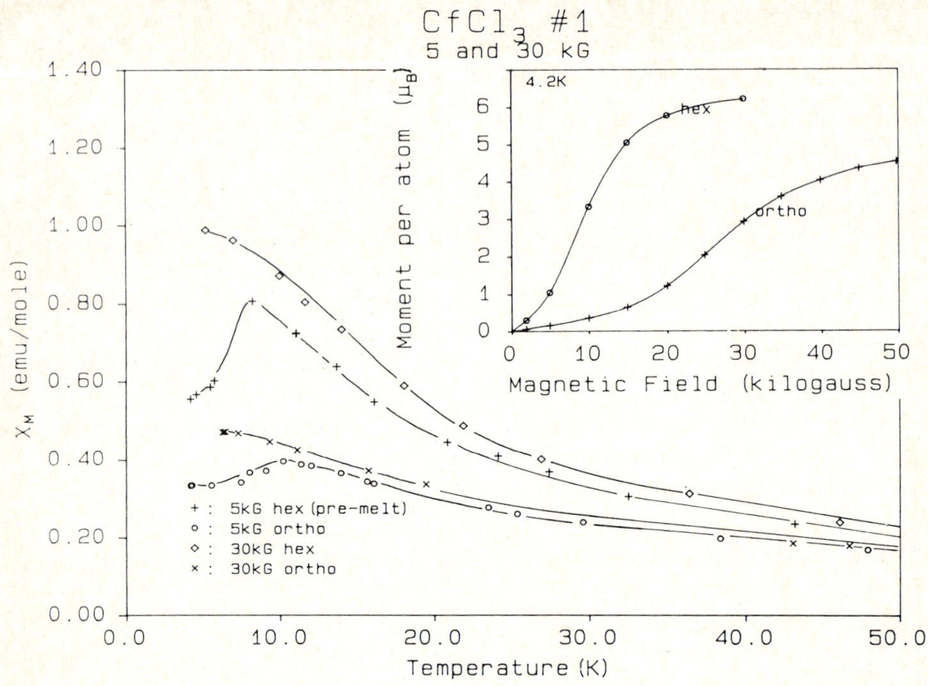

Fig. 27. Susceptibility versus temperature for the hexagonal and orthorhombic forms of $CfCl_3$ indicating the field-dependent transitions at low temperature. The inset gives the moment per atom versus field at 4.2 K for the two forms.

6.3.5. *Einsteinium halides*

Results of a preliminary study on EsF_3 have been reported by the authors (Huray et al. 1983). The measurements on a 3.25 μg sample were performed 10 days after its preparation, which meant 31% of the einstenium had decayed to berkelium. Results of the measurements are presented in fig. 24 and cover a limited temperature range (4.2–200 K). The effective moment obtained from a Curie–Weiss fit and corrected for the berkelium content has a value of $11.4\mu_B$.

6.4. *Pnictides*

Binary compounds of actinides or lanthanides with a pnictogen anion (group VA element) are semimetallic and have a simple cubic (NaCl) structure. For these compounds the ordering temperatures and effective moments vary with An–An spacing in a manner that cannot be explained using a simple dipole–dipole exchange-coupling model. For example, the uranium or neptunium pnictides have ordering temperatures that increase monotonically with increasing An–An separation, as seen in fig. 28. Since these compounds are semimetallic, it is postulated that in addition to direct overlap interactions, a Ruderman–Kittel–Kasuya–Yosida (RKKY) interaction is present. For uranium it has been postulated

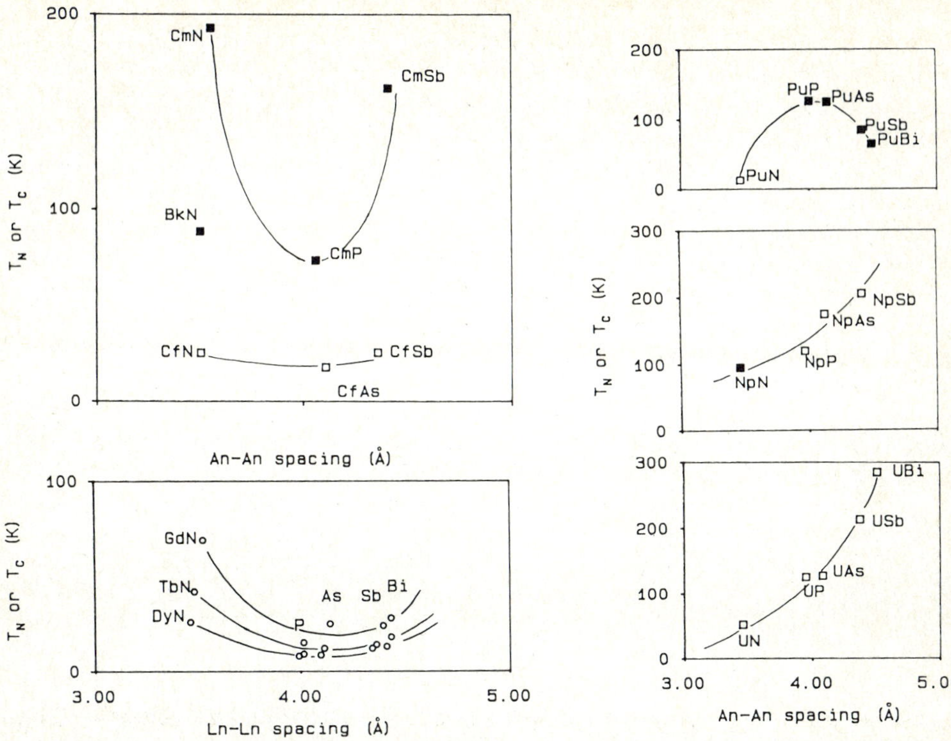

Fig. 28. Transition temperatures T_c (solid symbols) or T_N (open symbols) versus anion–anion separation for mono-pnictides of various lanthanides and actinides.

(Grunzweig-Genossar et al. 1968) that, in addition to the RKKY and the direct overlap interactions, there is also a superexchange coupling via the anions. The superexchange is necessary to account for the change from antiferromagnetism to ferromagnetism between a uranium mono-pnictide and the corresponding uranium monochalcogenide (group VIA element) which have almost identical An–An separation. Due to the localization of the 5f wavefunctions in the transamericium actinides there should be a reduction in contributions to the magnetic interactions from direct overlap and superexchange in pnictide compounds of these elements compared to the light actinide pnictides.

6.4.1. *Americium pnictides*

The first measurments of transamericium pnictide compounds involved the short-lived isotope americium-241. Dunlap (1971) observed TIP for AmSb and reported a value of 1250×10^{-6} emu/mol for χ_{TIP}. The TIP behavior agrees with an assumption of a trivalent americium ion in an f^6 configuration ($\mu_{eff} = 0$), but is higher than the 770×10^{-6} emu/mol value observed in oxide or fluoride compounds. Kanellakopulos et al. (1976), also using this isotope, performed measurements on AmN and AmAs. The nitride was also in agreement with an f^6 configuration but

TABLE 16
Pnictides.

	An–An spacing (Å)	μ_{eff} (μ_B)	θ_p (K)	$T_c(T_N)$	χ_0 $\left(10^{-6}\frac{emu}{mol}\right)$	Ref.[a]
^{241}AmN	3.532	—	—	—	780	[1]
^{241}AmAs	4.155	0.17	12	(13)	535	[1]
^{241}AmSb	4.412	0	—	—	1250	[2]
^{244}CmN	3.555	7.02	−109	109	—	[1]
^{248}CmN	—	6.93	−126	193	—	[3]
^{248}CmP	4.061	—	—	73	—	[1]
^{244}CmAs	4.168	6.58	−88	88	—	[1]
^{248}CmSb	4.414	—	—	162	—	[2]
^{249}BkN	3.50	7.87, 7.79	−42, −66	88	—	[4]
^{249}CfN	—	10.3	41	(25)	—	[5]
^{249}CfAs	4.108	10.1	29	(17.5)	—	[5]
^{249}CfSb	4.359	10.3	18.2	(25)	—	[5]

[a] References:
[1] Kanellakopulos et al. (1976).
[2] Dunlap (1971).
[3] Nave et al. (1981a).
[4] Nave et al. (1981b).
[5] Nave et al. (1986).

gave a $\chi_{TIP}=780\times10^{-6}$ emu/mol, in agreement with oxide and fluoride compounds. The arsenide, on the other hand, followed a modified Curie–Weiss law down to an antiferromagnetic transition at $T_N=13$ K. The value determined for μ_{eff} at 300 K was $1.14\mu_B$. Fournier and Troć (1985) analyzed their data and determined the values $\mu^*_{eff}=0.17\mu_B$ and $\theta_p=-12$ K with $\chi_{TIP}=535\times10^{-6}$ emu/mol and these are the values listed in table 16. Due to the small value for the moment and the trend of TIP in the other pnictides of americium this temperature-dependent paramagnetic component is most likely due to an impurity.

6.4.2. *Curium pnictides*

All but the bismuthide of the curium pnictides have been studied magnetically. Kanellakopolus et al. (1976) report measurements of CmAs and CmN using the short-lived isotope curium-244. Curie–Weiss behavior was observed in the region 100–300 K for CmAs and 150–300 K region for CmN. Both order ferromagnetically with $T_c=109\text{ K}\pm2$ K for CmN and $T_c=88\text{ K}\pm1$ K for CmAs as determined very accurately by extensive isomagnetization plots. Using these T_c values for $-\theta_p$ in the Curie–Weiss law, a good fit to the data is obtained in the limited regions of Curie–Weiss behavior with $\mu_{eff}=7.02\mu_B$ for CmN and $\mu_{eff}=6.58\mu_B$ for CmAs. It is remarkable that such a good fit to the data is obtained with $-\theta_p=T_c$ since this is not true for many ferromagnetic materials. Nave et al. (1981b) have reported measurements on CmP and CmSb, and for completeness the authors give results for CmN, all using the long-lived isotope ^{248}Cm. For this ^{248}CmN sample the transition temperature is much higher than that of the ^{244}CmN with $T_c=193$ K,

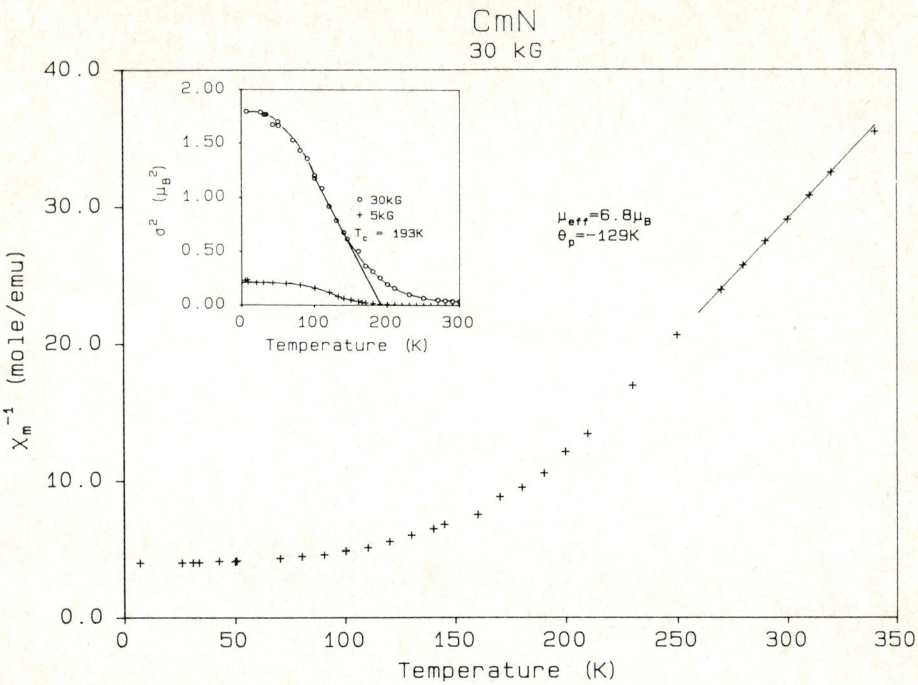

Fig. 29. Inverse susceptibility versus temperature for CmN. The solid line is the result of a Curie–Weiss fit to the data with the values for μ_{eff} and θ_p as indicated. The inset gives the square of the moment per atom versus temperature.

while the values $\mu_{eff} = 6.93\mu_B$ and $\theta_p = -126$ K are in reasonable agreement. This effect is not understood and the measurements must be repeated on additional samples. Figure 29 gives a plot of the inverse susceptibility versus temperature while the inset gives a plot of the square of the moment per atom (normalized to the 0 K value) as a function of temperature. A similar graph (fig. 30) is given for CmP and CmAs and summarizes the limited data available on these pnictides. The magnetic field was 1600 G for these measurements and the values $T_c = 73$ K for CmP, and $T_c = 162$ K for CmSb are determined as indicated. The maximum moment per atom was $0.1\mu_B$/atom for CmP and $0.4\mu_B$/atom for CmSb at 1600 G and $1.7\mu_B$/atom for CmN at 50 kG. From the data of Kanellakopulos et al. (1976) values of $1.1\mu_B$/atom for CmN and $2.2\mu_B$/atom for CmAs are determined at ~13 kG. These values are much smaller than the values for the theoretical saturation moment (table 7) and are small most likely due to a large magnetic anisotropy similar to that observed for the plutonium pnictides (Spirlet et al. 1983). Such an anisotropy would require significant deviation from the S state of the free-ion f^7 configuration. The paramagnetic effective moments are smaller than expected for Cm^{3+} (f^7 configuration, $\mu_{eff} = 7.55\,\mu_B$) even taking into account a negative spin polarization in the conduction electrons.

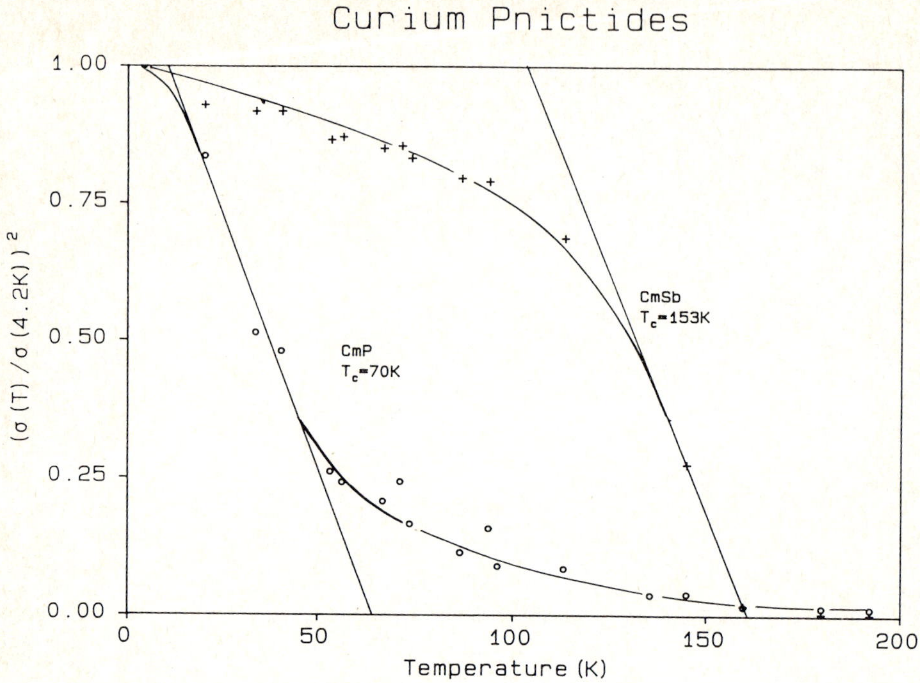

Fig. 30. The square of the moment per atom (normalized to the 4.2 K value) versus temperature for CmP and CmSb.

6.4.3. *Berkelium pnictides*

The authors (Nave et al. 1981b) have previously determined the magnetic susceptibility of two samples of BkN for magnetic fields below 1600 G. One had 3.9% Cf and the other 5.6% Cf at the time of measurement. Figure 31 gives a plot of the inverse susceptibility versus temperature, and the inset shows of a plot of the square of the moment per atom (normalized to the 0 K value) versus temperature. From the data one determines a transition temperature of 88 K for both, and a transition to a ferro- or ferrimagnetic state is indicated. The largest moment per atom is only $0.29\mu_B$, again probably indicating large magnetic anisotropy. The values $\mu_{eff} = 7.79\mu_B$, $7.87\mu_B$ and $\theta_p = -68$ K, -42 K, are determined from a Curie–Weiss fit to the limited temperature range of 80–150 K as indicated for one sample in fig. 31. The average value for μ_{eff} of $7.85\mu_B$ is close to that for and f^7 configuration ($7.55\mu_B$) and is considerably lower than the value of $9.3\mu_B$ expected for trivalent berkelium (f^8 configuration).

6.4.4. *Californium pnictides*

Recent measurements (Nave et al. 1986) on the californium pnictides CfN, CfAs and CfSb involved samples with masses in the range from 20 to 30 μg. Results of these measurements are given as graphs of inverse susceptibility versus temperature

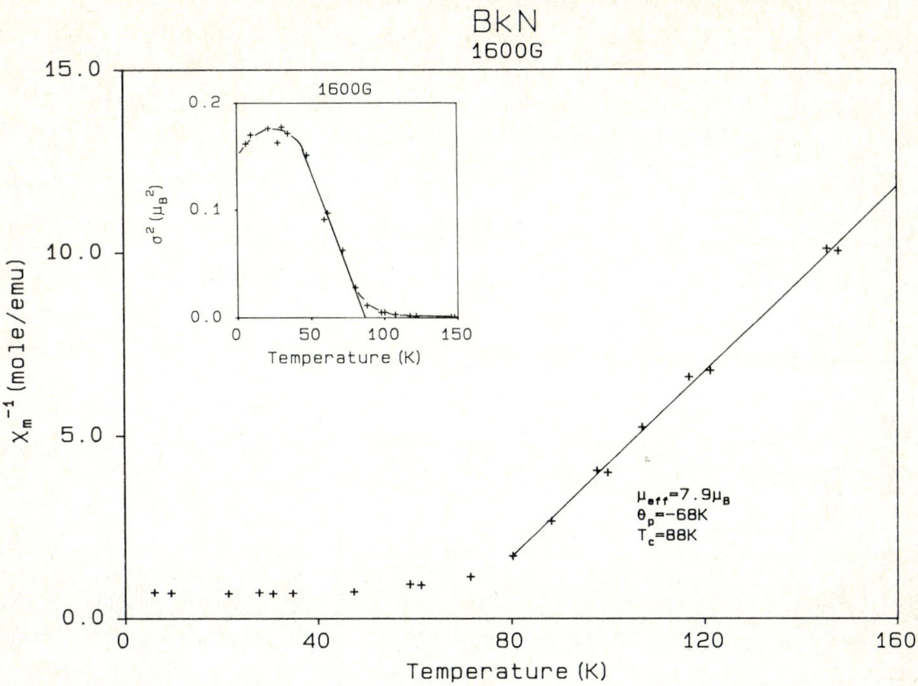

Fig. 31. Inverse susceptibility versus temperature for BkN. The solid line is the result of a Curie–Weiss fit to the data with the values for μ_{eff} and θ_p as indicated. The inset gives the square of the moment per atom versus temperature.

in figs. 32–34. Here it is seen that Curie–Weiss behavior extends to very low temperatures: ~50 K for CfN, ~20 K for CfAs and ~30 K for CfSb. The results of the Curie–Weiss fit to the data in the region from 80 K to 340 K were given in table 11. The values for the experimental μ_{eff} are in the range from 10.1μ_B to 10.3μ_B and are in accord with the intermediate coupling values for the f^9 (trivalent californium) or f^{10} (divalent californium) configurations which are 10.2μ_B and 10.3μ_B, respectively (table 11).

All samples had a low-temperature behavior indicating antiferromagnetic-type ordering, as illustrated by the maximum in the susceptibility plots shown in the insets of figs. 32–34. For CfN such a maximum is only present in the low-field data and there is a change in the magnetic character at higher fields as also observed in the GdN which is postulated to be a metamagnet (Wachter and Kaldis 1980).

The results of the measurements are summarized in fig. 28 where the ordering temperatures are plotted versus actinide–actinide separation. In the same figure, results for lanthanide homologues and light-actinide pnictides are given for comparison. The curium pnictides are more like the light-actinide pnictides in terms of exhibiting high transition temperatures and large variation across the series. On the other hand, the californium pnictides have low ordering temperatures

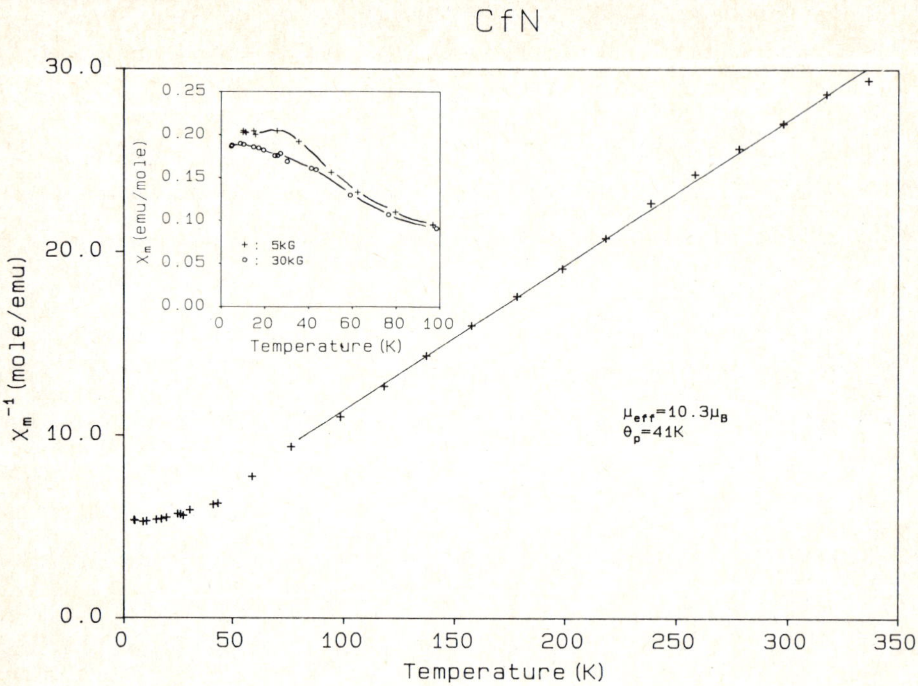

Fig. 32. Inverse susceptibility versus temperature for CfN. The solid line is the result of a Curie–Weiss fit to the data with the values for μ_{eff} and θ_p as indicated. The inset gives the susceptibility versus temperature in the transition region.

with little variation across the series like the homologues, the dysprosium pnictides. Since the only pnictide of berkelium which as been measured to date is the nitride, one cannot generalize about its series, but the BkN transition temperature is intermediate to CmN and CfN. This data implies that direct exchange and/or superexchange are decreasing with increasing atomic number as may be expected from increased localization. It should be noted that the nitrides may not fit in with the rest of the pnictide series based on the oversimplified assumption that the only property that varies from member to member in the series is the An–An spacing. For example, the type of magnetic order for the nitride is different from the remainder of the series for Pu and Np and due to the much smaller size of the nitrogen antion the An–An separation is almost the same as in the metal.

6.5. *Summary*

Magnetic moments of well-characterized lanthanide ions have been used to demonstrate the applicability of *L–S* coupling and Hund's rule to the 4f element series (section 2.4). By analogy, magnetic moments of lanthanide metals were compared to the predicted moments for a nominal $4f^n$ configuration to determine the 'effective' valency for atoms in the metal (all were 3+ except Sm and Yb which

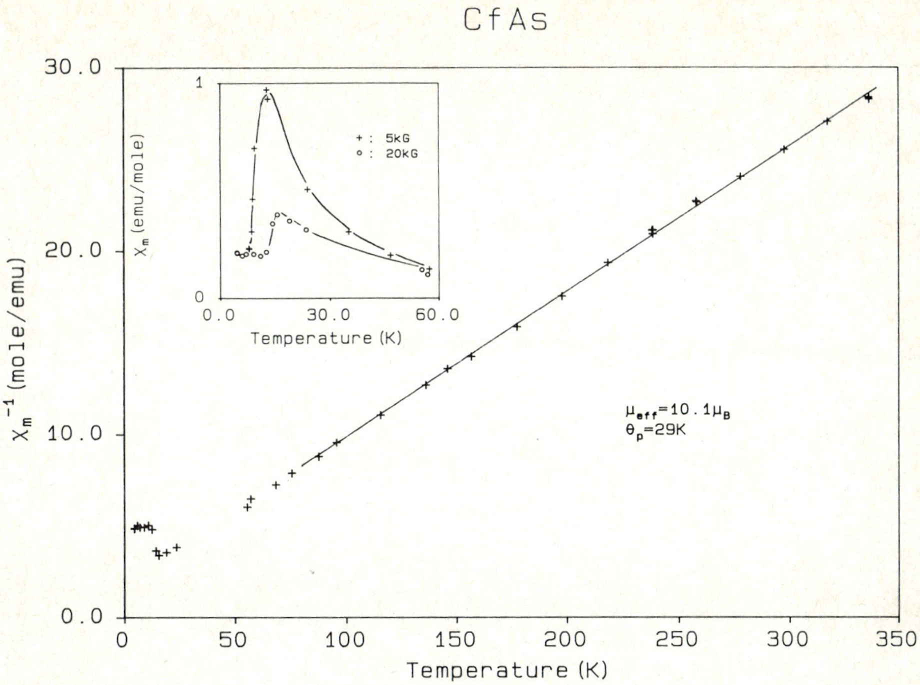

Fig. 33. Inverse susceptibility versus temperature for CfAs. The solid line is the result of a Curie–Weiss fit to the data with the values for μ_{eff} and θ_p as indicated. The inset gives the susceptibility versus temperature in the transition region.

were 'effectively' 2+). An examination of similar behavior for the more spatially extended orbitals of the 5f elements series has, until the last few years, been limited to the lighter members (actinium through plutonium). With the characterization of magnetic properties of the heavier ions and metals (americium through einsteinium) as presented in this chapter, a comparison can be made to the 4f homologues of most of the actinide series. As was observed in section 2.4, significant population of excited states, even for room-temperature measurements, is not expected and Van Vleck corrections should not be necessary. Because of the larger spin–orbit coupling for the 5f wavefunctions, however, the inclusion of intermediate coupling is expected to alter the character of the ground states and reduce the effective magnetic moments in the last half of the series (table 5).

Well-defined ionic compounds of the actinides are expected to have effective magnetic moments as given in table 5 if crystal-field effects are negligible. Am_2O_3, AmF_3, CmF_4, CmO_2 and $BaCmO_3$ should nominally correspond to the $5f^6$ configuration. Cm_2O_3, CmF_3, BkF_4, BkO_2 and $BaBkO_3$ should nominally correspond to the $5f^7$ configuration. Bk_2O_3, BkF_3, CfO_2 and $BaCfO_3$ should nominally correspond to the $5f^8$ configuration. Cf_2O_3, CfF_3, EsO_2 and $BaEsO_3$ should nominally act as ionic compounds corresponding to the $5f^9$ configuration.

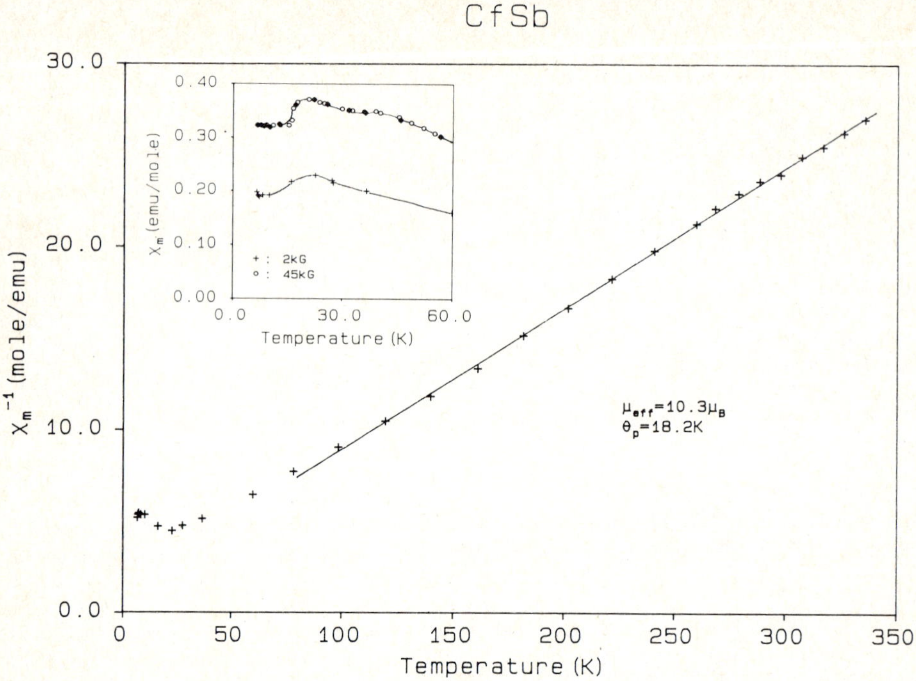

Fig. 34. Inverse susceptibility versus temperature for CfSb. The solid line is the result of a Curie–Weiss fit to the data with the values for μ_{eff} and θ_p as indicated. The inset gives the susceptibility versus temperature in the transition region.

And Es_2O_3 and EsF_3 should exhibit moments comparable with the nominal $5f^{10}$ configuration. The experimental effective magnetic moment values for these compounds are shown in fig. 35 as a function of the number of f electrons in their nominal electronic configuration. The solid line connects the theoretical values of table 4. As may be seen from this figure there is a reasonable correlation between the experimental and theoretical values except near the $5f^6$ configuration. This discrepancy has as yet not been proven conclusively by experiment but there is a strong indication that it is a real effect which warrants further theoretical consideration. One possible theoretical explanation lies in the crystal-field influence of neighboring anions upon the actinide ion which causes an admixture of different J multiplets (other than $J=0$) into the ground state. A more probably explanation is that the 5f electrons are participating in covalent bonding as described by Ellis (1985).

Since the actinide pnictides are semimetals, the conduction electrons are expected to play a strong role in screening the actinide ions and thus could modify the effective magnetic moment. These values are therefore not shown in fig. 35. For curium and berkelium the moments are considerably smaller than expected for an isolated 3+ ion, while for californium the agreement with such a model is very

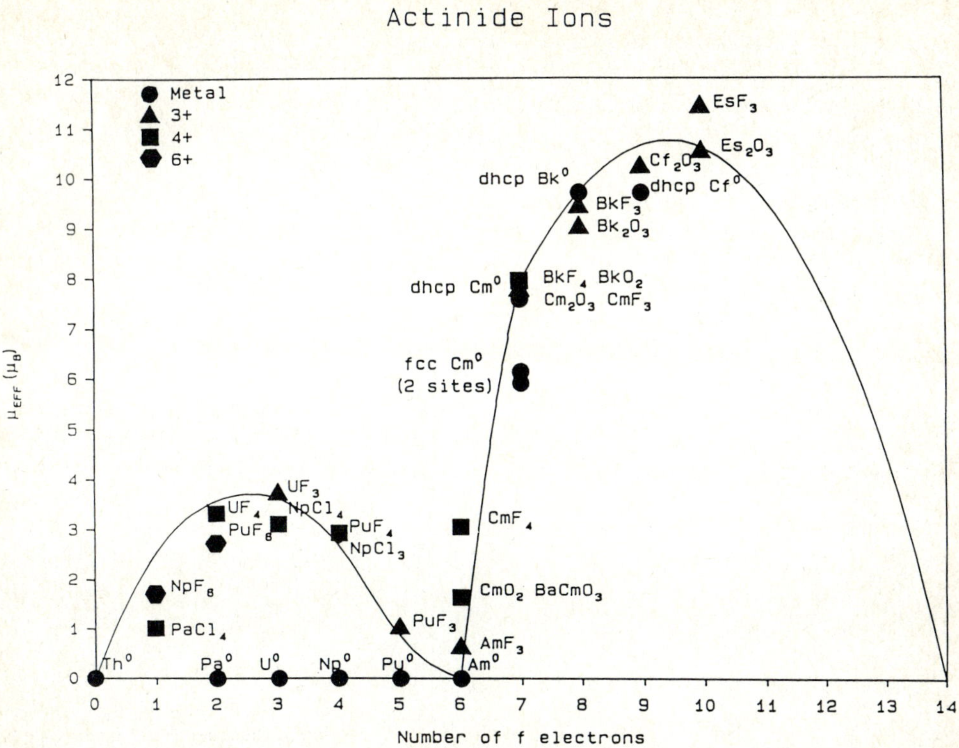

Fig. 35. The paramagnetic effective moments of actinide metals (solid circles) and compounds as a function of the number of f electrons. The solid line is from fig. 5.

good. It is clear that effects such as conduction electron polarization or covalency preclude treating the pnictides as isolated trivalent ions in all cases.

For the case of the lanthanide metals (section 4.2) it was found that an assignment of a tri-positive charge state gave an agreement between the measured effective magnetic moments and the theoretical values, except in the case of Eu and Yb metals which were better treated as di-positive. By analogy, we have plotted the effective magnetic moments of the actinide metals as circles in fig. 35 as if they were tripositive. Since curium metal has been measured in both an fcc as well as a dhcp phase, multiple circular data points are indicated in fig. 35 for the two phases of this metal. The complex thermal behavior of the magnetic susceptibility of fcc curium metal is consistent with ferrimagnetism with two different magnetic moments but could also be caused by a complex spin structure at low temperatures and significant crystal-field interactions at higher temperatures. The behavior will likely be confirmed only by neutron diffraction experiments.

The light-actinide metals (actinium through plutonium) do not exhibit a temperature-dependent magnetic susceptibility (Fournier and Troć 1985) and thus are consistent with zero effective magnetic moments. Clearly the treatment of metal

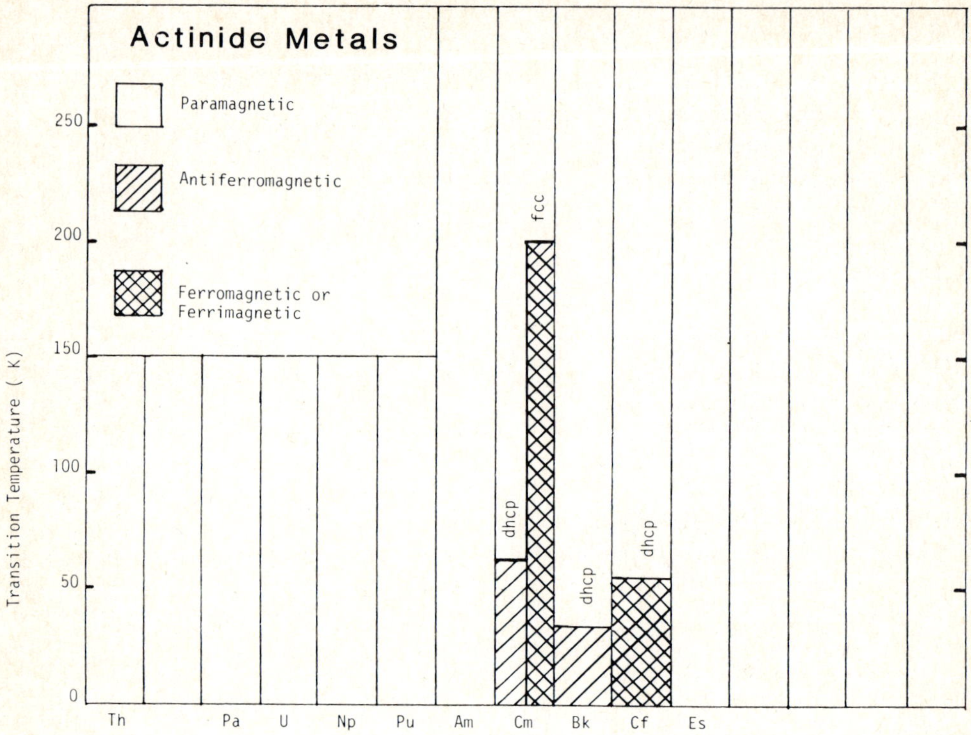

Fig. 36. Transition temperatures and regions of differing magnetic-ordering type for the actinide metals.

atoms for the light actinides (unlike the light-lanthanide homologues) as tri-positive ions is not possible. This failure is understood to exist from the nonlocalized or itinerant nature of the f electrons. Hence the picture of Smith and Kmetko (1983) that suggests americium is a transition element between the nonlocalized light actinides and the localized heavy actinides is confirmed by magnetic measurements.

Of the heavy actinide metals studied, all except einsteinium exhibit a transition to an ordered structure at low temperatures, as shown in fig. 36. Unlike the lanthanide metal homologues shown in fig. 7, the light actinides exhibit no ordering at low temperatures and the heavy actinides have much lower transition temperatures. The dhcp crystal structure of curium has an antiferromagnetic transition at about 65 K while the fcc structure has a ferrimagnetic or ferromagnetic transition at 205 K. Berkelium metal in the dhcp crystal structure likewise has a low-temperature transition at 34 K to an antiferromagnetic arrangement. Californium metal in the dhcp structure exhibits a ferromagnetic transition at 51 K but is not fully saturated by the largest fields applied (50 kG) and suggests a complex spin arrangement. The preliminary measurements on einsteinium metal are consistent with only paramagnetism and no low-temperature order.

Acknowledgements

Research of the authors is sponsored by the Division of Chemical Sciences, Office of Basic Energy Sciences, US Department of Energy under contracts DE-AS05-79ER10348 with the University of Tennessee, Knoxville and DE-AC05-840R21400 with the Martin–Marietta Energy Systems, Inc.

References

Amoretti, G., and J.-M. Fournier, 1984, J. Magn. & Magn. Mater. **43**, L217–L220.

Anderson, P.W., 1950, Phys. Rev. **79**, 350.

Bansal, B.M., 1966, Ph.D. Dissertation UCRL-16782 (University of California).

Baybarz, R.D., L.B. Aspray, C.E. Strousse and E. Fukushima, 1972, J. Inorg. Nucl. Chem. **34**, 3427–3431.

Bickel, M., 1982, Dissertation (Universität Karlsruhe).

Boeuf, A., J.-M. Fournier, J.F. Gueuegnon, L. Manes, J. Rebizant and F. Rustichelli, 1979, J. Phys. Lett. (France) **40**, 335.

Brodsky, M.B., 1971, in: Rare Earths and Actinides, Inst. Phys. Conf. Ser. No. 35 (Institute of Physics, London and Bristol) p. 75.

Brooks, M.S.S., B. Johansson and H.L. Skriver, 1984, in: Handbook on the Physics and Chemistry of the Actinides, Vol. 1, eds A.J. Freeman and G.H. Lander (North-Holland, Amsterdam) p. 153.

Carnall, W.T., and H.M. Crosswhite, 1985, Report ANL-84-90 (Argonne National Laboratory, Argonne, IL).

Casey, A.T., and S. Mitra, 1976, in: Theory and Applications of Molecular Paramagnetism, eds E.A. Boudreaux and L.N. Mulay (Wiley, New York) pp. 271–310.

Chan, S.-K., and D.J. Lam, 1974, in: The Actinides: Electronic Structure and Related Properties, Vol. 1, eds A.J. Freeman and J.B. Darby Jr (Academic Press, New York) pp. 1–4.

Chapin, P.S., M.C. Finn and J.M. Honig, 1964, Proc. 4th Rare Earth Research Conf., Phoenix, AZ (1964).

Dunlap, B.D., 1971, J. Appl. Phys. **42**, 1719.

Edelstein, N., 1985, in: Americium and Curium Chemistry and Technology, eds N. Edelstein, J.D. Navratil and W.W. Schulz (Reidel, Dordrecht, The Netherlands).

Ellis, D.E., 1985, in: Handbook on the Physics and Chemistry of the Actinides, Vol. 2, eds A.J. Freeman and G.H. Lander (North-Holland, Amsterdam) p. 1.

Fournier, J.-M., and R. Troć, 1985, in: Handbook on the Physics and Chemistry of the Actinides, Vol. 2, eds A.J. Freeman and G.H. Lander (North-Holland, Amsterdam) p. 29.

Fournier, J.-M., A. Blaise, W. Müller and J.-C. Spirlet, 1977, Physica B **86–88**, 30.

Fujita, D.K., 1969, Ph.D. Dissertation UCRL-19507 (University of California).

Fujita, D.K., T.C. Parsons, N. Edelstein, M. Noe and J.R. Peterson, 1976, The magnetic susceptibility of ^{244}Cm metal and ^{249}Cf metal, in: Transplutonium Elements, eds W. Müller and R. Linder (North-Holland, Amsterdam) p. 173.

Grünberg, P., and H.J. Maletta, 1982, in: Landolt–Börnstein: Numerical Data and Functional Relationships in Science and Technology, Vol. 12, eds K.-H. Hellwege and A.M. Hellwege (Springer, Berlin) pp. 67–158.

Grunzweig-Genossar, J., M. Kuznietz and F. Friedman, 1968, Phys. Rev. **173**, 562.

Haire, R.G., and R.D. Baybarz, 1973, J. Inorg. & Nucl. Chem. **35**, 489–496.

Haire, R.G., and R.D. Baybarz, 1974, J. Inorg. & Nucl. Chem. **36**, 1295.

Haire, R.G., U. Benedict, J.R. Peterson, C. Dufour and J.P. Itié, 1985, J. Less-Common Met. **109**, 71.

Hendricks, M.E., E.R. Jones Jr, J.A. Stone and D.G. Karraker, 1974, J. Chem. Phys. **60**, 2095.

Hessler, J., and W.T. Carnall, 1980, in: Lanthanide and Actinide Chemistry and Spectroscopy, ACS Symposium Series, Vol. 131, ed. N. Edelstein (American Chemical Society, Washington, DC) ch. 17.

Holtzberg, F., T.R. McGuire and S. Methfessel, 1970, in: Landolt–Börnstein: Numerical Data and Functional Relationships in Science and Technology, Vol. 4, eds K.-H. Hellwege and A.M. Hellwege (Springer, Berlin) pp. 94–109.

Hulliger, F., 1978, in: Handbook on the Physics and Chemistry of Rare Earths, Vol. 4, eds K.A. Gschneider Jr and L. Eyring (North-Holland, Amsterdam) pp. 153–236.

Huray, P.G., S.E. Nave, J.R. Peterson and R.G. Haire, 1980, Physica B **102**, 217.

Huray, P.G., S.E. Nave, R.G. Haire and J.R. Moore, 1983, Inorg. Chimica Acta, A **3**, 120.

Kalvius, G.M., S.L. Ruby, B.D. Dunlap. G.K. Shenoy, D. Cohen and M.B. Brodsky, 1969, Phys. Lett. B **29**, 489.

Kanellakopulos, B., and W. Müller, 1979, Re-

port AED-Conf-78-074-002 (AED, Berlin); INIS Atomindex **10**, 425659.
Kanellakopulos, B., A. Blaise, J.-M. Fournier and W. Müller, 1975, Solid State Commun. **17**, 713.
Kanellakopulos, B., J.P. Charvillat, F. Maino and W. Müller, 1976, in: Transplutonium Elements, eds W. Müller and R. Linder (North-Holland, Amsterdam) p. 181.
Kanellakopulos, B., E. Aderhold, E. Dornberger, W. Müller and R.D. Baybarz, 1978, Radiochim. Acta **25**, 89.
Karraker, D.B., 1975a, J. Chem. Phys. **63**(7), 3174.
Karraker, D.B., 1975b, J. Chem. Phys. **62**(4), 1444–1447.
Koehler, W.C., 1972, in: Magnetic Properties of Rare Earth Metals, ed. R.J. Elliot (Plenum, New York) pp. 81–128.
Lander, G.H., A.T. Aldred, B.D. Dunlap and G.K. Shency, 1977, Physica B+C **86–88**, 152.
MacChesney, J.B., H.J. Williams, R.C. Sherwood and J.F. Potter, 1964, J. Chem. Phys. **41**, 3177.
MacChesney, J.B., H.J. Williams, R.C. Sherwood and J.F. Potter, 1965, J. Appl. Phys. **37**(3), 1435.
Marei, S.A., 1965, Ph.D. Thesis UCRL-11984 (University of California).
Marei, S.A., and B.B. Cunningham, 1972, Inorg. & Nucl. Chem. **34**, 1203.
McWhan, D.B., 1961, Ph.D. Dissertation UCRL-9695 (University of California).
Moore, J.R., S.E. Nave, R.G. Haire and P.G. Huray, 1986, J. Less-Common Met. **121**, 187–192.
Morss, L.R., J. Fuger, J. Goffart and R.G. Haire, 1983, Inorg. Chem. **22**, 1983.
Nave, S.E., and P.G. Huray, 1980, Rev. Sci. & Instrum. **51**(5), 7.
Nave, S.E., and P.G. Huray, 1984, unpublished.
Nave, S.E., P.G. Huray and R.G. Haire, 1980, The magnetic susceptibility of ^{249}Bk metal, in: Crystalline Electric Field and Structural Effects in f-Electron Systems, eds J. Crow, R.P. Guertin and T. Mihalisin (Plenum, New York) p. 269.
Nave, S.E., P.G. Huray, J.R. Peterson, D.A. Damien and R.G. Haire, 1981a, Physica B **107**, 253.
Nave, S.E., R.G. Haire and P.G. Huray, 1981b, in: Actinides 1981, Pacific Grove, CA, LBL-12441, ed. L. Lizama (Lawrence Berkeley Laboratory) p. 144.
Nave, S.E., R.G. Haire and P.G. Huray, 1983, Phys. Rev. B **28**(5), 2317.
Nave, S.E., J.R. Moore, M.T. Spaar, R.G. Haire and P.G. Huray, 1985, Physica B **130**, 225.
Nave, S.E., J.R. Moore, R.G. Haire, J.R. Peterson, D.A. Damien and P.G. Huray, 1986, J. Less-Common Met. **121**, 319–324.
Nave, S.E., J.R. Moore, J.R. Peterson and R.G. Haire, 1987, J. Less-Common Met. **127**, 79–85.
Nestor, C.W., T.C. Tucker, T.A. Carlson, L.D. Roberts, F.B. Malik and C. Froese, 1966, Report ORNL-4027 (Oak Ridge National Laboratory, Oak Ridge, TN).
Roy, A., D.S. Buchanan, D.M. Ginsberg and P.R. Roach, 1982, Rev. Sci. & Instrum. **53**(7), 1053.
Russel, H.N., and F.A. Saunders, 1925, Astrophys. J. **61**, 38.
Schieber, M.M., 1967, Experimental Magnetochemistry (North-Holland, Amsterdam) p. 305.
Sinha, S.K., 1978, in: Handbook on the Physics and Chemistry of Rare Earths, Vol. 3, eds K.A. Gschneidner Jr and L. Eyring (North-Holland, Amsterdam) p. 489.
Smith, J.L., and E.A. Kmetko, 1983, J. Less-Common Met. **90**, 8.
Soderholm, L., N. Edelstein, L.R. Morss and G.V. Shalimoff, 1986, J. Magn. & Magn. Mater. **54–57**, 597–598.
Spirlet, J.C., J. Rebizant and O. Vogt, 1983, 13^{ièmes} Journées des Actinides (Elat, Israel).
Topp, N.E., 1965, The Chemistry of the Rare Earth Elements (Elsevier, Amsterdam) p. 11.
Van Vleck, J.H., 1932, The Theory of Electric and Magnetic Susceptibility (Oxford Press, London).
Wachter, P., and E. Kaldis, 1980, J. Magn. & Magn. Mater. **15–18**, 305, 306.

SUBJECT INDEX